SUPERNOVA 1987A:30 YEARS LATER - COSMIC RAYS AND NUCLEI FROM SUPERNOVAE AND THEIR AFTERMATHS

IAU SYMPOSIUM No. 331

COVER ILLUSTRATION: Celebrating the 30th anniversary of SN 1987A: a multiwavelength view

Upper left: ATCA 44 GHz from Zanardo, G., Staveley-Smith, L., Ng, C.-Y., et al. 2013, ApJ, 767, 98. *Upper right:* ALMA 870 μm from Indebetouw, R., Matsuura, M., Dwek, E., et al. 2014, ApJ, 782, L2. *Lower left:* ALMA CO(J=2-1) from Matsuura, M., Indebetouw, R., Woosley, S., et al. 2017, MNRAS in press (http://adsabs.harvard.edu/abs/2017arXiv170402324M). *Lower right:* ESO SINFONI 1.6 μm from Fransson, C., Larsson, J., Spyromilio, J., et al. 2016, ApJ, 821, L5. *Center image:* NASA's Chandra X-ray Observatory and Hubble Space Telescope (HST) composite image, Credit: X-ray: NASA/CXC/PSU/S. Park & D. Burrows; Optical: NASA/STScI/CfA/P. Challis. Figure courtesy of P. Cigan.

IAU SYMPOSIUM PROCEEDINGS SERIES

INTERNATIONAL ASTRONOMICAL UNION
UNION ASTRONOMIQUE INTERNATIONALE

International Astronomical Union

IAU

SUPERNOVA 1987A:30 YEARS LATER - COSMIC RAYS AND NUCLEI FROM SUPERNOVAE AND THEIR AFTERMATHS

PROCEEDINGS OF THE 331st SYMPOSIUM OF THE INTERNATIONAL ASTRONOMICAL UNION HELD IN SAINT-GILLES-LES-BAINS, LA REUNION ISLAND, FRANCE FEBRUARY 20–24, 2017

Edited by

ALEXANDRE MARCOWITH
LUPM, France

MATTHIEU RENAUD
LUPM, France

GLORIA DUBNER
IAFE, Argentina

ALAK RAY
TIFR, India

and

ANDREÏ BYKOV
IOFFE, Russia

CAMBRIDGE
UNIVERSITY PRESS

CAMBRIDGE UNIVERSITY PRESS
University Printing House, Cambridge CB2 8BS, United Kingdom
1 Liberty Plaza, Floor 20, New York, NY 10006, USA
10 Stamford Road, Oakleigh, Melbourne 3166, Australia

First published 2017

Printed in the UK by Bell & Bain, Glasgow, UK

Typeset in System $\LaTeX 2_\varepsilon$

A catalogue record for this book is available from the British Library Library of Congress Cataloguing in Publication data

ISBN 9781107192553 hardback
ISSN 1743-9213

Table of Contents

Preface

Supernovae, along with their remnants, are pivotal in many active fields of research as the key sources of cosmic high-energy particles and heavy elements, the privileged sites of cosmic dust formation, and the primary agents of chemical and dynamical evolution of galaxies. The precise observational studies of type Ia supernovae at cosmological distances recently challenged fundamental physics with the dark energy concept. The bright transient phenomenae associated with supernovae are among the rare celestial events visible with the naked eye. Yet, many open issues still need to be deeply explored such as the nature of the supernova progenitors and the final stages of stellar evolution, the explosion mechanisms and the associated nucleosynthesis, the shock acceleration processes and the long standing question of the origin of cosmic rays.

A bright "new star" appeared in the Large Magellanic Cloud on the 23rd of February, 1987, at 07:35 UT, manifesting itself firstly in the form of a burst of neutrinos, a few hours before its first light reached the Earth. Sanduleak -69° 202, a blue supergiant in the outskirts of the Tarantula Nebula in the Large Magellanic Cloud had just imploded. The resulting stellar explosion named SN 1987A, visible to naked eye, has been the nearest and brightest supernova witnessed since the invention of the telescope four centuries ago. Thanks to the wealth of multi-wavelength information collected in the course of thirty years of investigation, including the unique detection of MeV neutrinos associated with a core-collapse supernova, SN 1987A became the first case enabling to make significant progress in all of these above-mentioned fundamental topics in modern astrophysics.

At the occasion of the thirtieth anniversary of this "supernova of a lifetime", this IAUS 331 Symposium aimed at bringing together different communities in order to collate the accumulating knowledge on these explosive events and their remnants and shed further light on the open questions through an interdisciplinary approach. Such a meeting was then thought as the opportunity to make the link between the stellar progenitors and the multi-wavelength and multi-messenger manifestation of their aftermaths in terms of extreme sources of high-energy particles and nuclei. Therefore, several, interconnected, themes have been identified: (1) massive stars as supernova progenitors; (2) explosion mechanisms; (3) supernova properties, outcomes and impacts; (4) multi-wavelength and multi-messenger data on supernovae and their remnants, and (5) "SN 1987A, 30 years later" allowing for these diverse topics to be discussed at once, and hence for bridges among them to be built.

Following endorsement and sponsoring by IAU Commissions and Divisions and subsequent approval by the IAU Executive Committee in May 2016, speakers were invited and the different communities informed. Positive feedback translated into 19 invited review papers, 55 oral contributions and a dozen of posters, with the participation of nearly a hundred astronomers from 22 countries worldwide. The meeting location in La Réunion Island, a French overseas department in the Indian Ocean, has allowed a greater participation by scientists from Asia in particular (India, China, Japan, Taiwan), but also from South Africa and Australia, aside from the possibility it offered to stargaze within the wonders of the Southern Hemisphere.

It is a great pleasure to acknowledge the financial support of our sponsors listed on page *xxi* of these Proceedings, and the active support of the members of the Scientific and Local Organizing Committees in making this Symposium possible, and in the end, a great success.

To conclude, our hope would be threefold:
- may these Proceedings be found useful to anyone interested in supernovae and their remnants. As a healthy consequence of the rapid developments currently experienced in these lively fields of research,
- may their scientific contents become outdated in the horizon of the next decennial anniversary of SN 1987A. In particular, given that the probability of sighting the next Galactic supernova in our lifetimes is, though small, non null,
- may such an exceptional, eagerly awaited, event bring us as many unexpected discoveries and trigger as much excitement as SN 1987A and its historical predecessors in the Milky Way did.

Astonishing Light Burst At Night Emerges...

Matthieu Renaud and Alexandre Marcowith, SOC/LOC co-chairs
Les Avirons & Montpellier, France, June 9, 2017

THE ORGANIZING COMMITTEE

Scientific

A. Marcowith (France, co-chair)
G. Dubner (Argentina, co-chair)
A. Bykov (Russia, co-chair)
A. Bamba (Japan)
R. Diehl (Germany)
H.-T. Janka (Germany)
G. Meynet (Switzerland)
S. J. Smartt (UK)
M. Renaud (France, co-chair)
A. Ray (India, co-chair)
E. Amato (Italy)
R. Chevalier (USA)
F. Harrison (USA)
M. Lemoine-Goumard (France)
S. Safi-Harb (Canada)
P. A. Whitelock (South Africa)

Local

S. Colaïocco
A. Lèbre
A. Peyrot
M.-H. Grondin
A. Marcowith (co-chair)
M. Renaud (co-chair)

Acknowledgements

The symposium is sponsored and supported by the IAU Divisions G: Stars and Stellar Physics (coordinating division), D: High Energy Phenomena and Fundamental Physics and B: Facilities, Technologies and Data Science; and by the IAU Commissions G2 (Massive Stars), G3 (Stellar Evolution), D1 (Gravitational Wave Astrophysics), B1 (Computational Astrophysics), and B4 (Radio Astronomy).

The Local Organizing Committee operated under the auspices of the Laboratoire Univers et Particules de Montpellier (CNRS, Univ. Montpellier).

Funding by the
International Astronomical Union IAU,
Centre National de la Recherche Scientifique CNRS,
Université de Montpellier UM,
Laboratoire Univers et Particules de Montpellier LUPM,
Société Française d'Astronomie et d'Astrophysique SF2A,
Société Française de Physique SFP,
Programme National Hautes Energies PNHE,
Programme National de Physique Stellaire PNPS,
Programme National de Physique et Chimie du Milieu Interstellaire PCMI,
Programme National de Cosmologie et Galaxies PNCG,
LabeX Origines, Constituants et EVolution de l'Univers OCEVU,
Observatoire de Recherche Méditerranéen de l'Environnement OSU-OREME,
and
Regional Council of La Réunion,
is gratefully acknowledged.

CONFERENCE PHOTOGRAPH

Participants

Nayana **A.J**, NCRA-TIFR, Pune, India	nayan89deva@gmail.com
Niloufar **Afsariardchi**, University of Toronto, Toronto, Canada	afsariardchi@astro.utoronto.ca
Maarten **Baes**, Universiteit Gent, Gent, Belgium	maarten.baes@ugent.be
Jean **Ballet**, CEA/SAp, Saclay, France	jballet@cea.fr
Matthew **Baring**, Rice University, Houston, TX, USA	baring@rice.edu
Pooja **Bhattacharjee**, Bose Institute, Kolkata, India	pooja.bhattacharjee06@gmail.com
Michael **Bietenholz**, Hartebeesthoek Radio Astronomy Observatory, Krugersdorp, South Africa	michael@hartrao.ac.za
Artem **Bohdan**, Institute of Nuclear Physics PAS, Krakow, Poland	artem.bohdan@ifj.edu.pl
Gabriela **Castelletti**, IAFE, Buenos Aires, Argentina	castelletti.g@gmail.com
Poonam **Chandra**, NCRA-TIFR, Pune, India	poonam@ncra.tifr.res.in
Ryan **Chaves**, LUPM, Montpellier, France	ryan.chaves@umontpellier.fr
You-Hua **Chu**, ASIAA, Taipei, Taiwan	yhchu@asiaa.sinica.edu.tw
Phillip **Cigan**, Cardiff University School of Physics and Astronomy, Cardiff, United Kingdom	ciganp@cardiff.ac.uk
Francesco **de Palma**, INFN Bari, Bari, Italy	francesco.depalma@ba.infn.it
Antonio **de Ugarte Postigo**, IAA-CSIC, Granada, Spain	deugarte@iaa.es
Anne **Decourchelle**, CEA/SAp, Saclay, France	anne.decourchelle@cea.fr
Justine **Devin**, LUPM, Montpellier, France	justine.devin@etu.umontpellier.fr
Roland **Diehl**, Max Planck Institute (MPE), Garching, Germany	rod@mpe.mpg.de
Fiorenza **Donato**, Torino University, Torino, Italy	donato@to.infn.it
Elise **Egron**, INAF - Osservatorio Astronomico di Cagliari, Cagliari, Italy	elisegron@hotmail.fr
Claes **Fransson**, Stockholm University, Stockholm, Sweden	claes@astro.su.se
Chris **Fryer**, Los Alamos National Laboratory, Los Alamos, USA	fryer@lanl.gov
Yasuo **Fukui**, Nagoya University, Nagoya, Japan	fukui@a.phys.nagoya-u.ac.jp
Michael **Gabler**, MPI for Astrophysics, Munich, Germany	miga@mpa-garching.mpg.de
Yves **Gallant**, LUPM, Montpellier, France	gallant@in2p3.fr
Robert **Glas**, Max Planck Institute for Astrophysics, Munich, Germany	rglas@mpa-garching.mpg.de
Brian **Grefenstette**, Caltech, Pasadena, USA	bwgref@srl.caltech.edu
Jerome **Guilet**, Max-Planck Institut fuer Astrophysik, Munich, Germany	jguilet@mpa-garching.mpg.de
Alexander **Heger**, Monash Centre for Astrophysics, Notting Hill, Australia	alex.heger@gmail.com
Raphael **Hirschi**, Keele University, Keele, United Kingdom	r.hirschi@keele.ac.uk
Takuma **Ikeda**, Rikkyo University, Tokyo, Japan	t.ikeda@rikkyo.ac.jp
Adriano **Ingallinera**, INAF - OACT, Catania, Italy	ingallinera@oact.inaf.it
Hans-Thomas **Janka**, Max Planck Institute for Astrophysics, Garching, Germany	thj@mpa-garching.mpg.de
Vinay **Kashyap**, Department of Astronomy, Harvard University, Cambridge, USA	lvinayk@protonmail.com
Satoru **Katsuda**, Chuo University, Tokyo, Japan	katsuda@phys.chuo-u.ac.jp
Norita **Kawanaka**, Kyoto University, Kyoto, Japan	norita@kusastro.kyoto-u.ac.jp
Remi **Kazeroni**, Max Planck Institute for Astrophysics, Garching, Germany	remi.kazeroni@gmail.com
Charles **Kilpatrick**, University of California, Santa Cruz, United States	cdkilpat@ucsc.edu
Robert **Kirshner**, Gordon and Betty Moore Foundation, Palo Alto, CA, USA	robert.kirshner@moore.org
Sladjana **Knezevic**, Weizmann Institute of Science, Rehovot, Israel	sladjana.knezevic@weizmann.ac.il
Oleh **Kobzar**, Institute of Nuclear Physics PAS, Krakow, Poland	oleh.kobzar@ifj.edu.pl
Nukri **Komin**, University of the Witwatersrand, Johannesburg, South Africa	nukri.komin@wits.ac.za
Kei **Kotake**, Fukuoka University, Fukuoka, Japan	kkotake@fukuoka-u.ac.jp
Vladimir **Kulikovskiy**, CPPM, Marseille, France	kulikovskiy@cppm.in2p3.fr
Esha **Kundu**, Stockholm University, Stockholm, Sweden	esha.kundu@astro.su.se
Agnes **Lèbre**, LUPM, Montpellier, France	agnes.lebre@umontpellier.fr
Herman (Shiu-Hang) **Lee**, Kyoto University, Kyoto, Japan	shia520@gmail.com
Chuan-Jui **Li**, ASIAA, Taipei, Taiwan	cjli@asiaa.sinica.edu.tw
Bing **Liu**, Nanjing University, Nanjing, China	liubing1326@foxmail.com
Sara **Loru**, INAF - Osservatorio Astronomico di Cagliari, Cagliari, Italy	saraloru@oa-cagliari.inaf.it
Alexandre **Marcowith**, LUPM, Montpellier, France	Alexandre.Marcowith@umontpellier.fr
Pierrick **Martin**, IRAP, Toulouse, France	pierrick.martin@irap.omp.eu
Shu **Masuda**, Department of Physics, Kyoto University, Kyoto, Japan	masuda@cr.scphys.kyoto-u.ac.jp
Mikako **Matsuura**, Cardiff University School of Physics and Astronomy, Cardiff, United Kingdom	matsuuram@cardiff.ac.uk
Julie **McEnery**, NASA/GSFC, Greenbelt, USA	julie.mcenery@nasa.gov
Marco **Miceli**, Universita di Palermo, Palermo, Italy	miceli@astropa.unipa.it
Kuntal **Misra**, ARIES, Nainital, India	kuntal@aries.res.in
Giovanni **Morlino**, Gran Sasso Science Institute, Firenze, Italy	giovanni.morlino@gssi.infn.it
Jacek **Niemiec**, Institute of Nuclear Physics PAS, Krakow, Poland	Jacek.Niemiec@ifj.edu.pl
Evan **O'Connor**, North Carolina State University, Raleigh, USA	evanoconnor@gmail.com
Yukari **Ohtani**, National Astronomical Observatory of Japan, Tokyo, Japan	yukari.ohtani@nao.ac.jp
Salvatore **Orlando**, INAF - Osservatorio Astronomico di Palermo, Palermo, Italy	orlando@astropa.inaf.it
Po-Sheng **Ou**, ASIAA, Taipei, Taiwan	psou@asiaa.sinica.edu.tw
Alberto **Pellizzoni**, INAF - Osservatorio Astronomico di Cagliari, Cagliari, Italy	apellizz@oa-cagliari.inaf.it
Oleh **Petruk**, INAF - Osservatorio Astronomico di Palermo, Palermo, Italy	oleh.petruk@gmail.com
Georg **Raffelt**, Max Planck Institute for Physics (MPP), Munich, Germany	raffelt@mpp.mpg.de
Ninoy **Rahman**, Max Planck Institute for Astrophysics, Munich, Germany	nrahman@mpa-garching.mpg.de
Alak **Ray**, TIFR, Mumbai, India	akr@tifr.res.in
Soebur **Razzaque**, University of Johannesburg, Auckland Park, South Africa	srazzaque@uj.ac.za
Matthieu **Renaud**, LUPM, Montpellier, France	matthieu.renaud@lupm-univ-montp2.fr
Hidetomo **Sawai**, Research Organization for Information Science & Technology, Kobe, Japan	hidetomo.sawai@gmail.com
Ivo **Seitenzahl**, Australian National University, Weston Creek, Australia	ivo.seitenzahl@anu.edu.au
Susu **Shan**, National Astronomical Observatories, Chinese Academy of Sciences, Beijing, China	shansusu@nao.cas.cn
Rachel **Simoni**, Anton Pannekoek Institute, Amsterdam, The Netherlands	r.c.simoni@uva.nl
Noam **Soker**, Technion, Haifa, Israel	soker@physics.technion.ac.il
Niharika **Sravan**, Northwestern University, Evanston, USA	niharika@u.northwestern.edu
Georg **Stockinger**, Max-Planck-Institute for Astrophysics, Munich, Germany	georgsto@mpa-garching.mpg.de
Mingxu **Sun**, Beijing Normal University, Beijing, China	931855101@qq.com
Jorge Leonardo **Supan**, IAFE, Buenos Aires, Argentina	leo.supan@gmail.com
Iurii **Sushch**, DESY, Zeuthen, Germany	iurii.sushch@desy.de
Firoza **Sutaria**, Indian Institute of Astrophysics, Bangalore, India	fsutaria@gmail.com

Irene **Tamborra**, Niels Bohr Institute, Copenhagen, Denmark tamborra@nbi.ku.dk
Vincent **Tatischeff**, CSNSM, Orsay, France Vincent.Tatischeff@csnsm.in2p3.fr
Christina **Thöne**, IAA-CSIC, Granada, Spain cthoene@iaa.es
Diego F. **Torres**, ICREA & Institute of Space Sciences (IEEC-CSIC), Barcelona, Spain dtorres@ice.csic.es
Tomonori **Totani**, The University of Tokyo, Tokyo, Japan totani@astron.s.u-tokyo.ac.jp
Dmitry **Tsvetkov**, Sternberg Astronomical Institute, Moscow, Russia dmitry.tsvetkov@gmail.com
Stefano **Valenti**, UC Davis, Davis, United States stfn.valenti@gmail.com
Roland **Walter**, ISDC, University of Geneva, Versoix, Switzerland geminga@romandie.com
Annop **Wongwathanarat**, RIKEN, Wako, Japan annop.wongwathanarat@riken.jp
Giovanna **Zanardo**, International Centre for Radio Astronomy Research, West Perth, Australia giovanna.zanardo@gmail.com
Xiao **Zhang**, Nanjing University, Nanjing, China xiaozhang@nju.edu.cn

Supernova 1987A:30 years later - Cosmic Rays and Nuclei from Supernovae and their aftermaths
Proceedings IAU Symposium No. 331, 2017
A. Marcowith, M. Renaud, G. Dubner, A. Ray & A. Bykov, eds.

doi:10.1017/S1743921317004896

Progenitors of Core-Collapse Supernovae

R. Hirschi[1,2], D. Arnett[3], A. Cristini[1], C. Georgy[4], C. Meakin[3,5] and I. Walkington[1]

[1]Astrophysics Group, Keele University, Lennard-Jones Laboratories, Keele, ST5 5BG, UK
email: r.hirschi@keele.ac.uk

[2]Kavli IPMU (WPI), The University of Tokyo, Kashiwa, Chiba 277-8583, Japan
[3]Department of Astronomy, University of Arizona, Tucson, AZ 85721, USA
[4]Geneva Observatory, University of Geneva, Ch. Maillettes 51, 1290 Versoix, Switzerland
[5]Karagozian & Case, Inc., 700 N. Brand Blvd. Suite 700, Glendale, CA, 91203, USA

Abstract. Massive stars have a strong impact on their surroundings, in particular when they produce a core-collapse supernova at the end of their evolution. In these proceedings, we review the general evolution of massive stars and their properties at collapse as well as the transition between massive and intermediate-mass stars. We also summarise the effects of metallicity and rotation. We then discuss some of the major uncertainties in the modelling of massive stars, with a particular emphasis on the treatment of convection in 1D stellar evolution codes. Finally, we present new 3D hydrodynamic simulations of convection in carbon burning and list key points to take from 3D hydrodynamic studies for the development of new prescriptions for convective boundary mixing in 1D stellar evolution codes.

Keywords. convection, stars: evolution, interiors, rotation

1. Introduction

The progenitors of (iron) core-collapse supernovae (CCSNe) are by definition massive stars. Indeed, massive stars are stars massive enough to go through all the burning stages from hydrogen to silicon burning and form a core composed mostly of iron-group elements. Since these are the most stable elements, no more nuclear energy can be extracted to counter-balance gravity. Once the inner core is massive enough for electron degeneracy pressure to be overcome (Chandrasekhar mass limit), the iron core collapses and sometimes leads to a powerful explosion as in the case of SN1987A. In this paper, we review the general evolution of massive stars at solar metallicity as well as the impact of rotation and mass loss. We also discuss briefly the effects of metallicity. The models and plots presented in this paper are taken from Hirschi *et al.* (2004) unless otherwise stated. Other recent grids of models at solar metallicity can be found in Ekström *et al.* (2012); Chieffi & Limongi (2013); Sukhbold & Woosley (2014).

2. Evolution of Surface Properties (HR diagram) and Lifetimes

Figure 1 (*left*) shows the evolutionary tracks of different $20\,M_{\odot}$ models in the HR–diagram and thus how the surface properties of these stars evolve. The non-rotating model is representative of the lower end of massive stars, which keep an extended hydrogen-rich envelope, end as red supergiant and produce type II supernovae. The 300 $\mathrm{km\,s^{-1}}$ model is representative of the higher end of massive stars, for which most or all of the H-rich envelope is lost via stellar winds and the star ends as a hot star, generally a Wolf-Rayet star and produce a type Ib or Ic supernova depending on how much helium is left. Note

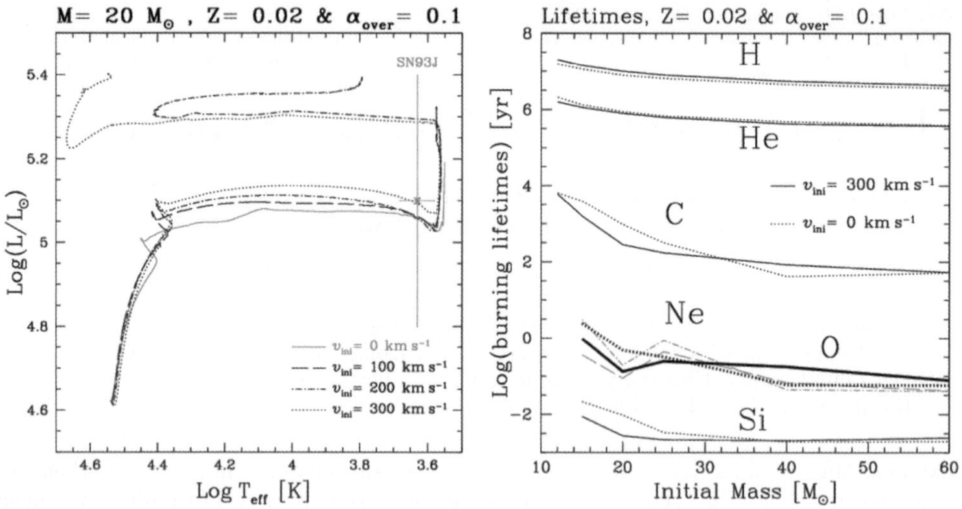

Figure 1. *Left*: HR–diagram for 20 $M_\odot$ models: solid, dashed, dotted-dashed and dotted lines correspond respectively to $v_{\rm ini}$= 0, 100, 200 and 300 km s^{-1}. We also indicate the position of the progenitor of SN1993J. *Right*: Burning lifetimes as a function of the initial mass and velocity. Solid and dotted lines correspond respectively to rotating and non–rotating models. Long–dashed and dotted–dashed lines are used for rotating and non–rotating Ne–burning lifetimes to point out that they are to be considered as estimates.

that type Ib and Ic supernovae also come from stars in multiple systems, in which the hydrogen-rich envelope is lost via Roche lobe overflow (see Langer 2012, for a review on the topic of binary interactions in massive stars). These two models also show the impact of rotation on the evolution of massive stars. The additional models with intermediate rotation ($v_{\rm ini}$= 100 and 200 km s^{-1}) show the smooth transition from non-rotating to fast rotating models (for HR diagrams covering the full IMF, see Ekström *et al.* 2012).

As mentioned above, massive stars go through 6 burning stages: H, He, C, Ne, O, and Si burning. The lifetimes of these stages are plotted in Fig. 1 (*right*). Whereas H and He-burning stages last for roughly 10^{6-7} and 10^{5-6} years, respectively, the lifetimes for the advanced phases is much shorter. This is due to neutrino losses dominating energy losses over radiation from C burning onwards. C, Ne, O ans Si burning phases last about 10^{2-3}, 1, 1 and 10^{-2} years respectively. Concerning the effects of rotation and mass loss, there is a mass range where rotational mixing ($M \lesssim 30 M_\odot$) or mass loss ($M \gtrsim 30 M_\odot$) dominates over the other process. For $M \lesssim 30 M_\odot$, rotation-induced mixing extends the H-burning lifetime and as a consequence shortens slightly He-burning lifetimes. For the advanced phases, rotation makes stars behave like more massive stars. This is clearly seen for C-burning lifetimes, which are shorter for rotating models. For $M \gtrsim 30 M_\odot$, strong mass loss leads to degeneracy in the lifetime and final properties.

3. Evolution of Central Properties in the Log $T_{\rm c}$–Log $\rho_{\rm c}$ Diagram and Lower Mass Limit for Massive Stars

Figure 2 (left) shows the tracks of the 15 and 60 $M_\odot$ models throughout their evolution in the central temperature versus central density plane (Log $T_{\rm c}$–Log $\rho_{\rm c}$ diagram). The 60 $M_\odot$ model is representative of the stars more massive than about 30 $M_\odot$, for which the evolution is mainly affected by mass loss. The 15 $M_\odot$ model is representative of the stars less massive than about 20 $M_\odot$, for which the evolution is mainly affected by rotational

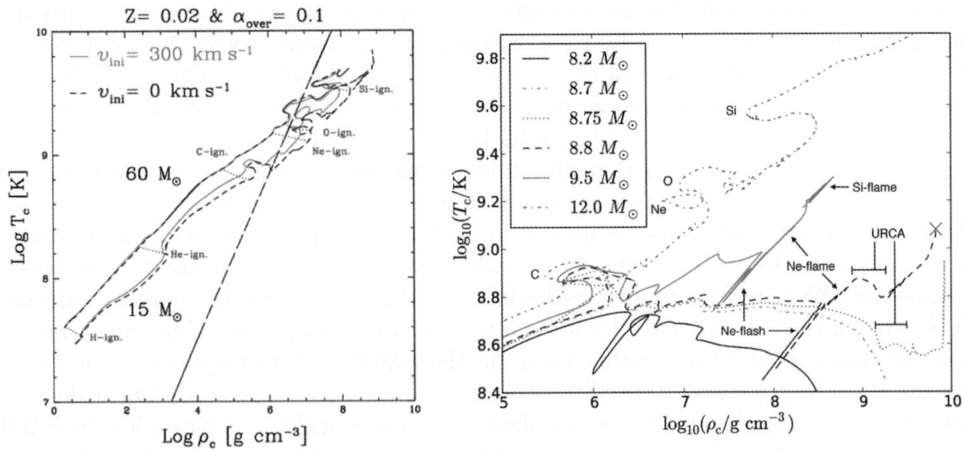

Figure 2. $\log T_c$ vs $\log \rho_c$ diagrams: Left: evolutionary tracks for the 15 and 60 $M_\odot$ models. Solid lines are rotating models and dashed lines are non–rotating models. The ignition points of every burning stage are connected with dotted lines. The additional long dashed line corresponds to the limit between non–degenerate and degenerate electron gas ($P^{\rm el}_{\rm perfect\ gas} = P^{\rm el}_{\rm degenerate\ gas}$). Figure taken from Hirschi *et al.* (2004). Right: The divergence of the models following C-burning at the transition between massive and intermediate mass stars. Figure taken from Jones *et al.* (2013).

mixing (already identified in Sect. 2). For the 15 $M_\odot$ model, the rotating tracks have a higher temperature and lower density due to more massive convective cores. The bigger cores are due to the effect of mixing, which largely dominates the structural effects of the centrifugal force. On the other hand, for the 60 $M_\odot$ model, mass loss dominates mixing effects and the rotating model tracks in the $\log T_c$–$\log \rho_c$ plane are at the same level or below the non–rotating ones (see Hirschi *et al.* 2004, for more details).

Recent models for the transition between massive and intermediate stars ($8 - 12\,M_\odot$ stars) can be found in Jones *et al.* (2013); Takahashi *et al.* (2013); Woosley & Heger (2015) and older models can be found in (Nomoto 1984, 1987; Ritossa *et al.* 1999). Evolutionary tracks from Jones *et al.* (2013) are shown in Fig. 2. Similar trends and conclusions are found in the other studies. The evolution and fate of stars in this mass range are sensitive to convective boundary mixing (CBM) treatment (e.g. overshooting), mass loss and CO core growth. Different choices of CBM lead to the transition mass being shifted up and down but we expect the same transitions and regimes to take place for different choices of CBM. The fate of super-AGB stars (SAGB, AGB stars that undergo carbon burning but not neon or subsequent burning stages) is highly sensitive to the mass–loss prescription on the SAGB and the rate at which the core grows (Poelarends *et al.* 2008). Mass loss and core growth compete against each other. At solar metallicity mass loss often wins and only a very narrow mass range at the top of the SAGB mass range will end as electron-capture supernovae (ECSN). The 8.7 and 8.75 $M_\odot$ models represent models in this narrow mass range. The 8.2 $M_\odot$ model represents models in the SAGB mass range for which mass loss wins and this model will end as a ONe white dwarf (WD). A temperature inversion develops in the core following the extinction of carbon-burning in both the 8.8 $M_\odot$ and 9.5 $M_\odot$ models. The neutrino emission processes that remove energy from the core are (over-)compensated by heating from gravitational contraction in more massive stars. However in these lower-mass stars the onset of partial degeneracy moderates the rate of contraction and hence neutrino losses dominate, cooling the central region. As a result, the ignition of neon in the 8.8 and 9.5 $M_\odot$ models takes place off center. These two models then go through neon(/oxygen; oxygen also burns via fusion in this situation)-flashes followed by the development of a neon(/oxygen)-flame. Owing to the high densities in the cores of these stars, the products of neon and oxygen burning are more neutron-rich than in more massive stars. This results in an electron fraction in the shell of as low as $Y_e \approx 0.48$. Due to its higher degeneracy, Y_e decreases faster in the $8.8\,M_\odot$ and it contracts faster than the time needed for the neon/oxygen flame to reach the centre of the star, both processes being helped by URCA pair processes. The core of the $8.8\,M_\odot$ model continuously contracts until the center reaches the critical density for electron captures by ^{24}Mg, quickly followed by further contraction to the critical density for those by ^{20}Ne and this model results in core collapse. The $8.8\,M_\odot$ model produces a ECSNe as for the $8.75\,M_\odot$ model but via a new evolutionary path coined "failed massive star" by Jones *et al.* (2013) rather than via the SAGB evolutionary path. The "failed massive star" path is also expected to take place for a narrow mass range but it does not critically depend on the uncertainties linked to mass loss, which is the case for the SAGB progenitors of ECSNe. Similarly to the $8.8\,M_\odot$ model starting neon burning off centre, the $9.5\,M_\odot$ model starts silicon burning off centre in a shell that later propagates toward the center. This is another example of the continuous transition towards massive stars, in which all the burning stages begin centrally. Although this model was not evolved to its conclusion, we expect that silicon-burning will migrate to the center, producing an iron core, and that it will finally collapse as an iron core-collapse SN (FeCCSN). The canonical massive star evolution (igniting C-, Ne-, O- and Si-burning centrally) leading to FeCCSN is expected to take place for stars with masses above $10\,M_\odot$. This mass range is represented by the $12\,M_\odot$.

The fate of models at the lower end of massive stars is studied further in Jones *et al.* (2014); Woosley & Heger (2015); Jones *et al.* (2016). At the other end of the IMF, very massive stars evolve far away from degeneracy. Very massive stars may encounter instead the pair-creation instability at very high temperatures (see Vink 2015, and references therein).

4. Structure Evolution and Pre-Supernova Properties

Figure 3 shows the evolution of the structure (Kippenhahn diagram) for $20\,M_\odot$ models. The y–axis represents the mass coordinate and the x–axis the time left until core collapse. The black zones represent convective zones. The abbreviations of the various burning stages are written below the graph at the time corresponding to the central burning stages. We note the complex succession of the different convective zones during the advanced phases. Sukhbold & Woosley (2014) study in detail the complex convective history in massive stars. In particular, their fig. 13 shows how the location in mass of the lower boundary of carbon burning convective shells play a key role in determining the compactness (O'Connor & Ott 2011) at the pre-supernova stage (see also Ertl *et al.* 2016). It is worth noting that a few physical ingredients of the stellar models influence carbon burning in general and thus the exact location of the convective shells and the compactness for a given initial mass. Carbon burning is sensitive to the amount of carbon (relative to oxygen) left at the end of helium burning. This in turn is influenced by the $^{12}C(\alpha,\gamma)^{16}O$ rate relative to the triple-α rate (Tur *et al.* 2009). Convective boundary criteria and mixing prescriptions also affect the carbon left over at the end of helium burning. Using Ledoux rather than Schwarzschild generally leads to smaller helium burning cores and more carbon left over. Extra mixing, especially towards the end of He-burning brings fresh α particles that can capture on ^{12}C and reduce its left over abundance. Finally rotation-induced mixing, as is clearly seen in Fig. 3, leads to

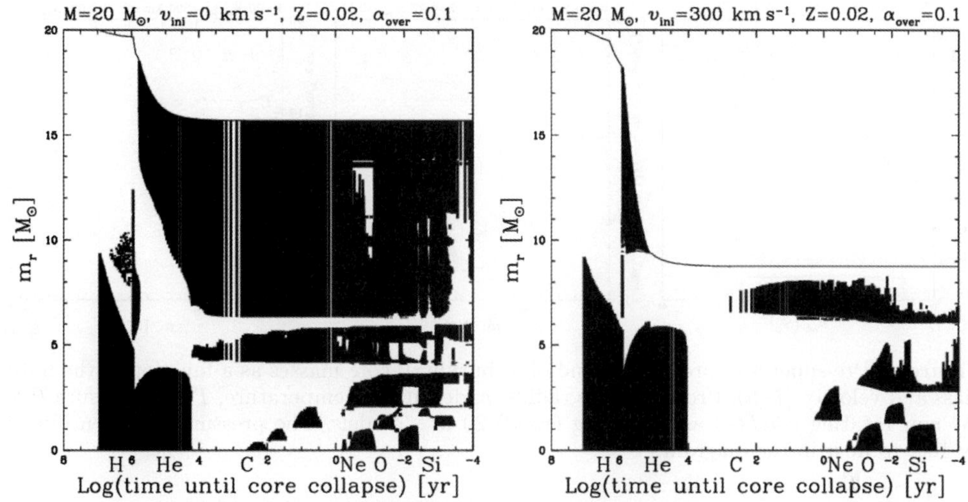

Figure 3. Kippenhahn diagrams for the non–rotating (*left*) and $v_{\rm ini}$= 300 km s^{-1} (*right*) 20 $M_\odot$ models. The black zones correspond to convective regions. Note that these plots are produced by drawing black vertical lines for a subset of the time steps of the model and thus the vertical white lines around log(time left until collapse) $\sim$ 3 are only due to the drawing technique and the models remain convective in between neighbouring black vertical lines.

significantly larger helium cores, less left over carbon and leads to radiative core carbon burning (*right* panel). The other differences between non-rotating and rotating models are the following. We can see that small convective zones above the central H-burning core disappear in rotating models. Also visible is the loss of the hydrogen rich envelope in the rotating models. The non-rotating 20 $M_\odot$ model is representative of the stars below 20 $M_\odot$, while the rotating 20 $M_\odot$ model is representative of the non–rotating and rotating models above 30 $M_\odot$. Above 30 $M_\odot$, all stars have very similar convective history after He–burning. They all lose their H-rich envelope and undergo core C burning under radiative conditions. The main difference between stars above 30 $M_\odot$ and the rotating 20 $M_\odot$ model is that stars above 30 $M_\odot$ have one large carbon convective shell that sits around 3 $M_\odot$ and thus does not influence much the final stages and the compactness at the pre-supernova stages. The complex history of convective zones and the uncertainties in the input physics mentioned here make it very hard to predict the exact explosion properties of a star of a given initial mass. Nevertheless, it is likely, as in the case of SAGB stars, that the same transitions would occur (e. g. from convective to radiative core carbon burning) even if the input physics changes. Convective boundary mixing (CBM) during carbon burning (and other stages), if able to change the extent of convective burning shells may affect the compactness of supernova progenitor significantly. 3D hydrodynamic simulations of convective boundary mixing will hopefully help constrain the 1D prescriptions used in stellar evolution codes (see Arnett *et al.* 2015, for a review on the topic). We present in Sect. 6 new 3D simulations of a carbon-burning shell in a 15 $M_\odot$ model (Cristini *et al.* 2017), which brings new light on CBM during this stage.

Figure 4 (*left*) shows the core masses as a function of initial mass for non–rotating (dotted lines) and rotating (solid lines) models. Since rotation increases mass loss, the final mass, $M_{\rm final}$, of rotating models is always smaller than that of non–rotating ones. Note that for stars with $M \gtrsim 50 M_\odot$ mass loss during the WR phase is proportional to the actual mass of the star. This produces a convergence of the final masses (see for instance Meynet & Maeder 2005). We can again see a general difference between the effects of

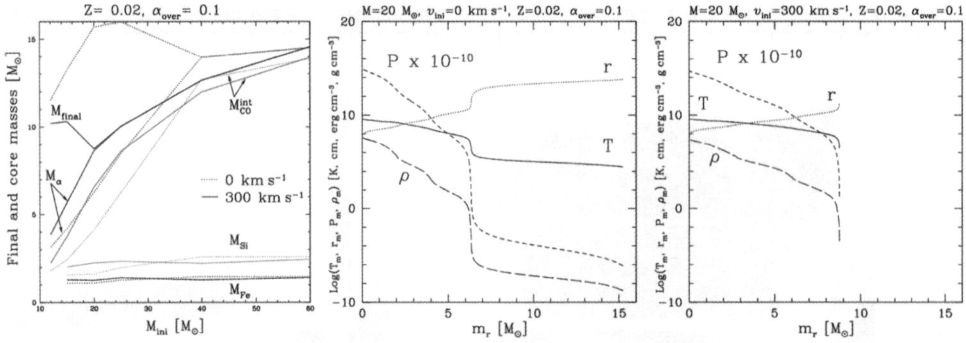

Figure 4. Pre-supernova properties (end of Si-burning): core masses as a function of the initial mass and velocity (*Left*). Profiles of the radius, r, density, ρ, temperature, T and pressure P for the non–rotating (*middle*) and rotating (*right*) 20 $M_\odot$ models. The pressure has been divided by 10^{10} to fit it in the diagram.

rotation below and above $30\,M_\odot$. For $M \lesssim 30\,M_\odot$, rotation significantly increases the core masses due to mixing. For $M \gtrsim 30\,M_\odot$, rotation makes the star enter the WR phase at an earlier stage. The rotating star spends therefore a longer time in this phase characterised by heavy mass loss rates. This results in smaller cores at the pre-supernova stage. We can see in Fig. 4 that the difference between rotating and non–rotating models is the largest between 15 and 25 $M_\odot$. As explained above, this will have an impact on the compactness in this sensitive mass range. Improvements in input physics may reconcile model predictions with observationally determined masses of type II supernova, with a maximum below 20 $M_\odot$, named the RSG problem by Smartt (2009) if the mass range of high compactness ends up covering the mass range between about 17 and 22 $M_\odot$, while more massive stars explode as type Ib or Ic supernova or fail to explode (see also Georgy *et al.* 2012; Walmswell & Eldridge 2012).

As well as the chemical composition (abundance profiles and core masses) at the pre-supernova stage, other properties, like the density profile, the neutron excess, the entropy and the total radius of the star, play an important role in the supernova explosion. Figure 4 (*middle*: non-rotating and *right*: rotating 20 $M_\odot$ models) shows the density, temperature, radius and pressure variations as a function of the Lagrangian mass coordinate at the end of the core Si–burning phase. The rotating model has lost its envelope, this truly affects the parameters towards the surface of the star. The radius of the star (BSG) is about one percent that of the non–rotating star (RSG). As said above this modifies strongly the supernova explosion. We also see that temperature, density and pressure profiles are flatter in the interior of rotating models due to the bigger core sizes.

5. Metallicity Effects

The effects of metallicity on stellar evolution are described in several studies (see for example Heger *et al.* 2003; Chieffi & Limongi 2004; Meynet *et al.* 1994). A lower metallicity implies a lower luminosity which leads to slightly smaller convective cores. A lower metallicity also implies lower opacity and lower mass losses (as long as the chemical composition has not been changed by burning or mixing in the part of the star one considers). So at the start of the evolution lower metallicity stars are more compact and thus have bluer tracks during the main sequence. The lower metallicity models also have a harder time reaching the red supergiant (RSG) stage (see Maeder & Meynet 2001, for a detailed discussion). Non–rotating models around $Z = 10^{-3}$ becomes a RSG only after the end of core He-burning and lower metallicity non–rotating models never reach the RSG stage. At even lower metallicities, as long as the metallicity is above about $Z = 10^{-10}$, no significant differences have been found in non-rotating models. Below this metallicity and for metal free stars, the CNO cycle cannot operate at the start of H–burning. At the end of its formation, the star therefore contracts until it starts He-burning because the pp–chains cannot balance the effect of the gravitational force. Once enough carbon and oxygen are produced, the CNO cycle can operate and the star behaves like stars with $Z > 10^{-10}$ for the rest of the main sequence. Shell H–burning still differs between $Z > 10^{-10}$ and metal free stars. Metal free stellar models are presented in Chieffi & Limongi (2004), Umeda & Nomoto (2005), Ekström *et al.* (2008) and Heger & Woosley (2010).

How does rotation change this picture? At all metallicities, rotation usually increases the core sizes, the lifetimes, the luminosity and the mass loss. Maeder & Meynet (2001) and Meynet & Maeder (2002) show that rotation favours a redward evolution and that rotating models better reproduce the observed ratio of blue to red supergiants (B/R) in the small Magellanic cloud. Rotating models around $Z = 10^{-5}$ become RSGs during shell He–burning. This does not change the ratio B/R but changes the structure of the star when the SN explodes. At even lower metallicities ($Z = 10^{-8}$ models presented in Hirschi 2007), the 20 $M_\odot$ models do not become RSG. However, more massive models do reach the RSG stage and the 85 $M_\odot$ model even becomes a WR star of type WO (see below). Maeder & Meynet (2001) also find that a larger fraction of stars reach break-up velocities during the evolution. The impact of rotation on nucleosynthesis and in particular the (not-so) weak *s*-process in low-Z massive rotating stars was studied in detail by Frischknecht *et al.* (2012, 2016).

Note that fast rotation at low metallicities may lead to energetic explosions such as gamma-ray bursts (see e.g. Hirschi *et al.* 2005; Heger *et al.* 2005; Yoon *et al.* 2006).

6. 3D Hydrodynamic Simulations of Convection in Carbon Burning

Due to the complex nature of stars, stellar models would ideally require three-dimensional (3D) (magneto-)hydrodynamic models that include all the relevant physics. 3D hydro models must use time steps that are at most days. The total lifetime of stars, however, is at least millions of years. This explains why most stellar evolution models are limited to (spherically-symmetric) one dimension (1D). The predictive power of 1D models, however, is crippled by 1D prescriptions of 3D phenomena containing free parameters, which need to be tuned to reproduce subsets of observations. A key uncertain prescription in 1D codes is that of convection, in particular convective boundary mixing (CBM). The mixing length theory (MLT) of convection used in most codes dates back to Böhm-Vitense (1958). The main deficiency of MLT (or MLT updates, e. g. Canuto & Mazzitelli 1991) is that it does not provide a treatment of the convective boundary. For the boundary placement, codes must use even simpler prescriptions based on linear analysis (Schwarzschild or Ledoux criterion) and *have to* add CBM to reproduce the main-sequence (MS) width (Brott *et al.* 2011; Ekström *et al.* 2012). The post-MS evolution is strongly affected by these choices, leading to uncertain predictions (Martins & Palacios 2013). Furthermore, SN progenitor structure is sensitive to the convection history as mentioned in Sect. 4. Asteroseismic observations are able to constrain further CBM choices (Mosser *et al.* 2014; Georgy *et al.* 2014). However, additional guidance and constraints are needed to improve CBM prescriptions in 1D codes and in turn the predictive power of stellar evolution. The computing power has finally reached the point where convective boundaries can be resolved in the largest simulations with $\geqslant 1024^3$ resolution (see e.g. Woodward *et al.*

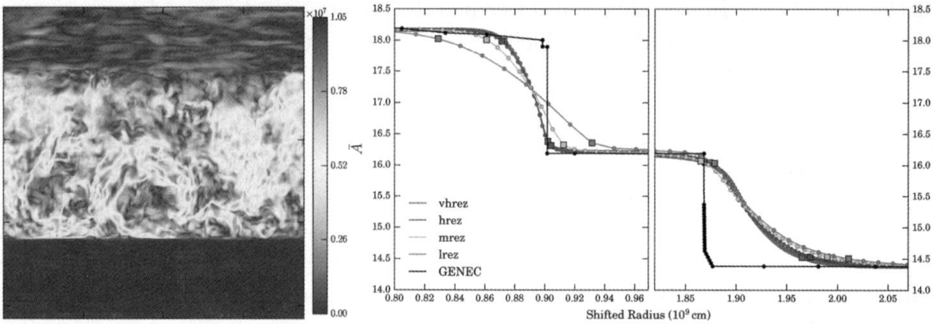

Figure 5. Vertical (2D) snapshot taken at 2,820 s into the 512^3 (hrez) simulation of the velocity magnitude in the plane of the snapshot, $\sqrt{v_x^2 + v_y^2}$, x being the vertical direction (*left*). The colour-map represents the velocity magnitude in cm s^{-1}. Radial compositional profiles at the lower (*middle*) and upper (*right*) convective boundary regions for the last time step of each model. The radius of each profile is shifted such that the boundary position coincides with the boundary position of the vhrez model. In this sense, it is easier to assess the convergence of each model's representation of the boundary at the final time-step. Individual mesh points are denoted by filled circles. Approximate boundary extent (width) is indicated by the distance between two filled squares for each resolution. The initial composition profile provided by the 1D GENEC code (Hirschi *et al.* 2004) is shown in black for a qualitative comparison only (see Cristini *et al.* 2017, for more details).

2015; Arnett *et al.* 2015; Jones *et al.* 2017). This means that now these simulations can provide the guidance and constraints to build the next generation of stellar evolution models including 3D-hydro-based (+asteroseismology-guided) prescriptions for CBM.

We summarise in this section the results of Cristini *et al.* (2017) who studied CBM in a carbon-burning convective shell in a 15 $M_\odot$ model using the PROMPI code (Meakin & Arnett 2007). A snapshot from these simulations is shown in Fig.5 (*left*). Entrainment of material at both convective boundaries was observed in all the simulations and was analysed in the framework of the entrainment law (Fedorovich *et al.* 2004; Garcia & Mellado 2014). The entrainment rate in our simulations was found to be roughly inversely proportional to the bulk Richardson number, $\mathrm{Ri_B}$ ($\propto \mathrm{Ri}^{-\alpha}$, $0.5 \lesssim \alpha \lesssim 1.0$). We also found that convective boundaries are broadened by shear mixing (Kelvin-Helmholtz instability) due to the fact that the flow has to do a U-turn at the boundaries. We estimated the boundary widths (see Fig. 5 *middle* and *right*) and found these to be roughly 30% and 10% of the local pressure scale height for the upper and lower convective boundary, respectively. While these widths are only estimates, they confirm that the lower boundary is narrower than the upper boundary. More importantly, the abundance profiles in the 3D simulations is much smoother than in the 1D model, which does not include any CBM.

7. Summary and Outlook

In this paper, we reviewed the general properties and evolution of CCSNe progenitors in the framework of 1D stellar evolution models. We also discussed the limitations of 1D models and how 3D hydrodynamic simulations can provide unique and crucial constraints for convective boundary mixing in massive stars, which is one of the major uncertainties affecting the progenitors of CCSNe. We then presented new 3D hydrodynamic simulations of convection in carbon burning. These new carbon burning simulations confirm the general findings of the oxygen shell simulations presented in Meakin & Arnett (2007).

This is promising for the long term goal of developing a convective boundary mixing prescription for 1D models which is applicable to all (or many) stages of the evolution of stars (and not only to the specific conditions studied in 3D simulations). The luminosity (driving convection) and the bulk Richardson number (a measure of the boundary stiffness) will be key quantities for such new prescriptions (also see Arnett *et al.* 2015).

The goal of 1D stellar evolution models is to capture the long-term (secular) evolution of the convective zones and of their boundaries, while 3D hydrodynamic simulations probe the short-term (dynamical) evolution. Keeping this in mind, the key points to take from existing 3D hydrodynamic studies for the development of new prescriptions in 1D stellar evolution codes are the following:

- Entrainment of the boundary and mixing across it occurs both at the top and bottom boundaries. Thus 1D stellar evolution models should include convective boundary mixing at both boundaries. Furthermore, the boundary shape is not a discontinuity in the 3D hydrodynamic simulations but a smooth function of radius, sigmoid-like, a feature that should also be incorporated in 1D models.
- At the lower boundary, which is stiffer, the entrainment is slower and the boundary width is narrower. This confirms the dependence of entrainment and mixing on the stiffness of the boundary.
- Since the boundary stiffness varies both in time and with the convective boundary considered, a single constant parameter is probably not going to correctly represent the dependence of the mixing on the instantaneous convective boundary properties. As discussed above, we suggest the use of the bulk Richardson number in new prescriptions to include this dependence. More in-depth discussions on this topic can be found in Arnett *et al.* (2015); Viallet *et al.* (2015)

Acknowledgements

The authors acknowledge support from EU-FP7-ERC-2012-St Grant 306901. RH acknowledges support from the World Premier International Research Centre Initiative (WPI Initiative), MEXT, Japan and the "ChETEC" COST Action (CA16117), supported by COST (European Cooperation in Science and Technology). This work used the Extreme Science and Engineering Discovery Environment (XSEDE), which is supported by National Science Foundation grant number OCI-1053575. CM and WDA acknowledge support from NSF grant 1107445 at the University of Arizona. The authors acknowledge the Texas Advanced Computing Center (TACC) at The University of Texas at Austin (http://www.tacc.utexas.edu) for providing HPC resources that have contributed to the research results reported within this paper. This work used the DiRAC Data Centric system at Durham University, operated by the Institute for Computational Cosmology on behalf of the STFC DiRAC HPC Facility (www.dirac.ac.uk). This equipment was funded by BIS National E-infrastructure capital grant ST/K00042X/1, STFC capital grants ST/H008519/1 and ST/K00087X/1, STFC DiRAC Operations grant ST/K003267/1 and Durham University. DiRAC is part of the National E-Infrastructure.

References

Arnett, W. D., Meakin, C., Viallet, M., *et al.* 2015, *ApJ*, 809, 30
Böhm-Vitense, E. 1958, *Z. Astrophys.*, 46, 108
Brott, I., de Mink, S. E., Cantiello, M., *et al.* 2011, *A& A*, 530, A115
Canuto, V. M. & Mazzitelli, I. 1991, *ApJ*, 370, 295
Chieffi, A. & Limongi, M. 2004, *ApJ*, 608, 405

—. 2013, *ApJ*, 764, 21
Cristini, A., Meakin, C., Hirschi, R., Arnett, D.and Georgy, C., & Viallet, M. 2017, *MNRAS*, submitted
Ekström, S., Georgy, C., Eggenberger, P., *et al.* 2012, *A& A*, 537, A146
Ekström, S., Meynet, G., Chiappini, C., Hirschi, R., & Maeder, A. 2008, *A& A*, 489, 685
Ertl, T., Janka, H.-T., Woosley, S. E., Sukhbold, T., & Ugliano, M. 2016, *ApJ*, 818, 124
Fedorovich, E., Conzemius, R., & Mironov, D. 2004, *J. Atmos. Sci.*, 61, 281
Frischknecht, U., Hirschi, R., Pignatari, M., *et al.* 2016, *MNRAS*, 456, 1803
Frischknecht, U., Hirschi, R., & Thielemann, F.-K. 2012, *A& A*, 538, L2
Garcia, J. & Mellado, J. 2014, *J. Atmos. Sci.*, 1935
Georgy, C., Ekström, S., Meynet, G., *et al.* 2012, *A& A*, 542, A29
Georgy, C., Saio, H., & Meynet, G. 2014, *MNRAS*, 439, L6
Heger, A., Fryer, C. L., Woosley, S. E., Langer, N., & Hartmann, D. H. 2003, *ApJ*, 591, 288
Heger, A. & Woosley, S. E. 2010, *ApJ*, 724, 341
Heger, A., Woosley, S. E., & Spruit, H. C. 2005, *ApJ*, 626, 350
Hirschi, R. 2007, *A& A*, 461, 571
Hirschi, R., Meynet, G., & Maeder, A. 2004, *A& A*, 425, 649
—. 2005, *A& A*, 443, 581
Jones, S., Andrassy, R., Sandalski, S., *et al.* 2017, *MNRAS*, 465, 2991
Jones, S., Hirschi, R., & Nomoto, K. 2014, *ApJ*, 797, 83
Jones, S., Hirschi, R., Nomoto, K., *et al.* 2013, *ApJ*, 772, 150
Jones, S., Röpke, F. K., Pakmor, R., *et al.* 2016, *A& A*, 593, A72
Langer, N. 2012, *ARA& A*, 50, 107
Maeder, A. & Meynet, G. 2001, *A& A*, 373, 555
Martins, F. & Palacios, A. 2013, *A& A*, 560, A16
Meakin, C. A. & Arnett, D. 2007, *ApJ*, 667, 448
Meynet, G. & Maeder, A. 2002, *A& A*, 390, 561
—. 2005, *A& A*, 429, 581
Meynet, G., Maeder, A., Schaller, G., Schaerer, D., & Charbonnel, C. 1994, *A& AS*, 103, 97
Mosser, B., Benomar, O., Belkacem, K., *et al.* 2014, *A& A*, 572, L5
Nomoto, K. 1984, *ApJ*, 277, 791
—. 1987, *ApJ*, 322, 206
O'Connor, E. & Ott, C. D. 2011, *ApJ*, 730, 70
Poelarends, A. J. T., Herwig, F., Langer, N., & Heger, A. 2008, *ApJ*, 675, 614
Ritossa, C., García-Berro, E., & Iben, Jr., I. 1999, *ApJ*, 515, 381
Smartt, S. J. 2009, *ARA& A*, 47, 63
Sukhbold, T. & Woosley, S. E. 2014, *ApJ*, 783, 10
Takahashi, K., Yoshida, T., & Umeda, H. 2013, *ApJ*, 771, 28
Tur, C., Heger, A., & Austin, S. M. 2009, *ApJ*, 702, 1068
Umeda, H. & Nomoto, K. 2005, *ApJ*, 619, 427
Viallet, M., Meakin, C., Prat, V., & Arnett, D. 2015, *A& A*, 580, A61
Vink, J. S., ed. 2015, Astrophysics and Space Science Library, Vol. 412, Very Massive Stars in the Local Universe (Springer International Publishing Switzerland)
Walmswell, J. J. & Eldridge, J. J. 2012, *MNRAS*, 419, 2054
Woodward, P. R., Herwig, F., & Lin, P.-H. 2015, *ApJ*, 798, 49
Woosley, S. E. & Heger, A. 2015, *ApJ*, 810, 34
Yoon, S.-C., Langer, N., & Norman, C. 2006, *A& A*, 460, 199

Supernova 1987A:30 years later - Cosmic Rays and Nuclei from Supernovae and their aftermaths
Proceedings IAU Symposium No. 331, 2017
A. Marcowith, M. Renaud, G. Dubner, A. Ray & A. Bykov, eds.

doi:10.1017/S1743921317004483

Evolution & Explosion of Massive Stars Leading to IIP-IIL SNe with MESA & SNEC

Sanskriti Das[1] and Alak Ray[2]

[1]Department of Physics, Indian Institute of Technology Bombay, Mumbai- 400076, India
email: dassanskriti@gmail.com

[2]Dept. of Astronomy & AstroPhysics, Tata Institute of Fundamental Research, Mumbai-400005,India
email: akr@tifr.res.in

Abstract. We show how the dense shells of circumstellar gas immediately outside the red supergiants(RSGs) can affect the early optical light curves of Type II-P SNe taking the example of 2013ej. The peak in V, R and I bands, decline rate after peak and plateau length are found to be strongly influenced by the dense CSM formed due to enhanced mass loss during the oxygen and silicon burning stage of the progenitor. We find that the required explosion energy for the progenitors with CSM is reduced by almost a factor of 2.

Keywords. Supernovae, hydrodynamics, radiative transfer, II-P, II-L, CSM, individual: 2013ej

1. Introduction

Massive stars (ZAMS mass $> 8M_\odot$) end up their lives through core-collapse supernovae (CCSNe). Type-II SNe, a class of CCSNe are identified by the P-cygni profile of hydrogen in their early spectra. Type II-P SNe cover a large fraction ($\sim 48\%$) of the whole type-II SNe population (Smith *et al.* 2011). These have pronounced plateaus in their visible band light curves that remain within 1 mag of maximum brightness for an extended period e.g. 60-100 rest frame days and is followed by exponential tail at late times. II-Ls were originally separately classified by their brighter peak and linearly falling luminosity after peak, but Sanders *et al.* 2015 shows that II-P and II-L form a continuous and statistically indistinguishable class of CCSNe. There has been search for a range of progenitors producing a continuous transition from II-P to II-L by arguing the qualitative similarity in their light curve pattern but so far it has not been possible to simulate II-L as an extreme case of II-P SNe (Morozova *et al.* 2015). Type II-n SNe show narrow emission lines of H-α due to the interaction of shock and the stellar ejecta with the circumstellar medium. So far, II-P/II-L and II-n have been kept into separate categories by virtue of their observed characteristics and inferred progenitor properties. Recently there has been considerations on the effect on the early light curves of II-P/II-L SNe due to dense circumstellar material immediately outside the progenitors (Valenti *et al.* 2015; Morozova *et al.* 2017; Yaron *et al.* 2017) and a subclass of *moderately interacting SNe* has been identified. This may act as an intermediate of II-P/II-L and II-n and fill the gap between observed ZAMS (Zero Age Main Sequence) mass range of their respective progenitors(Moriya *et al.* 2011).

We have evolved the stars in MESA (Modules for Experiments in Stellar Astrophysics)(Paxton *et al.* 2011,Paxton *et al.* 2013,Paxton *et al.* 2015) since their pre-ZAMS stage till the Fe core collapse with a history of enhanced mass loss rate in last few years, and constructed the CSM from that information. Our method is self-consistent in two

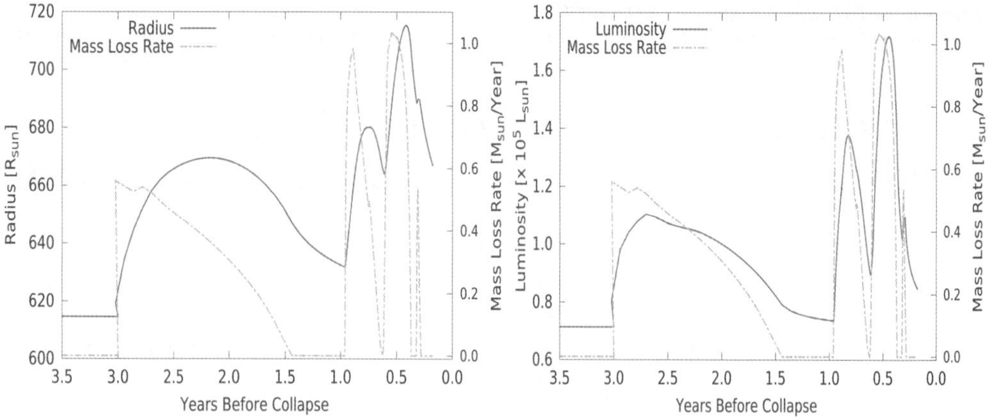

Figure 1. Variation in luminosity and radius with large mass loss rate in late stage of stellar evolution of a model of ZAMS mass 13 $M_{\odot}$.

ways: firstly, since the simulation of enhanced mass loss over a timescale of few years reveals the impact on surface properties (mainly luminosity and radius) naturally from the evolution (Fig. 1), the pre-SN progenitor carries the trace of the event with it and secondly, the modelling of the CSM also accounts for these changes in the star, and thus the CSM profile becomes more realistic. This is in contrast with the earlier works on the effect of CSM on II-P SNe where artificially designed CSM profile is stitched to the pre-SN star independently modelled from stellar evolution codes (e.g. KEPLER) without any huge mass loss history (Morozova *et al.* 2017; Moriya *et al.* 2017). The progenitor with CSM is then exploded via SNEC (SuperNova Explosion Code)(Morozova, Ott & Piro 2015) and the light curve is compared against the light curve of the progenitor with no excess mass loss history and hence no dense CSM in the immediate vicinity of the exploding star. We make a comparative study of the models best fitting the optical light curve of Type II-P SN2013ej (sometimes described as II-L because of fast decline of 1.74 mag/100 days in intermediate stage) and show in next section how the models with CSM improves the fitting.

2. Circumstellar medium structure and explosion characteristics

To study the effect of CSM in the context of SN2013ej we have taken $0.295Z_{\odot}$ metallicity, which is the equal to the metallicity of the nearby HII region number 197 of Cedrés *et al.* 2012 and ZAMS mass range of 12-19 $M_{\odot}$ for all of our MESA models. We use Ledoux criterion for convection, and take mixing length parameter $\alpha_{mlt} = 2$, exponentially decreasing overshooting (both for the core and the convective shells) with $f_{ov} = 0.025$ and $f_0 = 0.05$, and semi-convection efficiency $\alpha_{sc} = 0.1$. Near the center of the star we increase mass resolution with mesh_delta_coeff_for_highT = 1. During continuous mass ejection temporal resolution has been kept to default value with varcontrol_target $= 10^{-4}$. The resolution is increased by using max_timestep $= 0.05 \times 10^7$ seconds during the episodes of enhanced mass loss rate.The average mass loss rate has been kept to $\sim 10^{-6} M_{\odot} yr^{-1}$ (Vink's scheme for hot ($T_s > 10^4 K$) with wind η=1 and Dutch scheme for cool wind with η=1 and 0.5) constrained by X-ray studies which has probed mass loss rate in the range of 40-400 years before the explosion (Chakraborti *et al.* 2016). The enhanced mass loss rate was triggered by hydrogen stripping (remove_H_wind_mdot) and flashes (flash_wind_mdot) with an upper limit of $\sim 1 M_{\odot} yr^{-1}$ in last ~ 3 years of

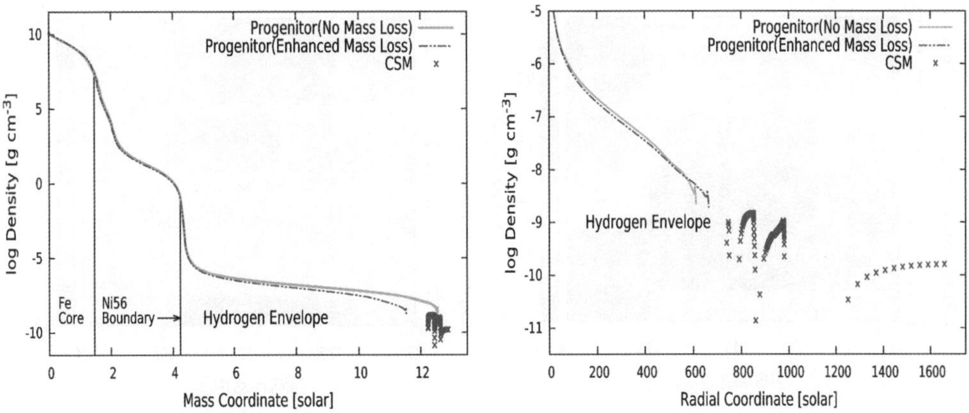

Figure 2. Pre-SN density profile of a progenitor of ZAMS mass 13 $M_{\odot}$. The model without enhanced mass loss and the one with shells of CSM have been shown in same plot for comparison. The radial profile of the central part of $\sim 4.5 M_{\odot}$ has been excluded from the plot on right side to show the CSM with better resolution.

evolution. Any heavy mass loss much before this would move away too far and become dilute to substantially affect the early light curves. Noting that the bolometric luminosity and radius vary with mass loss rate with certain correlation (Fig. 1), the CSM profile has been calculated from the history of mass loss rate, radius and surface temperature assuming asymptotic wind speed of 10 km/s. Although the launch velocity was found to be correlated with mass loss rate, we assume the ejected gas has had enough time to relax to the asymptotic speed during the episodic ejection. The composition in CSM is assumed to be the same as that of the surface. In contrast to a smooth continuous inverse square radial profile of density our constructed CSM is a collection of discrete shells of non-monotonically varying density. The resultant CSM has been stitched to the progenitor (e.g. Fig. 2) and exploded via SNEC using a thermal bomb. The explosion energy has been varied in the range of $0.4 - 1.4 \times 10^{51}$ erg. ^{56}Ni mass has been kept fixed at $0.0207 M_{\odot}$ and evenly spread from outside the Fe core to the middle of He core (Fig. 2). After boxcar smoothing this distribution gets modified with a tail extending upto the He core boundary.The mass of the NS has been taken to be equal to Fe core mass. The light curves have been calculated till 120 days to cover a part of the radioactive tail after plateau. The magnitude in different bands have been calculated using proper bolometric corrections with blackbody approximation.

The resultant light curves in V, R and I bands have been compared with the data given by Richmond 2014 and Yuan *et al.* 2016.† The reduced error of our SNEC outputs against the data has been defined following Morozova *et al.* 2017 as

$$\chi_{\nu}^2 = \sum_{V,R,I} \sum_{t_i \leqslant t_{pl}} \frac{(m_{obs}(t_i) - m_{th}(t_i))^2}{\sigma_i^2} \times \frac{1}{(n_V + n_R + n_I - 3)} \tag{2.1}$$

Here t_{pl} is plateau length which is $\sim$ 99 days ‡, $m_{th} = M_{th}$ + (distance and extinction correction), n_{λ} is no. of data points in a particular band. All sources of errors i.e. the photometric error in observational data, uncertainty in distance and extinction estimation

† Since SNEC does not calculate Fe line blanketing in U and B bands, these bands are excluded from goodness of fit analysis.

‡ By fitting the V-band light curve with the function $y(t) = -a_0/(1 + exp\frac{t-t_{pl}}{w_0}) + (p_0 \times t) + m_0$ as proposed in Valenti *et al.* 2016, the best fit value of the plateau duration is t_{pl}= 98.77 days

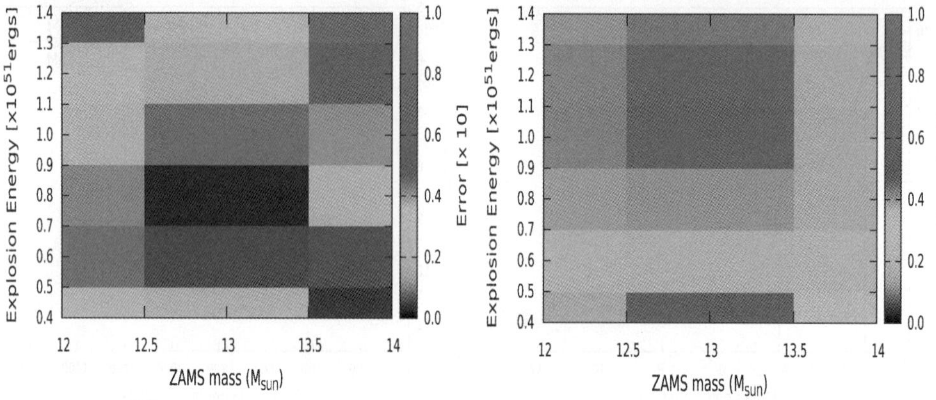

Figure 3. Error maps of the models of different ZAMS mass and explosion energy against V,R and I bands of SN2013ej (data: Yuan *et al.* 2016). The photometric error in observational data, errors in estimation of the distance of host galaxy and average galactic extinction (taken from Huang *et al.* 2015) have been included in the error analysis.

have been included in σ. We have used the distance and interstellar extinction correction quoted in Richmond 2014, Bose *et al.* 2015, Huang *et al.* 2015 and Yuan *et al.* 2016. We find that for the set of values quoted in Richmond the best fitted model with CSM has ZAMS mass 12- $13M_{\odot}$ and explosion energy $E_{exp} = 0.6 \times 10^{51}$ergs, while the values of Yuan and Bose predict best fitted ZAMS mass $= 13M_{\odot}$ and a range of $E_{exp} = 0.6 - 0.8 \times 10^{51}$ergs. Huang has considered a larger value of extinction by taking the reddening of host galaxy M74 into account. Using their values we can pinpoint a ZAMS mass $13M_{\odot}$ and $E_{exp} = 0.8 \times 10^{51}$ergs. The explosion of models without CSM deviate from the data with a different distribution (Fig. 3), and with an error larger by a factor of ~6-8 than explosions powered by the circumstellar medium (Table 1). The results have been cross-checked against the data from Valenti *et al.* 2014 and Bose *et al.* 2015 as well. It turns out that the model with CSM best fitted using Huang's values is visually closest to the data (Fig. 4). It strengthens the consideration of host extinction which is confirmed by recent analysis of massive star population around the site of explosion(Maund 2017). Although it was intended to fit upto the plateau only, since SNEC is not reliable beyond that region (Morozova, Ott & Piro 2015), the model with dense and nearby CSM fits the radioactive tail reasonably well in three bands. Also, the model predicts the peak magnitude in U and B bands correctly. On the other hand, the best-fit model without CSM hardly satisfies any feature of the light curve. This analysis is consistent with the spectroscopic observations of Bose *et al.* 2015 where a weak ejecta-CSM interaction was inferred from the high velocity $H\alpha - H\beta$ profiles. This suggests that the progenitor of SN2013ej was surrounded by dense CSM and supports the suggestion of Smith & Arnett 2014 of heavy mass loss immediately prior to the SN explosion.

3. Discussion

In the very late stage of post-main-sequence evolution of RSGs the mass loss rate can increase by several orders of magnitude (Quataert & Shiode 2012; Smith & Arnett 2014; Moriya 2014). This can give rise to a dense and slowly moving circumstellar gas just outside the star. Once the shock comes out of the star it hits onto the CSM and deposits a larger fraction of its energy to the CSM which gets diffusively radiated and brightens the early optical light curves by degrading the adiabatic loss.

Table 1. Overview of progenitor and explosion properties best-fitted for SN2013ej

ZAMS mass [$M_\odot$]	Pre-SN mass [$M_\odot$]	CSM mass [$M_\odot$][1]	Pre-SN radius [$R_\odot$]	CSM radius [$R_\odot$][2]	Energy [10^{51}ergs][3]	Goodness of Fit [χ^2_ν]
13	11.60	0.76	667	1650	0.6-0.8	0.0019-0.0028
13	12.36	—	617	—	1.0-1.2	0.0124-0.0227

Notes:
[1] lost mass during enhanced mass loss rate in last ($\sim 2-3$)few years.
[2] the external radius of the dense CSM formed in the last few years before explosion
[3] asymptotic energy of the shock after breakout
[4] ^{56}Ni mass was kept fixed at $0.021M_\odot$

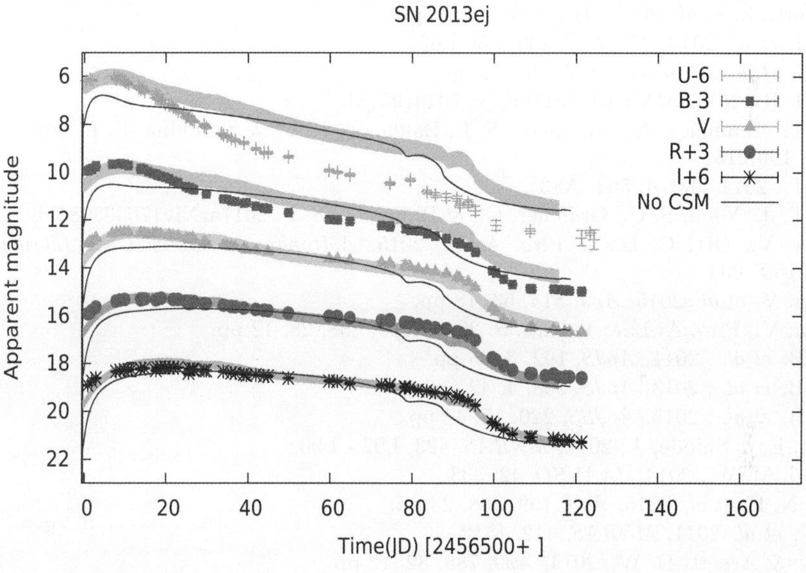

Figure 4. The UBVRI light curves of SN2013ej (data: Yuan *et al.* 2016). The curves for best-fitted model (fitted against VRI bands till the plateau of $\sim$ 99 days) with CSM are shown with the errors at each point. Values of distance and extinction estimation have been taken from Huang *et al.* 2015 [$d = 9.6 \pm 0.5$Mpc, $A_V = 0.37 \pm 0.19$ with relative extinction relation borrowed from Cardelli *et al.* 1989]. The best-fitted model without CSM is shown in thin lines for the sake of comparison.

Here we have discussed how the same star can pass through two evolutionary tracks differing in a timescale much less than the Kelvin-Helmholtz timescale and end up as different pre-SN progenitors, although occupying roughly the same position in H-R diagram. The density, velocity and temperature profiles of hydrogen envelope of the stars with no huge mass loss history are substantially different from those of the progenitors surrounded by dense and compact CSM (Fig.2); even though total mass (envelope + CSM) is same in both cases. So the rise and plateau of the light curve, which are functions of hydrogen profile are expected to be different. In these cases, although the pre-SN images cannot be distinguished, the light curves of explosion will differ. We note some particular differences in the presence of dense CSM: UV and optical bands are simultaneously bright in early phase, for a fixed shock energy the optical light curve is brighter at peak, flatter after peak and the plateau is longer. The dense and compact CSM behaves like an unbound extended part of the progenitor itself. The key difference between this scenario and usual II-n SNe is that the extended CSM in II-n is optically thin, while here the photosphere lies in the compact CSM for $\sim$40 days since explosion. In usual II-n the interactive phase lasts for $\sim$1 yr but in our case that is $\sim$ 2-3 days and the breakout happens when the shock has already traversed through the CSM. Radio and flash-ionization non-detection early on can constrain the radial extent of the CSM and confirm proximity and compactness (Yaron *et al.* 2017).

References

Bose, S. *et al.* 2015,*ApJ*, 806, 160, 18 pp.
Cardelli, J. A., Geoferey, G. C. & Mathis, J. S., 1989, *ApJ*, 345:245-256
Cedrés, B., Cepa, J., Bongiovanni, Á. *et al.* 2012, *A& A*, 545, A43
Chakraborti, S. *et al.* 2016, *ApJ*, 817, 22, 8 pp.
Fraser, M. *et al.* 2014, *MNRAS*, 439, L56-L60
Huang, F. *et al.* 2015, *ApJ*, 807, 59 , 12 pp.
Maund, J. R., 2017, *MNRAS*, 2017arXiv170401957M
Moriya, T., Tominaga, N., Blinnikov, S. I., Baklanov, P. V., & Sorokina, E. I., 2011, *MNRAS*, 415, 199-213
Moriya, T., 2014, *A& A*, 564, A83
Moriya, T. J., Yoon, S. C., Gräfener, G., & Blinnikov, S. I., 2017arXiv170303084M
Morozova, V., Ott, C. D., & Piro, A. L., 2015, *Astrophysics Source Code Library,* record ascl:1505.033
Morozova, V. *et al.* 2015, *ApJ*, 814, 63, 18 pp.
Morozova, V., Piro, A. L. & Valenti, S. 2017, *ApJ*, 838, 28, 12 pp.
Paxton, B. *et al.* , 2011, *ApJS*, 192, 3, 35 pp.
Paxton, B. *et al.* , 2013, *ApJS*, 208, 4, 42 pp.
Paxton, B. *et al.* , 2015, *ApJSS*, 220, 15, 44 pp.
Quataert, E. & Shiode, J., 2012, *MNRAS*, 423, L92 - L96
Richmond, M. W., 2014, *JAAVSO*, 42, 333
Sanders, N. E. *et al.* 2015, *ApJ*, 799, 208, 23 pp.
Smith, N. *et al.* 2011, *MNRAS*, 412, 1512
Smith, N. & Arnett, D. W., 2014, *ApJ*, 785, 82, 12 pp.
Valenti, S. *et al.* 2014, *MNRAS*, 438, L101-L105
Valenti, S. *et al.* 2015, *MNRAS*, 448, 2608-2616
Valenti S. *et al.* 2016, *MNRAS*, 459, 3939-3962
Yaron, O. *et al.* 2017, *Nature Physics*, 4025
Yuan, F. *et al.* 2016, *MNRAS* 461, 2003 - 2018

Supernova 1987A:30 years later - Cosmic Rays and Nuclei from Supernovae and their aftermaths
Proceedings IAU Symposium No. 331, 2017
A. Marcowith, M. Renaud, G. Dubner, A. Ray & A. Bykov, eds.

doi:10.1017/S1743921317004884

Unveiling the structure of the progenitors of type-IIP Supernovae through multi-waveband observations

F. K. Sutaria[1], Alak Ray [2], Subhash Bose[3] and Brijesh Kumar[4]

[1]Indian Institute of Astrophysics, Koramangala Block II, Bangalore 560034, India.
email: fsutaria@gmail.com

[2]Homi Bhabha Road, Navy Nagar, Colaba, Mumbai, Maharashtra 400005, India.
email:alak.ray@gmail.com

[3]Kavli Institute for Astronomy and Astrophysics, Peking University, Yi He Yuan Road 5, Hai Dian District, Beijing 100871, China.
email:email@subhashbose.com

[4]Aryabhatta Research Institute of Observational Sciences, Manora Peak, Nainital - 263001, Uttarakhand, India.
email:brij@aries.res.in

Abstract. Observational evidence from archival, pre-explosion images, suggests that progenitors of type-IIP SNe (SNe-IIP) have $8 \leqslant M_P \leqslant 17\ M_\odot$. However, the post-explosion temporal evolution of the event suggests that even in this mass range, the stellar evolutionary paths, the ensuing mass loss, and the eventual interaction of the supernova shock with the resulting CSM can show considerable diversity. Here we present the results from our program on multi-waveband (mainly optical) observations of SNe-IIP. Mass loss in their progenitors, with a massive and extended H-envelopes, is seen to occur via both strong stellar winds, or episodic mass ejections. Moreover, some type-IIP SNe also show unusually steep decline, characteristic of type-IIL (e.g. SN-IIP 2013ej). Our early and late-time spectrophotometry of these events shows CSM- shock interaction to varying degree among progenitors of comparable mass. Combined with X-ray data, our findings suggest that SNe-IIP progenitors can lose mass via strong stellar winds (e.g. SN2013ej, and SN2014cx), have episodic mass loss (SN2011ja), or have negligible mass loss (SN2012aw, SN2013ab).

Keywords. (stars:) supernovae: individual (SN2013ej, SN2013ab, SN2012aw)

1. Introduction

The advent of small and medium scale synoptic surveys over the last decade or so had greatly increased the number of supernovae events detected across all types and sub-types, while the ready accessibility of archival data had made it possible to serendipitously identify the progenitors for some of nearby events. The total number of events discovered to date stand at > 12000, and while most of them have been faint ($V > 18$) at discovery, $\sim 5\%$ of these have been bright enough to permit an optical and multi waveband study of the temporal evolution of their spectral and photometric properties, some over durations ranging from as early as $\sim$ few hr post explosion to few years. With this proliferation of observational data, the distinction between the various sub-types of type-II supernovae appears to be blurring. The spectral and temporal characteristics of supernovae being dependent on the progenitor properties, its evolutionary history, and on the interaction of the supernova shock with the environment (mainly, the circumstellar material ejected by the progenitor on it path towards the explosive finale), this has brought about renewed interest in post main-sequence evolutionary models of massive stars, and especially in the mechanisms and epochs of mass loss as a function of progenitor mass (M_{Prog}), ZAMS metallicity (Z_{Prog}) and spectral characteristics.

Core collapse supernovae (hereafter CC-SNe) in progenitors with a residual H-envelope have been traditionally classified (according to their light curves, and their early time spectra) in to types -IIP (-Plateau), -IIL (Linear), -IIn (Narrow emission lines) and -IIb (analogues of type-Ib, but with some H in the early spectra), with SNe-IIP being the most common type. The identification of the progenitors of 11 SNe-IIP, 2 SNe-IIL and 3 SNe-IIb from archival (mainly HST), imply ZAMS masses for SNe-IIP / -IIL of $8 \leqslant M_{Prog}/M_{\odot} \leqslant 15$ (see Smartt, 2015 and references therein), although early x-ray emission from some SNe-IIP suggests a possible upper limit of $M_{Prog} \leqslant 19\ M_{\odot}$. The direct observations set an overall upper limit of $\log L_{Prog}/L_{\odot} \leqslant 5.1$ dex is set for progenitors of all SNe-II. This is in contradiction with stellar evolutionary models, which can generate SNe-IIP in red super giants (RSG) with $16 < M_{RSG} < 30$ M$_{\odot}$. Moreover, the observed cosmic supernovae rate is significantly smaller than star formation rate. While there is some theoretical indication that this contradiction can be resolved if RSGs with compactness factor of $\zeta > 0.2$ collapse directly in to black holes (Horiuchi *et al.*, 2014), this solution than raises the question of the number of galactic, stellar mass, black holes – a problem which can only be resolved in the post-LIGO era. In the electromagnetic waveband, it becomes necessary to (a) increase the number of events with known or tightly constrained progenitors, and (b) to investigate both theoretically and observationally, all the possible scenarios of mass loss, including both continuous, as well as episodic (up to $\sim 1 M_{\odot}$) even in low mass ($8 < M_{RSG} < 16$ M$_{\odot}$) RSGs, occurring within a few months to a few decades immediately prior to the core-collapse.

As the supernova shock propagates through the star, and its overlying medium, its interaction illuminates the progenitor's history. As the shocked ejecta rams in to the circumstellar (or interstellar) medium (CSM or ISM), the event itself brightens and dims in the various electromagnetic bands. The optical photosphere lies within the ejecta, and moves inwards, along with the recombination front. Thus, the very early ($\leqslant 1$ d) UV/optical emission can be used to estimate the photospheric temperature of the progenitor (Rabinak *et al.*, 2011). M_{Prog} can be inferred indirectly, by modelling the light curves (see SN2013ej-opt and references therein) to estimate the mass of radioactive ^{56}Ni in the post-collapse core. Propagation of the recombination front in to the deeper layers of the ejecta, combined with the optical thinning of the ejecta itself, then reveals the inner structure of the stellar envelope, with metal lines [Fe, Ca, Sc] revealing themselves as the event reaches nebular stage. By contrast, thermal x-ray and synchrotron radio emission occur from SNe when the fast moving ($\sim 10^4$ km s^{-1}) ejecta slams into the slow moving ($\sim 10^2$ km s^{-1}) winds ejected by the progenitor (Chevalier *et al.*, 2006). While both forward and reverse shocks can emits x-rays, since the latter is embedded and propagating deep in to the dense ejecta profile, x-ray emission from this region is attenuated. Therefore, the effective x-ray photosphere lies within the CSM and moves outwards with the forward shock. The unfolding x-ray light curve therefore probes the mass-loss from progressively earlier pre-explosion epochs.

2. Observations

All of the events in the case studies presented here occurred in isolated progenitors with extended H-envelopes. All were detected early, within 0.6 (SN2013ej) to ~ 2.5 d (SN2012aw) post explosion. All were bright, with peak M_V in the range -17.3 – -18.5. While both SN2012aw and SN2013ej were x-ray bright, no early x-ray emission was

Table 1. UV/Optical/NIR characteristics of the various SNe-II discussed herein. Unless otherwise mentioned, data for SN2013ej is from SN2013ej-opt, for SN2013ab from SN2013ab-opt, and that for SN2012aw from SN2012aw-opt.

Object	Explosion date MJD	distance Mpc	$(M_V)_{peak}$	E(B-V)	$\log_{10} L_{bol}\|peak$ (UBVRI) erg s^{-1}	Plateau duration d	Plateau decline δM_V (100 d)$^{-1}$
SN2013ej	56497.3 $\pm$ 0.3	9.57 $\pm$ 0.57	-17.5	0.06 $\pm$ 0.001	42.37	$\sim$ 85	1.74
SN2013ab	56340.0 $\pm$ 1.0	24.0 $\pm$ 0.9	-17.3	0.044 $\pm$ 0.066	42.25	$\sim$ 78	0.92
SN2012aw	56002.6 $\pm$ 0.8	9.9 $\pm$ 0.1	-18.5	0.074 $\pm$ 0.008	42.12	$\sim$ 100	0.55

detected from SN2013ab. Since early radio emission is often correlated with x-ray, no search is reported in the literature for the radio counterpart of SN2013ab. Their notable optical characteristics are tabulated in 2, while the x-ray and radio properties are in 2.

The densely sampled light curves of SN2012aw (see figure 4, Bose *et al.*, 2013) reveal the presence of a local minima (followed by a sudden brightening) in the I at 31 d (42 d in V, 39 d in R), followed by a slow rise to a plateau at $\sim$ 73 d (51 d in V; 59 d in R). A comparision with the bolometric light curves shows that the VRI bands contribute $> 80\%$ to the total flux at these epochs, while the optical photospheric temperature (as inferred from the UBVRI curves) is $\sim$ 5000 K, i.e., well below that required for the onset of hydrogen - recombination. Thus, the local minima most likely signals a transition from a phase dominated by adiabatic cooling of the shocked ejecta, to one dominated by H-recombination, causing the characteristic plateaus of SNe-IIP. Since then, a similar dip has been seen observed in a few other events – notably, in SN2013ab (see figure 3, Bose *et al.*, 2015 I), and in SN2014cx around (53 d in V; Huang *et al.*, 2016). Such a minima is not seen in SN2013ej (figure 3, Bose etal.,2015 -II), which, while intrinsically brighter than 2012aw and 2013ab at discovery, declined steeply ($\delta M_V = 1.74$ (100 d)$^{-1}$) similar to a "typical" SNe-IIL rather than SNe-IIP, implying a more compact progenitor with a smaller, pre-explosion, H-envelope.

SNe-II light curves decline most rapidly in the UV bands. All 3 events in tab. 1 were detected in the *Swift*-UVOT bands for up to $>$ 100 d post-explosion. While the early light curves of SN2012aw and 2013ab decline steeply ($\sim$ 0.2 mag d^{-1} for 2013ab), by $\sim 30d$, they settle in to a plateau, as expected. By contrast, while the *Swift/UVOT* `uvw2` and `uvm2` light curves of SN-IIL 2013ej continued to decline until end of the observations at $\simeq$ 80 d, the NUV band `uvw1` light curve "plateaus" at 40 d, nearly concurrent with the `uvv` band light curve, and then sharply declines to the nebular tail powered by radioactive decay of $^{56}Ni \rightarrow^{56} Co \rightarrow^{56} Fe$ at 100 d

Temporal evolution of the spectra of all three events reveal interesting possibilities about the presence of anisotropic, inhomogeneities in the ejecta's density and abundance profiles. Weak absorption features at 4300 Åand 4850 Å, and stronger absorption features at 5500 Åand 4500 Å, associated with high velocity (hereafter HV) components of [He I] and H-β, were detected in the early spectra of SN2012aw (see Fig. 10, Bose *et al.*, 2013). The [He I] features at 17665 km/s on 7 d slowed down to 16643 km/s on 8 d, and does not appear in the spectrum taken 12 d post-explosion. A similar HV feature in the H-β is also seen to slow down from 21785 km s^{-1} to 21477 km s^{-1}, and then disappear over the same period. Finally, we note the reappearance of the HV feature in H-β near the start of the nebular phase at 104 d post-explosion, blended with [Ba II] and [Sc II] lines. While similar features have been detected in some other, well studied, luminous SNe-IIP, e.g. SN1999em, 1999gi and 2007od but they are by no means typical of SNe-IIP, and are not detected in the spectra of SN2013ab. Interestingly, these features are never seen in H-α, and their origins are yet disputed, although they may possibly be from dense

Table 2. Early radio and x-ray characteristics of our target type-IIP SNe. Unless otherwise mentioned, $3-\sigma$ upper limits are provided.

Event	x-ray		radio		
	days post-explosion	unabsorbed f_X (0.2-10 keV) erg cm^{-2} s^{-1}	days post-explosion	frequency GHz	flux f_r mJy
SN2013ej[1]	5.7	5.463×10^{-13}	7	6.7	< 35
SN2013ab[2]	4.11	$\leqslant 5.97\times10^{-13}$	–	–	–
SN202aw	4.35 [3]	$(7.8\pm2.1)\times10^{-14}$	7.65[3] 13.5[4]	20.8 21.2	0.160 ± 0.025 0.315 ± 0.01

Notes:
[1] From Swift detection (Margutti *et al.* 2013-II). Radio detection from 32m INAF-IRA in Medicina and Noto (Sokolovsky *et al.* 2013).
[2] Based on Swift observations (Margutti *et al.* 2013) with 10 keV thermal plasma model, and total column density of $N_H = 4.7\times10^{20}$ cm^{-2}.
[3] Swift detection (Immler & Brown, 2012.) and EVLA detection (Stockdale *et al.* 2012).
[4] EVLA detection (Yadav *et al.* 2012).

clumps within the ejecta itself. In this respect, the spectra of SN2013ej (see Figure 9, Bose etal.,2015 -II) are somewhat anomalous, in that we see the formation of an absorption feature in emission component of the H-α P-Cygni profile 97 d post-explosion onwards. Since 2013ej is also x-ray bright and showed x-ray emission at similar epochs, we interpret this as due to interaction of the ejecta with the CSM. Moreover, weak absorption features in the blue wing of H-α and H-β P-Cygni profiles in the early (12 d to 42 d) spectra of SN2013ej may also be early signs of CSM interaction, although they can also be modelled as HV features from [H I], or blended lines of [Si II] at 6347 and 6371 Å.

In addition to the early x-ray and radio characteristics of these events (see table 3), SN2013ej was also observed with Chandra over 5 epochs, from 28.9 d to 145.1 d post-explosion (see Table 2, Chakraborti *et al.*, 2016). The early x-ray emission is dominated by hard x-rays, generated via inverse Compton scattering of optical and UV photons in to x-ray energies. However, this component fades 28.9 d post explosion, and is replaced with a soft x-ray (thermal) component with median (0.5-8.0 keV) flux of 7.0×10^{-15} erg cm^{-2} s^{-1}. Treating the inverse Compton component as a power law, and the thermal component as a combination of bremsstrahlung and line emission from a hot plasma, x-ray spectral fitting was used to determine the column density of the cool shell embedded between the forward and the reverse shocks, as well as the temperature of the reverse shock itself. The mass of the surrounding shell, and hence the mass-loss rate as a function of look back time was then inferred as ranging from $(4.38\pm1.53)\times10^{-6}$ to $(2.61\pm0.50)\times10^{-6}$ $\dot{M}_\odot$ yr^{-1}, over look back times ranging from 48 ± 17 to 338 ± 19 yr. While no follow-up observations in x-ray or radio were carried out for the x-ray and radio-dim SN2013ab, Yadav *et al.* 2014 carried out extensive radio observations of SN2012aw at frequencies ranging from 1.3 to 32.0 GHz with the GMRT. Their radio light curves cover 8 d – 183 d post explosion, i.e. from the initial peak to early nebular stage of the optical light curve. Modelling the radio emission as synchrotron emission from relativistic electrons with a high energy cut-off, they determine a mass loss rate of 1.9×10^{-6} $\dot{M}_\odot$ yr^{-1} (see table 2) for the progenitor of this event.

3. Conclusions

The inferred properties of SN2012aw, 2013ab and 2013ej are given in table 2. Unless otherwise mentioned, M_P and R_P are extracted from the optical observations, using methods discussed in section I. The mass-loss rates for SN2013ej are inferred from

Table 3. Inferred properties of the supernova explosion and its progenitor from UV/optical/NIR , x-ray and radio observations

Event	Mass of ^{56}Ni $M_\odot$	Explosion energy 10^{51} erg.	M_P $M_\odot$	R_P $R_\odot$	$\dot{M}$ $\times 10^{-6}$ $M_\odot \mathrm{yr}^{-1}$	Comments Mass loss via stellar winds
SN2013ej	0.02	~ 2.3	13.7 ± 0.3 [1]	~ 450	4.38 ± 1.53 to 2.61 ± 0.50	Continuous mass-loss over look-back times ranging from 48 ± 17 to 338 ± 19 yr
SN2013ab	0.064	0.35	~ 9	~ 600	–	No evidence of a CSM seen.
SN2012aw	0.06 ± 0.01	0.9 ± 0.3	14 ± 5	337 ± 67	1.9	Assumes continuous mass-loss.

Notes:
[1] Note Based on MESA simulations of progenitors with a median mass loss rate of $(2.6 \pm 0.2) \times 10^{-6}$ $\dot{M}_\odot$ yr^{-1}, as inferred from Chandra observations (Chakraborti *et al.*, 2016). This is consistent with the optical value (from mass of ^{56}Ni) of ~ 14 $M_\odot$.

x-ray, and for SN 2012aw, from radio observations. As demonstrated, these observations provide far more stringent limits on the M_P than those set from models to the optical data alone. Archival HST exist only for fields of SN2012aw and SN2013ej, and set limits of $M_{prog}/M_\odot \leqslant 15.5$ and $8 < M_{prog}/M_\odot < 15.5$ respectively, consistent with the values derived from models to post-explosion observations.

We note that even within this very limited set of events, with $9 \leqslant M_P \leqslant 13.7$ $M_\odot$, there is a considerable diversity, both in the event characteristics, as well as in the progenitor-environment properties and their interaction. SN2012aw appears to be an unusually luminous SN-IIP, and with a progenitor mass-loss rate similar to that of SN2013ej. Moreover, the two events have similar M_P. Nevertheless, SN2013ej is closer to a "classic" type-IIL, while SN2012aw was similar in most optical aspects to the archetypal type-IIP SN1999em. It is to be noted that SN2013ej was initially identified as an SN-IIP based on its early time spectra. It is possible that it represents a transitionary event between the two classes – a feature which has been noted for several other SNe-II, as has been reported by other groups in the literature. The HV features near nebular stage may also suggest a possible bipolar distribution of ^{56}Ni, but detailed hydrodynamic models are not yet available to us, to confirm this. Events like SN2013ej demonstrate the need for spectral follow-up of SNe at several epochs, especially when classifying events for the determination of the cosmological SNe rate, as a function of the supernova type.

Next, comparing the inferred mass-loss rates of the progenitor of SN2013ej with that of SN2011ja (Chakraborti *et al.*, 2013), we note that while the former appears to have had strong stellar winds for the last ~ 400 yr of its lifetime, the latter appears to have either CSM bubbles caused by the hot progenitor wind, or was a progenitor with episodic mass loss. Finally, we note the occasional, brief, emergence of narrow line features, similar to those seen in the spectra of type-IIn events, in the H$-\alpha$ feature in the optical spectra of SNe-II, e.g., most prominently in the case of SN1999gi at 35 d and 82 d post-explosion, but also in SN2014cx (at 20.6 d and again at 27.8 d) and, to a lesser extent, in SN2016gkg. In the case of type-IIn, these features have been attributed to a dense, slow moving, clumpy CSM. Since typical $M_P/M_\odot \simeq 40$ for type-IIn, they are not expected in type-II with $M_P/M_\odot \leqslant 17$. While it is possible that in some cases this may be due to line blending from other atomic species (most notably, He), it may also imply the existence of a clumpy CSM due to episodic mass ejections from the progenitors. All of this suggests the need to explore multiple mechanisms for mass-loss from massive stars, as it could drastically change both their evolutionary track in the post main-sequence HR diagram, and the fate of the final, explosive, event.

References

Bose, S., Valenti, S., Misra, K., Pumo, M. L., Zampieri, L., Sand, D., Kumar, B., Pastorello, A., Sutaria, F., Maccarone, T. J., Kumar, B., Graham, M. L., Howell, D. A., Ochner., P., Chandola, H. C., & Pandey, S. B., 2015, *MNRAS*,450, 2373.

Bose, S., Sutaria, F., Kumar, B., Duggal, C., Misra, K., Brown, P. J., Singh, M., Dwarkadas, V., York, D. G., Chakraborti, S., Chandola, H. C., Dahlstrom, J., Ray, A., & Safonova, M., 2015, *ApJ*, 806, 160.

Bose, S., Kumar, B. , Sutaria, F., Kumar, B., Roy, R., Bhatt, V. K., Pandey, S. B., Chandola, H. C., Sagar, R., Misra, K., & Chakraborti, S., 2013, *MNRAS*, 433, 1871.

Chakraborti, S., Ray, A., Smith, R., Margutti, R., Pooley, D., Bose, S., Sutaria, F., Chandra, P. , Dwarkadas, V. V., Ryder, S. , & Maeda, K., 2016, *ApJ*, 817, 22.

Chakraborti, S., Ray, A., Smith, R., Ryder, S., Yadav, N., Sutaria, F., Dwarkadas, V. V., Chandra, P., Pooley, D., & Roy, R., 2013, *ApJ*, 774, 30.

Chevalier, R. A., Fransson, C., & Nymark, T. K., 2006, *ApJ*, 641, 1029.

Horiuchi,S., Nakamura, K., Takiwaki, T., & Kotake, K., Tanaka,M., 2014, *MNRAS*, 445, L99-L103.

Huang, F., Wang, X., Zampieri, L., Pumo, M. L. , Arcavi, I., Brown, P. J., Graham, M. L., Filippenko, A. V. , Zheng, W., Hosseinzadeh, G., Howell, D. A., McCully, C., Ruil, L., Valenti, S., Zhang T., Zhang J., Zhang K. & Wang, L., 2016, *ApJ*, 832, 139.

Immler, S. & Brown, P. J., 2012, *ATel* 3995.

Margutti, R., Soderberg, A. M. and Milisavljevic, D., 2013 *ATel* 4832.

Margutti, R., Chakraborti, S., Brown, P. J., & Sokolovsky, K. , 2013, *ATel* 5264.

Nagy, A. P., Ordasi A., Vink J., & Wheeler, J. C., 2014, *A & A*, 571, A77.

Rabinak, I. & Waxman, E., 2011, *ApJ*, 728, 63.

Stockdale, C. J., Ryder, S. D., Van Dyk, S. D., Bauer, F. E., Marcaide, J. M., Immler, S., Pooley D., Williams C. L. M., Weiler, K. W., & Horesh, A., 2012, *ATel* 4012.

Sokolovsky, K., Giroletti M., Stagni M. , Nanni M. & Mahabal, A., 2013, *ATel* 5264.

Smartt, S. J., 2015, *PASA*, 32, 16.

Valenti, S., Sand, D., Stritzinger, M., Howell, D. A., Arcavi, I., McCully, C., Childress, M. J., Hsiao, E. Y., Contreras, C., Morrell, N., Phillips, M. M., Gromadzki, M., Kirshner, R. P., & Marion, G. H., 2015, *MNRAS*, 448, 2608.

Yadav, N., Chakraborti, S., & Ray, A., 2012, *ATel* 4010.

Yadav, N., Ray, A., Chakraborti, S., Stockdale, C., Chandra, P., Smith, R., Roy, R., Bose, S., Dwarkadas, V., Sutaria, F., & Pooley, D., 2014, 782, 30.

Discussion

Supernova 1987A:30 years later - Cosmic Rays and Nuclei from Supernovae and their aftermaths
Proceedings IAU Symposium No. 331, 2017
A. Marcowith, M. Renaud, G. Dubner, A. Ray & A. Bykov, eds.

doi:10.1017/S174392131700521X

Radio and X-ray observations of supernovae in dense environments

Poonam Chandra

National Centre for Radio Astrophysics, Tata Institute of Fundamental Research
Pune University Campus, Ganeshkhind, Pune 411 007, India
email: poonam@ncra.tifr.res.in

Abstract. Here we discuss the observational properties of supernovae exploding in extremely dense environments, namely Type IIn supernovae (SNe IIn). In SNe IIn, the surrounding environments play significant role in the supernovae energetics and evolution. Thus they are different than other classes of core collapse supernovae, whose energetics are not significantly altered by their environments. Though high density of medium is a prerequisite for radio and X-ray emission, less than 10% on SNe IIn are bright in these bands. This has important implications for their progenitor models. I will discuss the radio and X-ray observations of SNe IIn, which are crucial to unravel their complex environments. We also discuss some individual supernovae belonging to this class and discuss as to how they have refined our understanding of SNe IIn. Finally the importance of well sampled long term light curves in radio and X-ray bands cannot be stressed enough.

Keywords. supernovae: general, circumstellar matter, radiation mechanisms: general, supernovae: individual(SN 1995N, SN 2006jd, SN 2010jl)

---

1. Introduction

Massive stars (mass $\geqslant 8M_{\odot}$) loose mass vigorously ($10^{-5} - 10^{-6} \mathrm{M}_{\odot} \mathrm{yr}^{-1}$ as compared to $\sim 10^{-14} \mathrm{M}_{\odot} \mathrm{yr}^{-1}$ in the Sun) from their outer layers towards the end stages of the evolution. Thus at the time of their demise as supernovae (SNe), they are surrounded by dense environments, created by the mass loss, termed as circumstellar medium (CSM). Usually the optical light curves and spectra in SNe are dominated by the explosion dynamics and the presence of the CSM has negligible effect. However, in a particular class of SNe, named Type IIn supernovae (SNe IIn), the mass loss is extremely high ($10^{-4} - 10^{-1} \mathrm{M}_{\odot} \mathrm{yr}^{-1}$), suggesting instability in the evolution of the progenitor just before the explosion. In these SNe, the CSM manifests itself in the form of narrow emission lines in their optical spectra. The bolometric and H-α luminosities become quite high due to the contribution from the interaction of the supernova (SN) ejecta with the CSM (Chugai 1990). Due to this extra source of energy, they are detectable up to cosmological distances. The highest redshift Typs IIn SN is at a redshift of 2.36 (Cooke *et al.* 2009). SNe IIn are rare and they make less than 10% of the total core collapse SNe (Taddia *et al.* 2013).

In a SN explosion, fast moving ejecta colliding with the CSM leads to a blast wave 'forward' shock with a velocitiy $v \sim 10{,}000$ km/s and temperature $T \sim 10^9$ K and a 'reverse' shock propagating into the stellar envelope with $v \sim 1000$ km/s *relative* to the *expanding* stellar ejecta, heating it to $T \sim 10^7$ K (see Fig. 1). The heated shocked plasma in the forward and reverse shocks give rise to X-ray emission. Additionally, the accelerated relativistic electrons in the forward shocked shell in the presence of the amplified magnetic field give rise to radio emission (Chevalier & Fransson 2016, Chevalier & Fransson 2003). Since the shock velocities are typically 100–1000 times the speed of the progenitor wind,

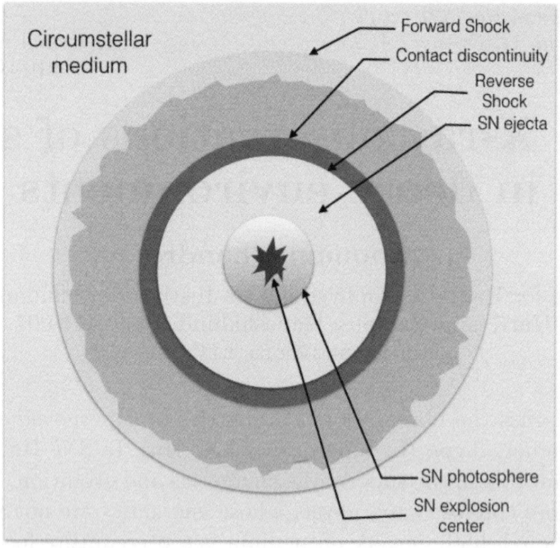

Figure 1. Schematic diagram of the ejecta interaction with the circumstellar medium. The diagram is not to scale.

the shock wave probes the past history of the star by sampling regions of the wind lost many hundreds–thousands of years ago.

In SNe IIn, the strong ejecta–CSM interaction leads to radiative shocks early on. In normal core collapse SNe, where the ejecta kinetic energy has only 1-2% contribution to the total radiated energy, in SNe IIn up to 50% of the ejecta kinetic energy can go to the radiated energy (Smith 2016). In this article, we discuss the radio and X-ray emission from SNe IIn. In §2, we discuss the SNe IIn in general, their progenitor models and diversity. Our understanding of SNe IIn through radio and X-ray observations are discussed in §3. Here we also discuss some important SNe IIn which have contributed significantly to our understanding. In §4, we summarize the main results.

2. Overview of SNe IIn

SNe IIn are not an intrinsic class of SNe exploding from a particular kind of progenitor stars. The prerequisite for their classification is a dense H-rich material surrounding the progenitor star. The diversity of the environment plays crucial role in the evolution and come to dominate the energetics of SNe IIn, e.g., a weak explosion in a dense CSM may appear very bright (Smith 2016). For this reason SNe IIn have displayed great deal of diversity. For example, SN 1998S was initially classified as a bright Type IIL SN. However, at a very early stage (< 10 days) characteristics of a IIn SN emerged. This also shows connection between IIn and IIL SNe. While SN 1988Z and SN 1986J remained bright for a decade after explosion, SN 1994W faded in only 130 days. After the discovery of unusually bright SN 2006gy with an absolute magnitude of about -21.5 (Smith *et al.* 2007), a new category of superluminous supernovae (SLSNe) has been identified. All SLSNe II are essentially IIn, except one Type IIL case, suggesting dense environments may be crucial in these SNe. A special class of SNe IIn has shown spectral similarities with Type Ia thermonuclear SNe. Some of these events are SNe 2002ic and 2005gj. They are considered hybrid between SNe Ia and IIn. Some SNe IIn also seem to be related to SNe Ib/c, e.g. SN 1995N was most likely a Type Ib/c exploding in a dense wind. SN 2001em and SN 2014C are unique SNe. They were identified at maximum light as a

SN Ib/c, later they revealed strong interaction with a dense CSM and were reclassified as SNe IIn (Chugai & Chevalier 2006, Milisavljevic *et al.* 2015).

While stellar models do not allow the kind of mass loss rates required to explain the SNe IIn, some theoretical stellar evolution models may incorporate enhanced mass loss shortly before explosion, these are super-asymptotic giant branch stars ($8 - 10M_{\odot}$), massive red supergiant stars ($17 - 25M_{\odot}$) and luminous blue variables (LBVs; $> 35M_{\odot}$). LBV progenitors are more favorable models for bright SNe IIn. However, LBV is a transient phase between O-type and Wolf-Rayet (WR) star and stars are not supposed to explode in this phase. Even if they explode, the fine tuning of the explosion at the time of enhanced episodic mass loss is not understood. In addition, Anderson *et al.* (2012) using H-α and near-UV data suggested that SN IIn progenitors come from low mass progenitors. Chevalier (2012) proposed a binary model where common envelope evolution can give rise to SNe IIn. In this model, the SN is triggered by the inspiral of a compact neutron star or black hole to the central core of the binary star.

However, archival data have suggested LBVs as progenitors in particular cases of Type IIn SN 2005gl (Gal-Yam *et al.* 2007) and SN 2010jl (Smith *et al.* 2011). In SN 2009ip, multiple pre-SN eruptions have been observed before the star exploding as SN IIn, strengthening the connection with LBV like progenitor (Mauerhan *et al.* 2013). Ofek *et al.* (2014) carried out analysis of Palomar Transient Factory archival data for more than 16 SNe IIn and inferred that more than 50% SNe IIn have precursors consistent with LBVs. In contrast, Bilinski *et al.* (2015) analysed 12 years of KAIT archival data at known SNe IIn locations and found no precusors, contradicting the above arguments. Recently Fox *et al.* (2017) presented HST/WFC3 imaging of the SN 2010jl field obtained in 2014, 2015, and 2016, where faint SN was seen 0.61" away from the blue source previously considered progenitor LBV star. Smith *et al.* (2016b) in late time HST observations found no trace of massive star formation at the SN 2009ip site. If LBVs are indeed transient phase between O-type and WR stars, they should be found in massive star forming regions. Smith & Tombleson (2015) concluded that LBVs are not found in O-type clusters and are statistically more isolated than O-type and Wolf-Rayet stars. Based on the relatively isolated environments of LBVs, they have challenged the traditional association of the LBVs and suggested that LBVs are product of binary evolution. The explained LBVs to be evolved massive blue stragglers. Smith *et al.* (2016b) suggested that the progenitor of SN 2009ip too was the product of a merger or binary mass transfer.

3. Radio and X-ray emission

SN CSM is best probed by studying the signatures of the interaction of SN ejecta with the CSM. In SNe IIn with very high CSM densities, the ejecta CSM interaction is expected to emit copiously in radio and X-ray bands. However, the statistics is contradictory. While ~ 400 SNe IIn are known, only 12 are shown to emit in X-ray bands; these include SNe 1978K, 1986J, 1988Z, 1995N, 1994W, 1996cr, 1998S, 2005ip, 2005kd, 2006gy, 2006jd and 2010jl (Ross & Dwarkadas 2017). The statistics is not much better in radio bands either. Less than half SNe IIn (~ 154) have been looked in the radio bands (mainly with the Very Large Array (VLA)) and $\sim 10\%$ have been detected.

However, there is a very interesting trend in radio and X-ray detected SNe IIn. While X-ray luminosities are towards the high end in SNe IIn, the radio luminosities are diverse and occupy 4 orders of magnitude (Fig. 2). Radio light curves are extremely diverse, the diversity translate to diversity in the stellar mass loss and could indicate the different mass loss history.

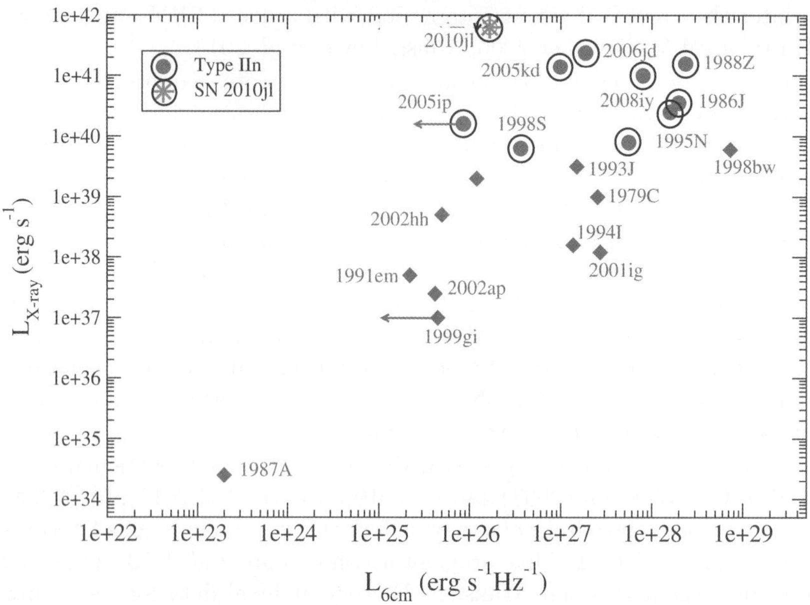

Figure 2. Peak radio and X-ray luminosities of some core collapse SNe. While radio luminosities occupy 4 orders of magnitude in SNe IIn and do not stand out from rest of the core collapse SNe, their X-ray luminosities are amongst the highest.

3.1. *X-ray Luminosity evolution*

Since reverse shock has higher density, the thermal X-ray emission from core collapse SNe, in general, has dominating contribution from the reverse shock. In SNe IIn, a radiative cooling shell forms between the reverse and forward shock. The radiative cooled shell absorbs most of the X-rays emanating from the reverse shock (Nymark *et al.* (2006). Observations have also shown that the detectable contribution from SNe IIn mainly comes from the dense forward shock (Chandra *et al.* 2012, Chandra *et al.* 2015).

While the thermal X-ray luminosity is expected to decline as $\sim 1/t$ for a steady wind, this relation is seen to be followed in hardly any core collapse SNe including SNe IIn (See 3 and (Dwarkadas & Gruszko 2012, Dwarkadas *et al.* 2016). This could be because other mechanisms like inverse-Compton scattering etc. may be at play. There is high probability that wind any not be steady $\rho \propto 1/R^2$ and has clumpy, asymmetric structure. However, a major role is also played by the constraints from the present X-ray instruments. While $\sim 1/t$ for a steady wind has been derived for a total X-ray luminosity (Fransson *et al.* 1996), current telescopes measure X-ray flux in a narrow X-ray range. This could give deviation from the expected luminosity profile, as well as may result in erroneous estimates of the parameters. It is crucial to have a wind band X-ray coverage to truly understand the nature of the X-ray emission. For example, in SN 2010jl, the joint *XMM-Newton* and *NuSTAR* spectrum (see Fig. 3) covering energy range 0.3–80 keV pinpointed the thermal X-ray emission temperature to be ~ 20 keV and established forward shock origin of the X-ray emission (Ofek *et al.* 2014, Chandra *et al.* 2015). This is the first time the (forward) shock temperature was observationally accurately measured in a SN.

3.2. *Absorption of radio emission*

On one hand high density CSM means efficient production of synchrotron radio emission, on the other hand higher density also translates into higher absorption. In classic core collapse SNe, the radio emission is absorbed either by the ionized CS medium (free-free

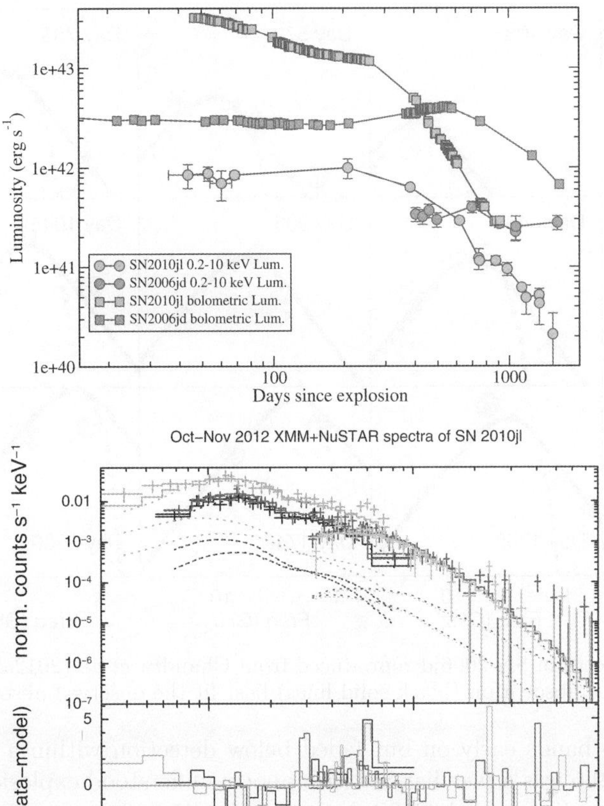

Figure 3. *U*pper panel: The blue curves are the 0.2–10 keV X-ray luminosity evolution of two SNe IIn 2006jd and 2010jl. None of them follow $\sim 1/t$ dependence. *L*ower panel: Joint *NuSTAR* and *XMM-Newton* spectrum of the SN 2010jl covering energy range 0.3–80 keV (reproduced from Chandra *et al.* (2015)).

absorption (FFA)) or by the in situ relativistic electrons (synchrotron self-absorption (SSA)). Observations when the SN is rising in the light curve evolution can disentangle the underlying absorption mechanism. The role of absorption mechanisms is different for different SNe (Chevalier *et al.* 2003). SNe IIn are expected to have external FFA due to their high density. However, in SNe IIn due to high density, radiative cooling shell is formed in the shocked region. The mixing of the cool gas in the forward shocked shell may cause some SNe IIn to undergo internal FFA. This indeed has been seen in SN 2006jd (Fig. 4 and Chandra *et al.* 2012), SN 1986J (Weiler *et al.* 1990) and SN 1988Z (van Dyk *et al.* 1993, Williams *et al.* 2002). Chandra *et al.* (2012a) have shown that modest amount of cool gas ($\sim 10^4$ K) i.e. $\sim 10^{-8} M_\odot$ can explain internal absorption in SN 2006jd.

The radio observations of SNe IIn also indicate that most of the SNe IIn are late radio emitters (e.g. Chandra *et al.* 2015). This could be partially due to the high absorption as well as partially due to observational biases. Some of the radio emitting SNe IIn were either discovered late or classified late, such as SN 1995N, 1986J, and 2006jd. SNe IIn which were discovered soon after the explosion, most of them did not have early radio observations within a month, except for SN 2009ip (Margutti *et al.* 2014). SN 2009ip was

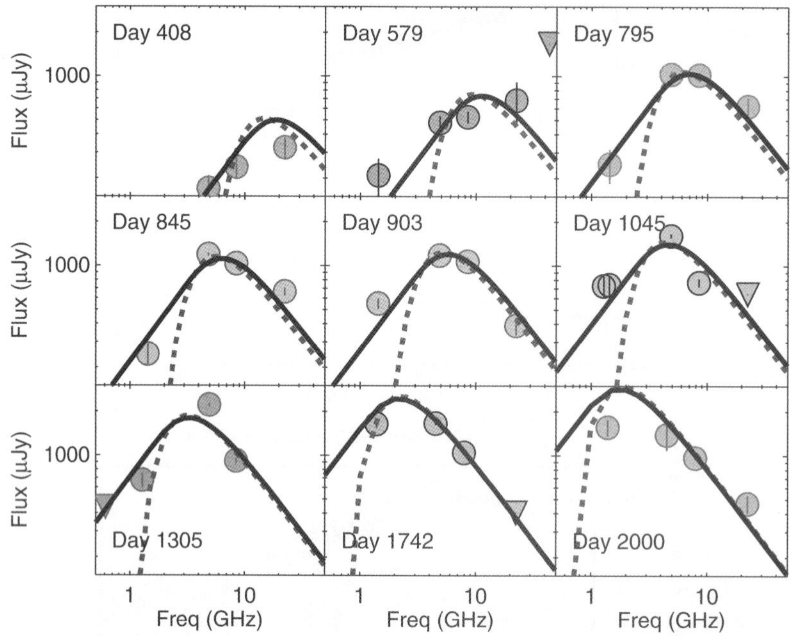

Figure 4. Spectra of SN 2006jd reproduced from Chandra *et al.* (2012a). Here internal free-free absorption (black solid lines) best fit the observed absorption.

detected in radio bands early on but faded below detection within a few tens of days. However, SN 2009ip was a peculiar SN with poorly understood explosion mechanism. In contrast the observations of SN 2010jl started by day 45 but the first detection happened only after 500 days (Chandra *et al.* 2015), clearly showcasing the efficient absorption of radio emission at early epochs.

The late radio turn on may also explain the overall low detection rates of SNe IIn. If a SN is not bright in radio bands in a first few epochs, the observational campaign for that SN is usually over. So even if a SN IIn was a potential radio emitter at later epochs, it would be missed in radio bands. van Dyk *et al.* (1996) observed 10 SNe IIn at an age of a few hundred days and still did not detect any; they set upper limits of 150 − 250 μJy. However, the current telescopes are able to reach at least an order of magnitude better sensitivity. In addition, the current upgraded VLA is extremely sensitive at higher frequencies where the radio absorption effects vanish much sooner ($\tau_\lambda \propto \lambda^2$). Strategic study of SNe IIn at higher frequencies at early times to low frequencies at late epochs is likely to result in the most complete sample of radio supernovae (Pérez-Torres *et al.* 2014, Wang *et al.*, 2015). SKA will be ideally suited for such studies at low frequencies. With an improved sensitivity level of 1 μJy with Square Kilometre Array (SKA) one can detect the brightest of radio SNe, such as the Type IIn SN 1988Z up to redshift of z=1 (Pérez-Torres *et al.* 2014),

3.3. *Evolution of column density*

SN 2010jl is the best studied X-ray supernova, where X-ray observations covered a period up to day 1500 with *Chandra*, *XMM-Newton*, *NuSTAR*, and *Swift* X-ray Telescope. This gave us opportunity to trace the evolution of column density from 40 to 1500 days. We witnessed three orders of magnitude change in column density over the span of X-ray observations (Chandra *et al.* 2012b, Chandra *et al.* 2015). At first epoch the column density associated with the SN was $N_H = 10^{24}$ cm^{-2} (for a metallicity of 0.3 solar, Stoll

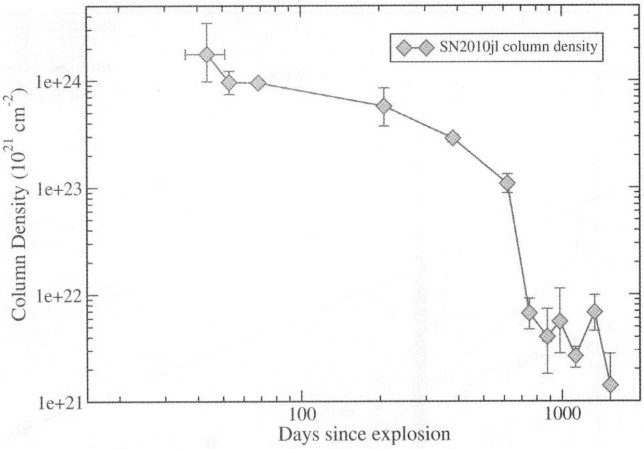

Figure 5. The column density evolution in SN 2010jl. The figure is reproduced from Chandra *et al.* (2015).

et al. 2011), 3000 times higher than the Galactic column density, which declined slowly by an order of magnitude upto day $\sim$ 650. Afterwards followed by a sharp decline, it settled to 10 times the Galactic value. The evolution of column density from Chandra *et al.* (2015) have been reproduced in Fig. 5.

The higher column density observed in the X-ray observations for SN 2010jl is not associated with the high host galaxy extinction. This indicates that the higher column density is due to the mass loss near the forward shock and thus is arising from the CSM. This was consistent with the variability of the column density too. This was the first time that external circumstellar X-ray absorption had been clearly observed in a SN and enabled us to trace the precise mass loss evolution history of the star.

3.4. *Episodic mass loss?*

It is becoming increasingly evident that the mass loss rate from some SNe IIn may not be steady. Unfortunately it is difficult to disentangle this effect because SNe IIn likely to remain highly absorbed for a long duration. By the time they become optically thin in radio bands, the strength of synchrotron emission is rather weak to see any such effects of mass loss. In addition, one needs a long term follow up to see the episodic mass loss. In X-ray observations, due to the lack of well sampled data, such trends are easy to miss. Yet, it seems that some SNe IIn seem to suggest episodic mass loss rate.

In Fig. 6, X-ray and radio light curves are plotted for a Type IIn SN 1995N (Chandra *et al.* 2009). The X-ray bands were chosen to mimic the *R*OSAT and *A*SCA bands. The radio data is at the four representative VLA frequencies. Around day $\sim$ 1300 onwards, bumps in the X-ray and radio light curves are seen. The long gap in the X-ray light curve does not allow to witness the evolution of this bump, however, in the radio data one can clearly see this effect. Due to the sensitivity limitation of the telescopes and lack of long term follow up, its not possible to explore the further bumps in time, if any. In SN 2006jd, one can also see a hint of such bumps near simultaneously in both X-ray and radio bands (Chandra *et al.* 2012a). The radio emission is too weak in SN 2010jl to note any such bumps. However, X-ray data seems to suggest a steady decline followed by a sudden decline settling down to steady decline again at $\sim$ 1000 days. (Chandra *et al.* 2015). These trends could be suggestive of the episodic mass loss rates in some SNe IIn and may support the LBV scenario. However, further study and much longer follow ups are needed.

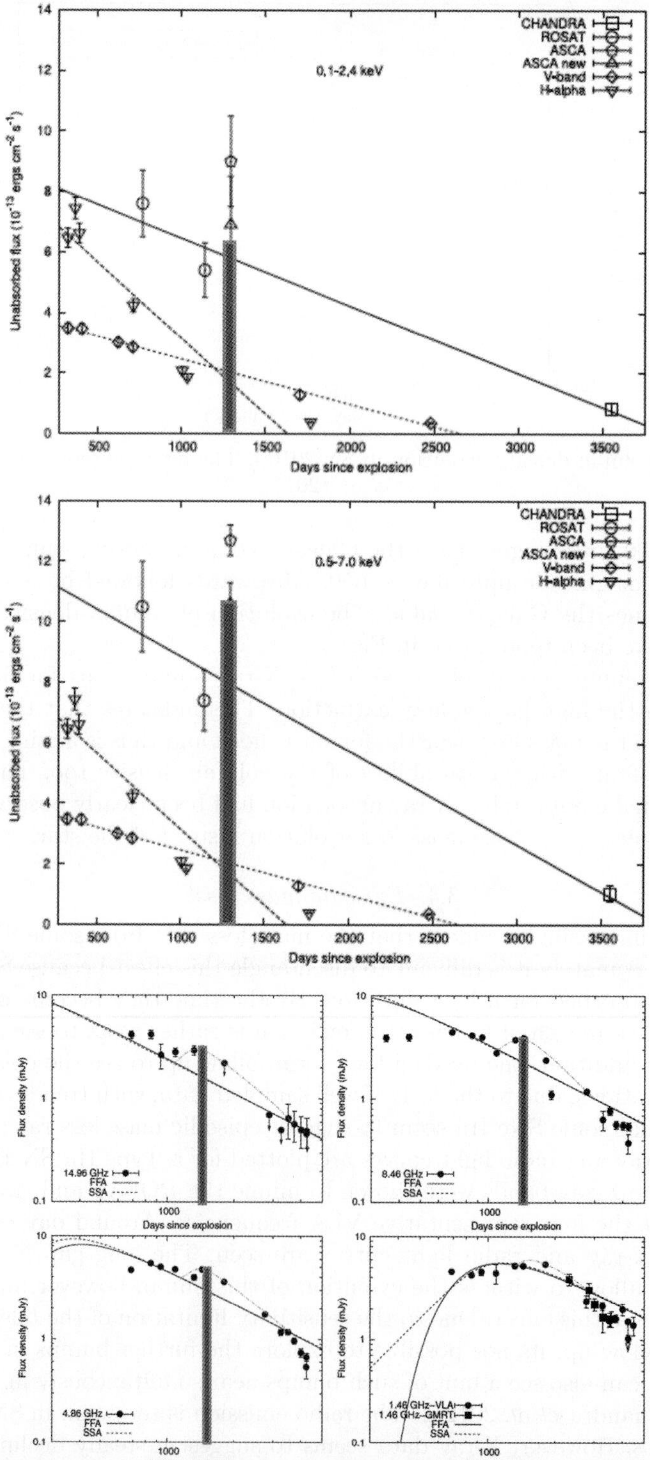

Figure 6. Radio and X-ray light curves of SN 1995N (Chandra *et al.* 2009). A hint of enhanced emission bump in radio and X-ray bands of a Type IIN SN 1995N around ~ 1300 days.

3.5. *Asymmetry in the explosion*

In optical bands, there has been various pieces of evidence suggesting asymmetry in the explosion. Polarization observations have indicated large-scale asymmetry, e.g., SN 1997eg (Hoffman *et al.* 2008), SN 1998S (Leonard *et al.* 2000), SN 2010jl (Patat *et al.* 2011). Many of the models explaining SNe IIn also require bipolar wind or disk like geometry (see Smith (2016a) for a review). In some SNe IIn, radio and X-ray interpretation of the data too seem to favour the asymmetric models.

SN 2010jl early optical spectra showed the presence of broad emission lines (Smith *et al.* 2012, Fransson *et al.* 2014, Ofek *et al.* 2014), explained by an electron scattering optical depth $> 1 - 3$, i.e., a column density $N_H \gtrsim 3 \times 10^{24}$ cm^{-2}, comparable with the column density seen in the first X-ray observations (Chandra *et al.* 2015). At later epochs ($t > 70$ day) X-rays column density declined to $N_H < 3 \times 10^{24}$ cm^{-2}. These constraints are difficult to reconcile with a spherical model of column density 3×10^{24} cm^{-2}. This may suggest that the X-rays escape the interaction region, avoiding the high column density CSM. This fits well in a scenario for the CSM having a bipolar geometry.

In SN 2006jd, it is difficult to reconcile the X-ray and radio data under spherical models (Chandra *et al.* 2012a) too. The column density of the matter to the X-ray emission was $\sim$ 50 times smaller than what was needed to produce the X-ray luminosity. The absorption column depends on the CSM wind density as $\propto \rho_w$ and the X-ray luminosity as $\propto \rho_w^2$. While clumpiness may explain this scenario, it lead to inconsistent shock velocity derived from the observed X-ray spectrum. The situation was explained if the CSM clouds interact primarily with the cool dense shell. This suggested that there is a global asymmetry in the CSM gas distribution that allowed a low column density in one direction, unaffected by the dense interaction taking place over much of the rest of the solid angle as viewed from the SN. This would also explain low column density to the radio emission too, because both radio and X-ray are from the same region. The lack of external FFA could also be explained in this model (Chandra *et al.* 2012a).

4. Summary

In this conference proceeding, we have summarized our current understanding of SNe IIn, mainly from the radio and X-ray observations. While X-ray luminosities seem to be occupying systematically high values, radio luminosities do not stand out. Due to their extreme densities, SNe IIn suffer absorption for a longer time and have tendency to turn up late in radio bands. Many SNe IIn radio spectra seem to be consistent with the internal free-free absorption, an effect negligible in other core collapse SNe. This is due to the formation of radiative shell. Cool gas (10^4 K) mixed into the hot shocked gas is efficient in doing the internal FFA. To disentangle the true nature of radio emission, multiple frequency sensitive radio observations spanning a large time span are needed.

The X-ray emission in SNe IIn is usually dominated by the forward shock, as radiative reverse shock emission is absorbed by the cool dense shell formed between the two shocks. None of the SNe IIn seem to follow X-ray luminosity evolution of $L \propto 1/t$ for steady spherical wind. Before getting into the physical reasons, one needs to understand telescope biases. Most of the X-ray telescopes works in narrow X-ray filters hence one gets spectral X-ray luminosity, whereas the above evolution is valid for the total X-ray luminosity. *NuSTAR* combined with other X-ray telescopes like *X*MM-Newton is a very promising telescope to obtain total X-ray luminosity. In addition, it is also emerging that the wind density profiles are far from standard $1/R^2$. There is certainly a hint of episodic mass ejection in some SNe IIn. SN 2010jl is an ideal example to follow the mass loss evolution due to its excellent X-ray coverage. Finally we note that since more than 50% O stars are in binary systems (Sana *et al.* 2012), one does need to take into account SN evolution under binary scenario. In addition disk like and bipolar asymmetries cannot be ignored while studying SNe IIn.

PC acknowledges support from the Department of Science and Technology via Swarana-Jayanti Fellowship award (file no. DST/SJF/PSA-01/2014-15).

References

Anderson, J. P., Habergham, S. M., James, P. A., & Hamuy, M. 2012, *MNRAS*, 424, 1372
Bilinski, C., Smith, N., Li, W., *et al.* 2015, *MNRAS*, 450, 246
Chandra, P., Stockdale, C. J., Chevalier, R. A., *et al.* 2009, *ApJ*, 690, 1839
Chandra, P., Chevalier, R. A., Chugai, N., *et al.* 2012, *ApJ*, 755, 110
Chandra, P., Chevalier, R. A., Irwin, C. M., *et al.* 2012, *ApJ* Letters, 750, L2
Chandra, P., Chevalier, R. A., Chugai, N., Fransson, C., & Soderberg, A. M. 2015, *ApJ*, 810, 32
Chevalier, R. A. & Fransson, C. 2003, *Supernovae and Gamma-Ray Bursters*, Edited by K. Weiler., Lecture Notes in Physics, 598, 171
Chevalier, R. A. 2012, *ApJ* Letters, 752, L2
Chevalier, R. A. & Fransson, C. 2016, *arXiv:*1612.07459
Chugai, N. N. 1990, *Soviet Astron.* (Letters), 16, 457
Chugai, N. N. & Chevalier, R. A. 2006, *ApJ*, 641, 1051
Cooke, J., Sullivan, M., Barton, E. J., *et al.* 2009, *Nature*, 460, 237
Dwarkadas, V. V. & Gruszko, J. 2012, *MNRAS*, 419, 1515
Dwarkadas, V. V., Romero-Cañizales, C., Reddy, R., & Bauer, F. E. 2016, *MNRAS*, 462, 1101
Fox, O. D., Van Dyk, S. D., Dwek, E., *et al.* 2017, *ApJ*, 836, 222
Fransson, C., Lundqvist, P., & Chevalier, R. A. 1996, *ApJ*, 461, 993
Fransson, C., Ergon, M., Challis, P. J., *et al.* 2014, *ApJ*, 797, 118
Gal-Yam, A., Leonard, D. C., Fox, D. B., *et al.* 2007, *ApJ*, 656, 372
Hoffman, J. L., Leonard, D. C., Chornock, R., *et al.* 2008, *ApJ*, 688, 1186-1209
Leonard, D. C., Filippenko, A. V., Barth, A. J., & Matheson, T. 2000, *ApJ*, 536, 239
Margutti, R., Milisavljevic, D., Soderberg, A. M., *et al.* 2014, *ApJ* , 780, 21
Mauerhan, J. C., Smith, N., Filippenko, A. V., *et al.* 2013, *MNRAS*, 430, 1801
Milisavljevic, D., Margutti, R., Kamble, A., *et al.* 2015, *ApJ*, 815, 120
Nymark, T. K., Fransson, C., & Kozma, C. 2006, *A& A*, 449, 171
Ofek, E. O., Sullivan, M., Shaviv, N. J., *et al.* 2014a, *ApJ*, 789, 104
Ofek, E. O., Zoglauer, A., Boggs, S. E., *et al.* 2014b, *ApJ*, 781, 42
Patat, F., Taubenberger, S., Benetti, S., Pastorello, A., & Harutyunyan, A. 2011, *A& A*, 527, L6
Pérez-Torres, M. A., Lundqvist, P., Beswick, R. J., *et al.* 2014, *ApJ*, 792, 38
Ross, M. & Dwarkadas, V. 2017, *arXiv:*1704.05866
Sana, H., de Mink, S. E., de Koter, A., *et al.* 2012, *Science*, 337, 444
Smith, N., Li, W., Foley, R. J., *et al.* 2007, *ApJ*, 666, 1116
Smith, N., Li, W., Miller, A. A., *et al.* 2011, *ApJ*, 732, 63
Smith, N., Silverman, J. M., Filippenko, A. V., *et al.* 2012, *AJ*, 143, 17
Smith, N. & Tombleson, R. 2015, *MNRAS*, 447, 598
Smith, N. 2016a, *arXiv:*1612.02006
Smith, N., Andrews, J. E., & Mauerhan, J. C. 2016b, *MNRAS*, 463, 2904
Stoll, R., Prieto, J. L., Stanek, K. Z., *et al.* 2011, *ApJ*,, 730, 34
Taddia, F., Stritzinger, M. D., Sollerman, J., *et al.* 2013, *A& A*, 555, A10
van Dyk, S. D., Weiler, K. W., Sramek, R. A., & Panagia, N. 1993, *ApJ* Letters, 419, L69
van Dyk, S. D., Weiler, K. W., Sramek, R. A., *et al.* 1996, *AJ*, 111, 1271
Wang, L., Cui, X., Zhu, H., & Tian, W. 2015, *Advancing Astrophysics with the Square Kilometre Array (AASKA14)*, 64
Weiler, K. W., Panagia, N., & Sramek, R. A. 1990, *ApJ*, 364, 611
Williams, C. L., Panagia, N., Van Dyk, S. D., *et al.* 2002, *ApJ*, 581, 396

Supernova 1987A:30 years later - Cosmic Rays and Nuclei from Supernovae and their aftermaths
Proceedings IAU Symposium No. 331, 2017
A. Marcowith, M. Renaud, G. Dubner, A. Ray & A. Bykov, eds.

doi:10.1017/S1743921317004859

Outbursts from evolved massive stars: SN 2015bh and its relatives

Christina C. Thöne[1], **Antonio de Ugarte Postigo**[1,2] **and Jose Groh**[3]

[1]IAA-CSIC, Glorieta de la Astronomía s/n, 18008 Granada, Spain
email: cthoene@iaa.es

[2]Dark Cosmology Centre, Niels-Bohr-Institute, Univ. of Copenhagen, Juliane Maries Vej 30, 2100 Copenhagen, Denmark

[3]Trinity College Dublin, The University if Dublin, Dublin 2, Ireland

Abstract. In their final stages, massive stars can show large eruptions which can resemble core-collapse IIn SNe. Here we present SN 2015bh in NGC 2770, a IIn/impostor, where archival data show variabilities for at least 21 years before the main event in 2015. Serendipitous spectra during an outburst are the only SN progenitor spectra available since SN 1987A and show an LBV with a fast, dense outflow. Analogues to SN 2015bh are SN 2009ip and SNhunt 248 while the SN 2000ch impostor could be equivalent to the outburst phase of SN 2015bh. It is currently unclear whether SN 2015bh (and SN 2009ip) were final core-collapse events. Alternatively, they might be large outbursts shedding the outer envelope and creating a Wolf-Rayet star in only a matter of decades. Future large-scale high-cadence surveys such as LSST will detect many more of these events and allow us a unique insight into the largely unknown late stages of massive stellar evolution.

Keywords. supernovae: individual (SN 2015bh, SN 2009ip), stars: evolution, stars: mass loss

1. Introduction

Luminous eruptions of massive stars are frequent, especially in the late stages of their evolution, but the mechanisms and processes are still poorly understood by theoretical models. So-called "SN impostors" are such massive outbursts that can reach almost the luminosity of a core-collapse SN, however, the event does not destroy the star. Eta Carinae, a luminous blue variable (LBV), is one of the most prominent examples with large documented outbursts in 1843 and 1892. Recently it had been suggested that it has experienced a major eruption once every about 300 years (Kiminki *et al.* 2016).

Another category of major outbursts are so-called "precursors" to SN events, which seem to be common in IIn SNe in the year prior to explosion (Ofek *et al.* 2014). The most typical case is an outburst with peaks several tens of days before the actual explosion and a luminosity several magnitudes below the final SN. In fact, they are so frequent that some relate this precursor to the actual SN and the later, brighter peak to the interaction with the circumstellar medium (Ofek *et al.* 2014). Possible precursors have also been observed in a SN IIb (Strotjohann *et al.* 2015) and in a Ic SNe (Corsi *et al.* 2014), although here they seem to be the exception. SNe IIb show some indication of extended envelopes or CSM interaction (indeed, the spectral evolution of IIb SNe change from pronounced hydrogen lines to He-dominated, Ib-like spectra and some of them show double-peaked lightcurves). The progenitors of SN IIn, IIb and Ic span a range of massive stars including LBVs, Wolf-Rayet (WR) stars and cool supergiants but all showing indications for different kinds of pre-explosion mass losses. Likewise, early observations of some SNe, so-called "flash-spectroscopy-SNe" (Gal-Yam *et al.* 2014), show narrow hydrogen lines from a dense CSM or envelope that disappear within days.

Here we present observations of the peculiar event SN 2015bh in NGC 2770 at a distance of ~27 Mpc which has hosted 3 other SNe prior to SN 2015bh (Thöne *et al.* (2017), see also Elias-Rosa *et al.* 2016). Archival observations over 21 years before the main event show variability of ~2 mag for two decades before a precursor with a peak in February 2015 and the brightest, so-called "main event" in May 2015. Our observations span the entire event from the precursor up to 200 days past maximum of the main event using the telescopes at Observatorio de Sierra Nevada (OSN), the Gran Telescopio Canarias (GTC) and *Swift* (Thöne *et al.* 2017). SN 2015bh shows many similarities to another famous event, SN 2009ip which also has several smaller outbursts documented prior to its possible explosion as IIn SN in 2012 (Pastorello *et al.* 2013, Mauerhan *et al.* 2013).

2. SN 2015bh from 1994 to 2015

Our search for the progenitor of SN 2015bh in archival data of NGC 2770 revealed that the star had been present in data at least back until 1994 and varying by about ~ 2 magnitudes over the years (Thöne *et al.* 2017). The progenitor was first detected during an outburst in 2013 by iPTF and named it PTF13efv, but it was not reported until the detection of the precursor in 2015 (Duggan *et al.* 2015). We obtained serendipitous imaging and spectra in Nov. 2013 while observing NGC 2770 and its satellite NGC 2770B just at the onset of the outburst later detected by iPTF, but did not realize the detection of an outburst until much later.

On Feb. 7, 2015, SNhunt reported a new transient later characterized as SN impostor or possible IIn (Elias-Rosa *et al.* 2015). This event was later known to be the precursor of SN 2015bh with a maximum ~ 40 days before the main event and a peak magnitude of R=–13.4 mag. We initiated a follow-up campaign with the 0.9 and 1.5m OSN telescopes which then lead to the discovery of a sudden rebrightening on May 16, 2015, the onset of the main event, which eventually reached an absolute magnitude of –17.5 mag around May 24. During the first month of the main event we obtained almost daily imaging and spectroscopy with OSIRIS at the GTC and *Swift* in UV and X-rays. Neither X-ray nor radio emission was detected from the SN and/or its progenitor (Thöne *et al.* 2017). By September 2015, the light curve had dropped to ~ –12.5 mag with a very shallow subsequent decay, implying that the material was not any more accelerated at this time. The light curve of the entire 21 years of observations is shown in Fig. 1.

The serendipitous spectra from the outburst in 2013 shows a typical LBV spectrum with a single P-Cygni profile and Balmer and Fe-lines, but no He-lines (this spectrum will be further discussed below). At the precursor in early 2015, a second absorption component emerges with a velocity of ~ –2000 km s^{-1} in addition to the absorption at –700 km s^{-1} already present in the LBV phase, suggesting a connection of the ejected material with the event that caused the precursor. At the onset of the main event, weak He lines furthermore start to appear. Around maximum of the main event the absorption components of the P-Cygni profile largely disappear but then later reemerge with a clear double P-Cygni profile in all lines while the He-lines had disappeared again. After the sun gap, the spectra have changed to broad nebular lines and the second, faster absorption now shows up as emission line accompanied by a small bump at the same velocity, but redshifted, suggesting some asymmetry in the ejecta. Our toy model for the explosion assumes a dense shell expelled some years before the main event which we see as the absorption component at –700 km s^{-1}. The faster material is expelled

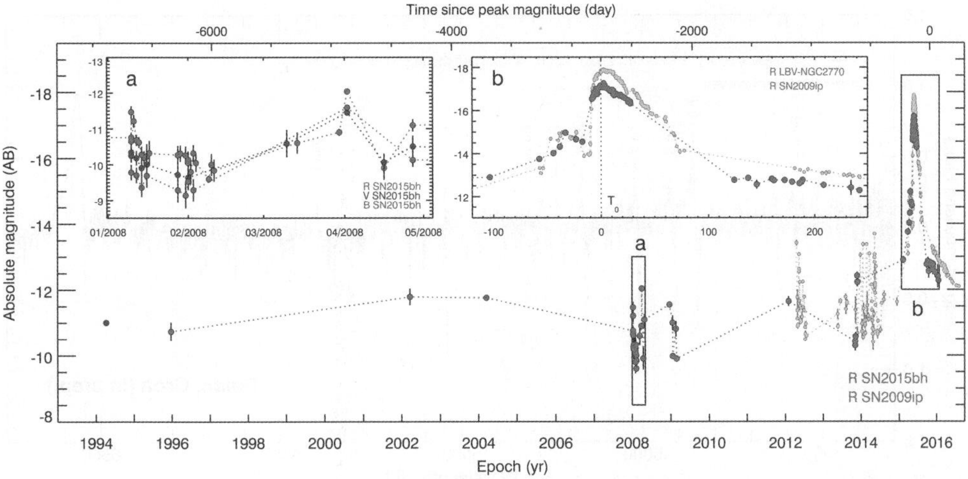

Figure 1. Lightcurve of SN 2015bh including the main event and precursor and the short term variations of its progenitor over 21 years. Inset a shows the dense monitoring of the progenitor during observations of SN 2008D and SN 2007uy in the same galaxy, inset b shows a blow-up of the precursor and main event. The grey dots are observations of the very similar event SN 2009ip shifted to the same T_0 as SN 2015bh (from Thöne *et al.* 2017)

during the precursor and interacts with the dense wind of the progenitor giving rise to the luminosity of the main event. Some time around 100 days post explosion it catches up with the first larger shell and shows up in emission.

The main event had a total energy release of $1.2 - 1.8 \times 10^{48}$ erg. We model the SED during the main event with a simple BB model and derive the temperature and radius evolution. Until about 20 days past maximum the BB has an expansion velocity of ~ 2200 km s^{-1} and a cooling rate of 660 K/day after which it slows down to 880 km s^{-1} and 150 K/day. The Hα EW behaves very similar to those of other IIn SNe with a drop around maximum below 100 Å and a subsequent rise to very high values of > 3000Å which could be indicative of a true core-collapse (Smith *et al.* 2014). The Balmer decrement (Hα / H β) drops below the value for case B recombination (2.76) around maximum which could be explained by a very dense plasma thermalizing the emission. During the "nebular phase" it rises to >10, which is also seen in the nebular phase of other types of events. An explanation by increasing extinction from the ejecta is excluded as the SED shows no signs of reddening and dust production.

3. The host galaxy NGC 2770 and the SN environment

NGC 2770 was observed with different 3D techniques to study the ISM in this "SN factory" as well as the surroundings of the SNe. However, only our drift-scan spectroscopy run to observe the satellite galaxy of NGC 2770 happened to also cover the centre of the galaxy and the site of the progenitor, which is also the dataset from which we obtained the serendipitous spectra of the SN 2015bh progenitor at the onset of the 2013 outburst. As the spectra are contaminated by the progenitor emission we can only study the neighboring HII regions. The star is located in one of the outer spiral arms of NGC 2770 and in the same arm as SN 2007uy. This is different from SN 2009ip which was found at a large distance from its host. The regions near the SN site have a metallicity of around half solar and a stellar population age of 7-10 Myr, consistent with single star LBV models.

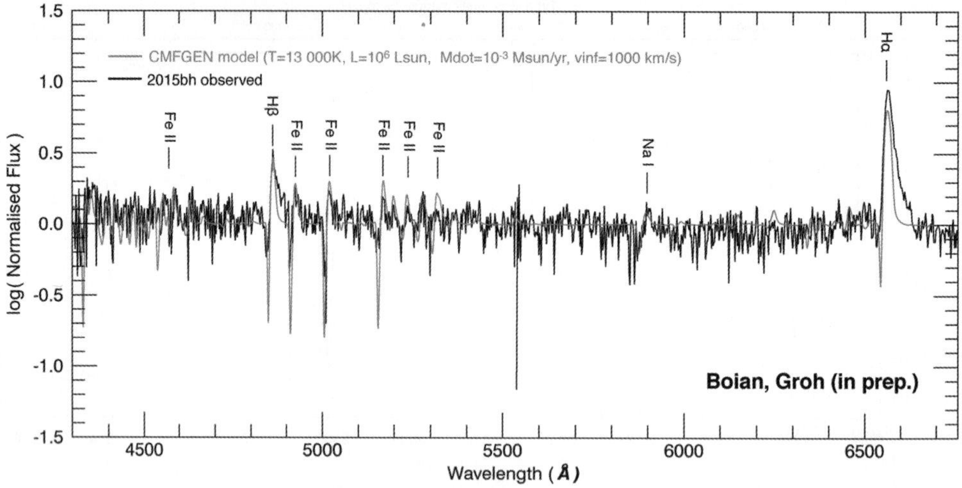

Figure 2. Modeling of the progenitor spectrum of SN 2015 at –550 d with an LBV showing a dense, fast outflow using the radiation transfer code CMFGEN Hillier & Miller 1998 The original spectrum is shown in black, the model spectrum in red.

Despite its elevated number of SNe, NGC 2770 does not show a particularly large star-formation rate, the only remarkable fact is a large HI mass (Thöne *et al.* 2009). Likewise, there are little indications for a major merger event triggering large star formation (Thöne *et al.* in prep.). The HI mass could be indicative of an inflow of gas onto the galaxy, giving rise to a recent enhancement in star-formation and hence SN occurrence.

4. The progenitor of SN 2015bh and its possible survivor

Our serendipitous spectra from one of the outbursts in Nov. 2013 are probably the only available spectra of a SN progenitor prior to explosion since SN 1987A. We model the spectrum at –550 d with template spectra of massive stars and find the best match with a luminous LBV ($L = 10^6 L_\odot$) with a temperature of 13,000 K showing a dense, fast outflow of $\dot{M} = 10^{-3} M_\odot$ and $v = 1000$ km s^{-1} (see Fig. 2). The large velocities indicate that the star is in outburst as this is too high for a usual stellar wind. Yellow hyper giant and WR models can be excluded with this model. Our model has a metallicity of about half solar, consistent with the metallicity of neighboring stellar populations in the host galaxy (see Sect. 3). Further details on this model will be given in Boian *et al.* in prep.

The progenitor of SN 2009ip has also been suggested to be a LBV star. Both SN 2015bh and SN 2009ip progenitors were always located red of typical LBVs in S-Dor outbursts at similar temperatures as Eta Carinae during its great eruption but at lower luminosities. During the pre-explosion event, the temperature stayed at similar values while the luminosity increased. The colors during the SN were typical of other IIn SNe. At late times, both SN 2009ip and SN 2015bh show a turn towards higher temperatures whose origin is still unclear. SN 2009ip now seems to be below the level of the proposed progenitor as observed in 1999 (Foley *et al.* 2011). At this point we cannot determine whether any of the two stars actually did explode. If they survived these outbursts, the stars might have lost most of their outer envelope and settle at a lower luminosity, while it seems they also turns towards higher temperatures similar to WR stars. LBVs have been suggested to be only a transitionary period on the evolution to a WR star (Maeder & Meynet 2000), but lately LBVs have also been observed to directly explode as core-collapse SNe (Gal-Yam & Leonard 2009). So far, it has been unclear how the star can loose enough mass in short time to evolve from a LBV to a WR star (Smith 2014). This kind of event might be the solution to this problem, expelling enough mass to transform a LBV in very short time into a WR star. In fact, many WR stars have been found to be embedded in SNR like shells. Another possibility is that these events happen in a binary system of which one star experiences core-collapse and the other, blue, companion emerges at late times. Late observations of the possible survivors will be crucial to confirm this possible scenario.

5. A new type of event?

SN 2015bh and its "twin" SN 2009ip are certainly not the only events of their type, in fact, more and more evidence appears that (very) massive stars at their late evolution stages show frequent, sometimes, massive outbursts. SN 2009ip also showed a complex absorption profile in the observed absorption lines indicating massive pre-explosion mass-loss (Margutti *et al.* 2014). Several other events have been suggested to be very similar to SN 2009ip (and hence SN 2015bh) such as SN 2010mc and LSQ13zm (see e.g. Elias-Rosa *et al.* 2016), however, most of them lack the extensive datasets of 2015bh and 2009ip, in particular at timescales long before the main event.

Two other interesting events in this context are the SNhunt 248 and SN 2000ch. The former has been largely neglected as possible member of this group, but shows a large number of similarities though at an overall lower luminosity: Variations of ~ 2 magnitude before the main event have been observed. The light-curve of the main event has been described as "triple-peaked". The first of these peaks corresponds to the precursor, the second is the peak of the main event. The third peak around +100 d could be a late rebrightening indicating a further interaction with the CSM. This late peak, albeit with a lower increase in luminosity has in fact also been observed in SN 2009ip and if it occured for SN 2015bh at a similar time post maximum, it would have been missed as the SN had been in the sun gap at that time. SNhunt 248 has less and weaker pronounced absorption components than SN 2009ip and SN 2015bh but shares some similarities (see Kankare *et al.* 2015). SN 2000ch has been detected in 2000 as impostor and showed new outbursts in 2009 – 2010 (Pastorello *et al.* 2010) at similar magnitude variations and timescales as SN 2015bh. Also the spectra resemble those of SN 2015bh during the outburst in 2013. SN 2000ch might turn into a similar event than 2015bh at some point in the (near?) future and should certainly be monitored.

Detecting these events and other variations of massive stars prior to their SN explosion is still difficult and requires high-cadency, large-scale surveys. PTF and iPTF have discovered several candidate precursors, but most lie at the very limit of its observing capabilities. Obtaining spectra at time before the explosion is even more complicated and has so far only been obtained for SN 2015bh and SN 2000ch. Future deep all-sky survey such as LSST will hopefully detect many more of these outbursts prior to explosion and allow for more thorough follow-up. This will be very important to gain insight in the so-far poorly understood late stages of massive stellar evolution and to improve our theoretical models.

References

Kiminki, M. M., Reiter, M., & Smith, N. 2016, *MNRAS*, 463, 845
Ofek, E. O., Sullivan, M., Shaviv, N. J., *et al.* 2014, *ApJ*, 789, 104
Strotjohann, N. L., Ofek, E. O., Gal-Yam, A., *et al.* 2015, *ApJ*, 811, 117

Corsi, A., Ofek, E. O., Gal-Yam, A., *et al.* 2014, *ApJ*, 782, 42
Gal-Yam, A., Arcavi, I., Ofek, E. O., *et al.* 2014, *Nature*, 509, 471
Khazov, D., Yaron, O., Gal-Yam, A., *et al.* 2016, *ApJ*, 818, 3
Thöne, C. C., de Ugarte Postigo, A., Leloudas, G., *et al.* 2017, *A&A*, 599, A129
Elias-Rosa, N., Pastorello, A., Benetti, S., *et al.* 2016, *MNRAS*, 463, 3894
Pastorello, A., Cappellaro, E., Inserra, C., *et al.* 2013, *ApJ*, 767, 1
Mauerhan, J. C., Smith, N., Filippenko, A., Blanchard, K. B. *et al.* 2013, *MNRAS*, 430, 1801
Duggan, G., Bellm, E., Leloudas, G., Kasliwal, M. M., & Kulkarni, S. R. 2015, *The Astronomers Telegram* 7515
Elias-Rosa, N., Benetti, S., Tomasella, L., *et al.* 2015, *The Astronomers Telegram*, 7042
Smith, N., Mauerhan, J. C., & Prieto, J. L. 2014, *MNRAS*, 438, 1191
Thöne, C. C., Michałowski, M. J., Leloudas, G., *et al.* 2009, *ApJ*, 698, 1307
Hillier, D. J. & Miller, D. L. 1998, *ApJ*, 496, 407
Foley, R. J., Berger, E., Fox, O., *et al.*, 2011, *ApJ*, 732, 32
Maeder, A. & Meynet, G. 2000, *A&A*, 361, 159
Gal-Yam, A. & Leonard, D. C. 2009, *Nature*, 458, 865
Smith, N. 2014, *ARA&A*, 52, 487
Margutti, R., Milisavljevic, D., Soderberg, A. M., *et al.* 2014, *ApJ*, 780, 21
Kankare, E., Kotak, R., Pastorello, A., *et al.* 2015, *A&A*, 581, L4
Pastorello, A., Botticella, M. T., Trundle, C., *et al.* 2010, *MNRAS*, 408, 181

Discussion

KIRSHNER: Is there any kind of relation in e.g. luminosity vs. timescale that would allow us to pick-up those pre-explosion outbursts from observations with LSST?

THÖNE: Interesting comment, this should certainly be given a thought for when LSST comes online.

VALENTI: Why do those events have the low Ni masses observed?

HIRSCHI: The low Ni masses are expected as these stars have massive cores that are hard to explode and hence little Ni is expected.

Supernova 1987A:30 years later - Cosmic Rays and Nuclei from Supernovae and their aftermaths
Proceedings IAU Symposium No. 331, 2017
A. Marcowith, M. Renaud, G. Dubner, A. Ray & A. Bykov, eds.

doi:10.1017/S1743921317004860

The diversity of GRBs and their SNe: Observations from the 10.4m GTC

Antonio de Ugarte Postigo[1,2], Christina Thöne[1], Zach Cano[1], David Alexander Kann[1], Luca Izzo[1], Rubén Sánchez Ramírez[1], Katarzyna Bensch[1] and Ana Sagues[1]

[1]Instituto de Astrofísica de Andalucía
Glorieta de la Astronomía s/n, 18008 Granada, Spain
email: deugarte@iaa.es

[2]Dark Cosmology Centre, Niels Bohr Institute, University of Copenhagen,
Juliane Maries Vej 30, 2100 Copenhagen Ø, Denmark

Abstract. Observing the supernovae (SNe) associated to the different types of gamma-ray bursts (GRBs) is one of the few means to study their progenitors. In the past years, it has become clear that GRB-like events are more heterogeneous than previously thought. There is a marked difference between long GRBs, which are produced by the collapse of very massive stars and are normally associated with broad-lined type Ic SNe, and short bursts, which occur when two compact objects merge and that, at least in some cases, can produce an associated kilonova. Moreover, the SNe associated with different sub-types of long GRBs are also seen to differ, especially those associated with ultra-long duration GRBs. To address this issue in a systematic way we started an observing programme in 2010 at the 10.4m GTC telescope. Here we present some results of our programme, including the detection of 12 new GRB-SNe. Highlights of our sample are the discovery of the first spectroscopic SN associated with a highly energetic ($E_{\gamma,iso} \sim 10^{54}$ erg) "cosmological" burst (GRB 130427A), the study of the SN associated with a shock-breakout GRB (GRB 140606B) and the SN associated with the peculiar ultra-long GRB 101225A at $z = 0.85$. The sample includes also the follow-up of several short GRBs in search for kilonovae emission (GRB 130603B and GRB 160821B are important examples). Amongst our latest results we present the photometric and spectroscopic observations of the SNe associated with GRB 150818A and GRB 161219B.

Keywords. gamma rays: bursts, supernovae: general

1. Introduction

GRBs are the most luminous explosions that we can witness in the Universe. They are produced during a cataclysmic stellar collapse, in which a black hole is formed and ultrarelativistic jets are released. Thanks to their extreme luminosity, they can be observed from the local Universe (the closest one recorded was localised at a redshift of 0.0085, Galama *et al.* 1998) to deep into the reionisation era, when the first stars had been just recently formed (the current record holder with a spectroscopic redshift happened at $z = 8.2$, Tanvir *et al.* 2009; Salvaterra *et al.* 2009, although there is a photometric redshift at $z = 9.4$, Cucchiara *et al.* 2011). They are typically classified into short-hard and long-soft GRBs (Kouveliotou *et al.* 1993) although other classes could also be there, pointing to different types of progenitor or environmental conditions (intermediate GRBs Horváth *et al.* 1998; ultra-long GRBs Levan *et al.* 2014). The progenitors have been proposed to be collapsing very massive stars (generally accepted for long GRBs) or the coalescence of two compact objects (model accepted for short GRBs). These different progenitors are

Table 1. Overview of the GRB-SN follow-up programme at the 10.4m GTC. The GRBs have been organised into ultralong, long, intermediate and short events.

GRB / SN	Redshift	T_{90}	Observations	References
101225A	0.847	> 1000	Photometry	Thöne *et al.* 2011
111211A	0.478	15	Spectroscopy	de Ugarte Postigo *et al.* 2012
111228A	0.716	101.2	Photometry	—
120729A	0.80	71.5	Photometry	Cano *et al.* 2014
130215A / SN 2013ez	0.597	65.7	Spectroscopy	Cano *et al.* 2014
130427A / SN 2013cq	0.34	163	Spectroscopy	Xu. *et al.* 2013
140606B	0.384	23.6	Spectroscopy	Cano *et al.* 2015
141225A	0.915	40.24	Photometry	—
150518A	0.256	165	Spectroscopy	—
150818A	0.282	123.3	Spectroscopy	de Ugarte Postigo *et al.* 2015
160131A	0.97	325	Photometry	—
100418A	0.6235	7	Photometry	de Ugarte Postigo *et al.* 2011
100816A	0.804	2.9	Photometry	—
120422A / SN 2012bz	0.283	5.35	Spectroscopy	Schulze *et al.* 2014
141004A	0.573	3.92	Photometry	—
161219B / SN 2016jca	0.1475	6.94	Spectroscopy	Cano *et al.* 2017
111117A	2.211	0.47	Photometry	Sakamoto *et al.* 2013; Selsing *et al.* 2017
130603B	0.356	0.18	Photometry	de Ugarte Postigo *et al.* 2014
160303A	—	5	Photometry	—
160821B	0.16	0.48	Photometry	—

expected to have either an associated supernova, in the case of the collapse of a massive star, or a kilonova, in the case of compact star mergers. To study these events and derive properties of their progenitors from a statistical approach, observations need to be performed in a systematic way. However, early studies were mainly focused on peculiar events, limiting the value of the GRB-SN samples.

2. The GTC GRB-SN programme

In 2010, shortly after the 10.4 m GTC began operations, we initiated our GRB-SN follow-up programme. Its aim is to systematically obtain photometric follow-up of all GRB-SNe at redshifts lower than 1.0 and spectroscopic follow-up of all GRBs at $z < 0.6$. If the GRB has $z < 0.3$ the follow-up consists of multiple spectroscopic observations. This follow-up is not restricted to long GRBs but we follow all classes of events. In the case of short GRBs we only expect to detect a kilonova component for the very nearby events ($z \sim< 0.15$).

During this time we have followed 10 long-soft GRBs, 4 short-hard, 5 intermediate and 1 ultra-long. The redshift range spans between $0.14 < z < 0.97$. We have obtained 7 spectroscopic detections. These observations are summarised in Table 1. In the following sections we give some examples of GRBs followed by our programme.

2.1. *GRB 100418A*

GRB 100418A was an intermediate-duration GRB at a redshift of $z = 0.6235$, with a luminous afterglow and a bright host galaxy (with a r'-band magnitude of 22.1), which complicated the study of its associated SN. Nevertheless, through the use of image-subtraction techniques, we detect SN emission peaking at $r' = 24.8$ mag, equivalent to an absolute magnitude of $M_{r'} \sim -17.0$ mag, 1.7 magnitudes fainter than the canonical SN1998bw.

2.2. *GRB 130215A / SN 2013ez*

GRB 130215A was detected at a redshift of $z = 0.597$, at the limit of the spectroscopic capabilities of our programme. A single spectrum was obtained using GTC during the

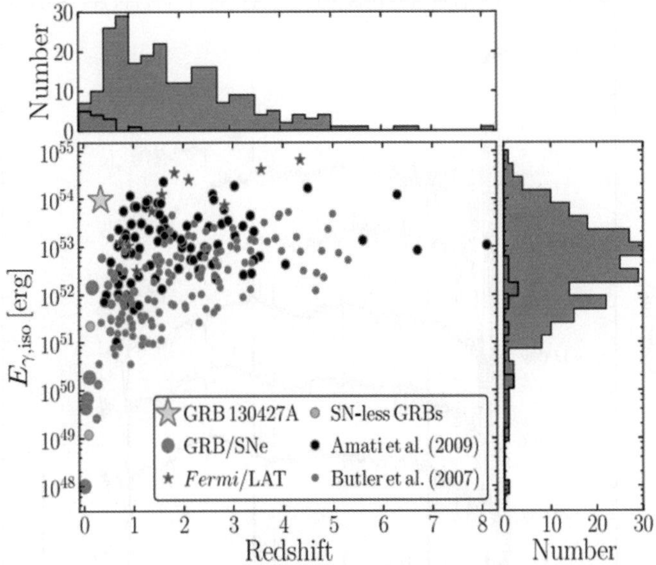

Figure 1. Comparison of isotropic equivalent energy release of different GRBs, showing that GRB 130427A was the first GRB-SN corresponding to the class of most energetic events, those with isotropic equivalent energy release in gamma rays of $E_{\gamma,iso} \sim 10^{54}$ erg. The histograms indicate the cumulative distributions of all GRBs (grey), and GRBs with identified supernova (red). The comparison also includes two GRBs that did not show a SN component in spite of deep searches (SN-less GRBs, Fynbo *et al.* 2006). From Xu *et al.* (2014).

maximum of the SN emission (de Ugarte Postigo *et al.* 2013a). This spectrum revealed broad absorption features typical of GRB-SNe. However, the ejecta was being expelled at a velocity of ~ 5000 km s^{-1}, significantly below the typical values for Ic-BL SNe at that time. This would mean that GRB 130215A / SN 2013ez would be the first non-broad-lined supernova associated with a GRB. The results of our study were published by Cano *et al.* 2014.

2.3. *GRB 130427A / SN 2013cq*

Most of the nearby GRB-SNe belong to a population of underluminous GRBs. This meant that our conclusions would be not representative of the overall population of GRBs, with a much larger range of gamma-ray energy releases. To overcome this, we studied several GRB-SNe that had intermediate luminosities in gamma-rays. However, it was not until GRB 130427A that we had the chance to observe the supernova associated with a highly energetic "cosmological" GRB, which is fully comparable to the bright GRBs usually only detected at $z > 1$. GRB 130427A had an isotropic equivalent gamma-ray energy release of $E_{\gamma,iso} = 9.6 \times 10^{53}$ erg, and was observed at a redshift of $z = 0.34$ (see Fig. 1). Its supernova was discovered by our team (de Ugarte Postigo *et al.* 2013b) as a typical broad-lined Ic, SN 2013cq, as was the case for all the other low- and intermediate-luminosity GRBs, implying that the properties of the supernovae associated with GRBs are independent of gamma-ray energetics. The data obtained by our programme were published by Xu *et al.* (2014).

2.4. *GRB 150818A*

GRB 150818A was detected at a redshift of $z = 0.282$. We performed a spectroscopic follow-up campaign covering the first month. 11 days after the GRB onset, we discovered the broad spectroscopic features characteristic of the Ic-BL SN. The discovery of the

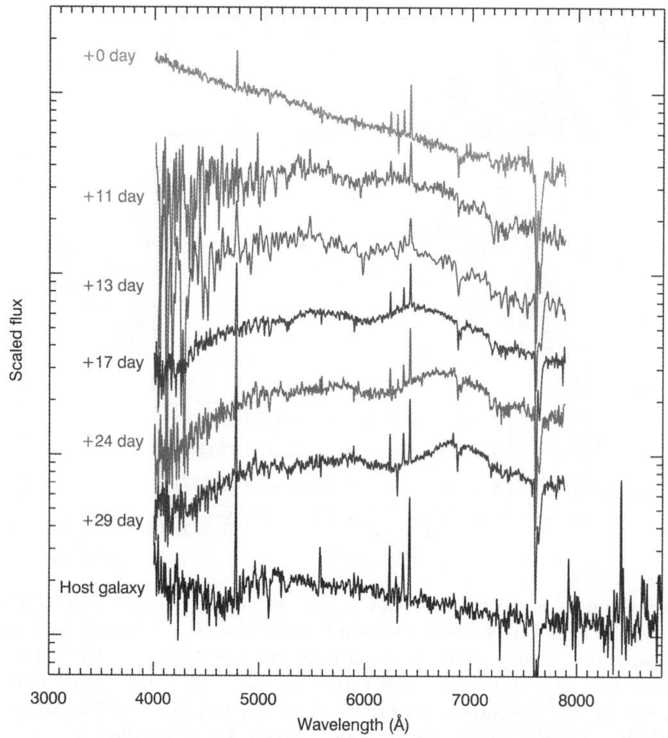

Figure 2. Spectroscopic follow-up of GRB 150818A, at a redshift of $z = 0.282$, ranging from the afterglow emission, shortly after the GRB, to 29 days later, showing the evolution of the supernova. At the bottom we show the host galaxy spectrum, obtained months later.

SN component associated with GRB 150818A was reported by de Ugarte Postigo *et al.* (2015). Figure 2 shows the spectral sequence obtained from GTC.

2.5. *GRB 161219B / SN 2016jca*

GRB 161219B is the latest GRB-SN studied by our programme at the time of this conference. At a redshift of $z = 0.1475$ it was amongst the closest GRB-SNe, and the nearest one since our programme began in 2010. The SN was spectroscopically discovered by our team (de Ugarte Postigo *et al.* 2016).We obtained multiple spectroscopic observations ranging over the first 3 months after the explosion. Our study has been presented by Cano *et al.* (2017), and it includes not only the study of the SN but also of its host galaxy, combining data from GTC and the Hubble Space Telescope.

2.6. *GRB 101225A*

GRB 101225A, also known as the *Christmas Burst*, was a very peculiar ultra-long GRB. It did not show a typical afterglow with a synchrotron emission, but instead, it presented what seemed to be a thermal emission, which evolved by cooling and expanding during the first days after the GRB. Starting 10 days after the event an additional component began to emerge, which we associate with a supernova that accompanied the burst. Although there was some confusion as to the origin of this event, which was reported as being of either Galactic (Campana *et al.* 2011) or extragalactic origin (Thöne *et al.* 2011), the later scenario was proved to be correct through host galaxy spectroscopy (Levan *et al.* 2013). At a redshift of $z = 0.847$ the supernova of GRB 101225A would be a luminous blue supernova, similar in brightness to SN 1998bw but with a bluer colour.

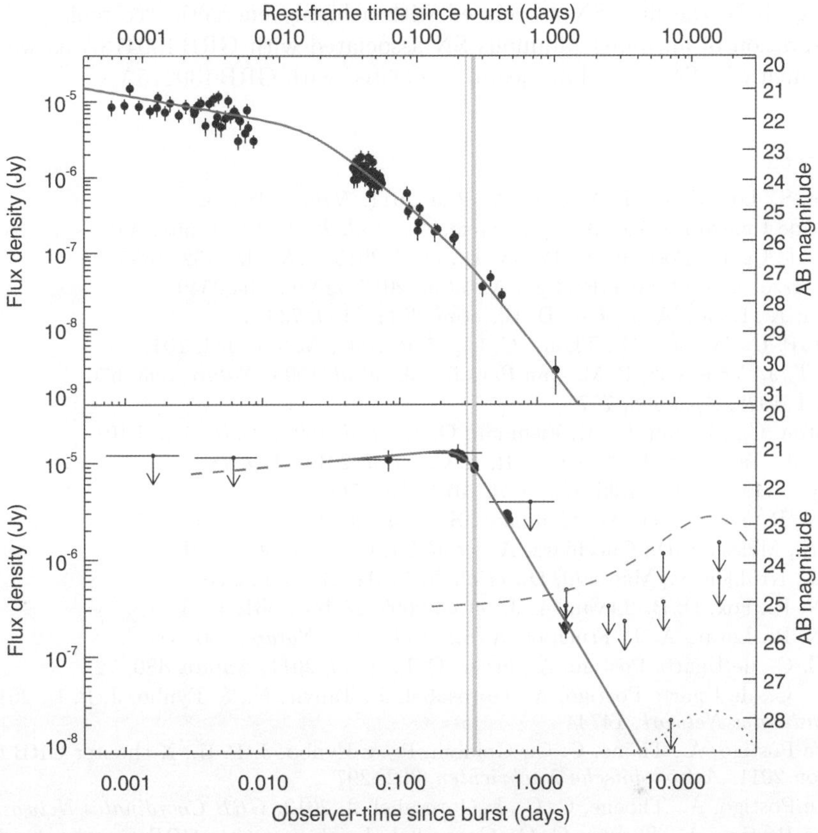

Figure 3. Light curve of the short GRB 130603B in X-rays (top) and optical (botom), showing the deep late limits compared to the prototypical GRB-SN, SN 1998bw marked by a dashed blue line. Our limits implied an emission dimmer than 100 times SN 1998bw (dotted blue line). From de Ugarte Postigo *et al.* (2014).

2.7. *GRB 130603B*

GRB 130603B was the first short GRB for which a useable afterglow spectrum was ever obtained (Thöne *et al.* 2013; de Ugarte Postigo *et al.* 2014). We measured for it a redshift of $z = 0.356$ and did a follow-up campaign to both study the afterglow emission and search for possible supernova or kilonova emission. Our deep observations set strong constraints to possible supernova emission (see Fig. 3), and constrained the afterglow evolution well enough to be able to later search for the kilonova with HST. Combining the GTC and the HST data it was, for the first time, possible to detect the emission of a kilonova emission associated with a short GRB (Tanvir *et al.* 2013). At present we are continuing to study in a similar way the emission of GRB 160821B at a possible redshift of $z = 0.16$.

3. Conclusions

During the early years of GRB-SN studies, the follow-up had been focused on peculiar events, which lead to biases in any possible statistical studies. To avoid this, in 2010 we began a systematic search for GRB-SNe at $z < 1.0$ using the 10.4m GTC telescope. Until now we have studied 20 objects (10 long, 5 intermediate, 1 ultra-long and 4 short). Amongst the accomplishments of our programme, we have the first SN detection for an ultra-long GRB, the first SN associated with a highly energetic "cosmological" event, the observation of an under-luminous SN associated with GRB 100418A as well as the measurement of a SN with slow ejecta, associated with GRB 130215A.

References

Campana, S., Lodato, G., D'Avanzo, P., *et al.* 2011, *Nature*, 480, 69

Cano, Z., de Ugarte Postigo, A., Pozanenko, A., *et al.* 2014, *A&A*, 568, A19

Cano, Z., de Ugarte Postigo, A., Perley, D., *et al.* 2015, *MNRAS*, 452, 1535

Cano, Z., Izzo, L., de Ugarte Postigo, A., *et al.* 2017, arXiv:1704.05401

Cucchiara, A., Levan, A. J., Fox, D. B., *et al.* 2011, *ApJ*, 736, 7

Fynbo, J. P. U., Watson, D., Thöne, C. C., *et al.* 2006, *Nature*, 444, 1047

Galama, T. J., Vreeswijk, P. M., van Paradijs, J., *et al.* 1998, *Nature*, 395, 670

Horváth, I. 1998, *ApJ*, 508, 757

Kouveliotou, C., Meegan, C. A., Fishman, G. J., *et al.* 1993, *ApJL*, 413, L101

Levan, A. J., Tanvir, N. R., Starling, R. L. C., *et al.* 2014, *ApJ*, 781, 13

Sakamoto, T., Troja, E., Aoki, K., *et al.* 2013, *ApJ*, 766, 41

Salvaterra, R., Della Valle, M., Campana, S., *et al.* 2009, *Nature*, 461, 1258

Schulze, S., Malesani, D., Cucchiara, A., *et al.* 2014, *A&A*, 566, A102

Selsing, J., Krühler, T., Malesani, D., *et al.* 2017, *A&A*, submitted

Tanvir, N. R., Fox, D. B., Levan, A. J., *et al.* 2009, *Nature*, 461, 1254

Tanvir, N. R., Levan, A. J., Fruchter, A. S., *et al.* 2013, *Nature*, 500, 547

Thöne, C. C., de Ugarte Postigo, A., Fryer, C. L., *et al.* 2011, *Nature*, 480, 72

Thone, C. C., de Ugarte Postigo, A., Gorosabel, J., Tanvir, N., & Fynbo, J. P. U. 2013, *GRB Coordinates Network*, 14744

de Ugarte Postigo, A., Thöne, C. C., Goldoni, P., & Fynbo, J. P. U., X-shooter GRB Collaboration 2011, *Astronomische Nachrichten*, 332, 297

de Ugarte Postigo, A., Thoene, C. C., & Gorosabel, J. 2012, *GRB Coordinates Network*, 12802

de Ugarte Postigo, A., Thoene, C. C., Gorosabel, J., *et al.* 2013a, *GRB Coordinates Network*, 14303

de Ugarte Postigo, A., Xu, D., Leloudas, G., *et al.* 2013b, *GRB Coordinates Network*, 14646

de Ugarte Postigo, A., Thöne, C. C., Rowlinson, A., *et al.* 2014, *A&A*, 563, A62

de Ugarte Postigo, A., Cano, Z., Perley, D. A., *et al.* 2015, *GRB Coordinates Network*, 18213

de Ugarte Postigo, A., Cano, Z., Izzo, L., *et al.* 2016, *GRB Coordinates Network*, 20342

Xu, D., de Ugarte Postigo, A., Leloudas, G., *et al.* 2013, *ApJ*, 776, 98

Supernova 1987A:30 years later - Cosmic Rays and Nuclei from Supernovae and their aftermaths
Proceedings IAU Symposium No. 331, 2017
A. Marcowith, M. Renaud, G. Dubner, A. Ray & A. Bykov, eds.

doi:10.1017/S1743921317004707

Radioactive decay of GRB-SNe at late-times

Kuntal Misra[1] and A. S. Fruchter[2]

[1]Aryabhatta Research Institute of observational sciencES,
Manora Peak, Nainital India 263 002
email: kuntal@aries.res.in

[2]Space Telescope Science Institute,
3700 San Martin Drive, Baltimore, MD USA 21210
email: fruchter@stsci.edu

Abstract. We present the late-time Hubble Space Telescope observations of two Gamma Ray Burst (GRB) associated supernovae (SNe), GRB 030329/SN 2003dh and XRF 060218/SN 2006aj. Using the multi-color data up to $\sim$ 320 days after the burst, we constrain the late-time decay nature of these SNe. The decay rates of SN 2003dh are steeper than SN 2006aj. A comparison with two other GRB SNe, GRB 980425/SN 1998bw and the SN associated with XRF 020903, shows that the decay rates of SN 2003dh are similar to XRF 020903 and those of SN 2006aj are similar to SN 1998bw. The late-time decay rates are steeper than the ^{56}Co$\rightarrow^{56}$Fe radioactive decay rate indicating that there is some leakage of gamma-rays. We also compare the late-time decay rates of nine type Ic SNe, including the SNe of long GRBs, Ic broad lined and normal Ics. The decay rates of the SNe sample show a remarkable similarity in I band at late-times with a scatter of $\sim$10%.

Keywords. gamma ray bursts, (stars:) supernovae: general, (stars:) supernovae: individual (SN 2003dh, SN 2006aj)

---

1. Introduction

Long duration Gamma Ray Bursts (GRBs) are thought to be associated to a special type of stripped envelope supernovae (SNe) that have lost most of their outer H and He envelope (type Ic). A few SNe out of the type Ic's show broad spectral lines with very high expansion velocities ($\sim$ 30, 000 km/s) and are known as 'broad lined' type Ic SNe. Till date all SNe associated with GRBs and X-Ray Flashes (XRFs) which have been spectroscopically typed have been found to be type Ic SNe. Because of the energetic nature of explosion, the SNe associated to GRBs are often termed as 'hypernovae'. While there is accumulating evidence that nearly all low redshift GRBs have an underlying broad lined type Ic SN, not all energetic broad lined type Ic's show a GRB association.

The first GRB to be associated with a SN, GRB 980425, was detected by the Gamma Ray Burst Monitor onboard the *BeppoSAX* satellite and further spectroscopic observations confirmed that the new optical transient in the error box of GRB 980425 was indeed a very luminous SN (named SN 1998bw) and was classified as a type Ic SN (Galama *et al.* 1998, Iwamoto *et al.* 1998). However, it was with the discovery of GRB 030329/SN 2003dh that the association of GRBs with SNe became incontrovertible and provided the first unambiguous evidence of a SN association with a long duration GRB. Since then about a dozen more SNe have been spectroscopically confirmed to have a GRB association with spectral features which are clearly of broad lined type Ic SNe (see Cano *et al.* 2016 for a review).

The SNe light curves are powered by the radioactive decay of ^{56}Ni $\rightarrow$ ^{56}Co $\rightarrow$ ^{56}Fe. The peak bolometric luminosity gives an estimate of ^{56}Ni synthesized in the explosion (Arnett 1982). The discovery of new GRB-SN associations and broad lined type Ic SNe is revealing a large diversity in terms of the ^{56}Ni production, ejecta mass and the explosion energy. The diversity in the ^{56}Ni produced comes in mainly because type Ic's show a lot of variability in the peak luminosity. Typically the broad lined type Ic without a GRB are found to be less luminous with smaller explosion energy and ejecta mass as compared to hypernovae associated to GRBs. The physical parameters for most of the GRB associated SNe are estimated from the information obtained using the early time light curve and spectrum. It is only in a few cases, like SN 1998bw where late-time data is available, that an estimation of physical parameters is possible (Patat *et al.* 2001) using the nebular phase data. In the case of GRBs it is not easy to construct late-time light curves of associated SNe because GRBs are typically distant objects and hence lack detailed study at late-times due to limited data.

In this paper, we examine the SNe associated with two long duration GRBs - SN 2003dh/GRB 030329 and SN 2006aj/XRF 060218 which are unique because of the availability of the *Hubble Space Telescope* (*HST*) data allowing us to constrain the decay nature of the late-time light curve of the SNe. Due to the proximity and nature of these GRB/XRF-SNe, the early afterglow and SN light curves of both SN 2003dh/GRB 030329 (z = 0.1685) and SN 2006aj/XRF 060218 (z = 0.033) have been studied by several authors in detail (Hjorth *et al.* 2003; Matheson *et al.* 2003; Stanek *et al.* 2003; Deng *et al.* 2005; Resmi *et al.* 2005; Cobb *et al.* 2006; Mazzali *et al.* 2006; Mirabal *et al.* 2006; Modjaz *et al.* 2006; Pian *et al.* 2006; Sollerman *et al.* 2006). The subset of GRB-SNe are compared with a sample of broad lined type Ic and normal type Ic supernovae which have published late-time light curves.

2. Data Reduction

HST observed the afterglow of GRB 030329 and the associated supernova SN 2003dh in *F606W* and *F814W* colors from $\sim$ 17 to 772 days after the burst. The last epoch *HST* observations at $\sim$ 772 and 422 days after the burst in *F606W* and *F814W* colors respectively are used to estimate the host galaxy flux. At late-times the contribution of the afterglow to the total flux becomes negligible and the supernova flux dominates. We therefore use the *HST* data beyond $\sim$ 72 days after the burst when the contribution from the afterglow becomes insignificant. The late-time observations of XRF 060218/SN 2006aj spanning $\sim$ 150–260 days after the burst were carried out at four epochs in four colors - *F435W, F555W, F625W* and *F814W*. We do not have SN 2006aj host galaxy images free from supernova flux to estimate the host galaxy flux.

The *HST* data were processed using the `multidrizzle` routine within the `stsdas` package in IRAF (Fruchter & Hook 2002). We used a `pixfrac` of 0.8 and a `scale` of 0.03333 to obtain the final drizzled image. Thus, the final images have a pixel size of 0.033 arcsec/pixel. The residual image in GRB 030329 is created by subtracting the last epoch image from the early epochs and is shown in Figure 1. In the subtracted image we accurately know the position of the optical transient and estimate the supernova flux in a fixed aperture. In the case of SN 2006aj/XRF 060218 subtracting the early epochs from the last epoch gives a clear detection of the optical transient and a very accurate position as shown in Figure 1. In the case of SN 2006aj the supernova flux is measured in a small aperture; the total flux thus contains the contribution of the host galaxy within this aperture. For both the sources, the corresponding AB magnitudes are estimated using the ACS zero points.

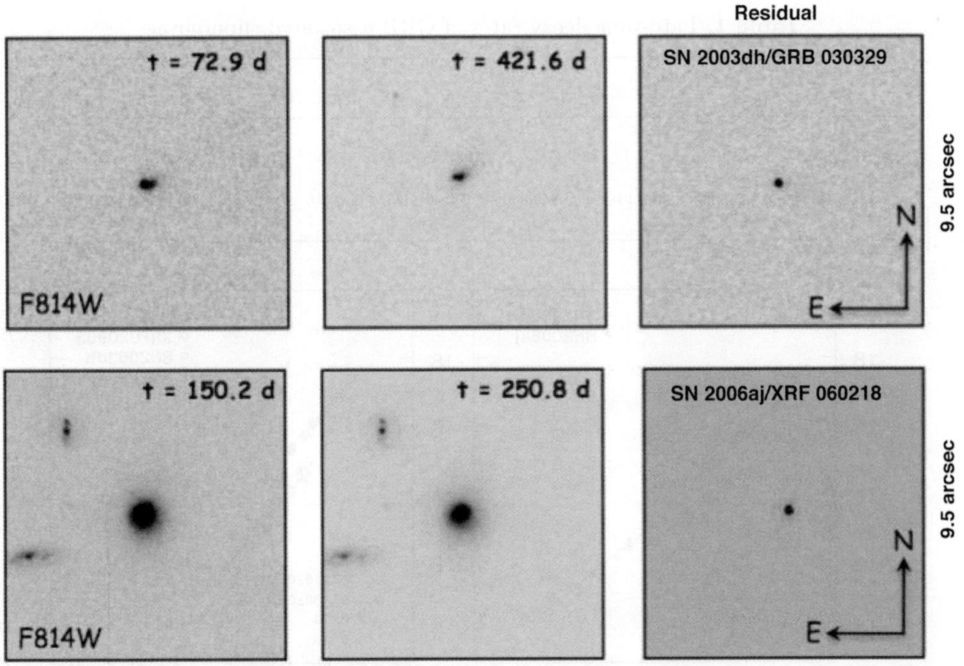

Figure 1. *HST* ACS imaging of SN 2003dh/GRB 03029 and SN 2006aj/XRF 060218 in F814W filter at two epochs as shown in the figure. The residual image is created by subtracting the last epoch image from early epochs.

3. Analysis and Preliminary Results

The unique data set of both SN 2003dh and SN 2006aj allows us to study the late-time behaviour of the light curve. The flux estimates of SN 2003dh are free of any host contamination as discussed in Section 2. To these fluxes we fit an exponential ($F(t) = Ae^{-\alpha t}$ where A is the normalization and α is the decay rate) and estimate the late-time decay rates. In the case of SN 2006aj since we do not have host galaxy flux measurements we add a constant C to the above mentioned exponential form in order to account for the host galaxy flux within the aperture. The decay rates for SN 2003dh and SN 2006aj as obtained from the exponential fits and the corresponding errors are listed in Table 1 and shown in Figure 2.

We also compare the late-time light curves of SN 2003dh and SN 2006aj to two other GRB associated SNe - GRB 980425/SN 1998bw (Patat *et al.* 2001) and the supernova associated with XRF 020903 (Bersier *et al.* 2006). The decay rates of these two supernovae are taken from the literature and are listed in Table 2 for a comparison with our results on SN 2003dh and SN 2006aj. The apparent magnitudes published in the literature are converted to AB magnitudes for a direct comparison with AB magnitudes derived from the *HST* observations. The absolute magnitudes have been derived from the observed photometry using a cosmology of H_0=70 km s^{-1} Mpc^{-1}, Ω_m=0.27 and Ω_Λ=0.73. For each SNe we plot the absolute magnitudes in Figure 2 corrected for the luminosity distance and total (Galactic+Host) extinction. We see that the decay rates of SN 2006aj are remarkably similar to those of SN 1998bw at late-times except for the B band. On the other hand we see that the decay rates are steeper in SN 2003dh and they are similar to the decay seen in the supernova associated to XRF 020903 (Table 2). In Figure 2 we have also shown the ^{56}Co$\rightarrow^{56}$Fe radioactive decay (0.0098 mag day^{-1}) in the case of complete

Table 1. Late-time decay rates of GRB associated supernovae

Object	B	V	R	I
GRB 030329/SN 2003dh	–	0.0218 ± 0.001	–	0.0189 ± 0.001
XRF 060218/SN 2006aj	0.0114 ± 0.004	0.0186 ± 0.001	0.0166 ± 0.0006	0.0170 ± 0.0008

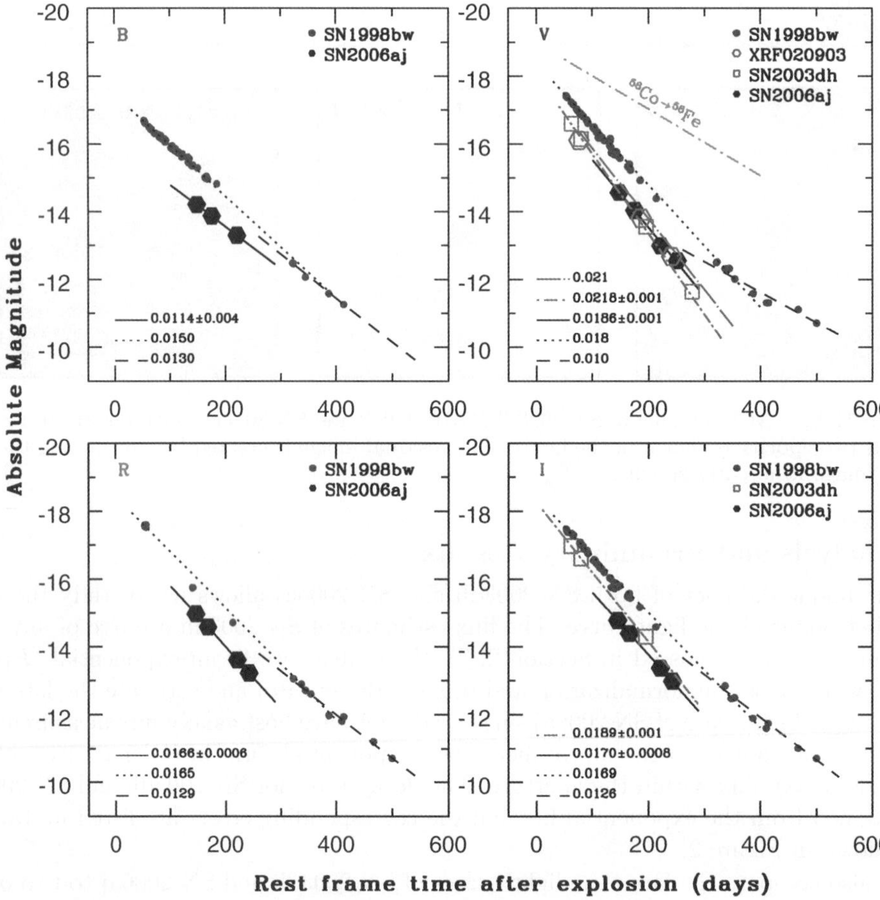

Figure 2. A comparison of multi-color absolute magnitudes of SN 1998bw (Galama *et al.* 1998; McKenzie *et al.* 1999; Sollerman *et al.* 2000; Patat *et al.* 2001), the supernova associated with XRF 020903 (Bersier *et al.* 2006), SN 2003dh (this work) and SN 2006aj (this work). The dotted and dashed lines represent the decay rates of SN 1998bw in the ranges 54-312 and 321-500 days respectively. The dash-dotted line corresponds to the ^{56}Co→^{56}Fe decay rate, as expected if there was no leakage of γ-rays.

γ-ray trapping. We see that the decay rates of all the supernovae shown in Figure 2 are steeper than the ^{56}Co→^{56}Fe decay rates indicating that there is some leakage of γ-rays. For a comparison, in SN 1998bw we also show the data in the time range of 324–504 days after the burst (Figure 2). At these late phases ($t > 300$ days), the multi-band light curves of SN 1998bw show a clear flattening and the light curve settles onto a ^{56}Co→^{56}Fe decay phase.

Table 2. A comparison of type Ic SNe late-time decay rates in the rest frame

Object	*B*	*V*	*R*	*I*	Reference
		GRB-SNe			
GRB 980425/SN 1998bw					
41-309 days	0.015 ± 0.001				Patat *et al.* 2001
41-323 days		0.018 ± 0.0006	0.0165 ± 0.0019	0.0169 ± 0.001	
308-400 days	0.013 ± 0.0008				
308-486 days		0.0101 ± 0.0008	0.0129 ± 0.0006	0.0126 ± 0.0008	
XRF 020903/SN					
76-240 days		0.021			Bersier *et al.* 2006
GRB 030329/SN 2003dh					
63-273 days		0.0218 ± 0.001			This work
63-195 days				0.0189 ± 0.001	
XRF 060218/SN 2006aj					
148-221 days	0.0114 ± 0.004				This work
148-250 days		0.0186 ± 0.001			
146-253 days			0.0166 ± 0.0006	0.0170 ± 0.0008	
		Type Ic-BL			
SN 2002ap					
135-230 days	0.017 ± 0.0004	0.021 ± 0.0002			Tomita *et al.* 2006
271-519 days	0.015 ± 0.002	0.014 ± 0.001			
136-513 days			0.016 ± 0.0001		
138-392 days				0.017 ± 0.0001	
SN 2003jd					
74-375 days	0.0144 ± 0.0013	0.0193 ± 0.0005	0.0152 ± 0.003	> 0.0163	Valenti *et al.* 2008
SN 2007ru					
>80 days		0.0154	0.0117		Sahu *et al.* 2009
		Type Ic			
SN 2004aw					
54-290 days	0.0137 ± 0.0018	0.0177 ± 0.0022	0.0138 ± 0.0014	0.0155 ± 0.0009	Taubenberger *et al.* 2006
SN 2007gr					
100-170 days	0.0153 ± 0.0003	0.0189 ± 0.0003	0.0128 ± 0.0004	0.0167 ± 0.0005	Hunter *et al.* 2009

We further increase our comparison sample by including the broad lined type Ic SNe that are not associated with any GRB/XRF and the normal type Ic SNe which have light curves published in the literature. We found three broad lined type Ic SNe (SNe 2002ap, 2003jd and 2007ru) without any GRB/XRF association and two normal type Ic SNe (SNe 2004aw and 2007gr) with published data extending out to reasonably late-times (nearly 170-500 days). Altogether we have a sample of nine type Ic SNe. In Table 2 we list the rest frame decay rates of the SNe in our sample in different colors at various time ranges. A comparison of the late-time decay rates of the SNe in our sample shows that the decay rates are fairly similar in the *I* band with a scatter of $\sim$ 10%. These decay rates are steeper than ^{56}Co$\rightarrow$^{56}Fe radioactive decay indicating that the γ-ray leakage seen in all type Ic SNe is standard.

4. Conclusions

We present the late-time *HST* ACS multi-color observations of two GRB associated SNe - GRB 030329/SN 2003dh and XRF 060218/SN 2006aj. Both these GRB associated supernovae are unique because of the availability of late-time data which allows us to constrain the decay behaviour at late-times. The SNe associated with GRB 980425/SN 1998bw and XRF 020903 are others for which late-time observations exist. The decay rates of SNe 2006aj and 1998bw are remarkably similar at late-times while those of SN 2003dh are steeper and similar to the SN associated with XRF 020903. In a larger sample of type Ic SNe including the broad lined type Ic with and without a GRB association and normal type Ic SNe, we find that the reported late-time decay rates (Table 2) are fairly similar in the I band with a scatter of $\sim$10%. It is also to be noted that the decay rates are steeper than ^{56}Co$\rightarrow$^{56}Fe radioactive decay indicating that the γ-ray leakage seen in all type Ic SNe is standard.

Acknowledgements

We thank the organizers of IAU Symposium 331 "SN 1987A, 30 years later - cosmic rays and nuclei from supernovae and their aftermaths" to provide an opportunity to present our work. KM acknowledges the financial support received from IAU to attend the symposium.

References

Arnett, W. D., 1982, *ApJ*, 253, 785
Bersier, D. *et al.*, 2006, *ApJ*, 643, 284
Cano, Z. *et al.*, 2016, arXiv: 1604.03549
Cobb, B. E. *et al.*, 2006, *ApJ*, 645, L113
Deng, J. *et al.*, 2005, *ApJ*, 624, 898
Fruchter, A. S. & Hook, R. N., 2002, *PASP*, 114, 144
Galama, T. J. *et al.*, 1998, *Nature*, 395, 670
Hjorth, J. *et al.*, 2003, *Nature*, 423, 847
Hunter, D. J. *et al.*, 2009, *A&A*, 508, 371
Iwamoto, K. *et al.*, 1998, *ApJ*, 534, 660
Matheson, T. *et al.*, 2003, *ApJ*, 599, 394
Mazzali, P. A. *et al.*, 2006, *Nature*, 442, 1018
McKenzie, E. H. *et al.*, 1999, *PASP*, 111, 964
Mirabal, N. *et al.*, 2006, *ApJ*, 643, L99
Modjaz, M. *et al.*, 2006, *ApJ*, 645, L21
Patat, F. *et al.*, 2001, *ApJ*, 555, 900
Pian, E. *et al.*, 2006, *Nature*, 442, 1011
Resmi, L. *et al.*, 2005, *A&A*, 440, 477
Sahu, D. K. *et al.*, 2009, *ApJ*, 697, 676
Sollerman, J. *et al.*, 2006, *A&A*, 454, 503
Sollerman, J. *et al.*, 2000, *ApJ*, 537, L127
Stanek, K. Z. *et al.*, 2003, *ApJ*, 591, L17
Taubenberger, S. *et al.*, 2006, *MNRAS*, 371, 1459
Tomita, H. *et al.*, 2006, *ApJ*, 644, 400
Valenti, S. *et al.*, 2008, *MNRAS*, 383, 1485

Supernova 1987A:30 years later - Cosmic Rays and Nuclei from Supernovae and their aftermaths
Proceedings IAU Symposium No. 331, 2017
A. Marcowith, M. Renaud, G. Dubner, A. Ray & A. Bykov, eds.

doi:10.1017/S1743921317004872

Properties of X-ray emission of an aspherical shock breakout

Yukari Ohtani[1], Akihiro Suzuki[2], Toshikazu Shigeyama[3] and Masaomi Tanaka[4]

[1]Center for Computational Astrophysics, National Astronomical Observatory of Japan, Mitaka, Tokyo 181-8588, Japan
email: yukari.ohtani@nao.ac.jp

[2]Yukawa Institute for Theoretical Physics, Kyoto University, Oiwake-cho, Kitashirakawa, Sakyo-ku, Kyoto, 606-8502, Japan
email: asuzuki@kusastro.kyoto-u.ac.jp

[3]Research Center for the Early Universe, Graduate School of Science, University of Tokyo, Bunkyo-ku, Tokyo 113-0033, Japan
email: shigeyama@resceu.s.u-tokyo.ac.jp

[4]Division of Theoretical Astronomy, National Observatory of Japan, Osawa 2-21-1, Mitaka, Tokyo 181-8588, Japan
email: masaomi.tanaka@nao.ac.jp

Abstract. We investigate the relation between the emission properties of supernova shock breakout in the circumstellar matter (CSM) and the behavior of the shock. Using a Monte-Carlo method, we examine how the light curve and spectrum depends on the asphericity of the shock and bulk-Compton scattering, and compare the results with the observed properties of X-ray outburst (XRO) 080109/SN 2008D. We found that the rise and decay time of the X-ray light curve do not significantly depend on the degree of shock asphericity and the viewing angle in a steady and spherically symmetric CSM. The observed X-light curve and spectrum of XRO 080109 can be reproduced by considering the shock with a radial velocity of 60% of the speed of light, and the wind mass loss rate is about $5 \times 10^{-4} M_{\odot}$.

Keywords. radiative transfer, scattering, methods: numerical, supernovae: general, supernovae: individual (SN 2008D), X-rays: individual (XRO 080109)

---

1. Introduction

A core-collapse supernova becomes bright in ultraviolet/optical bands when photons generated from the shock diffuse upstream. The timescale of the phenomenon, so-called "shock breakout", is determined by the light crossing time of the size of the emerging shock and the diffusion time of photons (e.g., Ensman & Burrows 1992; Matzner & McKee 1999). Since the emission properties of shock breakout strongly depend on the behavior of the shock, they can be used as probes of the stellar radius and the structure of the stellar surface.

X-ray outburst (XRO) 080109 is a fortunately detected shock breakout (e.g., Soderberg *et al.* 2008) associated with a Type Ib SN 2008D. The luminosity (in the energy range of [0.3,10] keV) reached its peak 6×10^{43} erg s^{-1} at ~ 100 sec (hereafter $\Delta t_{\rm p}$) from the onset and decreased until ~ 600 sec. The spectrum taken by the Swift/XRT can be well fitted by a power-law distribution.

The timescale of XRO 080109 depends on the shock radius at the moment of breakout. The rise time can be interpreted as the light crossing time of the shock radius at the moment of breakout (Soderberg *et al.* 2008) or the shock expansion timescale (Svirski & Nakar 2014). In both of the scenarios, the shock must break out at a radius of $\approx 10^{12}$ cm, which is a larger than the typical radius of a Wolf-Rayed star. For this reason, the shock breakout is believed to occur in a dense circumstellar matter (CSM). The scenario is also consistent with the fact that the duration of XRO 080109 corresponds to the diffusion timescale of photons in the unshocked CSM (Balberg & Loeb 2011). Since a wind mass loss event can cause uncertainties in the evolution of a massive star, studying shock breakout is also essential for stellar evolution shortly before the explosion.

The strong candidate of the origin of the observed power-law spectrum is repeated electron ("bulk-Compton") scattering across the shock. In fact, the observed spectrum of XRO 080109 can be reproduced by considering a shock with a radial velocity of $\sim 0.3c$, where c denotes the speed of light (Suzuki & Shigeyama 2010a).

In addition to the signatures of shock breakout in the CSM and bulk-Comptonization, the possibility of the asphericity of the shock has also been discussed. If the shock breaks out in the vicinity of the stellar surface (Suzuki & Shigeyama 2010a; Couch *et al.* 2011; Suzuki *et al.* 2016), the peak of the light curve is believed to be broader than that of a spherical symmetric supernova.

However, there have been no studies which reproduced both of the observed light curve and spectrum of XRO 080109 by an aspherical shock propagating into a thick CSM and bulk-Comptonization. In this study, we perform a radiative transfer calculation aiming to fully reconstruct the observation. In Section 2, we describe the model for the shock and how we calculate radiative transfer. In Section 3, we show the results and compare them with XRO 080109. In Section 4, we conclude this article.

2. Methods

We calculate the light curve and spectrum of the emission generated from the shock propagating into the CSM. In the following subsections, we describe our model for the shock (§2.1) and settings for radiative transfer calculations (§2.2).

2.1. *Model for shock*

To investigate the properties of X-ray emission, we use a simple model of an ellipsoidal shock in a steady and spherically symmetric wind as described in Figure 1 (left panel). The ejecta is regarded as a piston. The shocked ejecta is ignored because most photons are absorbed of scattered immediately at the contact surface where the supposed density is as high as 10^{20} cm^{-3}.

The matter is radiation dominated (thus the adiabatic index equals 4/3) behind the shock front. We calculate a pure radiation transfer in the shocked matter, so that the feedback from radiation is not taken into account. Here we consider a shock of which radial velocity does not change with time and follows the formula

$$v(f,\theta) = \frac{1-f}{[(1-f)^2\cos^2\theta + \sin^2\theta]^{1/2}} \times v(f,0), \tag{2.1}$$

where f is the oblateness of the shock front, θ the angle measured from the symmetric axis. If the kinetic energy of the ejecta is fixed, $v(f,0) = \sqrt{3} \times v_{f=0} \times (2f^2 - 4f + 3)^{-1/2}$. In this study, $v_{f=0} = 0.6c$. The shock radial velocity is displayed in Figure 1 (right panel) as a function of θ.

The shock radius $R_{\rm b}$ at the moment of breakout can be estimated by the observed rise time $\Delta t_{\rm p}$ of XRO 080109. For $f = 0$, we assumed that $R_{\rm b} = c\Delta t_{\rm p} = 3 \times 10^{12}$ cm. We also assumed that the distance between the shock front and contact surface is

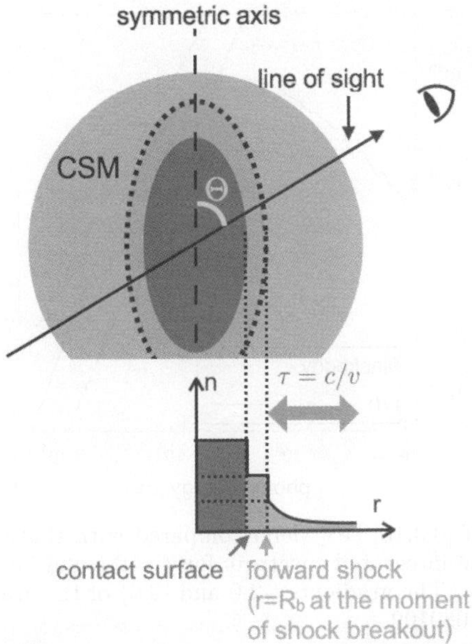

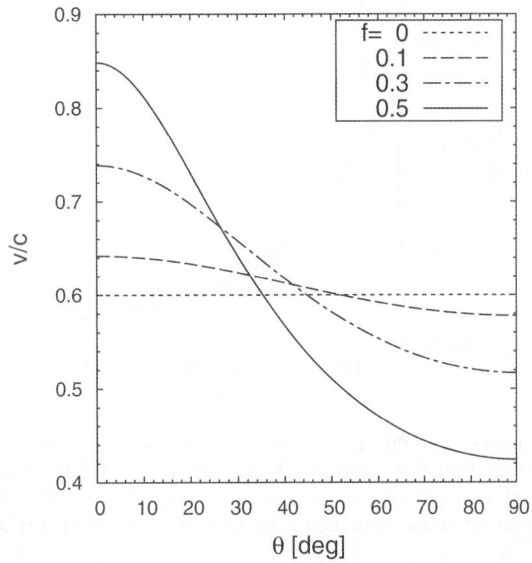

Figure 1. Left: Schematic view of an ellipsoidal shock driven by a piston moving in a steady, spherically symmetric CSM. Right: Angular distribution of the shock radial velocity for several values of oblateness f.

$R/7$, where R is the shock radius. In the unshocked CSM, the number density n_1 of electrons follows $n_1 = A \times r^{-2}$ where r is the radius. The constant A is determined so that the optical depth equals $c/v(f, \theta)$ at $r = R_{\rm b}$. Under the assumption of $v_{f=0} = 0.6c$, $A = 7.5 \times 10^{36}$ cm^{-1}. The wind is supposed to emanate from a carbon-oxygen layer, and become stationary when its velocity decreases to 1,000 km thus the mass loss rate becomes $5 \times 10^{-4} M_{\odot}$. From the above constructions, the number density n_2 of electrons and temperature $k_{\rm B}T$ ($k_{\rm B}$: the Boltzmann constant) become $\sim 10^{13}$ cm^{-3} and 0.11 keV (hereafter $k_{\rm B}T_{\rm b}$) at the moment of shock breakout.

2.2. *Monte-Carlo method*

We calculate the transfer of photons generated from the shock front by using a Monte-Carlo method. The basic construction of the code is the same as used in Ohtani *et al.* (2013). In this study, photons are assumed to be generated over a period of 0.5 sec with an energy distribution following the Planck function in the co-moving frame, and interact with matters via inverse Compton scattering and free-free absorption.

3. Results

First of all, we calculate the X-ray light curve and spectrum for a spherical shock and compare with the observation of XRO 080109. Then we investigate the dependence on the degree of shock asphericity and the viewing angle.

3.1. *Spherical shock*

The left panel of Figure 2 shows the X-ray light curve in the energy range of [0.3,10] keV for the wind mass loss rate of $5 \times 10^{-4} M_{\odot}$ yr^{-1} and the shock velocity of $0.6c$. The luminosity reaches its peak at ~ 40 sec and decreases for the for the subsequent several

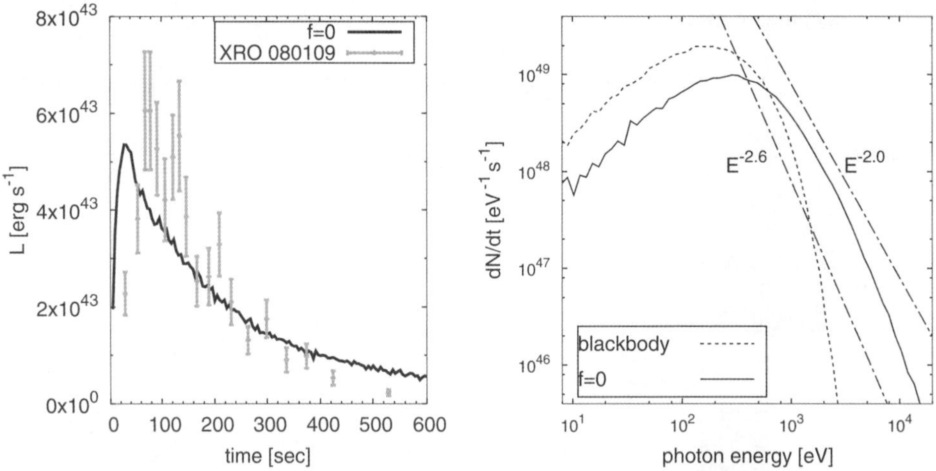

Figure 2. Left: Light curves in the energy range of [0.3,10] keV (left) compared with that of XRO 080109 (superposed bar) for $f = 0$. Right: Time-integrated spectrum for $f = 0$ (solid line) and a blackbody spectrum at a temperature of $k_\mathrm{B}T_\mathrm{b}$. The gradients (-2.0 and -2.6) of the gray straight lines show the $\pm 1\sigma$ values of those of XRO 080109.

hundred seconds. The rise time mainly depends on the light crossing time Δt_lc of the size of the emerging shock and weakly on the diffusion timescale $\Delta t'_\mathrm{diff}$ ($= \Delta R/v = 110$ sec) of photons in the shocked CSM. Due to weak relativistic effects, Δt_lc equals to $R_\mathrm{b}(1 - \cos\theta_\mathrm{rel})/c = 30$ deg, where $\theta_\mathrm{rel} = \tan^{-1}(c\gamma^{-1}v^{-1}) = 50$ deg and $\gamma = 1/\sqrt{1-(v/c)^2}$. The decay time is determined by the radius R_final where photons last scatter off electrons. We can roughly estimate R_final to be $\sim 10R_\mathrm{b}$ where the optical depth of the outer CSM decreases down to ~ 0.1. Therefore the decay time R_final/c equals to several hundred seconds. Note that we could not discuss the maximum luminosity because it depends on the assumed period of photon generation.

The right panel of Figure 2 shows the time-integrated spectrum (solid line) and a blackbody with a temperature of $k_\mathrm{B}T_{f=0}$ (dotted line). Due to the bulk-Comptonization, the maximum photon energy reaches $\sim m_\mathrm{e}v^2 \sim 10$ keV, where m_e is the electron mass. The high-energy component (1–7 keV) can be fitted by a power-law distribution, of which gradient lies in the 1σ error range of that of XRO 080109 (shown by straight lines in Figure 2).

3.2. *Aspherical shock*

We investigate the influence of the asphericity of the shock on the X-ray light curve and spectrum. In the previous works (Suzuki & Shigeyama 2010a; Couch *et al.* 2011; Matzner *et al.* 2013; Salbi *et al.* 2014; Suzuki *et al.* 2016) the properties of the emission generated near the stellar surface were strongly by the asphericity of the shock. Our results are fairly different from that because the shock breaks out simultaneously in all directions under the assumption of a shock of a constant velocity and a steady and spherical wind.

Figure 3 shows the X-ray light curves (0.3–10 keV) for aspherical shock waves. In the left panel, f has several values (0.1, 0.3, 0.5) and $\Theta = [80, 90]$ deg (the flux is averaged over this angular range), and in the right panel, f is fixed at 0.5 and $\Theta =$[0,10], [40,50] and [80,90] deg. (The graphs are shifted to the left side so that the time of onset correspond to those for $f = 0$.) The luminosity evolves with similar timescales (rise & decay times) regardless of f. As with the spherically symmetric case, the decay time is uniquely determined by the density distribution of the CSM. The rise time corresponds

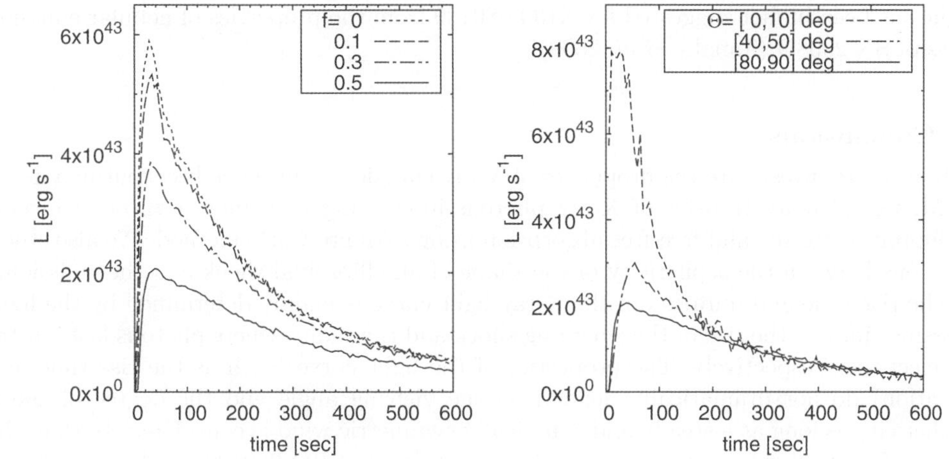

Figure 3. The same as the left panel of Figure 2 but for $f > 0$. Left: f-dependence (Θ =[80,90] deg). Right: Θ-dependence ($f = 0.5$). The graphs are shifted so that the times of onset correspond to those for $f = 0$.

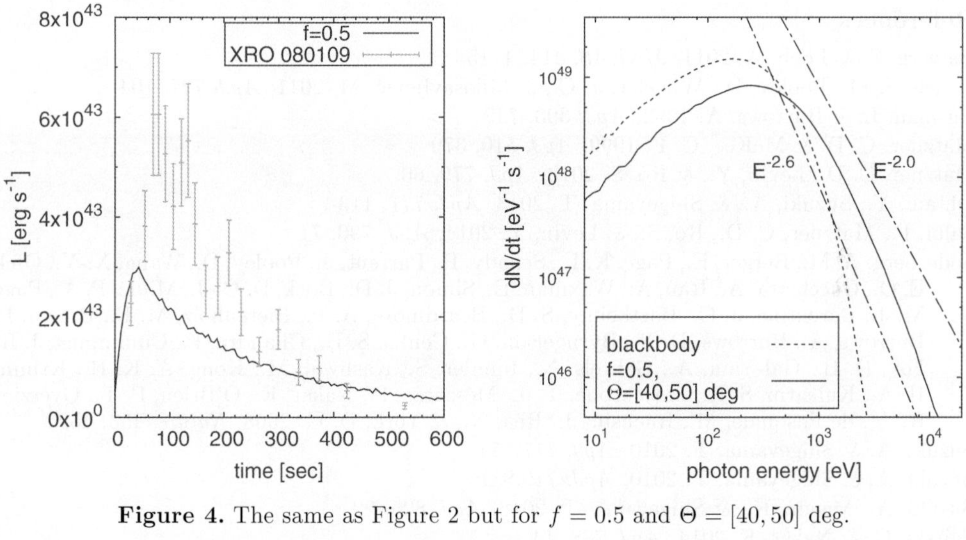

Figure 4. The same as Figure 2 but for $f = 0.5$ and Θ = [40, 50] deg.

to the light crossing time of the size of the emerging shock. If $f = 0.5$ and $\Theta \sim 0$ deg, for example, most photons traveling along the line of sight are generated near ($\theta \lesssim 30$ deg) the symmetry axis because of the high shock velocity ($\sim$ 0.8c) along the line of sight. Therefore the rise time equals to $\sim [R_{\mathrm{b},\theta=0} - R_{\mathrm{b},\theta=30^\circ} \cos 30^\circ]/c \sim 40$ deg, where $R_{\mathrm{b},\theta=0} = 4 \times 10^{12}$ cm and $R_{\mathrm{b},\theta=30^\circ} = 4 \times 10^{12}$ cm. If $\Theta = 90$ deg, the influence of relativistic effects on the motion of photons becomes weak due to the lower shock velocity of $0.5c$. As a result, the rise time equals ~ 40 sec, which is not significantly different from that for $\Theta = 0$. The peak luminosity decreases with increasing Θ, as well as the velocity of the shock propagating along the line of sight.

We compare the calculation for the ellipsoidal shock with the XRO 080109. Figure 4 shows the X-ray light curve (left) for $f = 0.5$ and Θ =[40,50] deg, of which overall shape is roughly consistent with the observation. The spectral gradient, shown in the right panel of Figure 4, is also consistent with that of XRO 080109. An off-axis line of sight has been in fact suggested for XRO 080109 from the properties of nebular emission lines of SN 2008D (Tanaka *et al.* 2009).

4. Conclusions

In order to investigate the properties of X-ray emission from shock breakout in a dense CSM, we calculate transfer of X-ray photons interacting with matter through inverse Compton scattering and free-free absorption using a Monte-Carlo method. We also study relations between the asphericity of the shape of an ellipsoidal shock and the emission.

The rise time and duration of the X-ray light curve is mainly determined by the light crossing times of the size of the emerging shock and the radius where photons last scatter off electrons, respectively. The properties of the light curve (such as the rise time and duration) do not dramatically depend on the viewing angle and the degree of shock asphericity as long as a steady and spherically symmetric wind is considered. Both of the observed X-ray light curve and power-law spectrum of XRO 080109 can be reproduced by a shock with a radial velocity of $0.6c$ and a (spherical) wind with a mass loss rate of $5 \times 10^{-4} M_{\odot}$.

References

Balberg, S. & Loeb, A. 2011, *MNRAS*, 414, 1715

Couch, S. M., Pooley, D., Wheeler, J. C., & Milosavljević, M. 2011, *ApJ*, 727, 104

Ensman, L. & Burrows, A. 1992, *ApJ*, 393, 742

Matzner, C. D. & McKee, C. F. 1999, *ApJ*, 510, 379

Matzner, C. D., Levin, Y., & Ro, S. 2013, *ApJ*, 779, 60

Ohtani, Y., Suzuki, A., & Shigeyama, T. 2013, *ApJ*, 777, 113

Salbi, P., Matzner, C. D., Ro, S., & Levin, Y. 2014, *ApJ*, 790, 71

Soderberg, A. M., Berger, E., Page, K. L., Schady, P., Parrent, J., Pooley, D., Wang, X.-Y., Ofek, E. O., Cucchiara, A., Rau, A., Waxman, E., Simon, J. D., Bock, D. C.-J., Milne, P. A., Page, M. J., Barentine, J. C., Barthelmy, S. D., Beardmore, A. P., Bietenholz, M. F., Brown, P., Burrows, A., Burrows, D. N., Byrngelson, G., Cenko, S. B., Chandra, P., Cummings, J. R., Fox, D. B., Gal-Yam, A., Gehrels, N., Immler, S., Kasliwal, M., Kong, A. K. H., Krimm, H. A., Kulkarni, S. R., Maccarone, T. J., Mészáros, P., Nakar, E., O'Brien, P. T., Overzier, R. A., de Pasquale, M., Racusin, J., Rea, N., & York, D. G. 2008, *Nature*, 453, 469

Suzuki, A. & Shigeyama, T. 2010, *ApJ*, 717,154

Suzuki, A. & Shigeyama, T. 2010, *ApJ*, 719,881

Suzuki, A., Maeda, K., & Shigeyama, T. 2016, *ApJ*, 825, 92

Svirski, G. & Nakar, E. 2014, *ApJ*, 788, 14

Tanaka, M., Yamanaka, M., Maeda, K., Kawabata, K. S., Hattori, T., Minezaki, T., Valenti, S., Della Valle, M., Sahu, D. K., Anupama, G. C., Tominaga, N., Nomoto, K., Mazzali, P. A., & Pian, E. 2009, *ApJ*, 700, 1680

Supernova 1987A:30 years later - Cosmic Rays and Nuclei from Supernovae and their aftermaths
Proceedings IAU Symposium No. 331, 2017
A. Marcowith, M. Renaud, G. Dubner, A. Ray & A. Bykov, eds.

doi:10.1017/S1743921317004537

Supernova 1986J: a Neutron Star or Black Hole in the Centre?

Michael F. Bietenholz[1,2]

[1]Hartebeesthoek Radio Observatory, PO Box 443, Krugersdorp, 1740, South Africa. [2]Also at Department of Physics and Astronomy, York University, Toronto, M3J 1P3, Ontario, Canada.

Abstract. Supernova 1986J is almost the same age as SN 1987A, but was Type IIn, and likely had a massive progenitor. Located at 10 Mpc in NGC 891, it is one of the few supernovae whose radio emission can be resolved using VLBI. We present a new 5-GHz global-VLBI image of SN 1986J from 2014 as well as broadband VLA flux-density measurements. SN 1986J is unusual in that a compact synchrotron radio-emitting component appeared in the centre of the expanding shell of ejecta ~14 yr after the explosion, which now dominates the VLBI image. The central component is stationary to within the uncertainties (< 570 km s^{-1}), and it has a marginally resolved HWHM radius of $(6.7^{+0.7}_{-3.7}) \times 10^{16}$ cm. The shell has expanded with average $v \simeq 5400$ km s^{-1}. The central component's 5-GHz flux density is still increasing with time, and at present it has a 5-GHz νL_ν luminosity of $\sim 4 \times 10^{35}$ erg s^{-1}, ~20 times that of the Crab Nebula. The central component may be due to a newly formed pulsar wind nebula, or an accreting black hole, or it may be due to interaction of the supernova shock with a highly structured environment left over from a progenitor which was in a close binary system. We discuss the newest observations and the constraints on its nature.

Keywords. (stars:) supernovae: individual (SN 1986J), radio continuum: stars)

SN 1986J was one of the most radio luminous supernovae (SNe) ever observed, and one of the few SNe still detectable more than $t = 30$ years after the explosion, thus we have been able to follow its evolution for longer than was possible for most other SNe. Its radio brightness and relatively close distance of $D = 10$ Mpc (Nasa/IPAC Extragalactic Database†) allowed both resolved very long baseline interferometry (VLBI) images and accurate measurements of its broadband radio spectral energy distribution (SED) to be made.

SN 1986J was unusual in that it was first discovered in the radio, some time after the explosion (Rupen *et al.* 1987). The best estimate of the explosion epoch is 1983.2 ± 1.1 ($t = 0$; see Bietenholz *et al.* 2002). Its optical spectrum showed prominent Hα lines but unusual narrow linewidths, which led Type IIn classification (Rupen *et al.* 1987). Due to its high radio brightness it was one of the first SNe to be observed with VLBI (e.g. Bartel *et al.* 1991), and we have continued the observations till the present. We discuss the VLBI observations in a series of papers, Bietenholz *et al.* (2002, 2010) and Bietenholz & Bartel (2017), which we will refer to as Papers I, II, and III respectively. Paper IV in this series is currently in progress.

We show our latest VLBI image at 5-GHz in Figure 1, observed on 2014 Oct 23, or at $t = 31.6$ yr. The observations were made using a global array of VLBI antennas from both the European VLBI Network and the National Radio Astronomy Observatory in the US (see Paper III, for details). A central component dominates the image. Such a central component has not been seen in any other SN (see, e.g. Bietenholz 2014; Bartel

† `https://ned.ipac.caltech.edu`

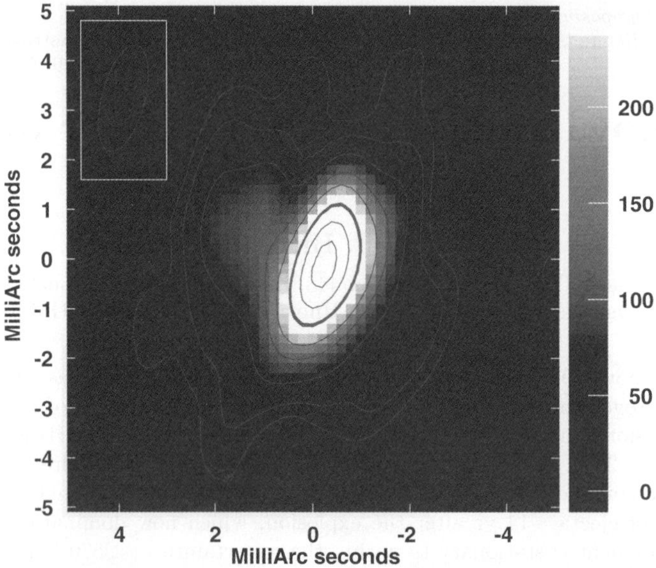

Figure 1. The 5-GHz VLBI image of SN 1986J, made from observations on 2014 Oct. 23, at age $t = 31.6$ yr, reproduced from Paper III. The contours are drawn at -3, 3, 5, 10, 15, 20, 30, **50**, 70 and 90% of the peak brightness, with the 50% contour being emphasized. The peak brightness was 617 μJy beam^{-1}, the total CLEANed flux density was 1622 μJy, and the background rms brightness was 5.9 μJy beam^{-1}. The colorscale is labelled in μJy beam^{-1}, and is saturated so as to better show the low-level emission. North is up and east to the left, and the FWHM of the convolving beam is indicated at upper left.

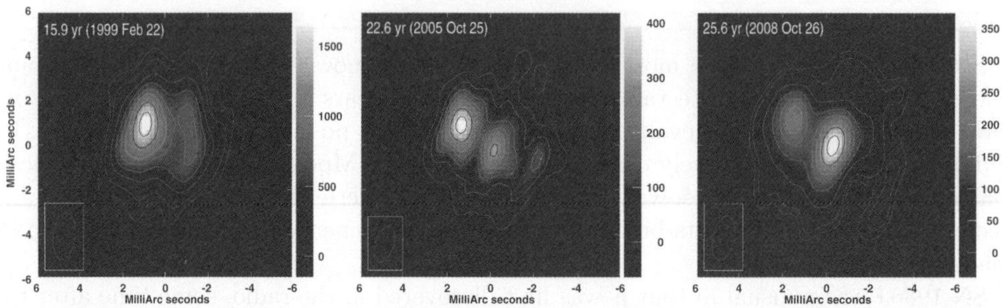

Figure 2. A sequence of 5-GHz VLBI images of SN 1986J, showing its evolution from $t = 15.9$ yr to 25.6 yr after the explosion, or 1999 to 2008. Each panel has the lowest contour at 3× the image background rms, and the remainder at 10, 20, 30, 40, **50**, 70, and 90% of the peak brightness, with the 50% one being emphasized. In each panel, the age and date are given at top left, and the FWHM size of the convolving beam at lower left. North is up and east to the left. The colorscale is in μJy beam^{-1}. For details of the images, see Paper II. An animation showing the evolution is included in the supplementary materials.

& Bietenholz 2014), and in particular, no central synchrotron emission has been seen in SN 1987A at cm wavelengths (Zanardo *et al.* 2014).

The central component is marginally resolved in this image, with a FWHM width of 900^{+100}_{-500} μas, corresponding to a radius of $r_{\rm comp} = 6.7^{+0.7}_{-3.7} \times 10^{16}$ cm (see Paper III, $D = 10$ Mpc). Its 5-GHz peak brightness in 2014 Oct. was 511 μJy beam^{-1}, corresponding to a spectral luminosity, L_ν, of 6×10^{25} erg s^{-1} Hz^{-1}, or ~20 times that of the Crab Nebula.

In Figure 2, we show a sequence of three 5-GHz images between $t = 15.9$ and 25.6 yr, showing the emergence of the central component at that frequency. The central component only begins to appear around $t = 20$ yr. In these images, the SN is clearly expanding. The expansion is well described by a powerlaw, with the radius, $r \propto t^{0.69\pm0.03}$ (Paper II). The average speed of the outer edge was 5400 km s^{-1} ($t = 0$ to 25.6 yr, for $D = 10$ Mpc).

A second hot-spot, located in the shell to the north-east, can be seen in the images in Figure 2. At $t = 15.9$ yr, it is the brightest point in the image. It is moving outward, but is also fading in time, and no longer clearly visible by $t = 31.6$ yr.

We measured the proper motions of the shell hotspot and the central component (Paper III), and found that the shell hotspot moves outwards with a mean projected velocity of 2810 ± 750 km s^{-1} to the NE, which is consistent with it taking part in the homologous expansion of the shell. The central component, on the other hand, is stationary to within the uncertainty of 570 km s^{-1} or 12 μas yr^{-1}.

Not only is the appearance of a central component unusual, but so is the evolution of SN 1986J's (SED). The radio SEDs of SNe usually evolve in a fairly predictable way: the flux density, S_ν, at frequency, ν, shows an optically-thick inverted spectrum, with positive spectral index $\alpha > 1$, ($S_\nu \propto \nu^\alpha$) up to a turnover frequency, and an optically-thin powerlaw above with $-0.8 < \alpha < -0.5$ above the turnover frequency. The absorption at low frequencies is either free-free absorption from intervening thermal material or synchrotron self absorption if the SN is still compact. The turnover frequency evolves downward in time, typically reaching 10 GHz at $t \sim 100$ d (see, e.g. Weiler *et al.* 2002). There is, however, a large variation in this timescale: for SN 1987A, for example, the turnover frequency reached 10 GHz at $t < 1$ d (Turtle *et al.* 1987) while for SN 2001em (e.g. Bietenholz & Bartel 2007), as well as for SN 1986J (Weiler *et al.* 1990), it did not reach 10 GHz till $t \sim 1000$ d.

SN 1986J's SED evolved in this fashion, albeit with slow evolution of the turnover frequency, until $t = 14.9$ yr, when an inversion appeared in the spectrum. We show the evolution of the SED in Figure 3. Using multi-frequency phase-referenced VLBI imaging at 5, 8.4 and 15 GHz, we showed in Bietenholz *et al.* (2004) that the spectral inversion was due to the central component, which was not yet promiment at 5-GHz but was already dominant at 15 GHz. Since that time, both the inflection point, where the spectrum becomes inverted and the higher frequency turnover point have been moving downward.

The emission from the expanding shell of ejecta in SN 1986J is expected to be optically thin at GHz frequencies, and to suffer no significant free-free absorption at this late stage (e.g. Weiler *et al.* 2002). The spectral inversion (at about 2 GHz in 2014) and the higher-frequency turnover (at about 15 GHz in 2014) therefore suggest a source of emission which is still being significantly absorbed by intervening thermal material. In Paper II, we argue that the absorption must be free-free and not synchrotron self-absorption.

We fitted the evolving SED of SN 1986J with a model which had 8 free parameters. The model consisted of two emitting components and some free-free absorption, and the intrinsic flux densities and the amount of absorption all vary with time in a powerlaw fashion. The first component is the emission from the shell, or the expanding shock front, which is unabsorbed and optically thin, and has flux-density

$$S_{\rm shell} = S_{0,\rm shell}(\frac{t}{20\ \rm yr})^{b_{\rm shell}}\ (\frac{\nu}{1\ \rm GHz})^{\alpha_{\rm shell}}$$

The second, due to the central component, has an intrinsic SED of

$$S_{\rm comp} = S_{0,\rm comp}(\frac{t}{20\ \rm yr})^{b_{\rm comp}}\ (\frac{\nu}{1\ \rm GHz})^{\alpha_{\rm comp}}$$

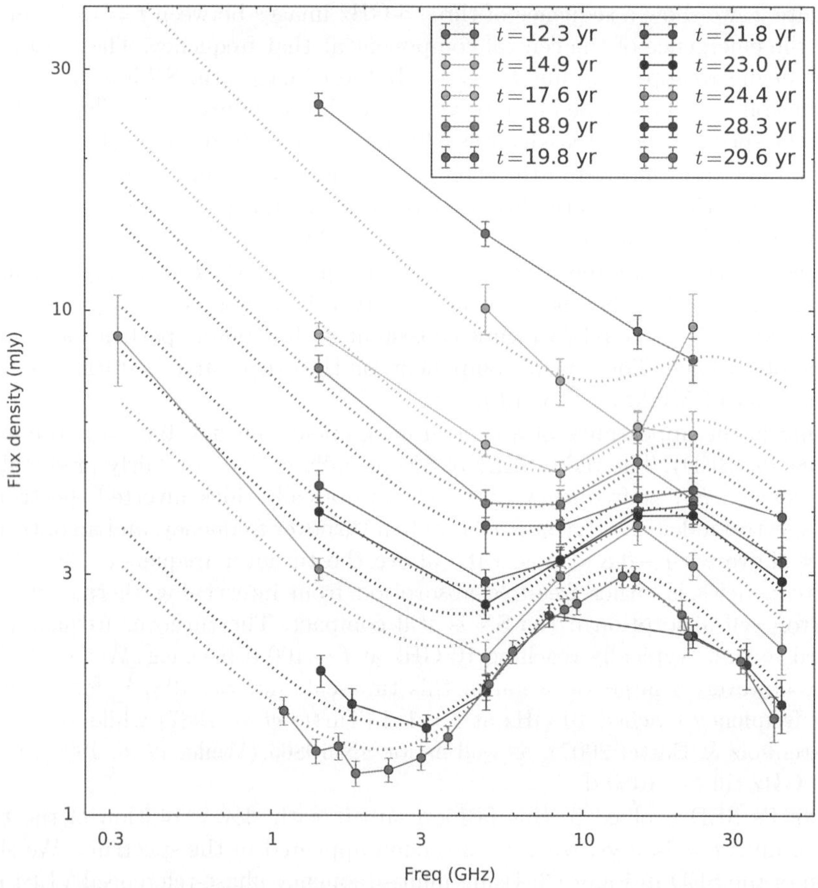

Figure 3. The evolving SED of SN 1986J. The coloured points and associated solid line shows the SED at the indicated age, with the earliest one at the top. The uncertainties are estimated standard errors, including statistical and systematic contributions combined. Data are taken from Paper III and Bietenholz *et al.* (2004), as well as from the Paper IV (currently in preparation). The dotted lines show a simple model fit to the evolving SED (see text) since the first appearance of the central (inverted spectrum) component at t = 14.9 yr. We plot the fitted spectra for each epoch using dotted lines of the same colours as were used for the data points. The model is fairly simple, and fits the SEDs at all epochs simultaneously with only eight free parameters, so the fit to the measurements at any one epoch is only approximate, but the model should give a good overall description of the evolution.

but is free-free absorbed by material with an emission measure, $\mathrm{EM} = \int N_e^2 \cdot dl$, where N_e is the electron number density and we assume an electron temperature of $T_e = 10^4$ K. In our model, the EM is also time-variable and is given by $\mathrm{EM}_0 \cdot (t/20\ \mathrm{yr})^{b_{\mathrm{EM}}}$ cm^{-6} pc. In the case of purely external absorption, the inverted part of the spectrum would have a steep frequency dependence, $\propto \nu^{2.1}$, which does not well match our observed SEDs (see Figure 3 and Paper II). Therefore, we take the emitting and the absorbing material to be mixed, in which case the inverted part of the spectrum is $\propto \nu^{(\alpha_{\mathrm{comp}}+2.1)}$.

We fitted this model to all the VLA flux density measurements after 2014 in Figure 3 (when the central component first appeared) using Bayesian statistics. Details will be given in our upcoming Paper IV. We obtained the following values for the eight free parameters: $S_{0,\mathrm{shell}} = 7.07 \pm 0.17$ mJy, $b_{\mathrm{shell}} = -3.92 \pm 0.07$, $\alpha_{\mathrm{shell}} = -0.63 \pm 0.03$, $S_{0,\mathrm{comp}} = 61 \pm 17$ mJy, $b_{\mathrm{comp}} = -2.07 \pm 0.16$, $\alpha_{\mathrm{comp}} = -0.76 \pm 0.07$, $\mathrm{EM}_0 = (1.64 \pm$

$0.20) \times 10^9$ cm^{-6} pc, and $b_{\rm EM} = -2.72 \pm 0.26$. We plot the model spectra as dotted lines in Figure 3. The model adequately reproduces the evolving SED, and therefore despite its relative simplicity, we feel gives us some insight into the physics.

The two fitted spectral indices, $\alpha_{\rm shell}$ and $\alpha_{\rm comp}$, are very similar, differing by only 1.7σ (formally). In view of the relatively simple nature of our model, we do not consider the difference to be significant†. Our measurements are therefore compatible with both the shell and the central component having the same optically-thin spectral index.

The intrinsic flux densities, $S_{\rm comp}$ and $S_{\rm shell}$ are both decreasing with time, with $S_{\rm shell}$ decreasing more rapidly, so that the central component becomes ever more dominant. The amount of absorption affecting the central component is also decreasing with time. At some frequencies, the *observed* flux density of the central component is increasing with time because the absorption is decreasing more rapidly than the $S_{\rm comp}$.

In light of these results, what can we say of the nature of the central component? Its high and still increasing brightness, coupled with its stationarity, and long lifetime argue that it really is in or near the three-dimensional centre of SN 1986J, and not associated with the shell and central only in projection.

There are a number of possibilities as to its nature: First, the central component could be due to a young pulsar wind nebula (PWN) formed around a newly born neutron star. Its central location and stationarity are readily explained. The size of the central component is consistent with the expected size of a PWN of 5% to 10% of the size of the forward shock. However, its steep intrinsic spectrum and the relatively rapid decay with time of its unabsorbed flux density are not consistent with what is expected of a PWN (see e.g. Gelfand *et al.* 2009; Chevalier & Fransson 1992).

Second, since SN 1986J had a massive progenitor, a black hole may have formed. The central component might therefore be due to accretion onto a newly-formed black hole. It is, however, far more radio-luminous than any known stellar-mass black hole system, having an observed $L_{\rm R} \simeq 3 \times 10^{35}$ erg s^{-1} (νL_ν at 5 GHz), with the unabsorbed value being about an order of magnitude larger. Known accreting stellar-mass black hole systems in X-ray binaries, by contrast, have $L_{\rm R} \lesssim 10^{32}$ erg s^{-1} (see, e.g. Körding *et al.* 2006). The radio luminosity of black hole systems depends on the accretion rate, which might be much higher in the only ∼30 yr old SN 1986J than in the much older X-ray binary systems, and thus produce SN 1986J's unusually high $L_{\rm R}$. However, in that case, SN 1986J would also be expected to have a high X-ray luminosity, $L_{\rm X}$. Houck (2005) found that the unabsorbed $L_{\rm X}$, at $t = 20.6$ yr was 1.9×10^{39} erg s^{-1} (0.5 to 2.5 keV), which is no higher than the brightest X-ray binaries (Körding *et al.* 2006).

There is a "fundamental plane" of black hole activity, with $L_{\rm R}$, $L_{\rm X}$, and the mass of the black hole all being correlated (see, e.g. Merloni *et al.* 2003). Even for high masses of 20 $M_\odot$, this fundamental plane predicts a ratio $L_{\rm R}/L_{\rm X}$ lower by several orders of magnitude than is observed for SN 1986J. Furthermore, in Galactic X-ray binaries, high accretion rates seem to *quench* the radio emission rather than enhance it (e.g. Fender *et al.* 2010). Therefore, SN 1986J's high radio luminosity, and in particular, its high ratio of $L_{\rm R}/L_{\rm X}$, seem discrepant with what is seen in other accreting black hole systems. Given, however, that little is known about newly-born accreting black hole systems inside SN ejecta, a black hole scenario cannot be conclusively ruled out.

Finally, the central component could be explained by SN 1986J's forward shock interacting with a highly anisotropic circumstellar medium (CSM), as was suggested by Chevalier (2012). Such an anisotropic CSM might arise as a result of SN 1986J being

† We also tried a model with only a single intrinsic spectral index characterizing both components, and it provided almost as good a fit with $\alpha = -0.64 \pm 0.03$

the second SN in a massive binary system, where there was a period of common envelope interaction between the compact remnant of the first, unobserved, supernova and SN 1986J's progenitor that resulted in very dense and anisotropic mass-loss concentrated in the equatorial plane of the binary. As a result of the dense CSM, the shock takes on something like an hourglass shape, and the central component would represent the narrow waist of the hourglass, the remainder of the shell is due to less equatorial parts of the shock which interact with the more normal supergiant wind of SN 1986J's progenitor. The density of the equatorial CSM could be an order of magnitude higher than that of the remainder (Chevalier 2012). The non-equatorial ejecta initially free-free absorb any emission from the narrow waist, and only after some time do the bulk of the ejecta expand or fragment to the point where the equatorial disk can be seen, at which point the radio central component "turns on." This hypothesis explains the size, central location, low proper motion of the central component, and the near equivalence of $\alpha_{\rm comp}$ and $\alpha_{\rm shell}$. Our fitted EM $(1.64 \pm 0.20) \times 10^9$ cm^{-6} pc is compatible with what would be expected from a few $M_\odot$ of ionized SN ejecta if the ejecta are somewhat centrally concentrated, as is expected to be the case.

At the moment none of the three hypotheses can be definitively ruled out. We are planning more observations to follow the evolution of this fascinating object and hopefully more decisively elucidate the nature of the central component.

References

Bartel, N. & Bietenholz, M. F. 2014, in IAU Symposium, Vol. 296, Supernova Environmental Impacts, ed. A. Ray & R. A. McCray, 53–57

Bartel, N., Rupen, M. P., Shapiro, I. I., Preston, R. A., & Rius, A. 1991, *Nature*, 350, 212

Bietenholz, M. 2014, in 12th European VLBI Network Symposium and Users Meeting (2014), published by SISSA, Trieste, ed. A. Tarchi, M. Giroletti, & L. Feretti, 51

Bietenholz, M. F. & Bartel, N. 2007, *ApJL*, 665, L47

—. 2017, *ApJ*, 839, 10, **Paper III**

Bietenholz, M. F., Bartel, N., & Rupen, M. P. 2002, *ApJ*, 581, 1132, **Paper I**

—. 2004, *Science*, 304, 1947

—. 2010, *ApJ*, 712, 1057, **Paper II**

Chevalier, R. A. 2012, *ApJL*, 752, L2

Chevalier, R. A. & Fransson, C. 1992, *ApJ*, 395, 540

Fender, R. P., Gallo, E., & Russell, D. 2010, *MNRAS*, 406, 1425

Gelfand, J. D., Slane, P. O., & Zhang, W. 2009, *ApJ*, 703, 2051

Houck, J. C. 2005, in X-Ray and Radio Connections (eds. L.O. Sjouwerman and K.K Dyer). Published electronically by NRAO, `http://www.aoc.nrao.edu/events/xraydio`, 4.6.

Körding, E., Falcke, H., & Corbel, S. 2006, *A&A*, 456, 439

Merloni, A., Heinz, S., & di Matteo, T. 2003, *MNRAS*, 345, 1057

Rupen, M. P., van Gorkom, J. H., Knapp, *et al.* 1987, *AJ*, 94, 61

Turtle, A. J., Campbell-Wilson, D., Bunton, J. D., *et al.* 1987, *Nature*, 327, 38

Weiler, K. W., Panagia, N., Montes, M. J., & Sramek, R. A. 2002, *ARAA*, 40, 387

Weiler, K. W., Panagia, N., & Sramek, R. A. 1990, *ApJ*, 364, 611

Zanardo, G., Staveley-Smith, L., Indebetouw, R., *et al.* 2014, *ApJ*, 796, 82

Supernova 1987A:30 years later - Cosmic Rays and Nuclei from Supernovae and their aftermaths
Proceedings IAU Symposium No. 331, 2017
A. Marcowith, M. Renaud, G. Dubner, A. Ray & A. Bykov, eds.

doi:10.1017/S1743921317004756

Constraining pulsar birth properties with supernova X-ray observations

Y. A. Gallant[1], R. Bandiera[2], N. Bucciantini[2] and E. Amato[2]

[1]LUPM, U. de Montpellier, CNRS/IN2P3, place E. Bataillon, 34095 Montpellier, France
email: gallant@in2p3.fr

[2]INAF – Osservatorio Astrofisico di Arcetri, Largo E. Fermi 5, I-50125 Firenze, Italy

Abstract. A large fraction of core-collapse supernovae are thought to result in the birth of a rotation-powered pulsar, which is later observable as a radio pulsar up to great ages. The birth properties of these pulsars, and in particular the distribution of their initial rotation periods, are however difficult to infer from studies of the radio pulsar population in our Galaxy. Yet the distributions of their birth properties is an important assumption for scenarios in which ultra-high-energy cosmic rays (UHECRs) originate in very young, extragalactic pulsars with short birth periods and/or high magnetic fields.

Using a model of the very young pulsar wind nebula's dynamical and spectral evolution, with pulsar wind and accelerated particle parameters assumed similar to those inferred from modeling young pulsar wind nebulae (PWNe) in our Galaxy, we show that X-ray observations of supernovae, a few years to decades after the explosion, constitute a favored window to obtain meaningful constraints on the initial spin-down luminosity of the newly-formed pulsar. We examine the expected emerging PWN spectral component, taking into account the X-ray opacity of the expanding supernova ejecta, and find that it is typically best detectable in $< 10\,\mathrm{keV}$ X-rays some years after the explosion. We use this framework to assess available X-ray observations and flux upper limits on supernovae, building on the work of Perna *et al.* (2008). We note that a resulting limit on spin-down luminosity corresponds univocally to a limit on the maximum magnetospheric acceleration potential, irrespective of the specific combination of magnetic field and rotation period that achieves it. We use available X-ray observations of supernovae to place constraints on the birth spin-down luminosity and period distribution of classical pulsars. We also examine the case of magnetars, born with much higher magnetic fields, and show that their much shorter initial spin-down time implies that any plausible signature of young magnetar wind nebulae can only be observed in harder X-ray or gamma-rays.

Keywords. pulsars, supernovae, acceleration of particles, X-rays, neutron stars

1. Introduction and motivation

The inferred Galactic radio pulsar birth rate is a large, and possibly dominant, fraction of the Galactic core-collapse supernova rate, implying that many, and possibly most, such supernovae result in the formation of a rotation-powered pulsar. Radio pulsar population synthesis studies (e.g. Faucher-Giguère & Kaspi 2006, and references therein) suggest that their surface magnetic fields B_* follow a log-normal distribution, with $\langle \log B_*/\mathrm{G} \rangle \approx 12.5 \pm 0.5$; the distribution of birth periods P_0 is however poorly constrained by such studies, as pulsars "lose memory" of P_0 after their initial spin-down time τ_0, which is of order kyr. Fast-spinning pulsars or magnetars have been proposed to power certain types of supernovae, but such scenarios currently lack an unambiguous observational signature (e.g., Chevalier 2011 and references therein). Here we propose to use a model of the "very young" wind nebulae powered by such pulsars to examine their potential observability.

2. Dynamical and spectral model of very young PWNe

Dynamical evolution. In the initial, so-called 'free expansion' phase of pulsar wind nebula (PWN) evolution, the wind of the newly-born pulsar blows a bubble of magnetized relativistic plasma inside the freely expanding ejecta of the supernova (e.g., van der Swaluw *et al.* 2001; Bucciantini *et al.* 2003). We assume that this wind carries most of the pulsar's spin-down power $\dot{E}$, and that the core of the ejecta, into which the PWN expands, has an approximately uniform density. In the first few years after the explosion, during which the pulsar's spin-down power remains approximately constant, the PWN's evolution can be described by an analytical solution in which its radius evolves as $R_{\mathrm{pwn}} \propto t^{6/5}$ (Chevalier & Fransson 1992 and references therein). We further assume that the magnetization fraction of the pulsar wind, η, remains constant and has a value similar to those inferred from models of 'young' PWNe with ages of order a few kyr; the median magnetization found in models of 9 young PWNe by Torres *et al.* (2014) was $\eta \approx 0.03$. Under the above assumptions, the PWN magnetic field in its early years is generally quite high:

$$B_{\mathrm{pwn}}\left(t_{\mathrm{age}}\right) \approx 0.4\ \mathrm{G} \times \sqrt{\frac{\eta}{0.03}} \left(\frac{E_{\mathrm{ej}}}{10^{51}\ \mathrm{erg}}\right)^{-0.45} \left(\frac{\dot{E}}{10^{38}\ \mathrm{erg/s}}\right)^{0.2} \left(\frac{t_{\mathrm{age}}}{\mathrm{yr}}\right)^{-1.3}, \tag{2.1}$$

where t_{age} is the supernova remnant's age and E_{ej} the initial kinetic energy of the ejecta, which are assumed uniform with a fiducial mass of $5\,M_{\odot}$. Strictly speaking, this expression assumes that radiative losses are dynamically unimportant, so that essentially all the pulsar's spin-down power can drive the expansion of the PWN. We will see below that this is not actually the case for very young PWNe, in which synchrotron losses are highly efficient. In the latter regime, the PWN magnetic field could be somewhat higher; estimates suggest that the correction factor remains of order unity, however, so we will use the magnetic field value from (2.1) in what follows.

Accelerated particle spectrum. Multi-wavelength observations of PWNe suggest that the relativistic electrons and positrons injected by the pulsar wind are accelerated to a broken power-law spectrum. In sufficiently young and magnetized PWNe, the maximum accelerated particle energy is reached when the synchrotron cooling time t_{cool} becomes shorter than the acceleration time t_{acc}, and observations of the high-energy cutoff of the Crab Nebula's synchrotron spectrum suggest that t_{acc} can be of order the particle's gyro-time (de Jager & Djannati-Ataï 2009 and references therein). In very young PWNe, the resulting acceleration time scale is much shorter than the dynamical time t_{age}:

$$t_{\mathrm{acc}}(\gamma_{\max}) \approx \frac{m_e\, \gamma_{\max}\, c}{e\, B_{\mathrm{pwn}}(t_{\mathrm{age}})} \approx 25\,\mathrm{s} \times \left(\frac{\eta}{0.03}\right)^{-0.75} \left(\frac{t_{\mathrm{age}}}{\mathrm{yr}}\right)^{1.95}, \tag{2.2}$$

with weaker dependencies on $\dot{E}$, the ejecta mass M_{ej} and E_{ej}. The acceleration process should thus operate efficiently, radiating synchrotron photons up to high-energy γ-rays.

Typical accelerated $e^{\pm}$ spectral indices, defined by $\mathrm{d}N_e/\mathrm{d}\gamma \propto \gamma^{-p}$ where γ is the particle Lorentz factor, are $p_1 \approx 1.5$ and $p_2 \approx 2.5$ for the low- and high-energy part of the (uncooled) spectrum, respectively, as shown by the median values of the PWN models of Torres *et al.* (2014). The break energy between the two power-law segments can differ significantly between PWNe, but a representative value is $\gamma_{\mathrm{br}} \simeq 3 \times 10^5$, the median value from the models of Torres *et al.* (2014). With the magnetic field (2.1), the corresponding (injection) break frequency in the synchrotron spectrum is given by:

$$h\,\nu_{\rm br} \approx 0.2\ {\rm keV} \times \left(\frac{\gamma_{\rm br}}{3\times 10^5}\right)^2 \sqrt{\frac{\eta}{0.03}} \left(\frac{E_{\rm ej}}{10^{51}\ {\rm erg}}\right)^{-0.45} \left(\frac{t}{\rm yr}\right)^{-1.3} . \tag{2.3}$$

Emitted synchrotron spectrum. The high magnetic field value implies rapid synchrotron cooling of the accelerated $e^{\pm}$; indeed, the spectrum injected in the nebula remains essentially unaffected by radiative losses only below the synchrotron cooling break frequency, defined by $t_{\rm cool} = t_{\rm age}$ and given by

$$\nu_{\rm cool} \approx 1.2\times 10^{10}\ {\rm Hz} \times \left(\frac{\eta}{0.03}\right)^{-1.5} \left(\frac{E_{\rm ej}}{10^{51}{\rm erg}}\right)^{1.35} \left(\frac{t}{\rm yr}\right)^{1.9} , \tag{2.4}$$

ignoring the weaker dependencies on $\dot{E}$ and $M_{\rm ej}$. As $\nu_{\rm cool} \ll \nu_{\rm br}$, the bulk of the synchrotron spectrum is 'cooled', with its peak in νF_ν at $\nu_{\rm br}$, and essentially all the accelerated particle energy is radiated away quasi-instantaneously (i.e. on a time scale short relative to the dynamical time $t_{\rm age}$, although long relative to the acceleration time $t_{\rm acc}$). For the particle spectral indices given above, the synchrotron spectral indices (defined by $F_\nu \propto \nu^{-s}$) are $s_1 \approx 0.75$ and $s_2 \approx 1.25$ below and above $\nu_{\rm br}$, respectively, and the flux value at $\nu_{\rm br}$ is given by $\nu_{\rm br}F_{\nu_{\rm br}} \approx \frac{1}{8}\dot{E}$ (assuming $\eta \ll 1$ and $\nu_{\rm cool} \ll \nu_{\rm br} \ll \nu_{\rm max}$). With all the above assumptions and fiducial parameter values, the fraction of the total pulsar spin-down power which is radiated in X-rays by the PWN is given by $L_{2-10\,{\rm keV}} \approx 0.09\ \dot{E}\ (t/{\rm yr})^{-0.325}$.

X-ray absorption by the ejecta. The X-rays emitted by the PWN are initially absorbed by the surrounding ejecta, and only become observable once these expanding ejecta become optically thin. The column density of the ejecta is equivalent to

$$N_{\rm H} = 4\times 10^{24}\ {\rm cm}^{-2} \times \left(\frac{M_{\rm ej}}{5\,M_\odot}\right)^2 \left(\frac{E_{\rm ej}}{10^{51}\ {\rm erg}}\right)^{-1} \left(\frac{t}{\rm yr}\right)^{-2} , \tag{2.5}$$

and we estimate the corresponding spectral absorption using the `phabs` model of interstellar X-ray absorption (Bałucińska-Church & McCammon 1992). Representative emerging PWN X-ray spectra at different ages, emitted as described earlier and absorbed as outlined above, are shown in Figure 1. It can be seen that PWN X-rays in the energy range below 10 keV typically only become observable a few years after the supernova, although harder X-rays can emerge earlier.

It should be noted that the X-ray absorption model we use is only an approximation, as it assumes the absorbing medium is neutral, while the ejecta are at least partially ionized at the ages of interest. The `phabs` model should nonetheless remain roughly applicable, as its dominant process is photoelectric absorption of K-shell electrons, which remains effective unless the ejecta ions are fully stripped. More importantly, in normalizing to the absorbing column (2.5) we implicitly assume a hydrogen-dominated composition and Galactic interstellar abundances; ejecta significantly enriched in heavy elements would yield correspondingly higher absorption.

3. Constraints on pulsar $\dot{E}_0$ and P_0 from observations

X-ray observations of supernovae. Perna *et al.* (2008) used archival X-ray observations of extragalactic core-collapse supernovae (SNe), generally performed years or decades after the explosion, to constrain the initial spin-down power $\dot{E}_0$ of any embedded pulsar. Here we improve the interpretation of their upper limits by using the physically-derived PWN spectral model described above instead of their empirical X-ray luminosity relation.

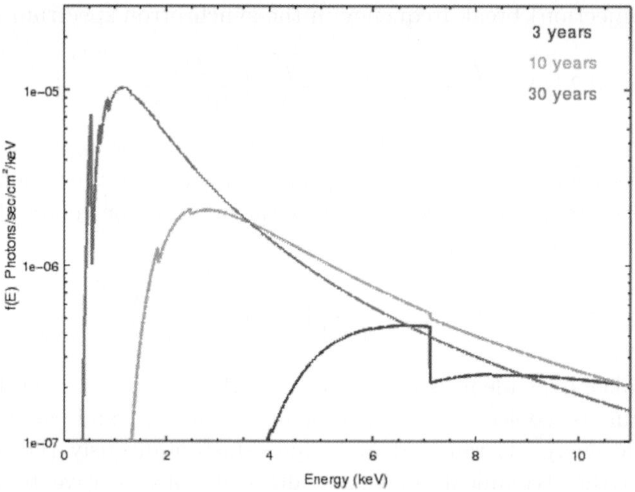

Figure 1. Estimated emerging PWN X-ray spectra, for the model and parameter values described in the text, at $t_{\rm age} = 3$, 10 and 30 years after the supernova explosion.

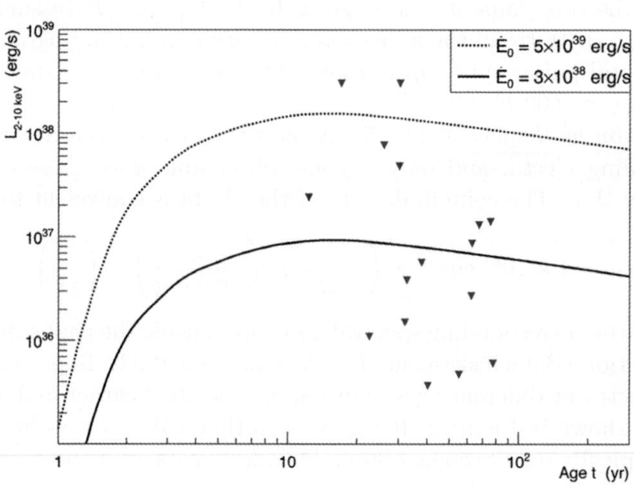

Figure 2. Estimated emerging PWN X-ray fluxes for the model and parameter values detailed in Sect. 2 and two representative values of $\dot{E}_0$, along with upper limits from Perna *et al.* (2008).

Figure 2 shows emerging PWN X-ray fluxes as a function of age, as predicted by the model described in Section 2, along with Perna *et al.*'s (2008) late-time (>10 yr) flux upper limits on SNe of type IIP. (As type IIP SNe make up the dominant fraction of core-collapse SNe, it is reasonable to suppose that at least a significant fraction of them give birth to pulsars.) As can be seen in Fig. 2, roughly 50% of the upper limits conflict with the predictions of the model for a spin-down luminosity of $\dot{E}_0 = 3 \times 10^{38}$ erg/s (similar to the Crab's), and ~90% conflict with $\dot{E}_0 = 5 \times 10^{39}$ erg/s (an order of magnitude higher).

Under the assumption that these SNe gave birth to pulsars, the above can be interpreted as upper limits on the fraction of pulsars born with high spin-down luminosities. A pulsar's spin-down luminosity can be related to its rotation period and surface magnetic field through the standard magnetic dipole radiation formula:

$$\dot{E} \equiv 3.3 \times 10^{40}\ \mathrm{erg/s} \left(\frac{B_*}{3 \times 10^{12}\ \mathrm{G}}\right)^2 \left(\frac{P}{10\,\mathrm{ms}}\right)^{-4} . \tag{3.1}$$

For a typical radio pulsar magnetic field value of 3×10^{12} G, the above limits would thus correspond to ~50% of such pulsars born with initial periods longer than ~30 ms, and ~90% with initial periods longer than ~15 ms.

Other constraints. An alternative approach exploits the fact that the wind nebulae of young and powerful extragalactic pulsars would contribute to the population of ultra-luminous X-ray sources (ULXs) to obtain constraints on their birth periods (Medvedev & Poutanen 2013, and references therein). These authors find, using the observed X-ray luminosity distribution of ULXs, that the average birth period P_0 of radio pulsars must be longer than ~10–40 ms, in agreement with the constraints derived above.

4. Applicability to magnetars

Magnetars are neutron stars whose surface magnetic fields, as inferred from their spin-down through the magnetic dipole formula (3.1), have typical values $B_* \sim 10^{14}$–10^{15} G, much higher than those of the rotation-powered, radio pulsars which we considered thus far. Magnetars further distinguish themselves by generally being radio-quiet, and by the fact that their observed emission is thought to be powered in large part by magnetic field decay rather than rotation.

Magnetars born with a very short rotation periods, of order a few milliseconds, have been proposed as the 'engine' of some classes of superluminous supernovae and gamma-ray bursts (e.g. Metzger *et al.* 2011, and references therein). It has also been proposed that ultra- high-energy cosmic rays (UHECRs) could originate in such fast-spinning magnetars (e.g., Blasi *et al.* 2000; Arons 2003; Fang *et al.* 2012; Lemoine *et al.* 2015). That magnetars should be born with short rotation periods is in fact expected in the dynamo theory of magnetar formation (Duncan & Thompson 1992). The question thus arises whether newly-formed, powerful magnetars formed in supernovae might yield a high-energy observational signature similar to that discussed above for very young pulsar wind nebulae.

Early spin-down of magnetars. Under the magnetic dipole spin-down assumption, a newly-formed pulsar or magnetar retains memory of its initial rotation period for a time of order its initial characteristic spin-down time,

$$\tau_0 \approx 180\,\mathrm{yr} \left(\frac{B_*}{3 \times 10^{12}\,\mathrm{G}}\right)^{-2} \left(\frac{P_0}{10\,\mathrm{ms}}\right)^{2}. \tag{4.1}$$

For the classical rotation-powered radio pulsars considered thus far, τ_0 is thus of order centuries or at least decades, implying that the initial spin-down luminosity remains approximately constant until the supernova ejecta become transparent to X-rays. For the much higher magnetic fields typical of magnetars, by contrast, τ_0 is of order days, or even less for very short birth periods P_0.

It follows that newly-born fast magnetars release their considerable rotational energy very early, well before the ejecta become optically thin to X-rays. For ages $t \gg \tau_0$, a neutron star's spin-down evolution is solely determined by its magnetic field B_*, higher B_* corresponding to lower $\dot{E}$ at a given age. Thus by the time the ejecta become transparent to X-rays, any observational signature of a magnetar's spin-down could no longer yield an upper limit on its initial rotation frequency.

It is moreover unclear whether one should expect magnetars to be surrounded by wind nebulae analogous to those described in Section 2; no 'magnetar wind nebulae' have been detected around the known, slower-spinning magnetars, with one plausible exception (Younes *et al.* 2016, and references therein). Nonetheless assuming that fast-spinning magnetars as well as classical pulsars give rise to similar wind nebulae, Murase *et al.* (2015) made predictions for their emerging spectra at supernova ages of order one year. They found that synchrotron emission from such nebulae should be detectable in the *hard X-ray* domain (at energies of several tens of keV), in which the ejecta become optically thin earlier, given the sensitivity of an instrument such as *NuSTAR*.

5. Summary

Classical radio pulsars ($B_* \sim 10^{12} - 10^{13}$ G). Extrapolating models of young pulsar wind nebulae and their parameter values back to the first several years after the supernova explosion, we find that such 'very young' PWNe efficiently radiate the newly- born pulsar's spin-down power $\dot{E}$. The emitted synchrotron radiation typically peaks (in νF_ν) at photon energies just below X-rays. At ages less than several years, this radiation is essentially fully absorbed in the supernova ejecta, and can thereby contribute to powering the supernova light curve.

At ages of several years to decades, the ejecta gradually become transparent to X-rays, in which the very young PWN can thus be directly observed. X-ray upper limits on supernovae observed more than a decade after the explosion then constrain the birth power $\dot{E}_0$ of pulsars. For typical magnetic field values, available limits are consistent with a median birth period $P_0 \gtrsim 30$ ms, and suggest that no more than ~10% of pulsars could be born with periods shorter than ~15 ms.

Magnetars ($B_* \sim 10^{14} - 10^{15}$ G). Their much stronger magnetic fields imply that newly-formed magnetars spin down much earlier after birth than classical radio pulsars, with characteristic spin-down times τ_0 of order days rather than centuries. They will thus have lost most of their initial rotation energy, and any memory of their birth period, well before the ejecta can become transparent to X-rays at energies $\lesssim 10$ keV. Very young magnetar wind nebulae, if they exist, may however be detectable in hard X-rays.

References

Arons, J. 2003, *ApJ*, 589, 871
Bałucińska-Church, M. & McCammon, D. 1992, *ApJ*, 400, 699
Blasi, P., Epstein, R. I., & Olinto, A. V. 2000, *ApJ* (Letters), 533, L123
Bucciantini, N., Blondin, J. M., Del Zanna, L., & Amato, E. 2003, *A&A*, 405, 617
Chevalier, R. A. 2011, *AIP Conf. Proc.*, 1379, 5 (`arXiv:1011.3731`)
Chevalier, R. A. & Fransson, C. 1992, *ApJ*, 395, 540
de Jager, O. C. & Djannati-Ataï, A. 2009, in: W. Becker (ed.), *Neutron Stars and Pulsars* (Berlin and Heidelberg: Springer-Verlag), p. 451 (`arXiv:0803.0116`)
Duncan, R. C. & Thompson, C. 1992, *ApJ* (Letters), 392, L9
Fang, K., Kotera, K., & Olinto, A. V. 2012, *ApJ*, 750, 118
Faucher-Giguère, C.-A. & Kaspi, V. M. 2006, *ApJ*, 643, 332
Lemoine, M., Kotera, K., & Pétri, J. 2015, *JCAP*, 07, 016 (`arXiv:1409.0159`)
Medvedev, A. S. & Poutanen, J. 2013, *MNRAS*, 431, 2690
Metzger, B. D., Giannios, D., Thompson, T. A., Bucciantini, N., & Quataert, E. 2011, *MNRAS*, 413, 2031
Murase, K., Kashiyama, K., Kiuchi, K., & Bartos, I. 2015, *ApJ*, 805, 82
Perna, R., Soria, R., Pooley, D., & Stella, L. 2008, *MNRAS*, 384, 1638
Torres, D. F., Cillis, A., Martín, J., & de Oña Wilhelmi, E. 2014, *J. High Energy Astrophys.*, 1, 31 (`arXiv:1402.5485`)
van der Swaluw, E., Achterberg, A., Gallant, Y. A., & Tóth. G. 2001, *A&A*, 380, 309
Younes, G., Kouveliotou, C., Kargaltsev, O., Gill, R., Granot, J., *et al.* 2016, *ApJ*, 824, 138

Supernova 1987A:30 years later - Cosmic Rays and Nuclei from Supernovae and their aftermaths
Proceedings IAU Symposium No. 331, 2017
A. Marcowith, M. Renaud, G. Dubner, A. Ray & A. Bykov, eds.

doi:10.1017/S1743921317005300

Constraints on environs around SN 2011fe and SN 2014J from radio modeling and observations

Esha Kundu[1], **Peter Lundqvist**[1] **and Miguel A. Pérez-Torres**[2,3]

[1]Department of Astronomy and The Oskar Klein Centre, AlbaNova, Stockholm University, SE-10691 Stockholm, Sweden
email: esha.kundu@astro.su.se, peter@astro.su.se

[2]Instituto de Astrofísica de Andalucía, Glorieta de las Astronomía, s/n, E-18008 Granada, Spain

[3]Visiting Scientist: Departamento de Física Teorica, Facultad de Ciencias, Universidad de Zaragoza, Spain
email: torres@iaa.es

Abstract. The radio non-detection of two Type Ia supernovae (SNe) SN 2011fe and SN 2014J has been modeled considering synchrotron radiation from shock accelerated electrons in the SN shock fronts. With 10% each of the bulk kinetic energy in electric and magnetic fields, a very low density of the medium around both the SNe has been estimated from the null detection of radio emission, around 1 and 4 years after the explosion of SNe 2014J and 2011fe, respectively. Keeping the fraction of energy in electrons fixed at 10%, a medium with particle density ~ 1 cm^{-3} is found when 1% of the post shock energy is in magnetic fields. In case of a wind medium, the former predicts the mass loss rate $\dot{M}$ to be $< 10^{-9}$ $M_\odot\,yr^{-1}$, and the latter gives an upper limit $\sim 10^{-9}$ $M_\odot\,yr^{-1}$, for wind velocity of 100 km s^{-1}, for both the SNe. The tenuous media obtained from this study favor the double degenerate as well as a spin up/down model for both SNe 2011fe and 2014J.

Keywords. stars: circumstellar matter, supernovae: general, supernovae: individual (SN 2011fe, SN 2014J)

---

1. Introduction

Type Ia supernovae (SNe) are accepted to be the thermonuclear explosion of white dwarfs (WDs) composed of carbon and oxygen (Hoyle & Fowler 1960). They are widely used as standard candles in cosmology (Riess *et al.* 1998, Perlmutter *et al.* 1999), and are one of the main sources of metals in the universe. Hence Type Ia SNe play an important role in terms of chemical evolution of galaxies. As it is understood so far, the successful disruption of a WD requires a companion star. In the single degenerate (SD) scenario the companion is a non-degenrate star, i.e., a main sequence or a asymptotic giant branch star, from which the WD accretes matter before explosion. Classically, the WD explodes when it reaches the Chandrashekar mass limit. Close to this limit, the central density is high enough for carbon to ignite, which obliterates completely the WD (Whelan & Iben 1973). A second possible formation channel consists of another WD as a companion, with the total mass of the WDs exceeding the Chandrasekhar mass. In this case, known as the double degenerate (DD) scenario, the two WDs spiral each other and emit gravitational waves. This causes the reduction of angular momentum from the system which reduces the orbital separation. Eventually, the two WDs merge and this results in a successful thermonuclear runaway (Iben & Tutukov 1984, Webbink 1984). The most likely type of progenitor of SNe Ia is not yet clear, which restrains their usage as cosmological standard candles.

The two formation channels are expected to yield significantly different circumbinary medium characteristics. Namely, in the SD scenario a fraction of the companion mass is usually lost to the surrounding medium via winds or Roche-lobe overflow. This results in a high density circumbinary medium. On the contrary, the merger model predicts a very clean, low-density environment around the progenitor star. The interaction of high velocity SN ejecta with this medium causes a strong shock wave to move through the circumstellar medium (CSM), which then sweeps the material up into a thin shell. The radio and X-ray emission from this shocked region thus allow to measure the density of the ambient medium, from which one could eventually distinguish the SNe Ia formation channels.

While radio emission has been detected from many core-collapse supernovae (e.g., SN 1993J (Bartel *et al.* 1994)) , SN 2011dh (Soderberg *et al.* 2012, Krauss *et al.* 2012)), no SN Ia has ever been detected at these wavelengths (Panagia *et al.* 2006, Chomiuk *et al.* 2016). Since the expected emitted radiation is weak, the large distance to most known SNe Ia makes it challenging to detect them. In this regard the two nearest young SNe Ia, SNe 2011fe and 2014J at distances of 6.4 (Shappee & Stanek 2011) and 3.4 Mpc (Marion *et al.* 2015), respectively, provide a unique opportunity to study these events in detail.

The non-detection of early-time radio and X-ray emission from SN 2011fe constrains the mass loss rate ($\dot{M}$) of the SN progenitor to be less than 6×10^{-10} $M_\odot\,yr^{-1}$ (Chomiuk *et al.* 2012) and 2×10^{-9} $M_\odot\,yr^{-1}$ (Margutti *et al.* 2012), respectively, considering a wind velocity (v_w) of 100 km s^{-1}. For a merger scenario they obtained a medium with density, $n_{\rm ISM}$, less than 6 and 150 cm^{-3}, respectively. Applying a similar formalism for the radio emission from SN 2014J, Pérez-Torres *et al.* 2014 estimated $\dot{M} < 7\times10^{-10}$ $(v_w/100 \text{ km s}^{-1})^{-1}$ $M_\odot\,yr^{-1}$ and $n_{\rm ISM} < 1.3$ cm^{-3}. Although the low density medium predicted from early observations favours a merger scenario over a SD channel for both SNe, the radio models contain a few poorly constrained shock parameters, the most relevant being the magnetic field energy density. In this situation, observations at late epochs ($t \gtrsim$ 1 yr after the explosion), in addition to the early time observations, are important in constraining the environment and hence the progenitor scenario.

2. Ejecta and CSM structure

The structure of the inner ejecta can be parametrized in different ways, but rather than assuming a power-law (Pérez-Torres *et al.* 2014) or an exponential density structure (Chomiuk *et al.* 2016), we have taken the density structure from explosion models, namely from N100 (Röpke *et al.* 2012, Seitenzahl *et al.* 2013) and a violent merger model (Pakmor *et al.* 2012). These two test the SD and DD formation channels of Type Ia, respectively. The outer ejecta profile is assumed to follow a power law distribution, i.e., $\rho_{ej} \propto r^{-n}$, where n is the power law index and r is the radius of the SN. In our study n can vary in the range 12 - 14. In this paper, we refer the N100 + power law structure for outer ejecta the N100 model and the violent merger + power law profile the merger model.

As discussed in § 1, in case of a SD channel, the winds from the progenitor system could result in a high density circumbinary medium. Therefore, for a constant value of $\dot{M}$ and v_w the density of the CSM could be written as $\rho(r) = \dot{M}/(4\pi r^2 v_w)$. For the DD channel the density of the ambient medium is constant and is characteristic of the interstellar medium (ISM). This implies that in case of DD scenario $\rho(r) = n_{\rm ISM}\,\mu$, where μ represents the mean atomic weight of the circumbinary medium. Therefore, for both media we can write $\rho(r) = Ar^{-s}$, where $s = 2$ (0) represents the wind (constant density) medium.

3. Radio modeling

Shocks are ideal places where particles can be accelerated efficiently. The interaction between SN ejecta and CSM launches a strong shock, called forward shock, into the ambient medium. The pressure behind this shock drives another shock, known as reverse shock, back into the ejecta. As a result a shocked shell is formed. This shell contains the shocked outer part of the ejecta and shocked circumstellar matter. When ejecta with a power law density structure interact with an ambient medium, having a power law profile for the density, the evolution of this shell can be described by a self-similar structure (Chevalier 1982). This implies that the shock radius evolves as $r_s \propto A^{1/(s-n)}\ t^{(n-3)/(n-s)}$, where t represents time since explosion.

We modeled the radio emission assuming synchrotron radiation from shock accelerated relativistic electrons. All electrons in the post shock region are considered to be accelerated and follow a power law distribution in energy, i.e., $dN/dE = N_0 E^{-p}$, where N_0 and E represent the normalization constant of the distribution and the electron energy, respectively. p is the power law index. For a non-relativistic shock, having a compression factor of $\eta = 4$, diffusive shock acceleration (DSA) predicts $p = 2$. However, as discussed in Ellison & Reynolds (1991), the presence of a sub-shock in a cosmic ray dominated shock could result in a particle spectrum, for low energy particles, with an index steeper than 2. Moreover, in case of Type Ib/c SNe it is found from the observations that $p = 3$ (Chevalier & Fransson 2006). We therefore assume $p = 3$ for our study.

The radio luminosity, with synchrotron self absorption (SSA) as the dominant absorption mechanism, from an optically thin shell having an outer radius r_s is given by

$$L_{\nu,\mathrm{thin}} = \frac{8\pi^2 k T_{\mathrm{bright}} \vartheta_\nu r_s^2}{c^2 f\left(\frac{\nu_{\mathrm{peak}}}{\nu_{\mathrm{abs}}}\right)} \nu_{\mathrm{abs},0}^{(p+3)/2}\ \nu^{-(p-1)/2}, \tag{3.1}$$

with

$$\nu_{\mathrm{abs},0} = \left(2\Delta r\ \varkappa(p)\ N_0\ B^{(p+2)/2}\right)^{2/(p+4)} \tag{3.2}$$

(Pérez-Torres *et al.* 2014), where T_{bright} and ϑ_ν being the brightness temperature and the ratio between the radio intensity from a given part of the shell to that from the centre part, respectively. Δr is the thickness of the radio emitting shell and it is $\propto r_s$. $\varkappa(p)$ and B represent the SSA coefficient and magnetic field strength, respectively. k is the Boltzmann constant and c represents the speed of light in vacuum. For a wind medium ($s = 2$), and with $p = 3$, the optically thin radio luminosity, eqn. 3.1, can be written as

$$L_{\nu,\mathrm{thin}} \propto \left(\frac{n-3}{n-2}\right)^{3.86} T_{\mathrm{bright}}\ \epsilon_{\mathrm{rel}}^{1.71}\ \epsilon_{\mathrm{B}}^{1.07} \left(\dot{M}/v_w\right)^{\frac{1.93n-8.43}{n-2}} t^{-\frac{(n+2.57)}{n-2}}\ \nu^{-1}, \tag{3.3}$$

and for a constant density medium ($s = 0$)

$$L_{\nu,\mathrm{thin}} \propto \left(\frac{n-3}{n}\right)^{3.86} T_{\mathrm{bright}}\ \epsilon_{\mathrm{rel}}^{1.71}\ \epsilon_{\mathrm{B}}^{1.07} \left(n_{\mathrm{ISM}}\right)^{\frac{1.93n-8.43}{n}} t^{\frac{(2.86n-25.3)}{n}}\ \nu^{-1} \tag{3.4}$$

(Kundu *et al.* 2017), where ϵ_{rel} and ϵ_{B} being the fraction of bulk kinetic energy in relativistic electrons and magnetic fields, respectively. If u_{th} represents the post shock thermal energy density, then $\epsilon_{\mathrm{B}} = u_B/u_{th} = B^2/(8\pi u_{th})$ and the fraction of post shock

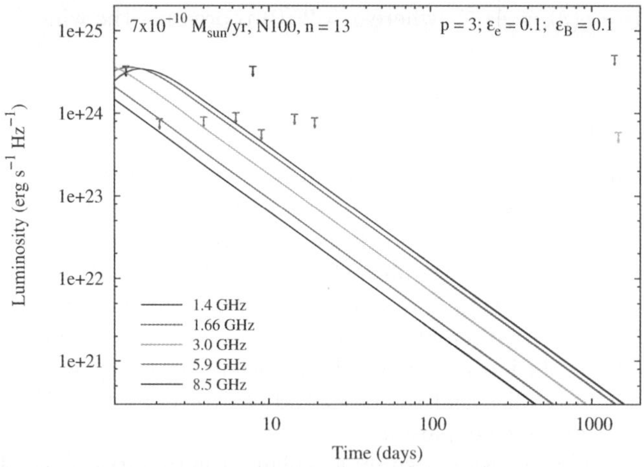

Figure 1. Light curves for SN 2011fe, in solid lines, predicted by the N100 model, for $n = 13$ and $\epsilon_{\rm e} = \epsilon_{\rm B} = 0.1$, when the SN plows through a wind medium. The upper limits are from Horesh *et al.* (2012), Chomiuk *et al.* (2012) and Kundu *et al.* (2017).

energy that has been channeled into electrons is $\epsilon_{\rm e} = \epsilon_{\rm nrel} + \epsilon_{\rm rel} = u_e/u_{th}$, where u_B and u_e are magnetic and electric field energy densities, respectively. Here, $\epsilon_{\rm nrel}$ is the fraction of post shock energy in non-relativistic electrons (non-relativistic electrons are those with kinetic energy $E_k < m_e c^2$ ($\equiv E_{\rm rest}$), where m_e is the mass of the electron). Just after the explosion the shock velocity is very high, which often implies $\epsilon_{\rm e} = \epsilon_{\rm rel}$. With time, as the shock gets decelerated, a part of the electron population becomes non-relativistic. This suggests $\epsilon_{\rm rel} = \epsilon_{\rm e}(\frac{E_{min}}{E_{\rm rest}})^{(p-2)}$ (Kundu *et al.* 2017), where E_{min} is the minimum energy of the electron population. Recent hybrid simulations have shown that in a non-relativistic parallel/quasi-parallel shock (i.e., the shock normal and the magnetic field vector are almost aligned) a maximum of 10 - 20% of the post shock energy can be converted into accelerated ions and the acceleration efficiency becomes negligible in a perpendicular shock (Caprioli & Spitkovsky 2014). Considering these facts we assume that for electrons the acceleration efficiency is 10%, i.e., $\epsilon_{\rm e} = 0.1$. For magnetic fields we consider two cases: (i) $\epsilon_{\rm B} = 0.1$ and (ii) $\epsilon_{\rm B} = 0.01$. This means that in one case it is assumed that the post-shock energy has been equally shared between electric and magnetic fields, i.e., $\epsilon_{\rm e} = \epsilon_{\rm B} = 0.1$ and for the second situation the energy in the magnetic fields is considered to be less than that in electric fields, i.e., $\epsilon_{\rm e} = 0.1$ and $\epsilon_{\rm B} = 0.01$.

Fig. 1 represents the radio light curves predicted by the N100 model for SN 2011fe, along with observational upper limits. As visible in the figure, the upper limit in luminosity measured around 2 days after the explosion at 5.9 GHz constrains the maximum mass loss rate of the system. This has been also noted by Chomiuk *et al.* (2012) and Pérez-Torres *et al.* (2014). With $n = 13$ and $\epsilon_{\rm e} = \epsilon_{\rm B} = 0.1$ it is found that $\dot{M} < 7 \times 10^{-10}$ $M_\odot$ yr^{-1} for $v_w = 100$ km s^{-1} (Kundu *et al.* 2017). Considering $n = 12$ and keeping other parameters fixed, the light curves from the merger model are shown in fig. 2. In this case, the radio upper limit at 3 GHz, obtained around 4 years after the explosion, limits $n_{\rm ISM}$ to be less than 0.23 cm^{-3} (Kundu *et al.* 2017). For our study we assume a constant value of $T_{\rm bright} = 5 \times 10^{10}$ K.

For SN 2014J the upper limits on $\dot{M}$ and $n_{\rm ISM}$ are derived from the observations done around 8 and 410 days after the explosion of the SN (Pérez-Torres *et al.* 2014, Kundu *et al.* 2017).

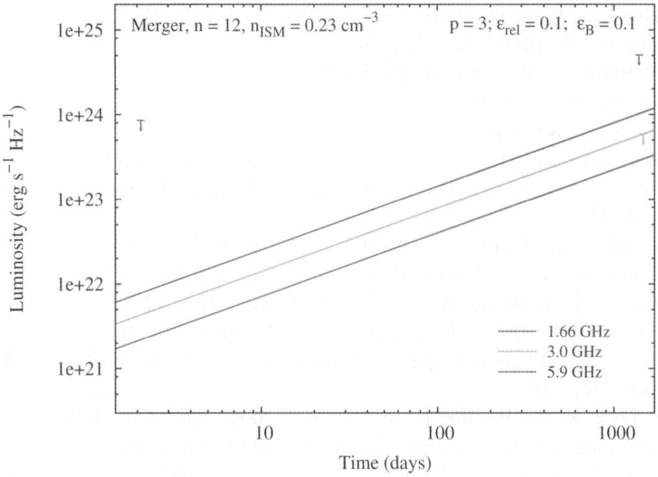

Figure 2. Light curve from a merger model for SN 2011fe when the SN expands in a constant density medium. The upper limits are from Chomiuk *et al.* (2012) and Kundu *et al.* (2017). (Only a few upper limits on radio luminosity are plotted here as these are the most constraining data points.)

4. Implications

It has been found from the radio modeling that the circumbinary media around both SNe 2011fe and 2014J are tenuous. The upper limit on mass loss rate of both SNe inferred from our modeling is 10^{-9} $(v_w/100 \text{ km s}^{-1})^{-1}$ $\text{M}_\odot \text{ yr}^{-1}$ when $\epsilon_\text{B} = 0.1$ and n varies in the range 12 - 14 (Kundu *et al.* 2017). For $\epsilon_\text{B} = 0.01$ the upper limit on $\dot{M}$ is $\sim 10^{-9}$ for $v_w = 100$ km s^{-1}. In case of a constant density medium, the upper limit on n_ISM, around both SNe, are estimated to be ~ 0.35 cm^{-3} and ~ 1 cm^{-3} when $\epsilon_\text{B} = 0.1$ and 0.01, respectively (Kundu *et al.* 2017). This implies that both SNe 2011fe and 2014J could be a result of mergers of two WDs, which predicts a very low density medium around the explosion site. On the other hand, the tenuous medium can also be due to recurrent novae. In this case, a cavity may exist close to the progenitor star. However, from optical nebular spectra this is found to be a less likely progenitor system for both SNe 2011fe and 2014J (Lundqvist *et al.* 2015). Another possibility within the SD scenario is the spin up/down model (Justham 2011). In this case, the WD spins down by losing angular momentum, and for some cases the spin down time could be more than 10^9 yrs (Ilkov & Soker 2012). During this time the CSM may diffuse into the ISM. Therefore, the surroundings could attain a density of the order of ISM density. However, if we assume n_ISM to be ~ 1 cm^{-3} (as indicated from other evidences in Kundu *et al.* (2017)), the radio non-detection at late time could point toward a low magnification efficiency of magnetic fields in SN shocks.

References

Bartel, N., Bietenholz, M. F., Rupen, M. P., Conway, J. E., *et al.* 1994, *Nature*, 369, 584

Caprioli, D. & Spitkovsky, A. 2014, *ApJ*, 783, 91

Chevalier, R. A. & Fransson C. 2006, *ApJ*, 651, 381

Chevalier, R. A. 1982, *ApJ*, 258, 790

Chomiuk, L., Soderberg, A. M., Moe, M., *et al.* 2012, *ApJ*, 750, 164

Chomiuk, L., Soderberg, A. M., Chevalier, R. A., *et al.* 2015, *ApJ*, 821, 119

Ellison, D. C. & Reynolds, S. P. 1991, *ApJ*, 382, 242

Horesh, A., Kulkarni, S. R., Fox, D. B., *et al.* 2012, *apj*, 746, 21
Hoyle, F. & Fowler W. A. 1960, *ApJ*, 132, 565
Iben, I., Jr. & Tutukov, A. V. 1984, *ApJS*, 54, 335
Ilkov, M. & Soker, N. 2011, *MNRAS*, 419, 1695
Justham, S. 2011, *ApJ*, 730, L34
Krauss, M. I., Soderberg, A. M., Chomiuk, l., *et al.* 2012, *ApJ*, 750, L40
Kundu, E., Lundqvist, P., Pérez-Torres, M. A., & Herrero-Illana, R., & Alberdi A. 2017, *arXiv:1705.04204*
Lundqvist, P., Nyholm, A., Taddia, F., *et al.* 2015, *A&A*, 577, 39
Margutti, R., Soderberg, A. M., Chomiuk, L., *et al.* 2012, *ApJ*, 751, 134
Margutti, R., Parrent, J., Kamble, A., *et al.* 2014, *ApJ*, 790, 52
Marion, G. H., Sand, D. J., Hsiao, E. Y., *et al.* 2015, *ApJ*, 798, 39
Pakmor, R., Kromer, M., Taubenberger, S., Sim, S. A., Röpke, F. K., & Hillebrandt, W. 2012, *ApJ* (Letters), 747, L10
Panagia, N., Van Dyk, S. D., Weiler, K. W., *et al.* 2006, *ApJ*, 646, 369
Perlmutter, S., Aldering, G., Goldhaber, G., *et al.* 1999, *ApJ*, 517, 565
Pérez-Torres, M. A., Lundqvist, P., Beswick, R. J., *et al.* 2014, *ApJ*, 792, 38
Riess, A., Filippenko, A. V., Challis, P., *et al.* 1998, *AJ*, 116, 1009
Röpke F. K., Kromer, M., Seitenzahl, I. R., Pakmor, R., *et al.* 2012, *ApJ* (Letters), 750, L19
Seitenzahl, I. R., Ciaraldi-Schoolmann, F., Röpke, F. K., *et al.* 2013, *MNRAS*, 429, 1156
Shappee, B. J. & Stanek, K. Z. 2011, *ApJ*, 733, 124
Soderberg, A. M., Margutti, R., Zauderer, B. A., *et al.* 2012 , *ApJ*, 752, 78
Webbink, R. F. 1984, *ApJ*, 277, 355
Whelan, J. & Iben, I., Jr. 1973, *ApJ*, 186, 1007

Supernova 1987A:30 years later - Cosmic Rays and Nuclei from Supernovae and their aftermaths
Proceedings IAU Symposium No. 331, 2017
A. Marcowith, M. Renaud, G. Dubner, A. Ray & A. Bykov, eds.

doi:10.1017/S1743921317004239

Spatial distribution of different subtypes of core-collapse and thermonuclear supernovae in the galaxies

Dmitry Yu. Tsvetkov and Nickolay N. Pavlyuk

Sternberg Astronomical Institute, Lomonosov Moscow State University,
Universitetsky pr. 13, Moscow 119234, Russia
email: tsvetkov@sai.msu.su

Abstract. The distributions of supernovae of different types and subtypes along the radius and in z coordinate of galaxies have been studied. We show that among SNe Ia in spiral galaxies, SNe Iax and Ia-norm have, respectively, the largest and smallest concentration to the center; the distributions of SNe Ia-91bg and Ia-91T are similar. A strong concentration of SNe Ibc to the central regions has been confirmed. In spiral galaxies, the supernovae of all types strongly concentrate to the galactic plane; the slight differences in scale height correlate with the extent to which the classes of supernovae are associated with star formation.

Keywords. supernovae: general, galaxies: general

1. Introduction

Despite the significantly increased interest in the studies of supernovae (SNe) in recent years and the rapidly growing number of discovered and thoroughly studied SNe, the problem of the nature of SN progenitors still remains unsolved for many classes of SNe. The most efficient method of solving this problem is the identification of SN progenitors in preexplosion images. About 20 SN progenitor stars have been detected in the last 15 years, and upper limits on the luminosity have been established for about 30 objects. Almost all of the detected SN progenitors are type IIP and IIb ones. It has been established that SNe IIP result from the explosions of red supergiants with initial masses from 8 to 15-18 $M_{\odot}$; the most probable precursors for SNe IIb are yellow supergiants, possibly, in binary systems (Smartt 2015).

In addition to the search for SN progenitors, other methods are also used to solve this problem: studies of the stellar population at explosion sites and the large-scale distribution of SNe in their host galaxies (see, e.g., Anderson *et al.* 2015a, Anderson *et al.* 2015b, Kuncarayakti *et al.* 2013a, Kuncarayakti *et al.* 2013b).

Anderson *et al.* (2015a), Anderson *et al.* (2015b) compared the distributions of supernovae of the main types with respect to the star formation rate in spiral galaxies (determined from the Hα flux) and showed SNe Ic to be most closely associated with star formation; the types Ib, IIP, IIn, and Ia follow next in order of decreasing association. The radial distributions of SNe and luminosity in galaxies were also compared. The corresponding distributions for SNe Ia and II turned out to be approximately identical and to show a deficit in the central parts of galaxies; SNe Ibc differ greatly from them and show a strong concentration to the galactic centers. Hakobyan *et al.* (2009) studied the radial distribution of core-collapse supernovae (CCSNe: types II and Ibc) in spiral galaxies. The surface density distribution of all CCSNe was shown to be exponential, with SNe Ibc concentrating more strongly to the galactic centers. No significant differences in the distributions of SNe Ib and Ic were found; there is some deficit of SNe II in the central regions of galaxies.

Although studies of the radial distribution of SNe cannot unambiguously indicate which stars explode as SNe of a certain type, they nevertheless can give important constraints on the nature of the stellar population of SN progenitors. The advantages of this method are a large statistical material and the possibility of a quick comparison of various SN types and subtypes between themselves. New species of SNe, which often become the "progenitors" of new types, are being continuously discovered at present. Note the introduction of subtypes of SNe Ia: types Ia-91bg, Ia-91T (e.g., Leibundgut 2000), the SN Iax type (Foley *et al.* 2013), the discovery of superluminous SNe (SLSNe, brighter than −21 mag) (Quimby *et al.* 2013) and low-luminosity SNe ($M \sim -14$ mag or fainter) (Zampieri *et al.* 2003).

An increase in the volume of observational data makes it possible to study the spatial distribution not only for the main SN types but also for some subtypes: SNe Iax, SNe Ia-91bg, SNe Ia-91T, IIb, and IIn. We have also attempted to compare the distributions of SNe in z coordinate in spiral galaxies for the first time.

2. Selection of input data

We took the samples of SNe to study the radial distributions from the Catalog of Supernovae of the Sternberg Astronomical Institute (Tsvetkov & Bartunov 1993, Tsvetkov *et al.* 2004), its updated version is accessible on the Internet†.

For our study, we selected only classified SNe. In those cases where the explosion occurred in a double or multiple system of galaxies, we selected only those SNe whose host galaxy was determined quite reliably. The intergalactic SNe or SNe without visible host galaxies were not considered. There should have been information about the morphological type, apparent sizes, and radial velocity for the host galaxies. If these data were available in both LEDA‡ and NED¶ preference was given to the former. Just as in our previous papers (Bartunov *et al.* 1992, Tsvetkov *et al.* 2004, Bartunov *et al.* 2007), we calculated the relative distance for each SN, $r = 2(\Delta\alpha^2 + \Delta\delta^2)^{0.5}/D_{25}$, where $\Delta\alpha$ and $\Delta\delta$ are the distances in arcseconds from the galactic center to the SN in right ascension and declination, D_{25} is the diameter of the galaxy to the 25 mag/arcsec^2 isophote. We also calculated the distances normalized to the galaxy's radius along the SN radius vector: $r_c = r/(\cos^2\theta + (b/a)^2 \sin^2\theta)^{0.5}$ where θ is the angle between the SN radius vector and the galaxy's major axis, a and b are the galaxy's apparent major and minor axes. For our study, we selected a total of 2341 SNe; among them there are 1048 SNe Ia, 984 SNe II; and 309 SNe Ibc. They were all discovered before February 16, 2013, with the addition of three 2014 supernovae whose host galaxies are seen exactly edge-on and can be used in considering the distributions in z coordinate. The fraction of supernovae discovered photographically is about 10-15%.

We introduced no radical changes into the technique of studying the radial distributions of SNe compared to our previous papers (Bartunov *et al.* 1992, Tsvetkov *et al.* 2004). We studied the dependence of the logarithm of the SN surface density σ in ring zones along the galactic radius. No smoothing of the data was applied. In those cases where the number of SNe within the ring zones dividing the galactic disk along the radius was

† http://www.sai.msu.su/sn/sncat/
‡ http://leda.univ-lyon1.fr
¶ http://ned.ipac.caltech.edu

not enough to construct the distribution, we increased the size of the corresponding rings beginning from a given radius and further out from the galactic center.

The considerable increase in the number of discovered SNe over the last 10 years has made it possible to study the distributions of SNe not only along the radius but also in a direction perpendicular to the galactic plane, in z coordinate. For this purpose, we selected the SNe that exploded in edge-on spiral galaxies. Our initial selection of galaxies was made according to the condition $i > 85°$, where the inclination i was determined from the formula $i = \arccos(1.04(b/a)^2 - 0.04)^{0.5}$. We then examined the photographs of the selected galaxies and included the galaxies that were undoubtedly seen almost exactly edge-on in our final sample. The sample contains 78 galaxies in which 26 SNe Ia, 44 SNe II, and 8 SNe Ibc exploded. For our study, we took the SN images stored at the "Bright Supernovae" site‖. These images were transformed until they coincided with the images of galaxies from the Digital Sky Survey (DSS)†† in three common stars using the IRAF software package‡‡.

Thus, we can determine the SN explosion site in the galaxy's DSS image. In this image, we drew the straight line corresponding to the galactic plane and determined the angular distances from SN to the galactic plane and from the galactic center to the base of the normal dropped from the SN on the galactic plane. These distances were then expressed in fractions of the galactic radius (z, r_z) or in kiloparsecs (Z, R_z).

3. Results of our study of the radial distribution of SNe

Previous studies showed that the observational selection effects, primarily the losses of SNe in the central parts of galaxies, play an important role in studying the radial distributions (Bartunov *et al.* 1992, Howell *et al.* 2000, Tsvetkov *et al.* 2004). This effect manifest itself most clearly in searching for SNe by the photographic method. However, whereas the fraction of SNe discovered on photographs was significant in previous studies, it is rapidly decreasing at present. It is of interest to estimate whether this effect is retained at present, when all SNe are discovered using CCD surveys. For this purpose, we studied the properties of the samples of SNe with respect to the radial velocities of their host galaxies. Obviously, the influence of selection effects on the results increases as we pass to higher redshifts. This is clearly demonstrated by Fig. 1, which shows the distribution of relative distances r_c as a function of the radial velocities of host galaxies V_r. There is a clear tendency for the minimum distance r_c to increase with V_r.

The influence of selection can also be estimated by dividing the samples of SNe into two parts, with radial velocities larger and smaller than some boundary value. When the boundary value of $V_r = 6000$ km s^{-1} is chosen, the numbers of SNe in the two parts of the samples turned out to be approximately equal. The results are shown in Fig. 1, where the "distant" (with larger radial velocities) and "nearby" (with smaller velocities) samples are denoted by the letters F and N. We may conclude that the influence of observational selection with this choice of the boundary velocity is insignificant for SNe Ia, while it is very large for SNe Ibc and especially for SNe II. Thus, a significant fraction of SNe in the central parts of galaxies are not detected even when searching for SNe with CCD detectors. This should be taken into account both when studying the distributions of SNe and when estimating the rate of SNe, especially when studying the redshift dependence of the SN rate. To completely eliminate the influence of selection, below we constructed

‖ http://www.rochesterastronomy.org

†† http://archive.stsci.edu

‡‡ IRAF is distributed by the NOAO, which is operated by AURA, Inc. under cooperative agreement with the NSF, http://iraf.noao.edu

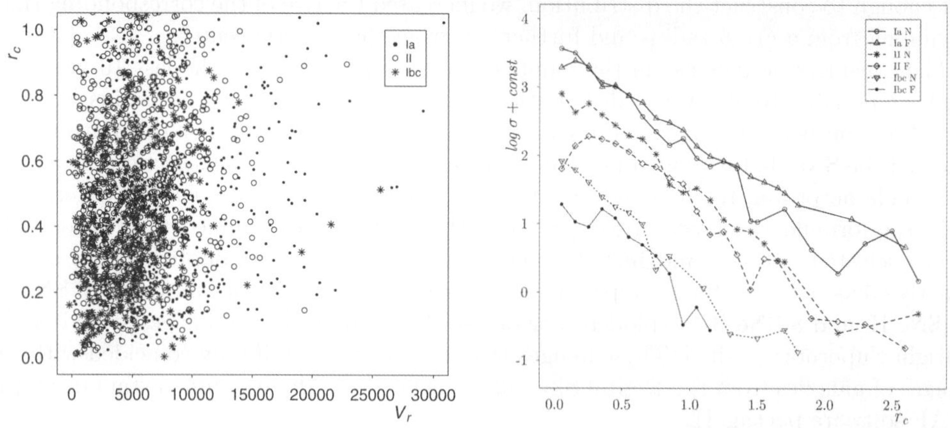

Figure 1. Selection effect on the discovery of SNe. Left panel: distributions of r_c as a function of the radial velocities of the galaxies V_r, right panel: distributions of surface density for "nearby" (N) and "distant" (F) samples.

all distributions with a radial velocity constraint, with the limiting velocity having been taken to be 6000 km s^{-1} for SNe II and Ibc and 8000 km s^{-1} for SNe Ia.

Despite the small number of SNe classified as subtypes of the Ia class, we nevertheless attempted to compare their radial distributions. The samples contained 167 SNe Ia-norm, 10 SNe Ia-91T, 14 SNe Ia-91bg, and 20 SNe Iax. The results for these subtypes are shown in Fig. 2. Probably the most interesting result is a reduced number of SNe Ia-norm in the central parts of galaxies. A strong concentration of SNe Iax to the central regions of galaxies seems quite unexpected. It is also very interesting that the radial distributions of SNe Ia-91bg and SNe Ia-91T agree better with one another than those for SNe Iax and SNe Ia-91T. Of course, these results are not yet quite reliable, and they should be repeated after increasing the sample size. Next, it is worth considering the situation with regard to SNe II and Ibc. The discovery of quite a few new objects has allowed us to separately study the radial distributions for their subtypes: IIP, IIL, IIn, IIb, Ib, and Ic (Fig. 2). Our samples contained 131 SNe IIP, 63 SNe IIn, 41 SNe IIb, 13 SNe IIL, 89 SNe Ib, and 60 SNe Ic. We restricted ourselves only to the SNe exploded in galaxies with radial velocities $V_r < 6000$ km s^{-1}. Given the small number of SNe IIL and the often unambiguous assignment of SNe to this class, we can say that their distributions do not differ significantly from the distribution of SNe IIP. A strong concentration of SNe Ib and Ic to the galactic centers is confirmed, with no difference between SNe Ib and Ic having been revealed. Interestingly, SNe IIb and IIn show a greater similarity to SNe IIP and Ibc, respectively.

4. Results of our study of the distributions of SNe in z coordinate

We constructed both relative and absolute distributions of SNe in z coordinate (Fig. 3). As expected, the SNe of all types strongly concentrate to the galactic plane (see, e.g., Hakobyan *et al.* 2009). The distributions can be fitted by exponentials with scale heights z_0, Z_0. For the relative distributions in z coordinate, $z_0 = 0.030 \pm 0.006$ for SNe Ia, slightly smaller $z_0 = 0.029 \pm 0.005$ for SNe II, and $z_0 = 0.024 \pm 0.006$ for SNe Ibc. Thus, the SNe Ibc are slightly closer to the galactic plane. A similar result is obtained when considering the absolute distributions: $Z_0 = 0.55 \pm 0.11$, 0.50 ± 0.08, and 0.40 ± 0.14 kpc for SNe Ia, II, and Ibc, respectively. We should note that some CCSNe occur at quite

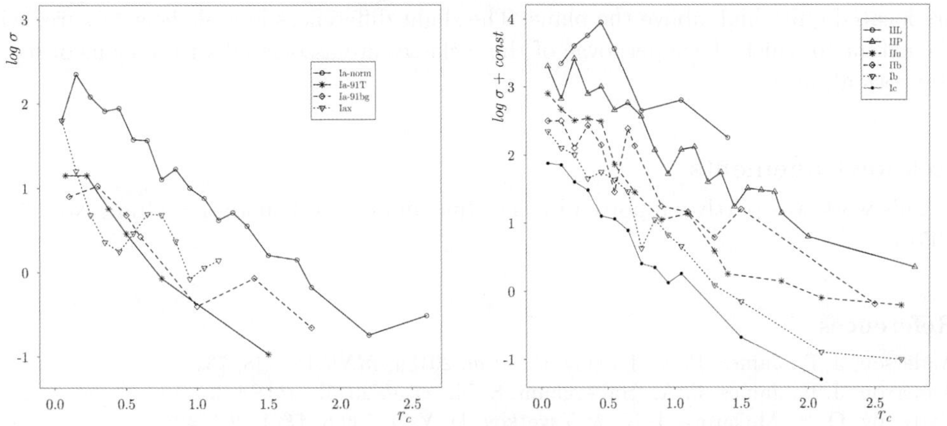

Figure 2. Radial distributions of subtypes of thermonuclear SNe (left panel) and CCSNe (right panel).

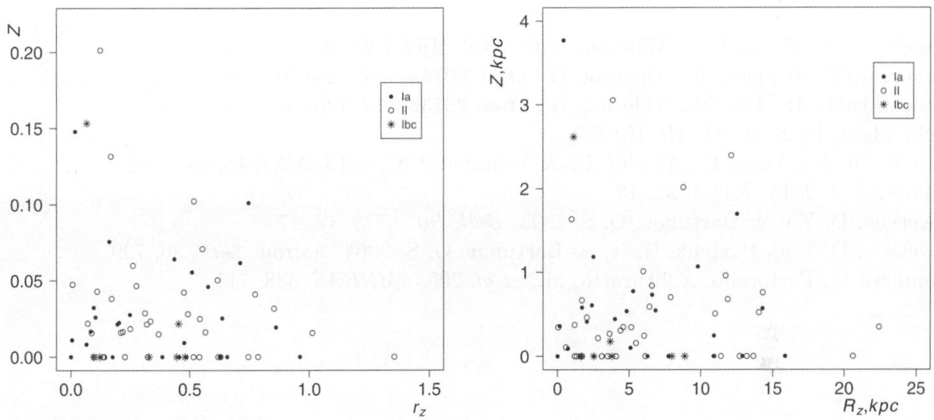

Figure 3. Distributions of SNe Ia, II and Ibc in z coordinate. Left panel: relative distributions, right panel: distributions of distances expressed in kiloparsecs.

high $Z \sim 3$kpc and there is a trend for SNe with high Z to be located in the central parts of the galaxies.

5. Conclusions

Based on data from the SN Catalog of the Sternberg Astronomical Institute, we studied the radial distributions of SNe in their host galaxies, with the radial distributions for subtypes of the main SN classes (Iax, Ia-91bg, Ia-91T, IIn, IIb, IIL, and IIP) having been compared for the first time. The distributions of SNe Ia, II, and Ibc in z coordinate have been studied for the first time. We showed a significant fraction of SNe in the central parts of galaxies to be lost for galaxies with $V_r > 10000$ km s^{-1} . The distributions of SNe Ia-91bg and Ia-91T in spiral galaxies turned out to be similar and different from those of SNe Ia-norm and Iax. SNe Iax show a strong concentration to the central regions of galaxies. We confirmed a stronger concentration of SNe Ib and Ic to the galactic centers than that for other types of core-collapse supernovae, with no difference between SNe Ib and Ic having been revealed. Our study of the distributions in z coordinate confirms a strong concentration of SNe of all types to the planes of spiral galaxies, but some CCSN are located quite high above the plane. The slight differences in scale heights agree with the extent to which the supernovae of these classes are associated with regions of recent star formation.

Acknowledgements

This work was partly supported by the Russian Science Foundation Grant No. 16-12-10519.

References

Anderson, J. P., James, P. A., Forster, F., *et al.* 2015a, *MNRAS*, 448, 732
Anderson, J. P., James, P. A., Habergham, S. M., *et al.* 2015b, *PASA*, 32, 19
Bartunov, O. S., Makarova, I. N., & Tsvetkov, D. Y. u. 1992, *A&A*, 264, 428
Bartunov, O. S., Tsvetkov, D. Y.u., & Pavlyuk, N. N. 2007, *Highlights Astron.*, 14, 316
Foley, R. J., Challis, P. J., Chornock, R., *et al.* 2013, *ApJ*, 767, 28
Hakobyan, A. A., Mamon, G. A., Petrosian, A. R., Kunth, D., & Turatto, M. 2009, *A&A*, 508, 1259
Howell, D. A., Wang, L., & Wheeler, J. C. 2000, *ApJ*, 530, 1666
Kuncarayakti, H., Doi, M., Aldering, G., *et al.* 2013a, *AJ*, 146, 30
Kuncarayakti, H., Doi, M., Aldering, G., *et al.* 2013b, *AJ*, 146, 31
Leibundgut, B. 2000, *A&AR*, 10, 179
Quimby, R. M., Yuan, F., Akerlof, C., & Wheeler, J. C. 2013, *MNRAS*, 431, 912
Smartt, S. J. 2015, *PASA*, 32, 16
Tsvetkov, D. Y.u. & Bartunov, O. S. 1993, *Bull. Inf. CDS*, 42, 17
Tsvetkov, D. Y.u., Pavlyuk, N. N., & Bartunov, O. S. 2004, *Astron. Lett.*, 30, 729
Zampieri, L., Pastorello, A., Turatto, M., *et al.* 2003, *MNRAS*, 338, 711

Supernova 1987A:30 years later - Cosmic Rays and Nuclei from Supernovae and their aftermaths
Proceedings IAU Symposium No. 331, 2017
A. Marcowith, M. Renaud, G. Dubner, A. Ray & A. Bykov, eds.

doi:10.1017/S1743921317004926

CSI in Supernova Remnants

You-Hua Chu

Institute of Astronomy and Astrophysics, Academia Sinica
email: yhchu@asiaa.sinica.edu.tw

Abstract. Supernovae (SNe) explode in environments that have been significantly modified by the SN progenitors. For core-collapse SNe, the massive progenitors ionize the ambient interstellar medium (ISM) via UV radiation and sweep the ambient ISM via fast stellar winds during the main sequence phase, replenish the surroundings with stellar material via slow winds during the luminous blue variable (LBV) or red supergiant (RSG) phase, and sweep up the circumstellar medium (CSM) via fast winds during the Wolf-Rayet (WR) phase. If a massive progenitor was in a close binary system, the binary interaction could have caused mass ejection in certain preferred directions, such as the orbital plane, and even bipolar outflow/jet. As a massive star finally explodes, the SN ejecta interacts first with the CSM that was ejected and shaped by the star itself. As the newly formed supernova remnant (SNR) expands further, it encounters interstellar structures that were shaped by the progenitor from earlier times. Therefore, the structure and evolution of a SNR is largely dependent on the initial mass and close binarity of the SN progenitor. The Large Magellanic Cloud (LMC) has an excellent sample of over 50 confirmed SNRs that are well resolved by Hubble Space Telescope, Chandra X-ray Observatory, and Spitzer Space Telescope. These multi-wavelength observations allow us to conduct stellar forensics in SNRs and understand the wide variety of morphologies and physical properties of SNRs observed.

Keywords. supernovae: general, supernova remnants, circumstellar matter, Magellanic Clouds

1. Introduction

There are two broad origins of supernovae (SNe): core-collapse SNe (CCSNe) originating from massive stars at the end of their evolution, and Type Ia SNe (SNe Ia) originating from white dwarfs in close binary systems. In the case of CCSN, the massive star progenitor had interacted with the ambient medium via UV ionising radiation and stellar winds before the star exploded as SN; thus the SN ejecta interacts with a surrounding that had been shaped and conditioned by the progenitor. The physical structure of the resulting supernova remnant (SNR) inevitably reveal imprints of the massive star progenitor, hence the study of CC SNRs may shed light on the nature of the massive progenitors.

SNe Ia are known to involve white dwarfs, but it is not known whether one or two white dwarfs are involved. The former, called single-degenerate (SD), requires a white dwarf accreting material from a close stellar companion whose envelope has expanded past the Roche limit (Whelan & Iben 1973; Nomoto 1982). The stellar companion is heated/shocked by the impact of SN ejecta, but survives the explosion. The latter, called double-degenerate (DD), is produced by the merger of two white dwarfs (Iben & Tutukov 1984; Webbink 1984), and after the explosion neither white dwarf remain. Therefore, if a surviving stellar companion is identified in a Type Ia SNR, the SD origin of the SN progenitor can be affirmed. The SD progenitor may produce a dense circumstellar medium (CSM) due to the mass loss of the stellar companion; therefore, a dense CSM detected in a Type Ia SNR also suggests a SD origin of the SN progenitor.

SNRs, particularly the young ones, provide sites where we can carry out CSI (circumstellar investigation, or crime scene investigation) to assess the nature of the SN progenitors. In this paper, I review how massive stars modify their surrounding medium and how this process affects the physical structures of their SNRs. I will use CCSNRs in the Large Magellanic Cloud (LMC) to demonstrate the imprints of their massive progenitors. I will also use Type Ia SNRs in the LMC to search for surviving companions and CSM for evidence of SD SNe Ia.

2. CSI in Core-Collapse SNRs

Massive stars are sources of UV radiation that can photoionize the ambient medium. They are also sources of stellar winds of velocities that vary along the stellar evolution. When a massive star is on the main sequence, it is surrounded by the interstellar medium (ISM). A main sequence O star has fast stellar winds with velocities of 1000-2000 km s^{-1} and mass loss rates of order of 10^{-7} $M_\odot$ yr^{-1}. The star not only photoionizes the ambient ISM into an HII region, but also sweeps up the ISM into a wind-blown bubble (Castor, McCray, & Weaver 1975; Weaver *et al.* 1977). As a massive star evolves off the main sequence to become a red supergiant (RSG) or a luminous blue variable (LBV), the fast stellar wind is replaced by a slow wind with velocity of 10–50 km s^{-1} and a mass loss rate of order of 10^{-4} $M_\odot$ yr^{-1}. This slow wind expands away from the central star to form a CSM within the central cavity of the interstellar bubble previously blown by the main sequence progenitor. As the massive star evolves further into the Wolf-Rayet (WR) phase, the slow wind is replaced by a copious fast wind of velocities 2000-3000 km s^{-1} and mass loss rate of $\sim 10^{-5}$ $M_\odot$ yr^{-1}. This fast wind sweeps up the CSM to form a circumstellar bubble.

It may have been puzzling that no interstellar bubbles around main sequence O stars as spectacular as the famous Bubble Nebula have been detected, especially since an O star's fast stellar wind *should* blow a bubble in the ISM. To definitively search for interstellar bubbles around main sequence O stars, Hubble Space Telescope (HST) images were obtained for the young HII regions N11B and N180B around OB associations LH10 and LH117, respectively; still, no bubbles can be identified morphologically in the Hα or [O III] images. The breakthrough came from the long-slit high-dispersion echelle spectra of these HII regions. Expanding shell structures around O stars are detected in these spectrograms; the shell sizes are $\sim$15 pc and the expansion velocities are 15–20 km s^{-1} (Nazé, Chu, Points, *et al.* 2001). It is conceivable that these expansion velocities produce shocks with a Mach number of 1.5–2, and such weak shocks do not cause strong compression to form dense interstellar shells that are identifiable morphologically.

The visibility of an interstellar bubble of an O star depends on the existence of other massive stars in its vicinity and the distribution of the ISM. For example, the interstellar bubble in the Orion Nebula is not clearly identified until XMM-Newton detected diffuse X-ray from the bubble interior, then it became obvious that the massive stars are close to the densest parts of the bubble and the less dense parts of the bubble extend further away from the stars (Güdel, Briggs, Montmerle, *et al.* 2008). Massive stars in OB associations or clusters will collectively form superbubbles with sizes larger than those of single-star interstellar bubbles. In an active star-forming region, such as the giant HII region 30 Doradus, the stellar energy feedback produce very complex structures, and effects of individual stars cannot be unambiguously identified.

The ambient medium of a massive star can be roughly characterized by the following physical parameters: interstellar bubble around a main sequence O star has density of 10–100 H-atom cm^{-3}, diameter of a few 10's pc, and expansion velocity $\sim$15 km s^{-1};

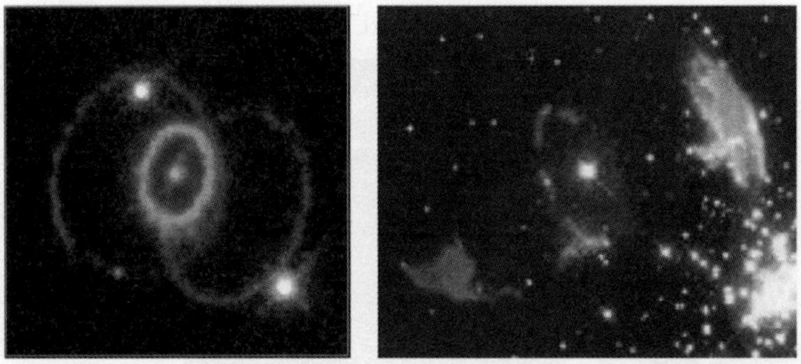

Figure 1. Left - Hubble Space Telescope image of SN1987A. Right - Hubble Space Telescope image of Sher 25.

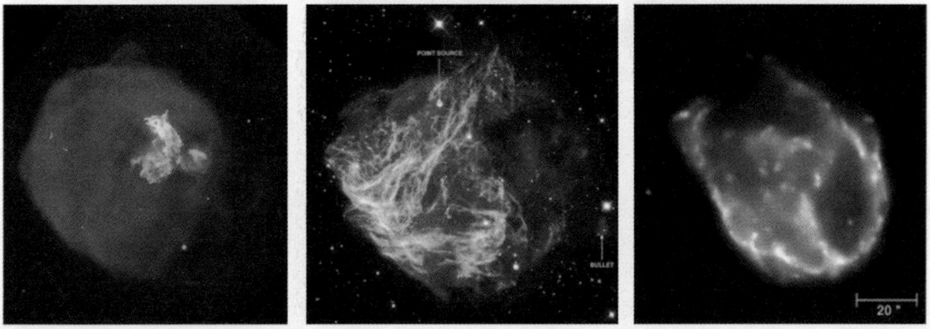

Figure 2. Left to right - Hubble Space Telescope images of N63A, N49, and Chandra X-ray image of N132D.

circumstellar nebula around a RSG has $>10^4$ H-atom cm^{-3}, 10^3 AU, and 10–50 km s^{-1}; circumstellar nebula around an LBV has 10^4 H-atom cm^{-3}, 1 pc, and 20–50 km s^{-1}; circumstellar bubble around a WR star has 10^2 H-atom cm^{-3}, a few pc, and 50–100 km s^{-1}. The detailed physical conditions of the ambient medium at the time of SN explosion depend on the exact evolutionary path and mass loss history. When a massive star explodes at the end of its life, the SN ejecta interacts successively with different layers of ambient medium, first CSM then ISM. The different conditions of the ambient medium produce a wide variety of physical structures in SNRs. Below I will use SNRs in the LMC to illustrate interactions between SN ejecta and the ambient medium.

SN1987A provides a clear example of SN ejecta interacting with a CSM (Figure 1). The progenitor of SN1987A, Sk−69 201, was a B3 I supergiant (Walborn, Lasker, Laidler, *et al.* 1988). The structure of Sk−69 202's CSM is almost identical to the CSM of Sher 25, a Galactic B1.5 I supergiant (Brandner, Chu, Eisenhauer, *et al.* 1997). Interestingly, the second youngest CCSNR in the LMC, 0540−69.3, also shows a [N II] ring around the SN that corresponds to a CSM similar to that around Sk−69 202 or Sher 25 (Caraveo, Mignani, & Bignami 1998; Sandin, Lundqvist, Lundqvist, *et al.* 2013).

As massive stars had blown interstellar bubbles during the main sequence phase, their SN ejecta will quickly sweep up the CSM and expand into the cavity of the bubble. Indeed, many young CCSNRs show indications of cavity explosion, for example N63A, N49, and N132D in the LMC (Figure 2). As noted earlier, interstellar bubbles of main sequence O stars do not show strong compression in the shell, so such bubbles are better described as interstellar cavities. When SN ejecta expands into the cavity, the expansion velocity is not dampened by dense material, so the X-ray image can show high luminosities in

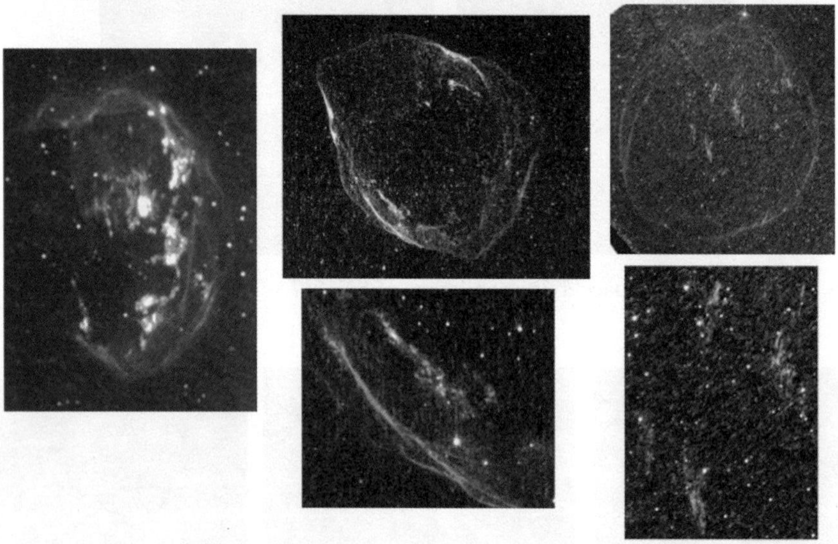

Figure 3. Hubble Space Telescope images of N103B, DEM L71, and 0548−70.4. The lower panels show close-ups of the knots.

CCSNRs that are already >20 pc in size (Warren, Hughes, Slane, *et al.* 2003; Park, Hughes, Slane, *et al.* 2012; Vogt & Dopita 2011).

3. CSI in Type Ia SNRs

Type Ia SNRs can be diagnosed by the over-abundance of Fe, Si, S, Ar, and Ca implied by the X-ray spectra of the SN ejecta (e.g., Hughes, Hayashi, Helfand, *et al.* 1995). Young Type Ia SNRs can also be diagnosed by their Balmer-dominated optical spectra due to collisionless shocks into a mostly neutral ISM (Chevalier, Kirshner, & Raymond 1980). They can also be unambiguously identified by spectra of their light echoes (Rest, Suntzeff, Olsen, *et al.* 2002; Rest, Matheson, Blondin, *et al.* 2005).

If CSM is detected in a Type Ia SNR, the SN progenitor is most likely of SD origin. Hubble Space Telescope images of Type Ia SNRs in the LMC have revealed dense knots in the interior of N103B, DEM L71, and 0548−70.4. These knots have densities a few thousand H-atoms cm^{-3}, which is more characteristic for CSM than ISM.

The Hubble Space Telescope Hα image of N103B shows a filamentary shell with dense knots in the shell interior *(Figure 3). High-dispersion long-slit echelle spectra of N103B show that the filamentary shell is Balmer-dominated while the dense knots show forbidden lines (Figure 4), clearly indicating that the knots belong to a CSM (Li, Chu, Gruendl, *et al.* 2017).

The dense CSM in N103B suggests that its SN progenitor has a SD origin. It is thus interesting to search for a surviving companion of the progenitor. A candidate has been identified near the explosion center defined by the Balmer-dominated shell (Li, Chu, Gruendl, *et al.* 2017). High-quality spectra of this star is needed to examine its radial velcoity, rotation velocity, and abundances in order to confirm its identification as the surviving companion.

This work was supported by the grant 104-2112-M001-044MY3 from the Ministry of Science and Technology of Taiwan.

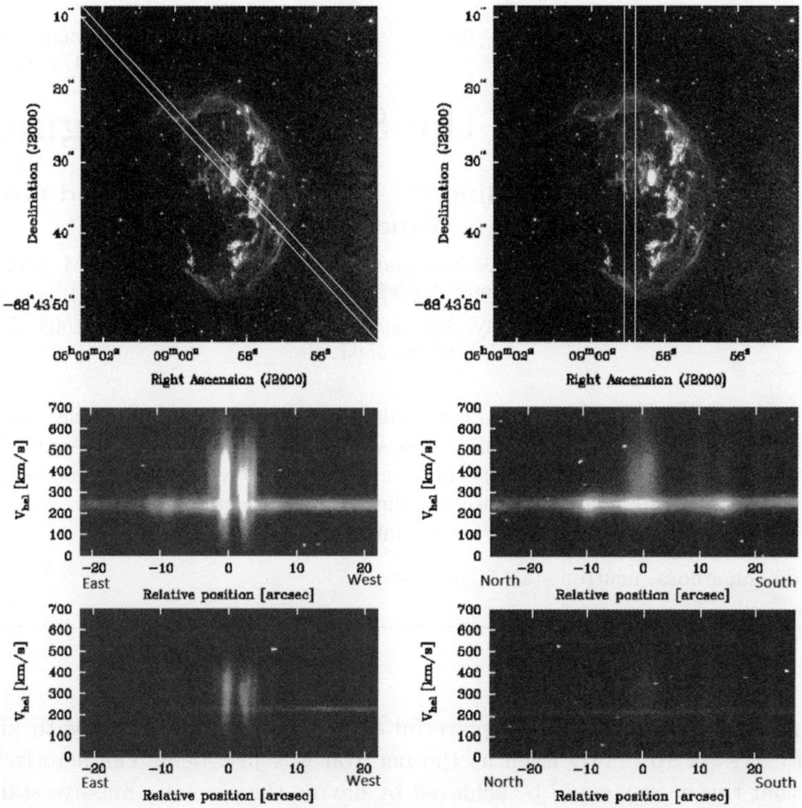

Figure 4. Top panel: Hubble Space Telescope Hα image of N103B. Bottom panels: long-slit echelle spectrograms of the Hα and [N II] λ6583 lines. It is clear that the filamentary shell has only Hα emission and no [N II], while the knots are shocked dense CSM with bright [N II] emission along with the Hα.

References

Brandner, W., Chu, Y.-H., Eisenhauer, F., Grebel, E. K., & Points, S. D. 1997, *ApJL*, 489, L153
Caraveo, P., Mignani, R., & Bignami, G. B. 1998, *Mem. Soc. Astronomia Italiana*, 69, 1061
Castor, J., McCray, R., & Weaver, R. 1975, *ApJL*, 200, L107
Chevalier, R. A., Kirshner, R. P., & Raymond, J. C. 1980, *ApJ*, 235, 186
Güdel, M., Briggs, K. R., Montmerle, T., *et al.* 2008, *Science*, 319, 309
Hughes, J. P., Hayashi, I., Helfand, D., *et al.* 1995, *ApJL*, 444, L81
Iben, I., Jr. & Tutukov, A. V. 1984, *ApJS*, 54, 335
Li, C.-J., Chu, Y.-H., Gruendl, R. A., *et al.* 2017, *ApJ*, 836, 85
Nazé, Y., Chu, Y.-H., Points, S. D., *et al.* 2001, *AJ*, 122, 921
Nomoto, K. 1982, *ApJ*, 257, 780
Park, S., Hughes, J. P., Slane, P. O., *et al.* 2012, *ApJ*, 748, 117
Rest, A., Matheson, T., Blondin, S., *et al.* 2008, *ApJ*, 680, 1137
Rest, A., Suntzeff, N. B., Olsen, K., *et al.* 2005, *Nature*, 438, 1132
Sandin, C., Lundqvist, P., Lundqvist, N., *et al.* 2013, *MNRAS*, 432, 2854
Vogt, F. & Dopita, M. A. 2011, *Ap. & Space Sci.*, 331, 521
Walborn, N. R., Lasker, B. M., Laidler, V. G., & Chu, Y.-H. 1987, *ApJL*, 321, L41
Warren, J. S., Hughes, J. P., & Slane, P. O. 2003, *ApJ*, 583, 260
Weaver, R., McCray, R., Castor, J., Shapiro, P., & Moore, R.1977, *ApJ*, 218, 377
Webbink, R. F. 1984, *ApJ*, 277, 355
Whelan, J. & Iben, I., Jr. 1973, *ApJ*, 186, 1007

Supernova 1987A:30 years later - Cosmic Rays and Nuclei from Supernovae and their aftermaths
Proceedings IAU Symposium No. 331, 2017
A. Marcowith, M. Renaud, G. Dubner, A. Ray & A. Bykov, eds.

doi:10.1017/S174392131700641X

Diagnostics of the Supernova Engine

Chris L. Fryer[1], Carola Ellinger[1], Patrick A. Young[2] and Gregory Vance[2]

[1]CCS-2, MS D409, Los Alamos Naitonal Laboratory, Los Alamos, NM 87544
email:*fryer@lanl.gov*

[2]Department of Physics and Astronomy, Arizona State University, P.O. Box 1504, Tempe, AZ 85287-1504

Abstract. The standard engine behind core-collapse supernovae is continuously evolving with increasingly detailed models. At this time, most simulations focus on an engine invoking turbulence above the proto-neutron star, sometimes termed the "convection-enhanced" engine. Here we review this engine and why it has become the standard for normal supernovae, focusing on a wide set of observations that provide insight into the supernova engine.

Keywords. supernova, neutron star, black holes

---

1. Introduction

Supernovae are among the most powerful explosions in the universe with kinetic energies in excess of 10^{51} erg†. Even as the neutron was just being characterized, it was realized that this energy could be achieved by having the core of a massive star collapse down to a core composed entirely of neutrons (Baade & Zwicky 1934). Ultimately, this collapse releases over 10^{53} erg of gravitational potential energy and only 1% of this energy is needed to drive the explosion.

This is easier said than done, and scientists have been working to achieve this 1% efficiency for the past 80 years. At this point in time, the arising favored model arguing that convection above the proto-neutron star increases the efficiency at which the potential energy released in the collapse is able to drive a strong mass ejection. The broad community now has a series of simulations that produce explosions. In this paper, we review this engine, comparing results of this engine with the current suite of observations. Leading alternative engines invoking magnetically dominated flows are much less sophisticated at the moment and hence, the comparison to observations is less constraining. But we will discuss these comparisons where possible. Section 2 provides a brief overview of the engine followed by a series of sections reviewing the supernova energetics (section 3), remnant masses (section 4), supernova asymmetries including observations of ejecta remnants(section 5).

2. Engine Models

Supernovae are produced at the end of a star's life when the central iron core collapses in on itself. The iron core is produced through a succession of burning phases where the ashes of the previous phase of nuclear burning becomes the fuel for the next phase as the core contracts and heats up. The iron core is supported by thermal and electron

† 10^{51} erg has been termed a FOE (for Fifty One Erg) by Hans Bethe. This unit is also termed a Bethe

degeneracy pressure. As silicon shell burning adds mass to the iron core, it contracts under its own weight. When the contraction creates sufficiently high densities and temperatures to allow electron capture onto protons, the collapse begins. Electron capture removes degeneracy pressure, causing the core to contract further that, in turn, increases the electron capture rate. In addition, the temperatures in the core become sufficiently high to cause the iron atoms to dissociate, removing thermal pressure. Very quickly, the stable neutron star evolves into a runaway collapse, imploding at nearly free-fall.

The collapse continues until the core reaches nuclear densities when nuclear forces and neutron degeneracy pressure dramatically increase the pressure, causing the core to bounce. Initial models argued this "bounce" would drive the explosion (Colgate & Johnson 1960). However, most of the energy in the bounce is stored in neutrinos and as the shock moves sufficiently outward to become optically thin to neutrinos, it quickly stalls. At this point, the interior of the star consists of a collapsed core (termed the proto-neutron star) $\sim 30-50\,\mathrm{km}$ in extent and the material from the failed bounce shock extending to $\sim 100-300\,\mathrm{km}$.

With the bounce shock failing, the proposed supernova engine evolved. Most of the gravitational potential energy released is deposited as thermal energy in the core. The photons at these densities have mean free paths smaller than a nanometer and can not transport energy. However, for the temperatures in the core, most of the energy is in neutrinos. Even with the extremely low matter-interaction cross-section of neutrinos, the neutrinos are also trapped, but the optical depth is low enough that the neutrinos can diffuse out, heating the failed shock. The next evolution of the supernova model argued that this neutrino heating could revive the shock (Colgate & White 1966).

Although simple prescriptions for this neutrino energy deposition did revive the shock, detailed 1-dimensional models were unable to produce a large enough neutrino flux to drive the explosion. To increase this flux, scientists looked into convective instabilities. Although the entropy gradient suggests a stable profile within the neutron star, the emission at the neutrinosphere (upper edge of the proto-neutron star) produces a composition gradient that is Ledoux unstable (Epstein 1979). This gradient arises as electron neutrinos from electron capture escape, allowing the electron fraction to decrease dramatically. This lepton instability criterion is:

$$(\partial\rho/\partial S)_{P,Y_l}dS/dr + (\partial\rho/\partial Y_l)_{P,S}dY_l/dr > 0, \tag{2.1}$$

where ρ is the density, S is the entropy, P is the pressure, and Y_l lepton fraction. As this instability builds, it can expand to cover a large fraction of the proto-neutron star(Burrows & Lattimer 1988, Keil *et al.* 1996), but the increase in flux is typically insufficient to drive an explosion.

While this was being studied, supernova 1987A exploded. This supernova both confirmed many of the theoretical models while also bringing new issues currently unexplained by the models. The detection of neutrinos from the core proved that the explosion occured during the collapse to a neutron star (Bionta *et al.* 1987, Hirata *et al.* 1987). But the ^{56}Ni decay gamma-rays appeared much earlier than scientists expected (Pinto & Woosley 1988). ^{56}Ni is produced in the innermost supernova ejecta where silicon is burned into ^{56}Ni as the shock plows through it. 1-dimensional models argued that it would take over 200 d for the ^{56}Ni to be uncovered enough such that the gamma-rays produced in its decay would leak out of the star. Observers detected those gamma-rays at 150 d. Scientists realized that this ^{56}Ni must be mixed out in the star. Additional evidence argued for extensive mixing, e.g. the broad lines seen in the iron lines iron lines produced from ^{56}Ni decay required that the ^{56}Ni be extensively mixed.

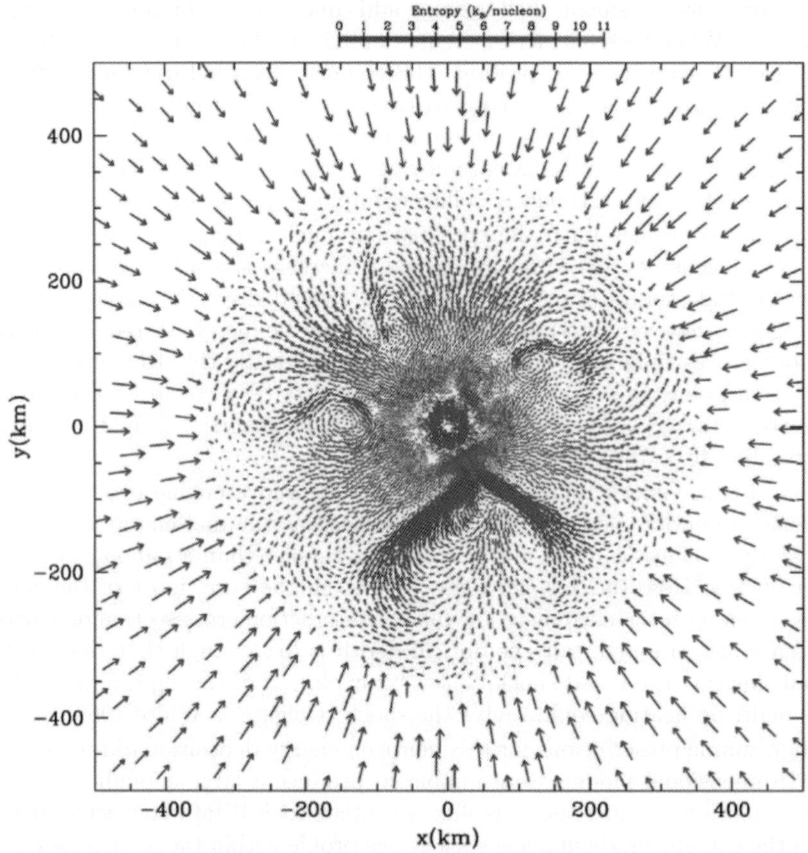

Figure 1. Slice of a 3-dimensional simulation of the core collapse engine. The colors denote the entropy and the length and direction of the arrows correspond to the magnitude and direction of the material. The neutron star is roughly 50 km in size and the bounce shock stalled at just above 200 km. Material from the infalling star piles up at the stalled shock until it drives a downflow - like dust motes on a pond surface. These downflows stream down to the proto-neutron star where they are either heated and begin to rise in an upflow or accrete onto the proto-neutron star (Fig. 2, Fryer & Warren 2002).

What causes this mixing? The supernova shock is unstable to Rayleigh-Taylor convection, but scientists studying this mixing in multi-dimensional codes were unable to achieve the high level of mixing observed (e.g. Herant *et al.* 1992). Scientists began to review the instabilities in the stalled shock region and the supernova engine began to evolve further. The bounce shock heats and compresses the star as it moves through it. In the strong shock limit, the pressure and density after being shocked are:

$$P_{\rm shock} = (\gamma+1)/2\rho_{\rm star}v^2 \tag{2.2}$$

and

$$\rho_{\rm shock} = (\gamma+1)/(\gamma-1)\rho_{\rm star} \tag{2.3}$$

where γ is the adiabatic index of the stellar material (we can assume a radiation dominated gas, $\gamma = 4/3$), $\rho_{\rm star}$ is the density of the imploding stellar material and v is the relative velocity of the shock and the infalling material: $v \approx v_{\rm shock} + v_{\rm free-fall}$. For a radiation dominated gas, the entropy (S) of material is:

$$S = S_0 \rho_{\rm star}^{-1/4} v^{3/2} \tag{2.4}$$

where we have subsumed any constants into S_0. If $\rho_{\rm star} \propto r^{-3}$ and $v \propto r^{-1/2}$, the entropy would be constant with respect to r and we would expect a stable stratification. However, typically the density falls of less steeply and the velocity decreases more steeply such that there is at least a portion of this failed shock region that is unstable to Rayleigh-Taylor instabilities. In addition, neutrino heating from the proto-neutron star would further drive the Rayleigh-Taylor instability. This convection, driven by limitations in the current engine and the evidence of mixing in SN 1987A, became the next step in the evolution of the supernova engine.

Figure 1 shows a slice of a 3-dimensional simulation of this convective engine. High entropy bubbles rise as low entropy downflows move from the top of the stalled shock down to the neutrino heating region (this region is termed the gain region because the net energy gained by neutrino deposition is large than that lost through neutrino emission). This convection helps the explosion in 2 ways: 1) by allowing material that has piled up at the top of the stalled shock to flow down toward the proto-neutron star, it reduces the pressure the convective region must overcome to drive an explosion, 2) by allowing the material heated by neutrinos to rise, its thermal energy is converted to kinetic energy, allowing it to cool before it begins to lose its energy through neutrino emission. With this engine, scientists began to produce explosions (Herant *et al.* 1994, Buras *et al.* 2006, Lentz *et al.* 2015).

As most groups confirming this engine with detailed models, there is little disagreement that this convection plays an important role. However, there are disagreements on the nature of the convection: for instance, is the matter moving because of Rayleigh-Taylor convection or the standing accretion shock instability (SASI). One way to decide this argument is based on the growth time of these instabilities. The Rayleigh-Taylor growth time ($\tau_{\rm conv}$) can be estimated by the Brunt-Väisäla frequency: $\tau_{\rm conv} = (|1/\omega^2|)^2$ where ω is the Brunt-Väisäla frequency:

$$\omega^2 = g/\rho(\partial\rho/\partial S)_P(\partial S/\partial r) \approx (1/S)(GM_{\rm enclosed}/r^2)(\Delta S/\Delta r) \tag{2.5}$$

with the gravitational acceleration $g = GM_{\rm enclosed}/r^2$, S is the entropy, P is the pressure, r is the radius and $M_{\rm enclosed}$ is the enclosed mass. For the conditions of our failed bounce, $\Delta S/S \approx 0.2$, $g \approx 1.5 \times 10^{12}\,{\rm cm\,s^{-2}}$, and $\Delta r = 10^7\,{\rm cm}$, the Rayleigh-Taylor growth time is 2 ms. The standing accretion shock instability has a much slower growth time $\sim 100\,{\rm ms}$ (Houck & Chevalier 1992, Blondin *et al.* 2003) and, unless the entropy gradient can be flattened (Rayleigh-Taylor could possible do this quickly), it should not dominate the material motion (Fryer & Young 2007).

Note that this supernova engine is not robust. It depends sensitively on our ability to model the convection as well as the details of the microphysics. But as we shall discuss in the next section, this appears to be what is needed to match nature.

Alternative engines exist. Among the most popular are engines driven by magnetic fields. The favored engine behind long-duration gamma-ray bursts argues that the collapsing core does not explode with this convection-enhanced engine and the core ultimately collapses to form a black hole. If the star is rotating fast enough, a disk will form around this black hole and this "collapsar" engine argues that strong magnetic fields will develop in this disk, producing a jet that drives the explosions (Woosley 1993). Scientists have also invoked this magnetically-driven disk engine for supernovae. One of the difficulties of this engine is getting enough rotation. The specific angular momentum of massive stars increases with radius. The collapsar engine leveraged this fact, allowing the

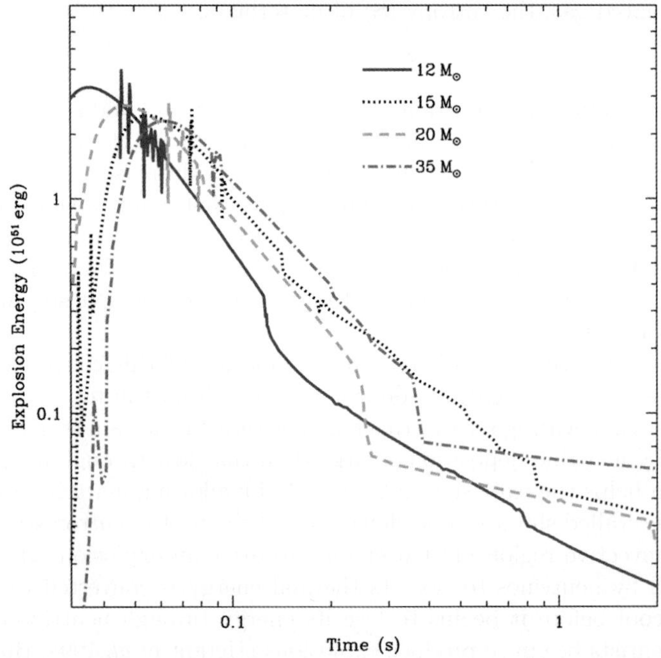

Figure 2. Supernova energy for 3 different models as a function of the time after collapse that the shock is launched. As the density of the infalling material decreases, less energy is needed to drive an explosion, but the resultant supernova explosion energy is less (Fig.1 of Fryer *et al.* 2012.)

low angular momentum material accrete onto the black hole before forming a disk. The specific angular momentum ($j_{\rm star}$)needed to form a disk is:

$$j = (GM_{\rm PNS} r_{\rm disk})^{1/2} \tag{2.6}$$

where G is the gravitational constant, $M_{\rm PNS}$ is the mass within the disk (roughly the proto-neutron star mass) and $r_{\rm disk}$ is the disk radius. For a 100 km disk around 1.5 M$_\odot$ proto-neutron star, the angular momentum need is 4×10^{16}cm^2 s^{-1}. This is only produced in rapidly spinning stars that neglect many physics processes that couple the neutron star (note that these stars would produced sub ms pulsars. Given that most pulsars are spinning at roughly 100 ms, such fast rotations are rare.

3. Supernova Energies

As we discussed in the introduction, the large, 1-3 foe, energies observed in supernovae is what prompted the first proposal of a core-collapse engine. But the distribution of the observed supernova energies place even stronger constraints on the engine and mark one of the strongest successes of the convective engine. The convective engine persists until the pressure at the edge of the failed shock that marks the edge of the convective region can overcome the ram pressure of the infalling star. The pressure in the convective region ($P_{\rm convection}$) assuming a roughly constant entropy region is Fryer *et al.* 2012:

$$P_{\rm convection}(r) = [0.25 S_0 M_{\rm NS} G S^{-1}(1/r - 1/r_{\rm shock}) + P_{\rm shock}^{1/4}]^4 {\rm erg\,g^{-1}} \tag{3.1}$$

where the pressure in the shock ($P_{\rm shock}$) is:

$$P_{\rm shock}(r) = 1/2 \rho_{\rm shock} v_{\rm free-fall}^2. \tag{3.2}$$

where $S_0 = 1.5 \times 10^{-11}$, $M_{\rm NS}$ is the mass of the neutron star, G is the gravitational constant, r and $r_{\rm shock}$ are the radii in the convective region and of the shock (edge of convective region) respectively, $\rho_{\rm shock}$ is the density of the shocked infalling materiall and $v_{\rm free-fall}$ is the free fall velocity. The specific energy ($u_{\rm convection}$) in this convective region is then:

$$u_{\rm convection}(r) = 3[4.7 \times 10^8 M_{\rm NS}(M_\odot)(S/10k_{\rm B}\,{\rm nucleon}^{-1}) \tag{3.3}$$

$$\times(r^{-1}(10\,{\rm km}) - r_{\rm shock}^{-1}(10\,{\rm km})) + 1.2 \times 10^6 M_{\rm NS}^{1/8}(M_\odot) \tag{3.4}$$

$$\times(\dot{M}_{acc}^{1/4}({\rm M_\odot s^{-1}})({\rm r_{shock}^{-5/8}}(200\,{\rm km})]^4\,{\rm erg\,cm^{-3}} \tag{3.5}$$

where $\dot{M}_{acc}$ is the accretion rate of the imploding star onto the convective region. The energy in this region depends both on this accretion rate, the entropy in the region, and the radius of the shock. If we assume the engine turns off once the explosion is launched because the density becomes too low to absorb any neutrinos, we can use simulation-guided values for the entropy and shock radius to determine the explosion energy as a function of explosion time (Fryer *et al.* 2012). Because the accretion rate decreases with time, the energy decreases with time. The supernova energy as a function of time and progenitor is shown in Figure 2. Note that this estimate is likely to be a lower limit because most simulated explosions are not complete; material continues to fall onto the neutron star after the supernova explosion. A fraction of this "fallback" material can absorb energy and become part of the explosion.

One of the energy requirements placed on the core-collapse model is to explain why most supernovae have energies that are 1% of the 10^{53} erg released in the collapse. This convective model provides a natural explanation for this limit - if the energy got more than a few foe, the shock would launch, turning off the engine. The energy from magnetic engine models can not explain this observational fact. Currently, practitioners of magnetic-field models argue the progenitor dictates the energy distribution, e.g. the distribution of rotations is such that most explosions extract 1% of the total energy released - the model is tuned to fit the answer. Despite the strengths of the convective engine, its limitation is that it can not explain extremely energetic supernova, e.g. hypernovae. In cases where there is considerable fallback, the convective engine can produce more energetic explosions, but it is unlikely that this is sufficient to explain hypernovae and a different engine, e.g. collapsar, is required to explain really powerful explosions.

4. Compact Remnants

The compact remnant mass distribution also places strong constraints on the supernova engine. When the convective engine was first proposed, compact mass distributions were believed to be delta functions, with neutron stars masses very close to $1.4\,{\rm M_\odot}$ and black hole masses of $\sim 7\,{\rm M_\odot}$ (Thorsett & Chakrabarty 1999). Defying observational evidence, the convective engine predicted a range of masses (Fryer & Kalogera 2001). We now know that there are a range of compact remnant masses, confirming this prediction of the convective engine. However, the convective engine did not predict the observed mass gap between neutron star and black hole masses. This gap, if it represents the birth mass distribution, has important implications on the convective engine and by constraining the engine to either very strong or very weak explosions (Fryer *et al.* 2012, Belczynski *et al.* 2012). Fig. 3 plots both predicted remnant distributions of remnants in binaries along with the observed distribution. Note, however, that scientists studying binary accretion have argued that the compact remnant can accrete considerable mass post-formation and

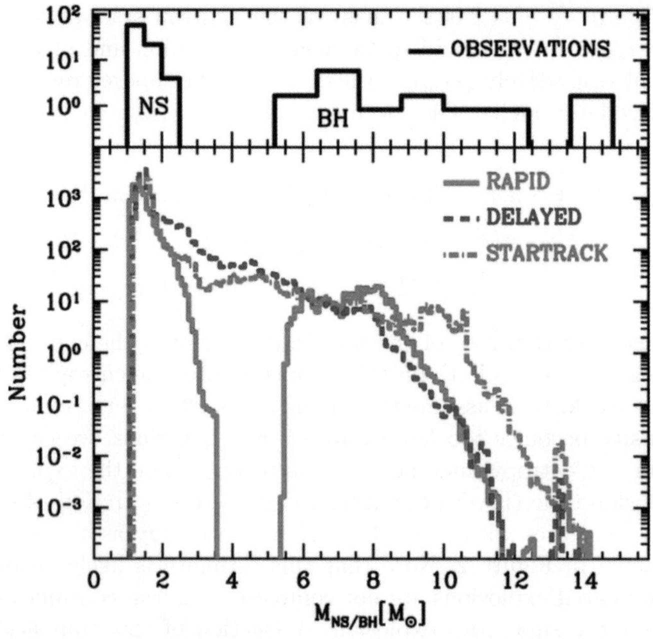

Figure 3. Mass distribution of compact remnants both from theoretical models and in observations. Compact remnant masses are only determined for systems in binaries and these theoretical models are produced through binary population studies (Fig. 1. of Belczynski *et al.* 2012.)

that no gap exists in the birth distribution of compact remnants (Fragos & McClintock 2015).

Magnetic field models have yet to explain these observations.

5. Supernova Asymmetries

For 1987A, the early emergence of the gamma-rays from ^{56}Ni decay and the spectral line widths drove the initial studies of the convective engine. But there is further evidence that supernovae are asymmetric. First and foremost, the pulsar velocity distribution suggests that neutron stars are born with a "kick" that, in some cases, can exceed $1000\,\mathrm{km\,s^{-1}}$. In addition, a straightforward explanation of the large polarization seen in some core-collapse supernovae is that the explosions driving these supernovae are asymmetric (see Hungerford *et al.* 2003 for a review). The ejecta remnants also show evidence for asymmetries. In the Cassiopeia A remnant, there is a visible "jet-like" structure in the silicon ejecta (Hwang *et al.* 2004). However, as we shall discuss below, the titanium distribution tells a very different story.

Before we discuss the latest results on Cassiopeia A, let's review the theoretical models. Not long after the first succesful supernova explosions were produced with the convective engine, scientists argued that the Rayleigh-Taylor convection would develop into low modes (as seen in many simulations of Rayleigh-Taylor convection) and, in some cases, could produce the large kicks (Herant 1995). The strongest convective cycles in the standing accretion shock instability are low mode and this model could produce asymmetric explosions to explain pulsar kicks. SN 1987A appears to be dominated by a single mode and such low modes are possible in the convective engine.

Jet-like explosions are much more difficult to explain. One way to make bimodal explosions is by collapsing a rotating star. The criterion for instability in a rotating star

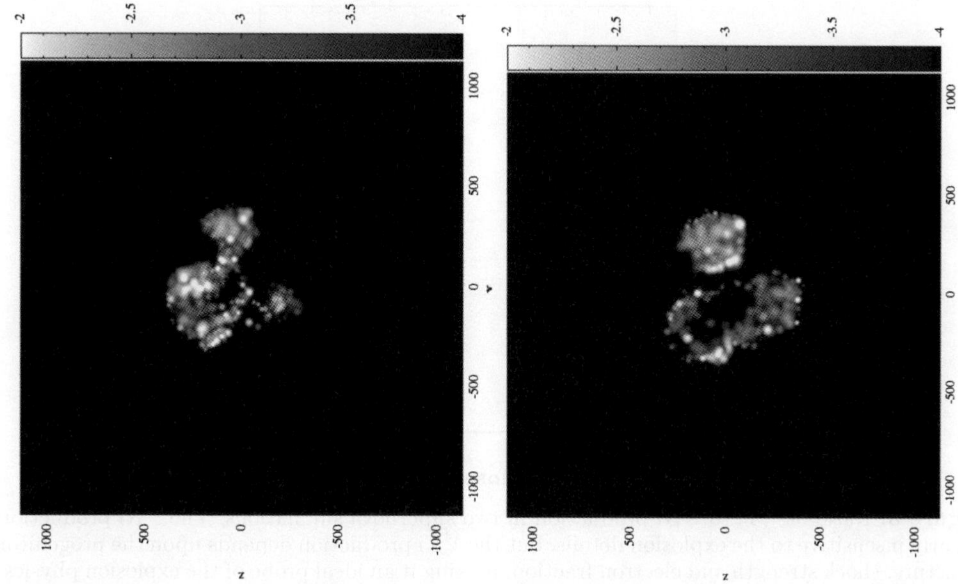

Figure 4. Color plots of two slices from a 3-dimensional supernova explosion of the ^{44}Ti/iron ratio. The low-mode convection produces low-mode clumps of ^{44}Ti similar to that observed in Cassiopeia A.

is:

$$g/\rho[(d\rho/dr)_S - d\rho/dr] - 1/r^3 dj^2/dr > 0. \quad (5.1)$$

where g is the gravitational acceleration, ρ is the density, j is the angular momentum, S is the entropy and r is the radius. If the angular momentum increases at larger radii (which it does for most stars), there is a stabilizing force against convection. The angular momentum is high along the spin equatorial plane, but convection can still occur along the rotation axis. The convective engine then produces an explosion that is much stronger along the rotation axis, even without magnetic fields (Fryer & Heger 2000). Of course, with magnetic fields, a range of jet models exist.

It is worth mentioning that the most extremeley asymmetric core-collapse events are hypernovae which not only appear jet-like, but also much too energetic to explain with the convective engine. It is likely that some mechanism, like the black-hole accretion disk magneto-hydrodynamic engines explain these explosions.

But let's return to what we believe to be normal supernova asymmetries - like the jet-like structure in Cassiopeia A. Although the remnant community argued, and in some cases still argues, that this jet structure must be caused by the engine (Hwang *et al.* 2004), analysis of the remnant can be difficult. Understanding the emission from excited lines requires that the material be heated, shocked by either the forward or reverse shock in the explosion. Because of this, the circumstellar medium can alter the shape and an alternate explanation for this jet-like feature is that the circumstellar medium is asymmetric (Blondin *et al.* 1996).

The NuSTAR satellite promised a more direct probe of the asymmetries in Cassiopeia A. ^{44}Ti is produced in the innermost ejecta of a supernova along with the ^{56}Ni. Its 60 year half-life make it ideally suited to probe the distribution of this ejecta in supernova remnants. Figure 4 shows the ^{44}Ti/iron ratio for a convectively-driven supernova. The low-mode convection produces an explosion with low-mode asymmetries which grow as the shock plows through the star.

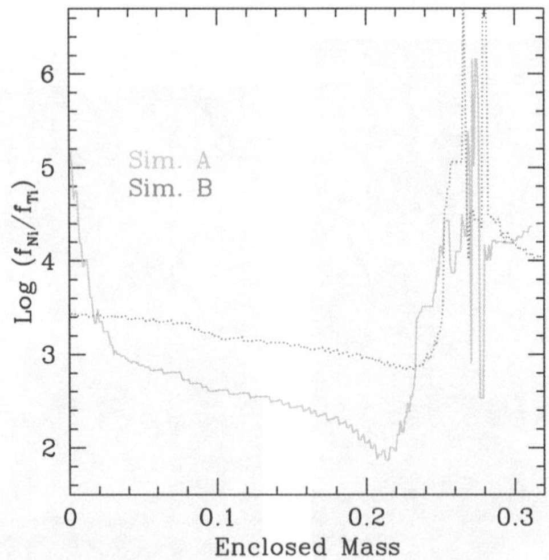

Figure 5. Ratio of ^{44}Ti to ^{56}Ni production in two supernova simulations. The ^{56}Ni production is fairly insensitive to the explosion details, but the ^{44}Ti production depends upon the progenitor structure, shock strength and electron fraction, making it an ideal probe of the explosion physics.

The NuSTAR satellite, with a wavelength coverage from a few to 80 keV, can directly observe 2 decay lines from ^{44}Ti. The strength of this probe is twofold: first, the ^{44}Ti is produced solely in the innermost ejecta with negligible mass fraction in the star or circumstellar medium (it is sometimes difficult to distinguish iron made in the supernova and iron already existing in the star) and second, we can observe all of the ^{44}Ti and, aside from small possible changes in the decay timescale due to excited states, the ^{44}Ti measurement is a clean, direct measurement of the ^{44}Ti. What we observed with a low mode, but not bimodal, ejection (Grefenstette *et al.* 2014, Grefenstette *et al.* 2017). The predictions of the convective model were correct. Although there are some arguments for jet jitter causing the observed structure, it appears that a magnetic-driven engine is not behind the explosion behind the Cassiopeia A remnant.

Figure 5 shows the ^{44}Ti/^{56}Ni ratio for two different supernova explosions. The strong sensitivity of the ^{44}Ti production to the characteristics of the supernova explosion make it an ideal probe of this explosion. We have only touched the surface of what we can learn from these explosions. Perhaps these comparisons will cause a new evolution in the model.

6. Summary

We have reviewed the evolution of the core-collapse engine into a convectively driven engine. Based primarily upon the fact that supernova are energetic and asymmetries seemed necessary to explain mixing, the convective engine explains naturally the preponderence of normal-energy, few foe, supernovae, predicted the range of remnant masses, and correctly predicted the asymmetries in the Cassiopeia A remnant. Alternative engine, e.g. magnetically driven engines, are needed to explain the most energy supernovae (e.g. hypernovae), but these engines have not been able to explain most of the observations for normal supernovae. Because of this, the convective-driven engine has become the favored mechanism for normal supernovae.

References

Baade, W. & Zwicky, F. 1934, *Phys. Rev.*, 46, 76
Bionta, R. M., Blewitt, G., Bratton, C. B., Casper, D., & Ciocio, A. 1987, *PRL*, 58, 1494
Blondin, J. M., Lundqvist, P., & Chevalier, R. A. 1996, *ApJ*, 472, 257
Blondin, J. M., Mezzacappa, A., & DeMarino, C. 2003, *ApJ*, 584, 971
Buras, R., Rampp, M., Janka, H.-T., & Kifonidis, K. 2006, *A&A*, 447, 1049
Burrows, A. & Lattimer, J. M. 1988, *Phys. Rep.*, 163, 51
Colgate, S. A. & Johnson, M. H. 1960, *PRL*, 5, 235
Colgate, S. A. & White, R. H. 1966, *ApJ*, 143, 626
Epstein, R. I. 1979, *ApJ*, 188, 305
Fragos, T. & McClintock, J. E. 2015, *ApJ*, 800, 17
Fryer, C. L. & Heger, A. 2000, *ApJ*, 541, 1033
Fryer, C. L. & Kalogera, V. 2001, *ApJ*, 554, 548
Fryer, C. L. & Warren, M. 2002, *ApJ*, 574, L65
Fryer, C. L. & Young, P. A. 2007, *ApJ*, 659, 389
Fryer, C. L., Belczynski, K., Wiktorowicz, G., Dominik, M., Kalogera, V., & Holz, D. 2012, *ApJ*, 749, 91
Grefenstette, B. W. *et al.* 2014, *Nature*, 506, 339
Grefenstette, B. W. *et al.* 2017, *ApJ*, 834, 19
Herant, M., Benz, W., & Colgate, S. 1992, *ApJ*, 395, 642
Herant, M., Benz, W., Hix, W. R., Fryer, C. L., & Colgate, S. A. 1994, *ApJ*, 435, 339
Herant, M. 1995, *Phys. Rep.*, 256, 117
Houck, J. C., & Chevalier, R. A. 1992 *ApJ*, 395, 592
Hirata, K., Kajita, T., Koshiba, M., Nakahata, M., & Oyama, Y. 1987, *PRL*, 58, 1490
Hungerford, A. L., Fryer, C. L., & Warren, M. S. 2003, *ApJ*, 594, 390
Hwang, U. *et al.* 2004 2004, *ApJ*, 615, L117
Keil, W., Janka, H.-T., & Müller, E. 1996, *ApJ*, 473, L111
Lentz, E. J., Bruenn, S. W., Hix, W. R., Mezzacappa, A., Messer, O. E. B., Endeve, E., Blondin, J. M., Harris, J. A., Marronetti, P., & Yakunin, K. N. 2015, *ApJ*, 807, L31
Pinto, P. A. & Woosley, S. E. 1988, *ApJ*, 329, 820
Thorsett, S. E. & Chakrabarty, D. 1999, *ApJ*, 512, 288
Woosley, S. E. 1993, *ApJ*, 405, 273

Supernova 1987A:30 years later - Cosmic Rays and Nuclei from Supernovae and their aftermaths
Proceedings IAU Symposium No. 331, 2017
A. Marcowith, M. Renaud, G. Dubner, A. Ray & A. Bykov, eds.

doi:10.1017/S1743921317004720

Aspherical Supernovae and Oblique Shock Breakout

Niloufar Afsariardchi and Christopher D. Matzner

Department of Astronomy and Astrophysics, University of Toronto, 50 St. George St, M5S 3H4, Toronto, Canada
email: afsariardchi@astro.utoronto.ca
email: matzner@astro.utoronto.ca

Abstract. In an aspherical supernova explosion, shock emergence is not simultaneous and non-radial flows develop near the stellar surface. Oblique shock breakouts tend to be easily developed in compact progenitors like stripped-envelop core collapse supernovae. According to Matzner *et al.* (2013), non-spherical explosions develop non-radial flows that alters the observable emission and radiation of a supernova explosion. These flows can limit ejecta speed, change the distribution of matter and heat of the ejecta, suppress the breakout flash, and most importantly engender collisions outside the star. We construct a global numerical FLASH hydrodynamic simulation in a two dimensional spherical coordinate, focusing on the non-relativistic, adiabatic limit in a polytropic envelope to see how these fundamental differences affect the early light curve of core-collapse SNe.

Keywords. supernovae: general, hydrodynamics, shock waves

1. Introduction

Shock breakout is the earliest observable phase of a supernova explosion and is associated with high energy radiation bursts in ultraviolet/X-ray frequencies and the release of high velocity ejecta (Klein & Chevalier 1978). This radiation dominated shock accelerates while approaching the stellar surface due to steeply decreasing density profile of the SN progenitor and sets the ejecta into motion (e.g., Sakurai 1960, Matzner & McKee 1999). The high energy photons diffuse out of the shock front when the optical depth of the progenitor τ drops below of the shock optical depth $\approx c/v_{sh}$, where v_{sh} is the shock velocity. The breakout flash follows by envelope cooling emission from the deeper layers of the expanding ejecta in optical/near-UV bands.

Studying the early light curve and spectrum of core collapse supernovae enables us to obtain invaluable information about the nature and dynamics of the supernova shock and the characteristics of its progenitor (e.g., radius, envelope composition, and mass loss history) which cannot be inferred in later times when the light curve is governed by the radioactive decay. In recent years, the advancement of modern supernova surveys (e.g., Kaiser *et al.* 2002, Kim *et al.* 2016) along with the development of sensitive high energy detectors in UV, X-ray, and γ-ray frequencies has enhanced both the number of early detected supernovae and the accuracy of constraining supernova progenitor with UV/X-ray observations.

The early light curve of observed SNe are generally interpreted by spherical explosion models (Matzner & McKee 1999, Nakar & Sari 2010, Rabinak & Waxman 2011), which have been extensively studied over past two decades. However, there has been growing evidence for aspherical scenarios when these models misinterpret the progenitor or cannot consistently explain the early lightcurve, spectrum, and polarization information. In the case of SN2011dh (Arcavi *et al.* 2011), for example, the progenitor was initially found to be a compact one based on the spherical models while it is now known as a yellow supergiant (YSG) from the stacked pre-SN images. This event also showed decreasing early polarization, suggesting aspherical structure in the fastest ejecta: (Mauerhan *et al.* 2015). For SN2008D, a Type Ib supernova (Soderberg *et al.* 2008), the polarization information and long 100 s detected X-ray flash supported aspherical breakout scenario (Couch *et al.* 2011).

Asymmetries seem theoretically inevitable in certain SN events. For instance, in the collapsar model of Gamma Ray Bursts (GRBs), the supernova signature is associated with the unbinding of the envelope by a highly asymmetric shock that is produced when a relativistic jet pierces a hole in the envelope (e.g., MacFayden & Woosley 1999). Moreover, asymmetries can be inherent to the core collapse mechanism, whether in the jet-driven explosions (Piran *et al.* 2017), or neutrino-driven ones (Utrobin *et al.* 2017). Finally, aspherical dynamics also arise when the progenitor shape deviates from a perfect sphere; examples include rapidly rotating progenitor in which the progenitors are oblate as well as Roche lobe binary systems in which they are tidally deformed.

Here, we provide an overview on the current state of aspherical shock breakout and early light curve modelling and analysis and briefly comment on our latest numerical simulation and its results.

2. Overview and Previous Work

The effect of asphericity on the SN observables have been studied in previous numerical and analytical articles. A general review of the previous works is as follows:

Early Work. Suzuki & Shigeyama (2010) investigate the effect aspherical energy deposition in a blue supergiant using a 2D hydrodynamical simulation. They derive the early light curve assuming a black-body radiation with a constant temperature along the progenitor. To treat the asymmetries, each patch of the progenitor is considered to radiate when the shock arrives and continue radiating for a constant amount of time. This has the drawback of ignoring the non-radial flows, which can completely change the shock emergence dynamics.

Jet-driven Hydrodynamic Simulations. The 2D hydrodynamic simulations of Couch *et al.* (2011) are targeted to reproducing the early light curve and the spectra of SN2008D. The explosion is invoked by jets that are active for about few seconds. The light curve and spectra modelling are done semi-analytically in post-processing. Unlike Suzuki & Shigeyama (2010), they take into account the effect of non-radial flows by integrating black-body photon emission along thermalization front (i.e., where radiation and matter temperatures are equal). Their analysis results indicate that their reproduced light curve fits the observed light curve well, but their spectra poorly matched the observed one. As they point out, this can be due to simple black-body assumption that does not allow the photon color temperature to go above the thermalization temperature. Furthermore, the luminosity must have been set where the photons are able to diffuse upstream, not at thermalization front; this condition is met when dynamical time is equal to photon diffusion time, i.e., $t_{\mathrm{dyn}} \sim t_{\mathrm{diff}}$. The depth of thermalization front and the diffusion front may substantially differ, particularly in compact progenitors like the one considered in this study.

Oblique Shock Breakout. More recently, Matzner *et al.* (2013) analytically found that if the pattern speed of the shock progression along the stellar surface v_ϕ in aspherical explosions is about the shock speed v_{sh}, the shock dynamics are different from spherical case. Such "oblique shock breakouts" are predicted to suppress the direct breakout flash, limit the shock velocity, and cause non-radial flows to collide outside the star – the

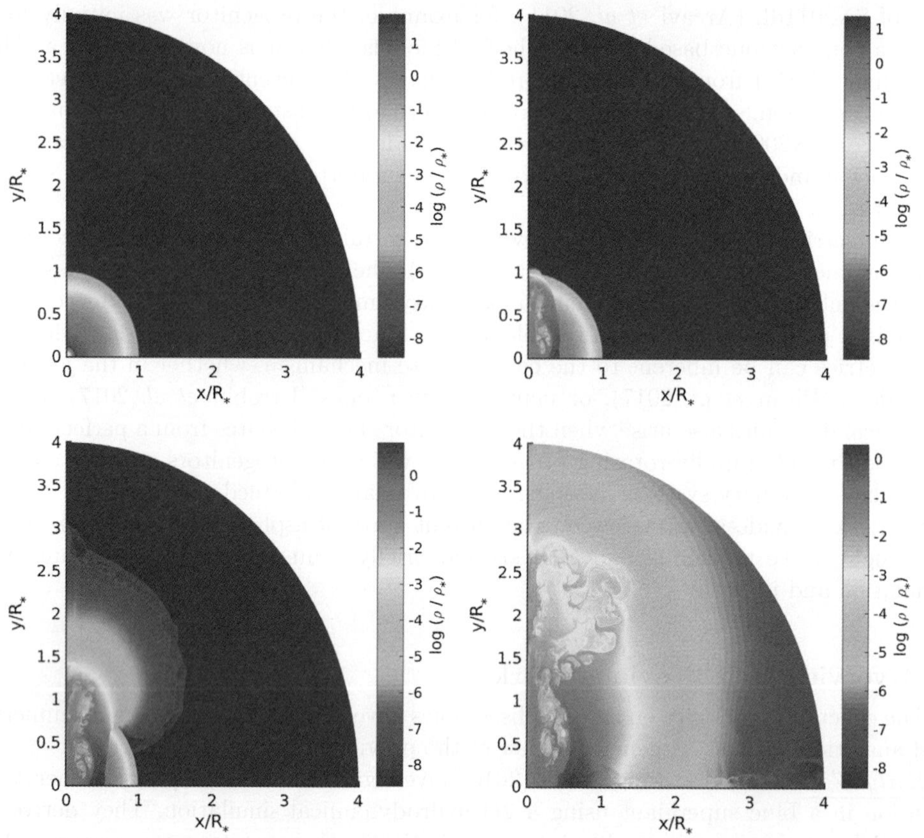

Figure 1. The snapshots of the density distribution of our scale-free simulations are shown. The top left panel shows the initial density distribution. In top right plot, the bipolar shock is emerged from the pole; the shock breaks normal to the stellar surface. The bottom left panel depicts the spray of the ejecta outside the stellar surface; at this stage the shock has become oblique. The bottom right panel marks the time when the shock has completely traversed the progenitor and the circumstellar collisions is happening along the equator.

predictions that are verified in the zoomed-in local simulations of Salbi *et al.* (2014). Despite providing the qualitative description of distinctive early light curve, these studies do not quantitatively model the early light curves or the spectrum.

Radiation Transfer Simulation. Recently, Suzuki *et al.* (2016) simulate aspherical explosion in a BSG progenitor using a single-group radiative transfer code and provide bolometric light curves for the simulation results. They, however, did not observe oblique shock breakout due to the insufficient degree of asphericity in their simulation.

3. Simulation and Results

We construct a global numerical FLASH (Fryxell 2000) hydrodynamic simulation in a two-dimensional uniform spherical-polar coordinate. The explosion is first triggered spherically in the center of a $n = 3$ polytrope with energy deposited in few central cells where, resulting in a Sedov explosion with a sharp peak in the density and pressure profiles. The outcome of the spherical Sedov explosion is saved at time at early time when the shock is at about $0.2R_*$. The saved velocity profile is then multiplied by $\cos^k(2\phi)$ to gain axisymmetrical momentum. A few density snapshots from our simulations are shown in

Model	M_* [2×10^{33}g]	R_* [6×10^{10}cm]	E_* [10^{51} erg]	X	κ [cm^2/g]
RSG	14	500	1	0.7	0.34
BSG	15	49	1	0.7	0.34
Ic	5	0.2	1	1	0.2

Table 1. Simulations parameters that represent various explosions. The dynamical model is a complete $n = 3$ polytrope.

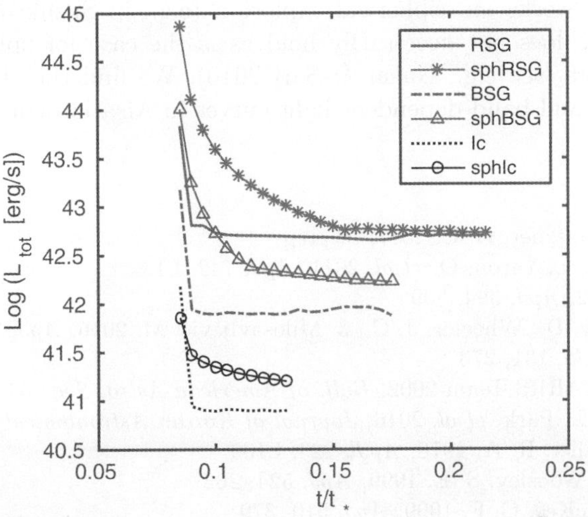

Figure 2. Angle-averaged apherical bolometric early light curve for three progenitors: RSG, BSG, and Type Ic model are shown. For comparison, the spherical early light curve of Nakar & Sari (2010) is also shown for all three models. The time axis is normalized to the characteristic time $t_* = R_*\sqrt{M_*/E_*}$, where R_*, M_*, and E_* are progenitor radius, progenitor mass, and explosion energy respectively shown in Table 1.

Figure 1. In our simulations, we employ adiabatic flow and ignore gravity and relativity, rendering them scale free. However we post-process them to check these approximations (see Afsariardchi & Matzner 2017), and to account or radiative transfer (absorptive emission and diffusion). In this process, we scale the simulations to three physical contexts: Red Supergiant (RSG), Blue supergiant (BSG), and compact Ic model, described in Table 1. The extent of diffusion front, has revealed that our simulations are mainly relevant to compact progenitors and BSGs. For RSGs, the photon diffusion starts early on in the hydrodynamic evolution and thus our adiabatic assumption breaks. Moreover, we calculated the obliquity scales of our simulations based on their definition in Matzner *et al.* 2013 to see when and in which progenitors a normal shock turns to oblique shock.

In order to obtain bolometric luminosity light curves for each progenitor model, we find diffusion front and use diffusion approximation (details in Afsariardchi & Matzner 2017) to calculate luminosity using a method similar to Chevalier (1992). In Figure 2, the early bolometric light curve of three different progenitors is shown for our aspherical simulation. For comparison, the spherical analytical models of Nakar & Sari (2010) are also plotted. The aspherical light curves are all angle-averaged to show overall evolution of the energy release rate as oppose to observer-dependent light curves. As shown in Figure 2, the light curve of all three progenitors have a narrow peak in the early bolometric light curve. This peak is associated with shock breakout from the poles, when the shock is still normal to the stellar surface, but as the shock becomes oblique the luminosity significantly declines and the light curve enters a phase of constant luminosity. This is because the direct breakout flash is obscured under an optically thick spray of ejecta when the shock is oblique. Furthermore, comparing the constant part of aspherical light curves to the corresponding spherical ones, we find that the asherical light curves are few (i.e., < 3) times less luminous than their spherical counterparts. This indicates that the oblique shock breakout emission is almost indistinguishable from the expanding ejecta phase of spherically symmetric SNe.

Deriving photon energies is more complicated; the assumption of thermalization that was made in prior works on aspherical explosion (e.g., in Suzuki & Shigeyama 2010, Couch *et al.* 2011) does not necessarily hold as is the case for spherical explosion in compact progenitors (see e.g., Nakar & Sari 2010). We find the thermalization front, color temperature, and band-dependent light curves in Afsariardchi & Matzner (2017).

References

Afsariardchi, N. & Matzner, D. C. 2017, *in prep*

Arcavi, I., Gal-Yam, A., Yaron, O. *et al.* 2011, *ApJ*, 742, L18

Chevalier, R. A. 1992, *ApJ*, 394, 599

Couch, S. M., Pooley, D., Wheeler, J. C., & Milosavljević, M. 2016, *ApJ*, 727, 104

Fryxell, B. 2000, *ApJS*, 131, 273

Kaiser, N. & Pan-STARRS Team 2002, *Bull. of American Astro. Soc.*, 34, 1304

Kim, S.-L., Lee, C.-U., Park, *et al.* 2016, *Journal of Korean Astronomical Society*, 49, 37

Klein, R. I. & Chevalier, R. A. 1978, *ApJl*, 223, L109

MacFadyen, A. I. & Woosley, S. E. 1999, *ApJ*, 524, 262

Matzner, C. D. & McKee, C. F. 1999, *ApJ*, 510, 379

Matzner, C. D., Levin, Y., & Ro, S. 1999, *ApJl*, 779, 60

Mauerhan, J. C., Williams, G. G., Leonard, D. C., *et al.* 2015, *MNRAS*, 453, 4467

Nakar , E. & Sari, R. 2010, *ApJ*, 725, 904

Piran, T., Nakar, E., Mazzali, P., & Pian, E. 2017, *ArXiv e-prints*

Rabinak, I. & Waxman, E. 2011, *ApJ*, 728, 63

Sakurai, A. 1960, *Comm. Pure Appl. Math*, 13, 353

Salbi, P., Matzner, C. D., Ro, S., & Levin, Y. 2014, *ApJ*, 790, 71

Soderberg, A. M., Berger, E., Page, K. L., *et al.* 2008, *Nature*, 453, 469

Suzuki, A. & Shigeyama, T. 2010, *ApJl*, 717, L154

Suzuki, A., Maeda, K., & Shigeyama, T. 2016, *ApJ*, 825, 92

Utrobin, V. P., Wongwathanarat, A., Janka, H.-T., & Mueller, E. 2017, *ArXiv e-prints*

Discussion

HIRSCHI: Do you also consider hydrogen poor progenitors?

AFSARIARDCHI: Yes, our type Ic model is hydrogen poor.

FRYER: Are your light curves sensitive to simulation resolution?

AFSARIARDCHI: We did a resolution study, and saw the same overall evolution, we, however, have not generated the light curve for our low resolution runs.

RAY: How does asphericity affect the relativistic flow?

AFSARIARDCHI: Asphericity put a new lower limit on the maximum speed of the flow. So, this new limit may prevent the ejecta to become relativistic.

Supernova 1987A:30 years later - Cosmic Rays and Nuclei from Supernovae and their aftermaths
Proceedings IAU Symposium No. 331, 2017
A. Marcowith, M. Renaud, G. Dubner, A. Ray & A. Bykov, eds.

doi:10.1017/S1743921317004409

Linking three-dimensional core-collapse supernova simulations with observations

Annop Wongwathanarat[1,2]

[1]RIKEN, Astrophysical Big Bang Laboratory, 2-1 Hirosawa, Wako, Saitama 351-0198, Japan
email: annop.wongwathanarat@riken.jp

[2]Max-Planck-Institut für Astrophysik, Karl-Schwarzschild-Str. 1, 85748 Garching, Germany

Abstract. To understand a wide variety of properties of young core-collapse supernova (CCSN) remnants being revealed by modern observations three-dimensional simulations of CCSNe starting from the initiation of the explosion until the expanding stellar debris transform into gaseous remnants are needed. We briefly review recent progress in modeling CCSNe on a long time scale. A current effort to model bolometric light curves based on 3D CCSN explosion models for comparison with observational data from SN 1987A is also discussed.

Keywords. supernovae: general, supernovae: individual (Supernova 1987A), hydrodynamics, instabilities

1. Introduction

Multi-wavelength observations of nearby young CCSN remnants, e.g. Crab, Cassiopeia A, and SN 1987A, reveal large scale asymmetries and complex structures of different chemical elements in the expanding stellar debris. Because interactions between ejecta and circumstellar materials are still minimal for young remnants these observations immediately suggest that the explosions were not spherically symmetric. Origins of asymmetries which influence the final observed morphology of remnants include asymmetries in the progenitor core prior to core collapse (Arnett & Meakin 2011; Couch *et al.* 2015; Müller *et al.* 2016), hydrodynamic instabilities such as convection (e.g., Herant *et al.* 1994) and the standing accretion shock instability (SASI, Foglizzo 2002; Blondin *et al.* 2003; Blondin & Mezzacappa 2006) which occur during the shock revival phase, and secondary mixing (e.g, Rayleigh-Taylor and Richtmyer-Meshkov) instabilities triggered by the SN shock propagation through the onion shell structure of the progenitor star (Chevalier 1976; Kifonidis *et al.* 2006). Therefore, to make a direct comparison of the ejecta asymmetries, which is one of the many CCSN observables besides neutrinos, gravitaional waves, nucleosynthesis yields, remnant properties, light curves, and spectra, between CCSN models and observations it is unavoidable that the models must consider the development of asymmetries in different stages of CCSN explosions and their complex interplays consistently.

Undoubtedly, performing a numerical simulation of CCSN explosions which considers the core collapse through the shock revival phase until the SN ejecta expand and transform into a gaseous nebula is an extremely difficult task. Apart from big hurdles in modeling the core of massive stars before collapse and the initiation of the explosions in multi dimensions, carrying out long-time simulations tracing the explosion asymmetries from the moment after the SN shock resumes its outward propagation until it breaks out of the surface of the progenitor star also requires a huge computational demand because of the massive length- and timescales needed to be covered by these simulations. Here, we give a short review of progress in simulating three-dimensional (3D) long-time CCSN explosions and discuss how the outcomes of these simulations be used in connection with one-dimensional (1D) light curve modeling to constrain the structure of the progenitor star in the case of SN 1987A.

2. Progress on long-time simulations of CCSNe

After the explosion of SN 1987A on February 23, 1987 a large number of two-dimensional (2D) simulations studying the growth of Rayleigh-Taylor instabilities (RTIs) while the SN shock plows through shells of elements inside the progenitor star have been performed. An extensive list of references to such works was already given in, e.g., Wongwathanarat *et al.* (2015) and Mao *et al.* (2015). One of the main goals of these simulations was to explain the mechanism which facilitates outward mixing of the radioactive nickel synthesized deep inside the SN core into the hydrogen rich envelope and inward mixing of hydrogen into the helium rich layer. Such large-scale radial mixing was implied following analyses of observational data obtained from SN 1987A such as early detections of hard X-ray (Dotani *et al.* 1987; Sunyaev *et al.* 1987) and gamma-ray lines (Matz *et al.* 1988), a high-velocity feature in the [Fe II] line at ∼3500 km/s (Haas *et al.* 1990), and the 1D light curve modeling by radiation hydrodynamic calculations (Woosley 1988; Shigeyama & Nomoto 1990; Blinnikov *et al.* 2000; Utrobin 2004). It turned out that RT mixing seeded by small-scales random pertubations in 2D models initiated by point-like or piston-driven explosions was insufficiently vigorous to carry out nickel to high velocities beyond ∼3000 km/s as required by observations, unless an asymmetric or jet-like explosion is assumed (e.g., Nagataki *et al.* 1998; Nagataki 2000). This problem was termed the "nickel discrepancy" by Herant & Benz (1992).

The number of attempts to perform 3D calculations investigating the growth of RTIs in the SN envelope is much smaller than 2D studies. Nagasawa *et al.* (1988) made the first advance to the third dimension by performing 3D models of point-like explosions in $n = 3$ polytropes using smoothed particle hydrodynamics (SPH). A motivation for Nagasawa *et al.* (1988) was that non-axisymmetric modes of RTIs which could not be captured by 2D calculations can be present in 3D. They reported fast growth of RTIs which can explain the early rise of X-rays observed in SN 1987A, but Muller *et al.* (1989) who did a similar 3D calculation with a grid-based code contradicted their results. Kane *et al.* (2000) made a comparison of RT growth rate of single-mode perturbation in 2D and 3D, and found that the perturbation grows ∼30% faster in a 3D calculation. This result was later tested by Joggerst *et al.* (2010). Nevertheless, the conclusion made by Joggerst *et al.* (2010) was different. They concluded that RTIs grow faster in 3D only initially. But, soon after RT fingers begin to interact and merge the growth rate decreases. As a result the extent of the fluid mixing zone is similar in 2D and 3D calculations. Hungerford *et al.* (2003) and Hungerford *et al.* (2005) re-investigated effects of asymmetric explosions on the extent of RT mixing and the resulting emergence of gamma-rays lines. They claimed that single-lobe explosions can explain the redshifted high-velocity component of the iron line observed in SN 1987A. The 3D calculation which followed the development of RTIs for the longest time was reported by Ellinger *et al.* (2012). They studied properties of clumped ejecta caused by fragmentaions due to RTIs and their evolution until ∼30 years after the explosion.

All of the studies mentioned above, however, neglected detailed explosion physics which is, in fact, very crucial in determining the ejecta asymmetries at late times. The calculations instead relied on simple methods to drive explosions. Using a different approach to initiate the explosion Kifonidis *et al.* (2003) and Kifonidis *et al.* (2006) were the first to model both the neutrino-driven explosion engine and the subsequent evolution of the SN shock until shock breakout in 2D. Modeling the explosions starting from core bounce in multi-D allowed them to capture the development of low-mode convective instabilities and SASI in their calculations. These explosion asymmetries act as large-scale large-amplitude perturbations for secondary RTIs at composition shell interfaces to grow strongly.

The importance of the 3D geometry and the development of low-mode instabilities in the explosion engine on the growth of secondary RTIs were both taken into account in the work by Hammer *et al.* (2010). They demonstrated that a 3D calculation of a neutrino-driven explosion starting from shortly after core bounce can provide a solution for the nickel discrepancy. They reported the asymptotic velocities of nickel-rich clumps up to ~4500 km/s in their 3D model, which are much higher velocities than in 2D calculations that they performed. They explained the 2D-3D difference in the velocity distribution of nickel at shock breakout by a simple geometrical argument. In 2D geometry a nickel rich clump in the simulation assumes a torus-like structure because of the axially symmetric constraint. Thus it experiences larger drag force in comparison with a spherical-like clump in 3D geometry.

The result obtained by Hammer *et al.* (2010) was different from that of Joggerst *et al.* (2010). However, their results, in fact, do not contradict each other. Because Joggerst *et al.* (2010) initiate the explosions using pistons and let RTIs develop from small-scale random perturbations, numerous small RT fingers grow and interact with each other. In contrast, only a few nickel-rich RT fingers develop in the 3D calculation of Hammer *et al.* (2010). These fingers grow from large-scale large-amplitude perturbations created by neutrino-driven convection and SASI during the first second of the explosion, and they propagate independently in different directions.

The work by Hammer *et al.* (2010) was a big step toward long-time modeling of neutrino-driven CCSNe in 3D. Still, they only investigated one 15 $M_\odot$ blue supergiant progenitor by Woosley *et al.* (1988). The growth of RTIs at shell interfaces depends greatly on the progenitor structure too, as shown by Herant & Benz (1991) in their 2D calculations in two different progenitor stars. Wongwathanarat *et al.* (2015) continued along the line of the work by Hammer *et al.* (2010), adding a progenitor dependence study. They performed a set of 3D simulations of CCSNe until ~1 day after the explosions for four different progenitor stars; two 15 $M_\odot$ red supergiant stars by Woosley & Weaver (1995) and Limongi *et al.* (2000), a 20 $M_\odot$ blue supergiant progenitor (Shigeyama & Nomoto 1990), and also the 15 $M_\odot$ blue supergiant star by Woosley *et al.* (1988) which was used in the study of Hammer *et al.* (2010).

Indeed, Wongwathanarat *et al.* (2015) found great differences in the morphology of the nickel-rich ejecta at shock breakout between cases of different progenitors. In red supergiants, RTIs develop extremely vigourously at the He/H composition interface due to a very steep density gradient there. In addition, a very strong reverse shock forms at that interface. The reverse shock travels backward in mass coordinates and compresses the trailing nickel-rich ejecta into a very thin layer. After being compressed the nickel-rich ejecta fragments into numerous small clumps.

The dynamics of the forward shock, reverse shock, and nickel-rich ejecta is different in the case of blue supergiant progenitors. In fact, it is also different even among the two blue supergiant stars that were considered. The simulations of the 20 $M_\odot$ star by Shigeyama & Nomoto (1990) showed much less vigorous growth of RTIs, at both C+O/He and He/H interfaces. In contrast, RTIs grows strongly at C+O/He interface in the 15 $M_\odot$ by Woosley *et al.* (1988). Consequently, the nickel-rich ejecta penetrate through the He shell quickly and thus later can avoid being drastically decelerated by the reverse

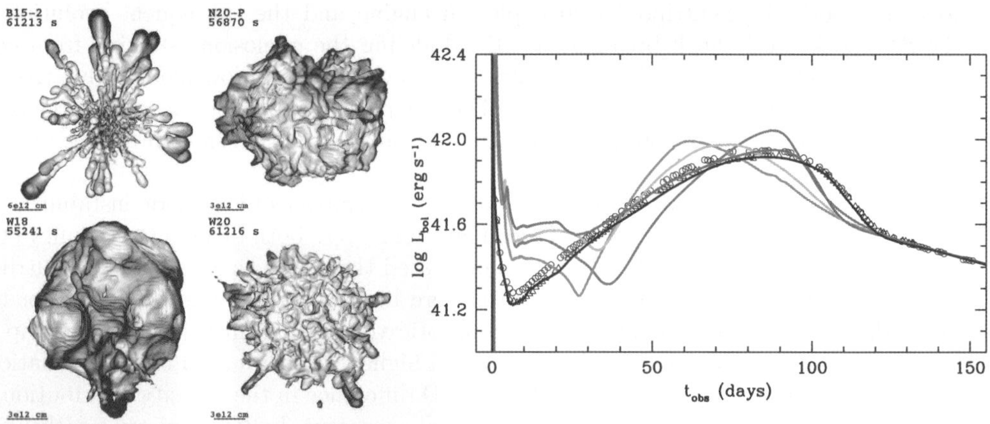

Figure 1. Morphological structures of nickel-rich ejecta in 3D simulations of Wongwathanarat *et al.* (2015) and Utrobin *et al.* (2015) and their corresponding bolometric light curves calculated by 1D radiation Lagrangian hydrodynamic code CRAB. The model B15-2 (upper left) yields a light curve with a wide dome shape (blue line) which is in good agreement with the observed data from SN 1987A during day $\sim 40-100$. The black solid line corresponds to the bolometric light curve of an optimal model calculated with a non-evolutionary progenitor star by Utrobin (2005).

shock, partially because of a mild acceraltion of the SN forward shock. As a result the morphology of the nickel-rich ejecta resulting from this model looks unique, consisting of many elongated RT fingers. In summary, the ejecta asymmetries at shock breakout can depend very sensitively on the density structure of the progenitor star.

3. 1D light curve modeling based on 3D explosion models: the case of SN 1987A

All previous studies on light curve comparison with observations of SN 1987A have relied on models of CCSN explosions that are initiated in 1D. To reproduce the smooth rise and the wide dome-shaped light curve of SN 1987A during days $\sim 10-120$ outward mixing of radioactive nickel and inward mixing of hydrogen have to be introduced artificially (e.g., Utrobin 2005). Recently, Utrobin *et al.* (2015) made the first attempt to calculate bolometric light curves in 1D based on angular-averaged 3D explosion models which are evolved from shortly after core bounce until shock breakout. This light curve modeling approach eliminates the need to introduce artificial mixing within the SN ejecta because mixing is already captured in 3D hydrodynamic calculations of the SN explosions.

Utrobin *et al.* (2015) consider 3D explosion models of four different blue supergiant progenitors: two of which were calculated by Wongwathanarat *et al.* (2015). The computed bolometric light curves are compared to observational data of SN 1987A. They found that one of their models, model B15-2, yields a light curve with a broad dome shape which is in good agreement with observational data during day $\sim 40-120$. In all other models, the resulting light curves are either peak-like or a dome-shaped not wide enough. The failure to produce a wide enough dome-like light curve in these models means too short diffusion time of photons from the radioactive nickel decay chain powering the light curve, which results from insufficient outward mixing of nickel into the hydrogen envelope. The model B15-2 is based on the $15\,M_{\odot}$ blue supergiant by Woosley *et al.* (1988) whose density structure allows for strong RT mixing near the C+O/He interface (Wongwathanarat *et al.* 2015). Disappointingly, all models considered by Utrobin *et al.* (2015) do not produce a good match with the initial light curve peak observed during the first ∼10 days of the SN explosion. This is caused by a too large radius of the progenitor star used for the calculations. Note that all of the considered progenitors are single stars, while SN 1987A may have been the result of a binary star.

Nevertheless, the results obtained by Utrobin *et al.* (2015) demonstrated an intriguing possibility to probe the density structure inside the progenitor star by light curve modeling.

4. Conclusions and Outlook

During the past decade considerable progress has been made in modeling CCSN explosions on a long time scale in three dimensions. 3D models which follow the time evolution of CCSNe from shortly after core bounce until shock breakout have become feasible (e.g., Hammer *et al.* 2010; Wongwathanarat *et al.* 2015). They revealed that morphology of the ejecta, especially the ejecta that are rich in heavy elements synthesized during the explosions, can depend sensitively on the density structure of the progenitor star. These 3D models already allow for a direct link between CCSN models and observations, e.g., through the 1D bolometric light curve modeling (Utrobin *et al.* 2015). Evolving 3D explosion models until decades or centuries after the explosion is currently a work in progress (Gabler *et al.*, these proceedings).

One of the major uncertainties in modeling CCSNe is the lack of a 3D model of the core of a massive star before collapse. So far, 3D simulations of CCSNe have relied solely on 1D models computed by stellar evolution codes. Asymmetries in convective shell burning during last minutes before core collapse can influence the explosion dynamics and thus may also affect the explosion asymmetries (e.g., Müller *et al.* 2016). 3D modeling of the core of massive stars is still in its infancy stage. There have been only two such calculations by Couch *et al.* (2015) and Müller *et al.* (2016) for a $15\,M_\odot$ and a $18\,M_\odot$ progenitor. More progenitor models are urgently needed.

References

Arnett, W. D. & Meakin, C. 2011, *ApJ*, 733, 78
Blinnikov, S., Lundqvist, P., Bartunov, O., Nomoto, K., & Iwamoto, K. 2000, *ApJ*, 532, 1132
Blondin, J. M. & Mezzacappa, A. 2006, *ApJ*, 642, 401
Blondin, J. M., Mezzacappa, A., & DeMarino, C. 2003, *ApJ*, 584, 971
Chevalier, R. A. 1976, *ApJ*, 207, 872
Couch, S. M., Chatzopoulos, E., Arnett, W. D., & Timmes, F. X. 2015, *ApJ* (Letters), 808, L21
Dotani, T., Hayashida, K., Inoue, H., Itoh, M., & Koyama, K. 1987, *Nature*, 330, 230
Ellinger, C. I., Young, P. A., Fryer, C. L., & Rockefeller, G. 2012, *ApJ*, 755, 160
Foglizzo, T. 2002, *A&A*, 392, 353
Haas, M. R., Erickson, E. F., Lord, S. D., *et al.* 1990, *ApJ*, 360, 257
Hammer, N. J., Janka, H.-T., & Müller, E. 2010, *ApJ*, 714, 1371
Herant, M. & Benz, W. 1991, *ApJ* (Letters), 370, L81
Herant, M. & Benz, W. 1992, *ApJ*, 387, 294
Herant, M., Benz, W., Hix, W. R., Fryer, C. L., & Colgate, S. A. 1994, *ApJ*, 435, 339
Hungerford, A. L., Fryer, C. L., & Rockefeller, G. 2005, *ApJ*, 635, 487
Hungerford, A. L., Fryer, C. L., & Warren, M. S. 2003, *ApJ*, 594, 390
Joggerst, C. C., Almgren, A., & Woosley, S. E. 2010, *ApJ*, 723, 353
Kane, J., Arnett, D., Remington, B. A., *et al.* 2000, *ApJ*, 528, 989
Kifonidis, K., Plewa, T., Janka, H., & Müller, E. 2003, *A&A*, 408, 621

Kifonidis, K., Plewa, T., Scheck, L., Janka, H.-T., & Müller, E. 2006, *A&A*, 453, 661
Limongi, M., Straniero, O., & Chieffi, A. 2000, *ApJS*, 129, 625
Mao, J., Ono, M., & Nagataki, S., et al. 2015, *ApJ*, 808, 164
Matz, S. M., Share, G. H., Leising, M. D., Chupp, E. L., & Vestrand, W. T. 1988, *Nature*, 331, 416
Muller, E., Hillebrandt, W., Orio, M., *et al.* 1989, *A&A*, 220, 167
Müller, B., Viallet, M., Heger, A., & Janka, H.-T. 2016, *ApJ*, 833, 124
Nagasawa, M., Nakamura, T., & Miyama, S. M. 1988, *PASJ*, 40, 691
Nagataki, S. 2000, *ApJS*, 127, 141
Nagataki, S., Shimizu, T. M., & Sato, K. 1998, *ApJ*, 495, 413
Shigeyama, T. & Nomoto, K. 1990, *ApJ*, 360, 242
Sunyaev, R., Kaniovsky, A., Efremov, V., *et al.* 1987, *Nature*, 330, 227
Utrobin, V. P. 2004, *Astron. Lett.*, 30, 293
Utrobin, V. P. 2005, *Astron. Lett.*, 31, 806
Utrobin, V. P., Wongwathanarat, A., Janka, H.-T., & Müller, E. 2015, *A&A*, 581, A40
Wongwathanarat, A., Müller, E., & Janka, H.-T. 2015, *A&A*, 577, A48
Woosley, S. E. 1988, *ApJ*, 330, 218
Woosley, S. E., Pinto, P. A., & Ensman, L. 1988, *ApJ*, 324, 466
Woosley, S. E. & Weaver, T. A. 1995, *ApJS*, 101, 181

Supernova 1987A:30 years later - Cosmic Rays and Nuclei from Supernovae and their aftermaths
Proceedings IAU Symposium No. 331, 2017
A. Marcowith, M. Renaud, G. Dubner, A. Ray & A. Bykov, eds.

doi:10.1017/S1743921317004586

Core-Collapse Supernova Simulations including Neutrino Interactions from the Virial EOS

Evan O'Connor[1,2], C. J. Horowitz[3], Zidu Lin[3] and Sean Couch[4,5,6,7]

[1]Department of Astronomy and Oskar Klein Centre, Stockholm University, AlbaNova, SE-106 91 Stockholm, Sweden email: evan.oconnor@astro.su.se

[2]NCSU, Department of Physics, Campus Code 8202, Raleigh, NC, 27695, USA

[3]Center for Exploration of Energy and Matter and Department of Physics, Indiana University, Bloomington, IN 47408, USA

[4]Department of Physics and Astronomy, MSU, East Lansing, MI 48824, USA

[5]Department of Computational Mathematics, Science, and Engineering, Michigan State University, East Lansing, MI 48824, USA

[6]National Superconducting Cyclotron Laboratory, MSU, East Lansing, MI 48824, USA

[7]Joint Institute for Nuclear Astrophysics, MSU, East Lansing, MI 48824, USA

Abstract. Core-collapse supernova explosions are driven by a central engine that converts a small fraction of the gravitational binding energy released during core collapse to outgoing kinetic energy. The suspected mode for this energy conversion is the neutrino mechanism, where a fraction of the neutrinos emitted from the newly formed protoneutron star are absorbed by and heat the matter behind the supernova shock. Accurate neutrino-matter interaction terms are crucial for simulating these explosions. In this proceedings for IAUS 331, *SN 1987A, 30 years later*, we explore several corrections to the neutrino-nucleon scattering opacity and demonstrate the effect on the dynamics of the core-collapse supernova central engine via two dimensional neutrino-radiation-hydrodynamics simulations. Our results reveal that the explosion properties are sensitive to corrections to the neutral-current scattering cross section at the 10-20% level, but only for densities at or above $\sim 10^{12}\ \mathrm{g\,cm^{-3}}$.

Keywords. (stars:) supernovae: general, stars: neutron, methods: numerical, radiative transfer, neutrinos, hydrodynamics, scattering

---

1. Introduction

The observation of neutrinos from SN 1987A confirmed the basic nature of core collapse and highlighted the important role neutrinos play. Since then, 30 years ago, the community has made tremendous progress in elucidating this process. While neutrinos are an extreme *sink* of energy, and part of the reason that the supernova shock initially stalls, they are also thought to be crucial in reviving the supernova shock via the *neutrino mechanism*. The neutrino mechanism relies primarily on charged-current heating by electron neutrinos and antineutrinos on the neutrons and protons located behind the shock. This heating gives rise to the so-called gain region, where the heating by the background neutrino field being emitted from deeper regions is larger than the neutrino cooling from the matter. The net result is an increasing internal energy and, if sufficient, the development of an explosion. The amount of neutrino heating depends sensitively on the spectra of the background neutrino radiation fields, both the overall normalization (the luminosity) and its shape (mean energy, root mean squared energy, etc.).

One of the major difficulties in modeling core-collapse supernovae is the treatment of the neutrino radiation fields. Soon after core bounce, in the core of the protoneutron star, the neutrinos are completely trapped. The mean free path is much shorter than the typical length scales and these neutrinos are in equilibrium with the surrounding matter. However, as one moves away from the core, the densities drop rapidly, the mean free path increases, and the neutrinos decouple from the matter. By $\sim$100 km, most neutrinos are essentially free streaming and only a small percentage ($\sim 5 - 10\%$) will have interactions before exiting the star. It is critical to model the transition away from equilibrium since it is this transition which sets the spectrum of the radiation field responsible for the neutrino heating further out. To model this transition precisely one needs spectral neutrino transport and accurate neutrino opacities.

In this proceedings for IAU Symposium 331, "SN 1987A, 30 years later", we present a series of 2D core-collapse supernova simulations with spectral neutrino transport where we investigate in detail the effect of varying the neutrino opacities on the core-collapse supernova explosion mechanism. In particular, we use `FLASH` (Fryxell *et al.* 2000; Couch and O'Connor 2014; O'Connor and Couch 2015) which is outfitted with an energy-dependent M1 neutrino transport scheme (see below) and an effective general relativistic treatment of gravity. We explore recent corrections to the neutral-current scattering cross sections proposed by Horowitz *et al.* (2017). We do this generally for four progenitor models, and in detail for one model in particular. Our main conclusion is that the core-collapse supernova explosion mechanism is sensitive to the details of the neutral current cross section at densities of $\sim 10^{12} - 10^{13}\,\mathrm{g\,cm^{-3}}$. This is currently the limit of model-independent calculations of the neutrino interactions, therefore care and caution must be used when developing models for higher densities.

2. Methods

Our `FLASH` simulations follow the methods of O'Connor & Couch (2017; *in prep*), a significantly updated version of O'Connor and Couch (2015). Our velocity-dependent neutrino transport scheme is based on the M1 formalism, where we evolve the first two angular moments of the neutrino distribution function, the energy density and the momentum density. We evolve these moments as a function of neutrino species (3 species in total, ν_e, $\bar{\nu}_e$, and $\nu_x = \{\nu_\mu, \bar{\nu}_\mu, \nu_\tau, \bar{\nu}_\tau\}$) and neutrino energy (our simulations use 12 energy groups for each species). We fully take into account gravitational red shifting. We use the LS220 EOS (Lattimer and Swesty 1991). Our neutrino opacities come from `NuLib`, an open-source neutrino interaction library (O'Connor 2015). The base rates are based on Bruenn (1985) and include corrections for weak magnetism and nuclear recoil (Horowitz 2002). The specific corrections we explore here are based on Horowitz *et al.* (2017). In Horowitz *et al.* (2017), we use the Virial equation of state (EOS) to derive model-independent expressions for the axial response function (S_A) of the neutral-current, neutrino-nucleon scattering cross section, which is given as

$$\frac{1}{V}\frac{d\sigma}{d\Omega} = \frac{G_F^2 E_\nu^2}{16\pi^2}\left[g_a^2(3-\cos(\theta))(n_n+n_p)S_A + (1+\cos(\theta))n_n S_V\right]. \tag{2.1}$$

The overall effect of the corrections, due to many-body effects, is to lower the axial response function, S_A, and therefore lower the total neutral-current scattering cross section. Since the Virial EOS is only valid at low densities, we transition to a model-dependent expression for S_A at high densities, which is based on the random phase approximation (RPA) work of Burrows and Sawyer (1998). We empirically fit our results with the following expression for use in our simulations.

$$S_A^f(n,T,Y_e) = \frac{1}{1 + A(1 + Be^{-C})}, \tag{2.2}$$

where

$$A(n,T,Y_e) = A_0 \frac{n(1 - Y_e + Y_e^2)}{T^{1.22}}\ ;\ B(T) = \frac{B_0}{T^{0.75}}\ ;\ C(n,T,Y_e) = C_0 \frac{nY_e(1 - Y_e)}{T^{0.5}} + D_0 \frac{n^4}{T^6},$$

where n is the baryon density in units of nucleons/fm^3, T is the matter temperature in MeV, and Y_e is the electron fraction. We take as constants, $A_0 = 920$, $B_0 = 3.05$, $C_0 = 6140$, and $D_0 = 1.5 \times 10^{13}$. We note that setting $B_0 = 0$ removes the Virial EOS contributions to the neutrino response and effectively reduces our expressions for the axial response function to the values from Burrows and Sawyer (1998). As noted in Horowitz *et al.* (2017), at low densities, where the Virial EOS is valid, the Virial contributions decrease S_A up to a factor of 2 more than the contributions from Burrows and Sawyer (1998). We refer the reader to Figure 2 of Horowitz *et al.* (2017) for a graphical display of S_A and S_A^f for different ρ, T, and Y_e values and for various neutrino interaction assumptions.

The goal of these proceedings is to explore the effect of these Virial contributions, as well as the many-body effects in general, on the core-collapse supernova dynamics. To this end, we perform a large number of simulations, broken into two parts. First, we simulate core-collapse in four progenitor models from Woosley and Heger (2007): s12WH07, s15WH07, s20WH07, and s25WH07. For each progenitor we run a simulation with and without the corrections from Horowitz *et al.* (2017). Second, for model s20WH07 we simulate an additional 15 simulations, split into three sets: without corrections, with only high density corrections (i.e. $B_0 = 0$), and with full corrections. In an attempt to remove the effect of stochastic motions on our results and interpretations, for each of the three sets we do five simulations, each one starting with different random perturbations (at the 0.1% level) to the density field. We apply these perturbations everywhere at the start of the 2D simulation. Each simulation is started at 15 ms after bounce. We take the evolution up to 15 ms from a 1D `GR1D` simulation (O'Connor 2015).

3. Results

3.1. *Progenitor Dependence*

We simulate core-collapse in four progenitors, with and without the neutral-current scattering cross section corrections discussed above. We show the evolution of the shock radius in Figure 1. Dark colours represent the simulations without the corrections, while the lighter colours are the corresponding simulations with corrections. In all cases the shock radius propagates out to ~150 km, stalls, and recedes. For models s15WH07, s20WH07, and s25WH07, the reduced neutral-current scattering cross sections lead to earlier explosions. Explosions times are reduced by ~100 ms −150 ms. For model s12WH07, the corrections lead to a late-time explosion where there was not one observed in the model without corrections. The impact of these corrections is more modest than determined by Burrows *et al.* (2016), where the inclusion of these corrections dramatically decreased the explosion times in several models.

3.2. *s20WH07*

Two dimensional simulations of core-collapse supernovae are sensitive to stochastic variations in the matter motions during the accretion phase. This can lead to differences in

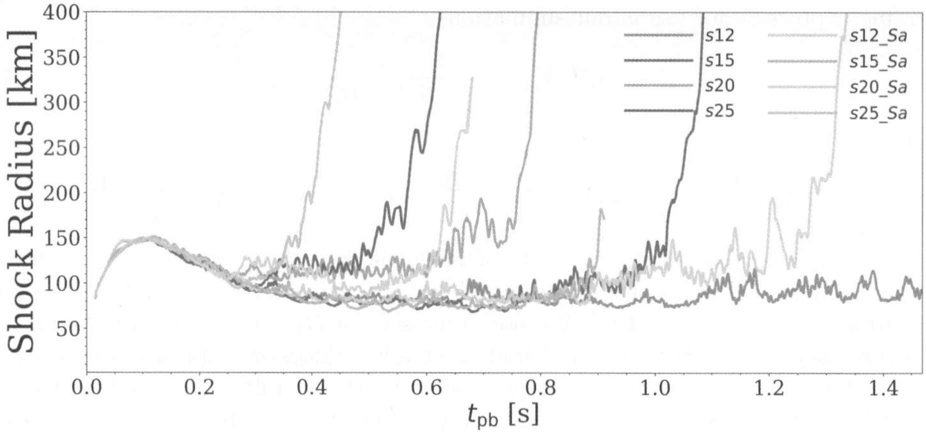

Figure 1. Shock radius evolution for four different progenitors, with and without the explored corrections from Horowitz *et al.* (2017). The corrections lead to earlier explosions.

explosion times which could potentially cloud interpretations. To overcome this, and to probe the effect of the corrections discussed in §2 more deeply, we perform many simulations with the s20WH07 model. As mentioned above, we run 15 simulations in total and explore three unique setups. Each simulation is started with a different set of random perturbations on the density field ($\rho \to \rho \times (1+r)$; where r is a different random number for each zone between -0.001 and 0.001). The first set of five simulations ignores any corrections to the neutral-current scattering cross section from the axial response, i.e. $S_A = 1$ in Eq. 2.1. The second set of simulations uses S_A as shown in Eq. 2.2 but sets $B_0 = 0$ and the third set used Eq. 2.2 directly. We show the evolution of the shock radius in these 15 simulations in Figure 2. We colour-code each set, but do not distinguish between the five runs of each set for clarity. We see the same results as the previous section. The reduced value of S_A due to many-body effects leads to earlier explosions. We define the explosion time to be when the shock crosses 400 km. With this definition, the average explosion times, and the variances are determined to be:

$$t_{\rm exp}^{S_A=1} = (723 \pm 63)\,{\rm ms}; \qquad t_{\rm exp}^{S_A^f,B_0=0} = (597 \pm 37)\,{\rm ms}; \qquad t_{\rm exp}^{S_A^f} = (583 \pm 42)\,{\rm ms}.$$

The key observation we make here is that the Virial EOS contributions to the axial response, which lower S_A at lower densities, do not have a significant effect on the explosion dynamics. Statistically the explosion times are indistinguishable from those that only include the reduction due to the effective RPA corrections. We explore this in more detail by examining the neutrino quantities.

In Figure 3, we show graphs of various neutrino quantities over the first 600 ms of evolution for the three different sets of simulations. For each set we average the quantities for the five different simulations. This significantly reduces the scatter and allows us to see the effect of each neutrino opacity change in detail. In the upper left panel we show the luminosity for each of the three neutrino species and for each set of simulations. The largest effect we see is the increase in the ν_x luminosity, by ~10%, when the many-body corrections to the axial response function are considered. The lower total scattering rates that result from the lower S_A allow the ν_x neutrinos to escape more easily and results in more cooling. Consequently, this leads to higher ν_x average energies (upper right panel) and faster contracting protoneutron star radii (bottom right panel). The faster contraction has the side effect of increasing the electron type neutrino and anti-neutrino luminosities and average energies which leads to increased neutrino heating (bottom left

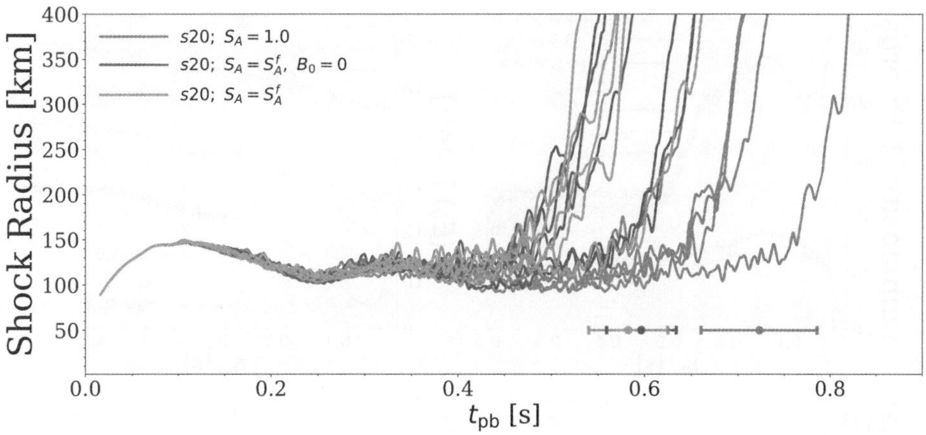

Figure 2. Shock radius evolution for 15 simulations of s20WH07 using three different neutrino-nucleon neutral current scattering cross sections based on Eq. 2.1. Five use $S_A = 1$ (blue), five take $S_A = S_A^f; B_0 = 0$ (red), the remaining five use $S_A = S_A^f$ (green). We include the average explosion time and the variance within the five simulations.

panel). This is the same effect observed in Melson *et al.* (2015) when they reduced the neutral-current scattering by considering relatively large strange-quark contributions.

Here we notice that the Virial contributions do indeed lead to more cooling, higher energies, and faster contraction compared to the $B_0 = 0$ simulation, as expected. However, the additional effect is relatively small compared to the impact the Virial contributions have on S_A (upwards of a factor of 2 more important than the RPA corrections at low densities). We can resolve this apparent conflict by considering the properties of the region where the ν_x's are emitted from and scatter through and comparing those to the region where the Virial contributions dominate. Unlike ν_e and $\bar{\nu}_e$, ν_x do not undergo charged-current interactions. The dominant emission process is via pair production. Due to the temperature dependence, the luminosity build up occurs deeper (and at higher densities, $\rho \sim 10^{12} - 10^{13}\,\mathrm{g\,cm^{-3}}$, and temperatures $T \sim 10\,\mathrm{MeV}$) in the protoneutron star. At these densities and temperatures the corrections to S_A in Eq. 2.2, which are of order ~20%, are either dominated by the effective RPA (at the highest densities) or of similar value between the Virial and effective RPA. These two reasons explain why the Virial EOS contributions to the axial response do not significantly affect the core-collapse supernova dynamics.

4. Conclusions

In this proceedings for IAUS 331, we present a detailed look at the effect of corrections to the neutral-current neutrino-nucleon scattering cross sections on the core-collapse supernova explosion dynamics. In particular, we look at recent model-independent corrections to the axial response (S_A) of low-density nuclear matter described by the Virial EOS. We argue here that corrections to S_A at the level of ~20% can influence the dynamics, however, at the relevant densities ($10^{12} - 10^{13}\,\mathrm{g\,cm^{-3}}$) and temperatures (~10 MeV) our model-independent calculation begins to break down and we must resort to model-dependent, RPA calculations to estimate S_A. Due to the important role these corrections may play in the evolution of the central engine of core-collapse supernovae, we advocate for a better understanding of the neutrino response at these densities.

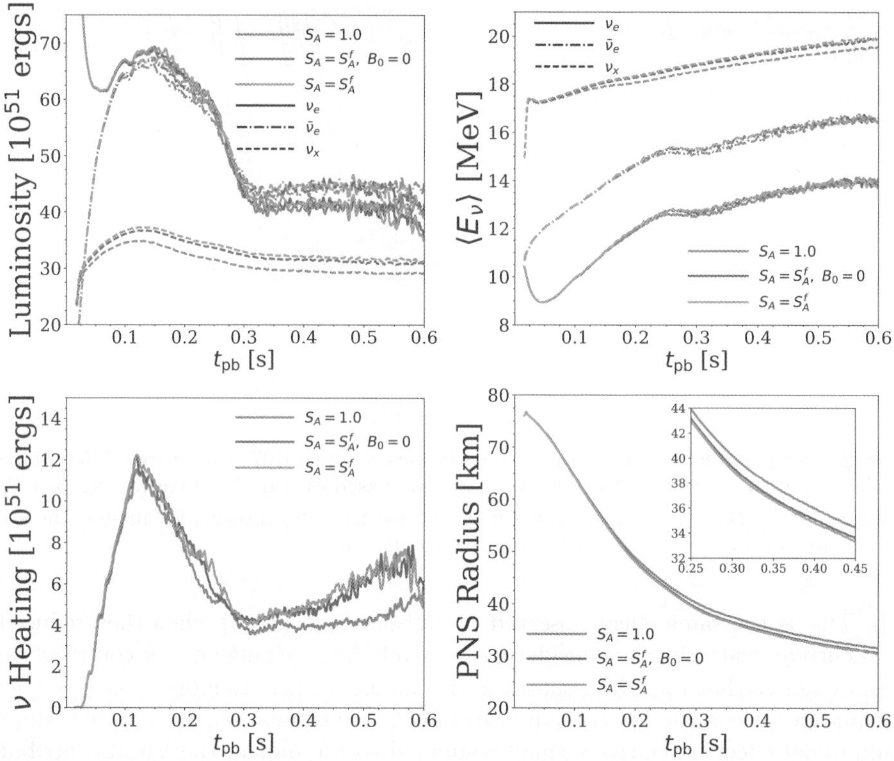

Figure 3. Neutrino related quantities for model s20WH07. Here we show the neutrino luminosity (top left) and average energy (top right) for each neutrino species. We also show the net neutrino heating (bottom left) and the protoneutron star radius (bottom right; defined as the radius of $\rho = 10^{11}\,\mathrm{g\,cm^{-3}}$). For each of the three configurations studied, we average the quantities from the five simulations in order to reduce scatter and better ascertain the overall effect.

Acknowledgements: We thank O.L. Caballero and A. Schwenk for collaborative work on the Virial EOS. Support was provided by NASA through Hubble Fellowship grant #51344.001-A awarded by the STScI, which is operated by the AURA, for NASA, under contract NAS 5-26555. Computations were performed on the Zwicky cluster at Caltech, supported by the Sherman Fairchild Foundation and by NSF award PHY-0960291, and on XSEDE which is supported by National Science Foundation grant number ACI-1548562.

References

Bruenn, S. W. (1985). *ApJS*, 58:771–841.
Burrows, A. & Sawyer, R. F. (1998). *PRC*, 58:554.
Burrows, A., Vartanyan, D., Dolence *et al.* (2016). *ArXiv e-prints: 1611.05859.*
Couch, S. M. & O'Connor, E. P. (2014). *ApJ*, 785:123.
Fryxell, B., Olson, K., Ricker, P., *et al.* (2000). *ApJS*, 131:273.
Horowitz, C. J. (2002). *PRD*, 65(4):043001.
Horowitz, C.J., Caballero, O.L., Lin, Z., *et al.* (2017). *PRC*, 95(2):025801.
Melson, T., Janka, H.-T., Bollig, R., Hanke, F., Marek, A., & Müller, B. (2015). *ApJL*, 808:L42.
O'Connor, E. (2015). *ApJS*, 219:24.
O'Connor, E. & Couch, S. M. (2015). *submitted to ApJ; arXiv:1511.07443.*
Lattimer, J. M. & Swesty, F. D. (1991). *Nuc. Phys.*, A535:331.
Woosley, S. E. & Heger, A. (2007). *Phys. Rep.*, 442:269.

Supernova 1987A:30 years later - Cosmic Rays and Nuclei from Supernovae and their aftermaths
Proceedings IAU Symposium No. 331, 2017
A. Marcowith, M. Renaud, G. Dubner, A. Ray & A. Bykov, eds.

doi:10.1017/S1743921317004471

Incidence of stellar rotation on the explosion mechanism of massive stars

Rémi Kazeroni,[1,2] Jérôme Guilet[1] and Thierry Foglizzo[2]

[1]Max-Planck-Institut für Astrophysik, Karl-Schwarzschild-Str. 1, D-85748 Garching, Germany
email: kazeroni@MPA-Garching.MPG.DE[2]

[2]Laboratoire AIM, CEA/DRF-CNRS-Université Paris Diderot, IRFU/Département d'Astrophysique, CEA-Saclay, F-91191, France

Abstract. Hydrodynamical instabilities may either spin-up or down the pulsar formed in the collapse of a rotating massive star. Using numerical simulations of an idealized setup, we investigate the impact of progenitor rotation on the shock dynamics. The amplitude of the spiral mode of the Standing Accretion Shock Instability (SASI) increases with rotation only if the shock to the neutron star radii ratio is large enough. At large rotation rates, a corotation instability, also known as low-T/W, develops and leads to a more vigorous spiral mode. We estimate the range of stellar rotation rates for which pulsars are spun up or down by SASI. In the presence of a corotation instability, the spin-down efficiency is less than 30%. Given observational data, these results suggest that rapid progenitor rotation might not play a significant hydrodynamical role in the majority of core-collapse supernovae.

Keywords. hydrodynamics, instabilities, shock waves, stars: neutron, stars: rotation, supernovae: general

1. Introduction

A core-collapse supernova represents the death of a massive star and the birth of a neutron star. The initial spin period of the neutron star provides some insight on the explosion mechanism as well as the rotation of the progenitor. Neutron stars are expected to rotate slowly at birth with spin periods covering a range from tens to hundreds milliseconds (Faucher-Giguère & Kaspi 2006; Popov & Turolla 2012). This is at odds with 1D stellar evolution calculations which predict a maximum period of $10-15$ ms (Heger *et al.* 2005). However such calculations seem to overestimate the angular momentum contained in the cores of low-mass red giants (Cantiello *et al.* 2014).

Asymmetric explosions. The favored framework to explain the explosion of moderately rotating stellar cores relies on the delayed neutrino-driven mechanism proposed by Bethe & Wilson (1985). A spherical accretion shock stalls at a radius of about 150 km for a few hundreds milliseconds after the formation of a proto-neutron star (PNS) and is eventually revived thanks to neutrino absorption. Numerical simulations have shown that transverse motions induced by hydrodynamical instabilities play a decisive role in the explosions of progenitors above ~ 10 $M_{\odot}$. Two types of instabilities are able to break the spherical symmetry of the collapse: the neutrino driven convection (Herant *et al.* 1994; Foglizzo *et al.* 2006) and the Standing Accretion Shock Instability (SASI) (Blondin *et al.* 2003). Asymmetric explosions are supported by a number of observational evidence including a large pulsar kick that may be imparted by a global deformation $l = 1$ of the shock wave due to SASI (Scheck *et al.* 2004; 2006; Wongwathanarat *et al.* 2010).

Impact of rotation. Even if the first 3D explosions from ab initio models have been obtained in the last years, explosions are still slightly under-energetic and not yet routinely achieved (Janka *et al.* 2016). Moderate rotation may be an important ingredient to foster explosions. It reduces the critical neutrino luminosity required to revive the stalled shock (Iwakami *et al.* 2014). A SASI spiral mode is more easily destabilized by rotation (Blondin & Mezzacappa 2007; Yamasaki & Foglizzo 2008). In a non-rotating progenitor, such a spiral wave could spin the neutron star up to periods compatible with the observations (Blondin & Mezzacappa 2007; Foglizzo *et al.* 2012; Guilet & Fernàndez 2014). On the contrary, a SASI spiral mode may spin a pulsar down when the core is rotating (Blondin & Mezzacappa 2007). Such a scenario might lead to a counter-rotating neutron star depending on the efficiency of the angular momentum redistribution. When the rotation rate is high enough, a more vigorous one-armed spiral mode related to a corotation instability may arise and help the shock revival (Takiwaki *et al.* 2016). In this work, a simplified setup is used to investigate the role of progenitor rotation, regarding its impact on the shock wave dynamics and on the initial neutron star spin period.

2. A simplified model

Physical model. The stalled shock phase is studied by the mean of an idealized model which is dedicated to SASI in its simplest form. The numerical domain is restricted to the equatorial plane of a massive star described in cylindrical geometry, which enables non-axisymmetric motions in 2D. The flow, modeled by an idealized gas with an adiabatic index $\gamma = 4/3$, is decelerated by a shock wave of initial radius $r_{\mathrm{sh}0}$ and an incident Mach number $\mathcal{M}_1 = 5$. The matter accretes subsonically onto the surface of the PNS at a radius r_*. No heating function is employed in order to suppress the neutrino-driven convection. A cooling function is used to imitate the neutrino emission due to electron capture in the region surrounding the PNS with the approximation $\mathcal{L} \propto \rho P^{3/2}$ (Blondin & Mezzacappa 2006) where ρ and P respectively stand for the density and the pressure.

Rotation rates. Once the flow has relaxed on the numerical grid, a constant specific angular momentum j is injected through the outer boundary. The numerically relaxed 1D rotating flow is then perturbed by two counter-rotating density perturbations to trigger two spiral modes $m = \pm 1$ where m is the azimuthal wavenumber (see Kazeroni *et al.* 2017). The range of rotation rates considered in this study covers three orders of magnitude: $10^{13}\,\mathrm{cm}^2\,\mathrm{s}^{-1} \leqslant j \leqslant 10^{16}\,\mathrm{cm}^2\,\mathrm{s}^{-1}$. If the angular momentum were conserved down to the surface of a pulsar of 10 km radius, $j = 10^{16}\,\mathrm{cm}^2\,\mathrm{s}^{-1}$ would correspond to a spin period of 0.6 ms. The rotation rates considered are not high enough for the centrifugal force to greatly impact the dynamics. The stationary shock radius increases by less than 10% due to the centrifugal force (Fig. 1 & 2). Note that higher rotation rates are often considered in global simulations (e.g. Nakamura *et al.* 2014; Kuroda *et al.* 2014; Takiwaki *et al.* 2016). The angular momentum of the accreted material is poorly known since stellar evolution calculations predict values of $j \sim 10^{14} - 10^{15}\,\mathrm{cm}^2\,\mathrm{s}^{-1}$ if the magnetic field is taken into account and $j \sim 10^{16} - 10^{17}\,\mathrm{cm}^2\,\mathrm{s}^{-1}$ otherwise (Heger *et al.* 2005).

Non-rotating cases. Using non-rotating simulations of this model, Kazeroni *et al.* (2016) showed that a spiral mode prevails non-linearly only if the radii ratio $R \equiv r_{\mathrm{sh}0}/r_*$ is such that $R \gtrsim 2$. If this condition is fulfilled, a spiral mode emerges after three to five oscillations induced by a SASI sloshing mode. In the opposite case, the non-linear dynamics is dominated by a sloshing mode, irrespective of the shape of the initial perturbation.

Parametric simulations. A total of 140 simulations are used to cover the parameter space and the following radii ratio are used: $R = \{1.67,\ 2,\ 2.5,\ 3,\ 4,\ 5\}$. Periodic boundary

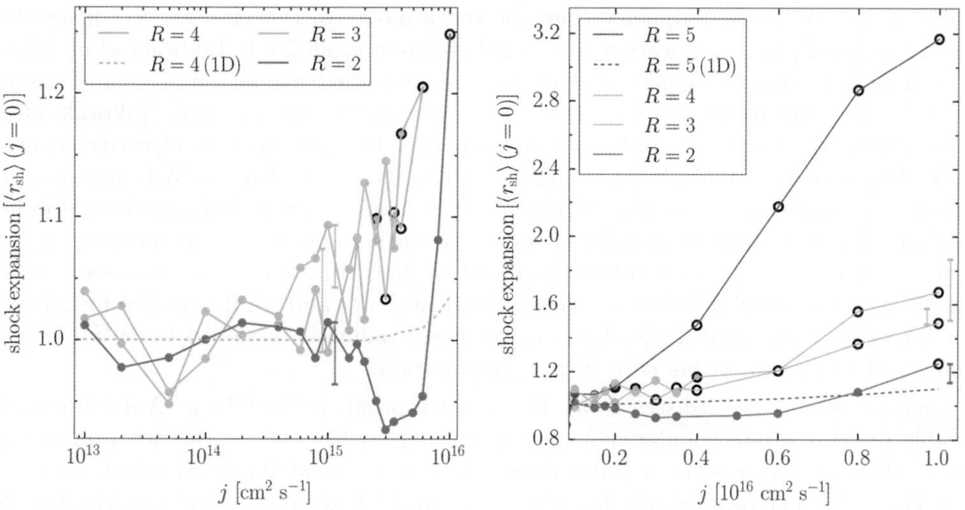

Figure 1. Influence of rotation on the non-linear regime of SASI (left panel) and the corotation instability (right panel) for different values of R. The solid lines represent the mean shock radius as a function of j in the 2D cases whereas the dashed lines show the effect of the centrifugal force in 1D simulations. The values are averaged in the non-linear regime and normalized by the ones measured in the non-rotating cases. The black circled points denote that a corotation emerges in the linear regime (empty circles) or in the non-linear regime (full circles). To reduce the variability of the results, the non-rotating simulations are repeated five times with slightly different initial perturbation amplitudes. The simulations where $j = 10^{15}$ cm^2 s^{-1} (left) and $j = 10^{16}$ cm^2 s^{-1} (left) are also repeated five times to assess the stochasticity of our results. The vertical bars indicate the minimum and maximum values obtained for each value of R.

conditions are employed in the azimuthal direction, a reflexive condition at the inner boundary and a constant inflow at the outer boundary which is placed far enough to prevent any interaction with the shock wave. The numerical resolution is 1600 cells in the azimuthal direction and from 1600 to 2730 in the radial one.

3. Influence of rotation

SASI dominated regime. The simulations of the linear regime reproduce the growth rates computed by Yamasaki & Foglizzo (2008) with a perturbative analysis with discrepancies of less than 10% (Kazeroni *et al.* 2017). The linear increase of the growth rates with rotation does not seem to depend on the geometry and the dimensionality (Blondin *et al.* 2017). In the non-linear regime of the instability, the shock wave expands with rotation only if the radii ratio is such that $R \gtrsim 2$. In that case, the shock radius is increased by 10% to 20% compared to situations where rotation is absent (Fig. 1, left panel). If $R < 2$, the averaged shock radius diminishes by 10% in the range of rotation rates in which spiral modes dominate the non-linear regime. This surprising result may be a consequence of the influence of rotation on the competition between the fundamental mode and higher harmonics. Note that this radii ratio threshold is similar to the critical ratio below which a sloshing mode dominates systematically the non-linear dynamics in non-rotating simulations (Kazeroni *et al.* 2016). The case-to-case stochasticity of the results may be as large as the effect of rotation on the shock dynamics (vertical bars in Fig. 1).

Overlap with a corotation instability. In cases with a large enough rotation rate, a corotation radius can emerge above the PNS surface. It delimits the region in which the flow has a higher rotational frequency than the spiral mode. In this case another instability known as low-T/W † can overlap with SASI (Kuroda *et al.* 2014; Takiwaki *et al.* 2016). This additional instability can produce a strong acoustic wave in the post-shock region with an open one-armed spiral pattern and may ease the shock revival (Takiwaki *et al.* 2016). While the mechanism of SASI has clearly been interpreted as an advective-acoustic cycle (Foglizzo *et al.* 2007; Guilet & Foglizzo 2012), the origin of low-T/W is uncertain. In the case of a differentially rotating neutron star, it has been suggested that the instability could be due to an energy transfer occurring across a band around a corotation radius where acoustic waves are amplified (Watts *et al.* 2005; Passamonti & Andersson 2015). In our study, an overlap between a corotation instability and SASI could be responsible for a more robust spiral mode. Indeed, the shock radius is increased by 50% to 200% compared to a non-rotating case (Fig. 1, right panel).

Angular momentum redistribution. The rotation profile induced by a SASI spiral mode is able to redistribute angular momentum and impact the spin of the neutron star. In an idealized scenario, some angular momentum is stored in the SASI spiral wave and expelled in the explosion while the opposite quantity is accreted onto the surface of the neutron star. If the progenitor is not rotating, this mechanism can induce a spin-up of the neutron star (Blondin & Mezzacappa 2007; Fernàndez 2010) which may reach periods of a few tens milliseconds (Guilet & Fernàndez 2014). This process has been reproduced experimentally in a shallow-water analogue of SASI where the dynamics of a hydraulic jumps is similar to the one of a shock wave in the equatorial plane of a massive star (Foglizzo *et al.* 2012; 2015). Concerning rotating cases, (Blondin & Mezzacappa 2007) proposed that a SASI spiral mode could reduce the neutron star spin compared to an estimate based on angular momentum conservation during the collapse. If enough angular momentum is redistributed, this process could lead to a counter-rotating neutron star compared to its progenitor.

Neutron star spin. The efficiency of the spin-up and the spin-down generated by spiral modes can be assessed by comparing the idealized angular momentum redistribution to a spherically-symmetric collapse leading to a rotational frequency $f_{\rm CORE}$ (Fig. 2). If the radii ratio is such that $R \lesssim 2$, the saturation amplitude of the spiral is too low to drive a significant redistribution and the neutron spin is barely affected by SASI. For higher radii ratios, rotation impacts the initial neutron star spin throughout the range of rotation rates considered. If the rotation is low, i.e. $f_{\rm CORE} \lesssim 10 - 30$ Hz, the angular momentum redistribution proceeds very similarly to the non-rotating case. The neutron star periods obtained are close to the ones derived from non-rotating cases and the direction of rotation of the spiral mode is stochastic, so is the one of the neutron star. The amount of angular momentum required to balance the quantity redistributed by SASI is similar to the amount stored in a spiral wave of a non-rotating case. If $f_{\rm CORE} \lesssim 100$ Hz, the neutron star spin is mostly set by SASI which can induces a significant spin-down. Finally, if $f_{\rm CORE} \gtrsim 100$ Hz, a corotation instability can develop. The overlap between the two instabilities do not lead to a strong spin-down effect, the efficiency of the process is less than 30% (Kazeroni *et al.* 2017). Observational data indicate that $f_{\rm CORE} = 100$ Hz may represent an upper limit to initial neutron star spins. Our results suggest that a corotation instability may not be compatible with such a constraint, at least when magnetic fields are neglected.

† where T and W respectively stand for the rotational kinetic energy and the gravitational potential energy.

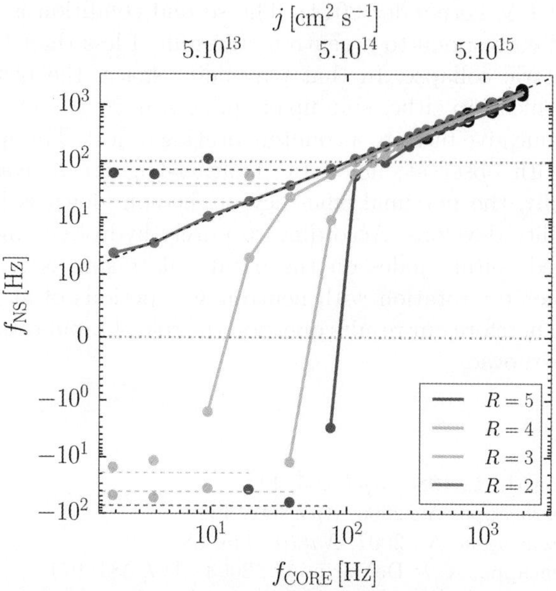

Figure 2. Relationship between the rotational frequency inferred from a spherically-symmetric collapse ($f_{\rm CORE}$) and the one which accounts for the development of spiral modes ($f_{\rm NS}$). The black dashed line illustrates the absence of angular momentum redistribution ($f_{\rm NS} = f_{\rm CORE}$). Neutron stars above and below this line are respectively spun up and down by SASI. The upper gray dot-dashed line indicates the maximal rotation frequency which seems compatible with the observations ($f \lesssim 100\,$Hz). Counter-rotating neutron stars correspond to the cases below the lower grey dot-dashed line. The horizontal dashed lines refer to the rotational frequencies obtained in simulations without rotation. The angular momentum redistribution at low rotation rates (isolated points) is very similar to the non-rotating cases.

4. Conclusion

A parametric study based on numerical simulations of an idealized model has been conducted to investigate the interplay between SASI and stellar rotation. Our approach enables to isolate the development of hydrodynamical instabilities from the complexity of physical effects at play in ab initio simulations of core-collapse supernovae and to perform a wide coverage of the parameter space.

Non-linear dynamics. Rotation is able to boost a SASI spiral mode only if the shock to the PNS radii ratio is large enough. This threshold is similar to the one above which a spiral mode prevails in a non-rotating progenitor (Kazeroni *et al.* 2016). The non-monotonic influence of the rotation rate on the shock radius as well as the stochasticity of the dynamics suggest that a single self-consistent 3D simulation including rotation may not fully capture the role of rotation on the shock dynamics. Above a specific angular momentum of $10^{15} - 10^{16}$ cm^2 s^{-1}, the emergence of a corotation above the PNS surface can dramatically impact the dynamics. The development of a corotation instability can lead to an increase of the shock radius of $50 - 200\%$ compared to a non-rotating case whereas SASI alone only induces an increase of $10 - 20\%$. The corotation instability seems to have a similar influence in the more realistic model of Takiwaki *et al.* (2016). However the mechanism of the instability remains unclear.

Pulsar spin. SASI has the potential to significantly impact the initial neutron star spin period if two conditions are satisfied (Kazeroni *et al.* 2017). The shock to the PNS radii ratio should be large enough, i.e. $R \geqslant 3$, to lead to a high saturation amplitude of the spiral mode. This quantity sets the amount of angular momentum redistributed by a spiral mode (Guilet & Fernàndez 2014). The second condition is that the progenitor rotation rate should correspond to a neutron star spin of less than 100 Hz inferred from a spherically-symmetric collapse. In that parameter space, the birth period is largely impacted by SASI which can either spin up or spin down the nascent neutron star and in the low rotation regime give birth to a counter-rotating object. The spin periods obtained are in agreement with observational data (Faucher-Giguère & Kaspi 2006; Popov & Turolla 2012). Finally, the maximal efficiency of the spin-down is less than 30% when a corotation instability develops. According to purely hydrodynamical simulations, the impact of one armed spiral modes on the natal pulsar spin is not strong enough to reconcile rapid progenitor rotation with neutron star periods of at least 10 ms derived from observations. Therefore our results question the role of rapid rotation in the majority of core-collapse supernovae.

References

Bethe, H. A. & Wilson, J. R., 1985, *ApJ*, 295, 14
Blondin, J. M. & Mezzacappa, A., 2006, *ApJ*, 642, 401
Blondin, J. M. & Mezzacappa, A., 2007, *Nature*, 445, 58
Blondin, J. M., Mezzacappa, A. & DeMarino C., 2003, *ApJ*, 584, 971
Blondin, J. M., Gipson, E., Harris, S. & Mezzacappa, A. 2017, *ApJ*, 835, 170
Cantiello, M., Mankovich, C., Bildsten, L. Christensen-Dalsgaard, J., & Paxton, B., 2014, *ApJ*, 788, 93
Faucher-Giguère, C.-A. & Kaspi, V. M., 2006, *ApJ*, 643, 332
Fernández, R., 2010, *ApJ*, 725, 1563
Foglizzo, T., Scheck, L. & Janka, H.-T., 2006, *ApJ*, 652, 1436
Foglizzo, T., Galletti, P., Scheck, L. & Janka, H.-T., 2007, *ApJ*, 654, 1006
Foglizzo, T., Masset, F., Guilet, J., & Durand, G., 2012, *Phys. Rev. Lett.*, 108, 051103
Foglizzo, T., *et al.*, 2015, *PASA*, 32, 9
Guilet, J. & Fernández, R., 2014, *MNRAS*, 441, 2782
Guilet, J. & Foglizzo, T., 2012, *MNRAS*, 421, 546
Heger, A., Woosley, S. E., & Spruit, H. C., 2005, *ApJ*, 626, 350
Herant, M., Benz, W., Hix W. R., Fryer C. L., & Colgate, S. A., 1994, *ApJ*, 435, 339
Iwakami, W., Nagakura, H. & Yamada, S., 2014, *ApJ*, 793, 5
Janka, H.-T., Melson, T. & Summa, A., 2016, *Annual Review of Nuclear and Particle Science*, 66, 341
Kazeroni, R., Guilet, J. & Foglizzo, T., 2016, *MNRAS*, 456, 126
Kazeroni, R., Guilet, J. & Foglizzo, T., 2017, *arXiv*, 1701.07029
Kuroda, T., Takiwaki, T. & Kotake, K., 2014, *Phys. Rev. D*, 89, 044011
Nakamura, K., Kuroda, T., Takiwaki, T. & Kotake, K., 2014, *ApJ*, 793, 45
Passamonti, A. & Andersson, N., 2015, *MNRAS*, 446, 555
Popov, S. B. & Turolla, R., 2012, *Ap&SS*, 341, 457
Scheck, L., Plewa, T., Janka H.-T., Kifonidis, K. & Müller, E., 2004, *Phys. Rev. Lett.*, 92, 011103
Scheck, L., Kifonidis, K., Janka H.-T., & Müller, E., 2006, *A&A*, 457, 963
Takiwaki, T., Kotake, K. & Suwa, Y., 2016, *MNRAS*, 461, L112
Watts, A. L., Andersson, N. & Jones, D. I., 2005, *ApJL*, 618, L37
Wongwathanarat, A., Janka, H.-T., & Müller, E., 2010, *ApJL*, 725, L106
Yamasaki, T. & Foglizzo, T., 2008, *ApJ*, 679, 607

Supernova 1987A:30 years later - Cosmic Rays and Nuclei from Supernovae and their aftermaths
Proceedings IAU Symposium No. 331, 2017
A. Marcowith, M. Renaud, G. Dubner, A. Ray & A. Bykov, eds.

doi:10.1017/S1743921317004732

How to form a millisecond magnetar? Magnetic field amplification in protoneutron stars

Jérôme Guilet[1,2,3], Ewald Müller[1], Hans-Thomas Janka[1], Tomasz Rembiasz[4], Martin Obergaulinger[4], Pablo Cerdá-Durán[4] and Miguel-Angel Aloy[4]

[1]Max-Planck-Institut für Astrophysik, Karl-Schwarzschild-Str. 1, D-85748 Garching, Germany

[2]Max-Planck Princeton Center for Plasma Physics

[3]Laboratoire AIM, CEA/DRF-CNRS-Université Paris Diderot, IRFU/Département d'Astrophysique, CEA-Saclay, F-91191, France

[4]Departamento de Astronoma y Astrofsica, Universidad de Valencia C/ Dr. Moliner 50, 46100 Burjassot, Spain

Abstract. Extremely strong magnetic fields of the order of 10^{15} G are required to explain the properties of magnetars, the most magnetic neutron stars. Such a strong magnetic field is expected to play an important role for the dynamics of core-collapse supernovae, and in the presence of rapid rotation may power superluminous supernovae and hypernovae associated to long gamma-ray bursts. The origin of these strong magnetic fields remains, however, obscure and most likely requires an amplification over many orders of magnitude in the protoneutron star. One of the most promising agents is the magnetorotational instability (MRI), which can in principle amplify exponentially fast a weak initial magnetic field to a dynamically relevant strength. We describe our current understanding of the MRI in protoneutron stars and show recent results on its dependence on physical conditions specific to protoneutron stars such as neutrino radiation, strong buoyancy effects and large magnetic Prandtl number.

Keywords. instabilities, magnetic fields, MHD, stars: neutron, stars: magnetic fields, stars: rotation, supernovae: general

1. Introduction

The delayed injection of energy due to the spin down of a fast rotating, highly magnetized neutron star is the most popular model to explain a class of superluminous supernovae (e.g. Kasen & Bildsten 2010; Inserra *et al.* 2013). The birth of such millisecond magnetars is furthermore a potential central engine for long gamma-ray bursts and hypernovae (e.g. Duncan & Thompson 1992; Metzger *et al.* 2011; Obergaulinger & Aloy 2017). One of the most fundamental and open question in these models is the origin of the strong magnetic field that is invoked. The most studied mechanism capable of generating such a strong magnetic field is the growth of the magnetorotational instability (MRI) in the protoneutron star (Akiyama *et al.* 2003; Masada *et al.* 2007; Obergaulinger *et al.* 2009; Sawai & Yamada 2014; Guilet *et al.* 2015; Guilet & Müller 2015; Rembiasz *et al.* 2016a,b; Mösta *et al.* 2016). In the following we review several aspects of the physics of the MRI in the physical conditions relevant to protoneutron stars, by describing its linear growth (Section 2) and its non-linear dynamics (Section 3).

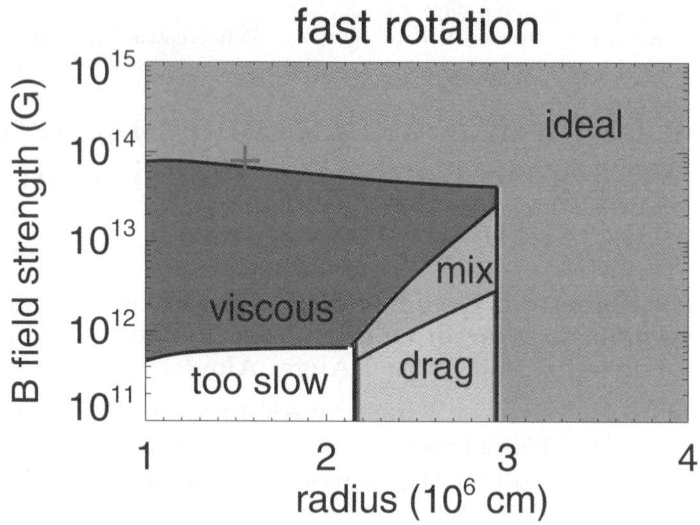

Figure 1. Different regimes of MRI growth as a function of radius and magnetic field strength in a fast rotating PNS model at $t = 170\,\mathrm{ms}$ post-bounce (from Guilet *et al.* 2015). See the text for a description of the different regimes. The parameter range used in the simulations by Obergaulinger *et al.* (2009) is shown in red. The regime labelled as "too slow" corresponds to growth rates lower than $10\,s^{-1}$.

2. Linear growth of the MRI

Although it is too often neglected in numerical simulations, neutrino radiation can have a dramatic impact on the linear growth of the MRI. This was shown by Guilet *et al.* (2015) by applying analytical results for the linear growth of the MRI to a numerical model of a protoneutron star (PNS). In Guilet *et al.* (2015), we have shown that, depending on the physical conditions, the MRI growth can occur in three main different regimes (Fig. 1):

• Viscous regime (dark blue color in Fig. 1) : on length-scales longer than the neutrino mean free path, neutrino viscosity significantly affects the growth of the MRI if $E_\nu \equiv \frac{v_A^2}{\nu\Omega} < 1$, where v_A is the Alfvén velocity, ν the viscosity induced by neutrinos and Ω the rotation angular frequency. The growth of the MRI is then slower and takes place at longer wavelengths compared to the ideal regime. In the viscous regime, the wavelength of the most unstable mode is independent of the magnetic field strength, while the growth rate is proportional to the magnetic field strength (Fig. 2). As a result, a minimum magnetic field strength of $\sim 10^{12}$ G is required for the MRI to grow on sufficiently short time-scales.

• Drag regime (light blue color in Fig. 1) : on length-scales shorter than the neutrino mean free path, neutrino radiation exerts a drag force $-\Gamma \mathbf{v}$, where $\mathbf{v}$ is the fluid velocity perturbation and $\Gamma \simeq 6 \times 10^3\,(T/10\,\mathrm{MeV})^6\,\mathrm{s}^{-1}$ is a damping rate. This drag has a significant impact on the MRI if the damping rate is larger than the rotation angular frequency ($\Gamma > \Omega$). In this regime, the growth rate of the most unstable mode is independent of the magnetic field strength, but is reduced by a factor of Γ/Ω compared to the ideal regime (Fig. 2). The wavelength of the most unstable mode is not much affected by the neutrino drag.

• Ideal regime (orange color in Fig. 1) : this is the classical MRI regime which occurs when neutrino viscosity or drag are negligible. The growth rate of the MRI is then a fraction of the angular frequency, independent of the magnetic field strength, while the most unstable wavelength is proportional to the magnetic field strength.

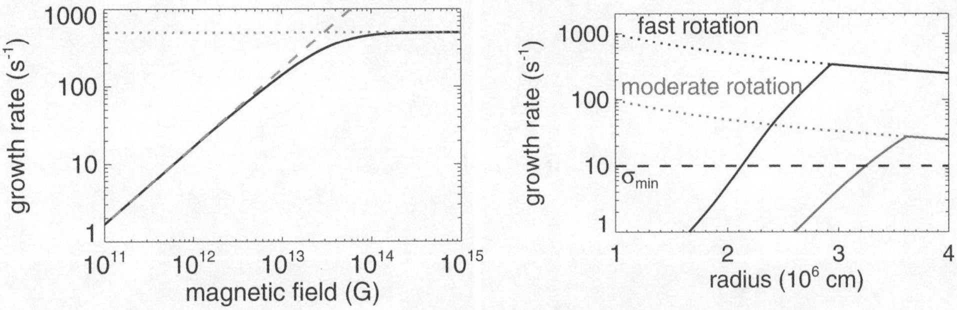

Figure 2. Maximum growth rate of the MRI in the viscous regime (left panel) and in the drag regime (right panel). *Left panel:* the solid line shows the MRI growth rate taking into account neutrino viscosity, while the dotted line shows the ideal MHD limit. The dashed line shows equation (15) of Guilet *et al.* (2015), which is valid in the very viscous regime ($E_\nu \ll 1$). *Right panel:* Solid lines show the MRI growth rate including the impact of neutrino drag, while dotted lines show the growth rate in the ideal MHD limit. Black and red colors represent, respectively, a profile of fast rotation and a ten times slower rotation (see Guilet *et al.* 2015 for details). Finally, the dashed line shows the (somewhat arbitrary) minimum growth rate below which the growth is labelled as "too slow" in Fig. 1.

Fig. 1 shows where these three regimes apply, as a function of magnetic field strength and radius in the PNS (note that qualitatively similar results were obtained by Guilet *et al.* (2017) in the context of neutron star mergers). Three regions in the PNS can be distinguished:

- Deep inside the PNS, the relevant MRI regime is the viscous MRI. The MRI can grow on sufficiently short time-scales if the initial magnetic field is above a critical strength.
- At intermediate radii, MRI growth can take place both in the viscous regime at wavelengths longer than the neutrino mean free path, and in the drag regime at length-scales shorter than the mean free path. Since the growth rate in the viscous regime is proportional to the magnetic field strength, the growth is faster in the viscous regime above a critical magnetic field strength while it is faster in the drag regime for weaker magnetic fields. In addition, in-between these regimes, the MRI growth occurs in a mixed regime (shown in green in Fig. 1) where electron neutrinos are diffusing and thus induce a viscosity, while the other species are free streaming and exert a drag.
- Near the PNS surface, the viscous regime is irrelevant because the neutrino mean free path is longer than the wavelength of the MRI. Furthermore, in this region the neutrino drag does not affect much the growth of the MRI because the damping rate is smaller than the angular frequency, i.e. $\Gamma < \Omega$. As a consequence the MRI growth takes place in the ideal regime without much impact of neutrino radiation.

For the sake of simplicity and in order to show the impact of neutrinos in a clear way, this analysis neglected the effect of buoyancy. As shown analytically by Menou *et al.* (2004) and Masada *et al.* (2007) and confirmed by numerical simulations in Guilet & Müller (2015), entropy and composition gradients can act against the MRI, but this is alleviated by the diffusion due to neutrinos.

3. Non-linear dynamics of the MRI

In order to know the efficiency of magnetic field amplification, numerical simulations of the non-linear dynamics driven by the MRI are necessary. These simulations have so far

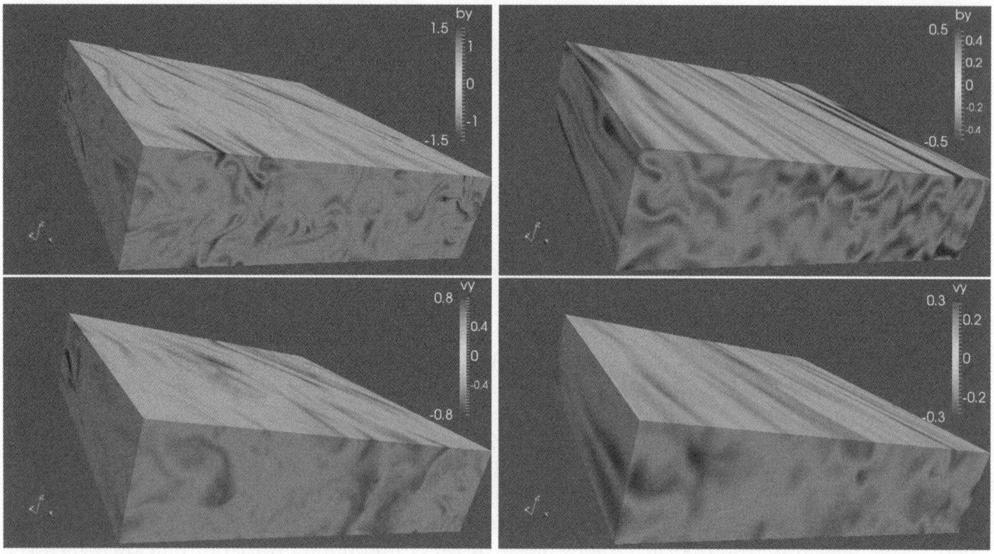

Figure 3. Snapshots of two simulations of a local model of the MRI in the viscous regime using the Boussinesq approximation (from Guilet & Müller 2015). The left column corresponds to a buoyantly unstable flow with $N^2/\Omega^2 = -1$ while the right column corresponds to a buoyantly stable flow with $N^2/\Omega^2 = 10$. The two rows show the azimuthal magnetic field (top) and azimuthal velocity (bottom). Note that we adjusted the color scales to the level of turbulence, i.e. they are different for the two columns.

mostly been performed in local or semi-global models describing a small portion of the PNS (e.g. Obergaulinger *et al.* 2009; Masada *et al.* 2012; Guilet *et al.* 2015; Rembiasz *et al.* 2016a,b). The first phase of MRI growth is often dominated by channel flows, which are the fastest growing MRI modes in the presence of a poloidal magnetic field. These modes, which are uniform in the horizontal direction, are approximate non-linear solutions and are therefore potentially able to grow well into the non-linear regime. Their growth is terminated when parasitic instabilities (either the Kelvin-Helmholtz instability or resistive tearing modes) are able to destroy their structures (Rembiasz *et al.* 2016a,b). By studying in detail this process with two different numerical codes, Rembiasz *et al.* (2016b) have been able to show that channel modes can only amplify the magnetic field by a factor $\sim$10 under the physical conditions prevailing in a PNS. A further process such as a turbulent MRI-driven dynamo is therefore necessary to reach very strong magnetic fields.

In Guilet & Müller (2015), we have studied the turbulent phase following the disruption of the channel modes, taking for the first time into account both the viscosity and diffusion due to neutrinos and the buoyancy force due to entropy/composition gradients. This was made possible owing to the use of the Boussinesq approximation, which we showed to be well suited to a local model of a PNS. In addition to drastically reducing the computing time this approximation avoids artifacts at the boundaries caused by global gradients in fully compressible simulations. Fig. 3 shows snapshots of the turbulent phase in a buoyantly unstable case (left panel) and a buoyantly stable case (right panel). While the buoyantly unstable dynamics is turbulent and largely non-axisymmetric, the buoyantly stable dynamics contains more large-scale axisymmetric structures. A channel flow at the largest scale available in the box can be distinguished even during the turbulent phase, which is further illustrated in Fig. 4 by the horizontally averaged space-time diagram (left

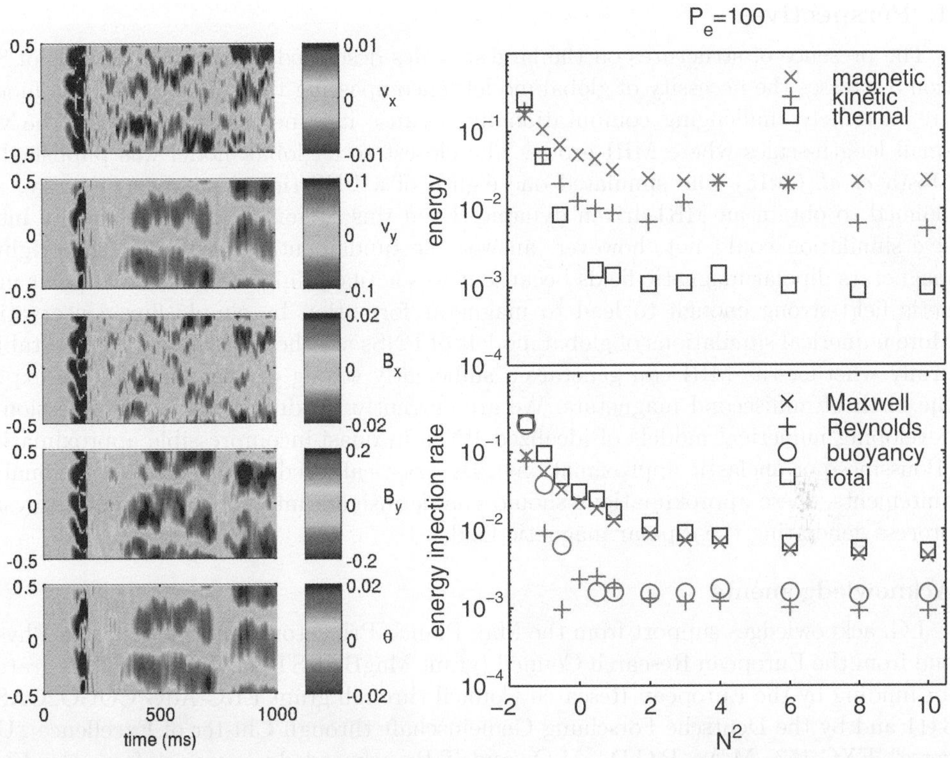

Figure 4. *Left panel:* Quasi-stationary channel flow for $N^2/\Omega^2 = 10$ and high thermal diffusion Pe $= 100$. The five panels show space time diagrams (from top to bottom) of the horizontally averaged radial velocity, azimuthal velocity, radial magnetic field, azimuthal magnetic field, and buoyancy variable. *Right panel:* Time and volume averages of the turbulent energy densities (in units of 2×10^{29} erg cm^{-3}), and energy density injection rates (in units of 2×10^{32} erg s^{-1} cm^{-3}) as a function of the dimensionless buoyancy parameter N^2/Ω^2. They show the magnitude of magnetic (red $\times$), kinetic (blue $+$), and thermal (black squares) energies in the top panel, and energy injection rates due to Maxwell stress ($\times$ symbols), Reynolds stress ($+$ signs), buoyancy force work (circles), and sum of the three (squares) in the bottom panel. Positive injection rates are shown in red color, while negative ones (i.e. energy is removed from turbulent motions) are shown in blue.

panel). Fig. 4 (right panel) furthermore shows the kinetic, magnetic and thermal energies and energy injection rates as a function of the dimensionless buoyancy parameter N^2/Ω^2 (where N is the Brunt-Väisälä frequency). The magnetic energy decreases with N^2/Ω but, interestingly, it becomes almost constant in the stable buoyancy regime ($N^2 > 0$) suggesting an efficient magnetic field amplification even in the presence of a strong, stable stratification.

While these simulations were able to use realistic values for the viscosity due to neutrinos, they have the shortcoming of vastly overestimating the resistivity (like all other simulations, be it by including it explicitly like here or by suffering from a large numerical resistivity (Rembiasz *et al.* 2016)). Preliminary results show that this is likely to cause a significant underestimate of the efficiency of magnetic field amplification, thus confirming the strong dependence of the MRI on the magnetic Prandtl number (the ratio of viscosity to resistivity) known in the context of accretion disks (e.g. Fromang *et al.* 2007).

4. Perspectives

The presence of structures on the largest scales described by the local models of Section 3 stresses the necessity of global models encompassing the whole PNS. Such models are extremely challenging computationally because it is necessary to resolve the very small length-scales where MRI grows. The closest to a global model was published by Mösta *et al.* (2015) who simulated one eighth of a PNS (in an octant symmetry) and claimed to obtain an MRI-driven dynamo. Even this extremely computationally intensive simulation could not, however, answer the fundamental question of the origin of magnetars dipolar magnetic fields because it was actually initialized with a dipolar magnetic field strong enough to lead to magnetar formation by simple flux conservation. More numerical simulations of global models of PNSs are therefore necessary to establish firmly whether the MRI can generate a sufficiently strong dipolar magnetic to explain the birth of millisecond magnetars. We are currently making steps in this direction by developing numerical models of idealized PNSs in quasi-incompressible approximations (Boussinesq or anelastic approximations). By drastically reducing the computational requirements, these approximations should enable insights into the fundamental physical process generating the dipolar magnetic field.

Acknowledgements

J.G. acknowledges support from the Max-Planck-Princeton Center for Plasma Physics and from the European Research Council (grant MagBURST–715368). H.T.J. is grateful for funding by the European Research Council through grant ERC-AdG COCO2CASA-3411 and by the Deutsche Forschung Gemeinschaft through Cluster of Excellence "Universe" EXC-153. M.A., P.C.D., M.O. and T.R. acknowledge support from the European Research Council (grant CAMAP-259276) and from grants AYA2013-9340979-P, AYA2015-66899-C2-1-P and PROME-TEOII/2014-069.

References

Akiyama S., Wheeler J. C., Meier D. L., & Lichtenstadt I., 2003, *ApJ*, 584, 954
Duncan R. C. & Thompson C., 1992, *ApJL*, 392, L9
Foglizzo, T., Kazeroni, R., Guilet, J., Masset, F., González, M. *et al.*, 2015, *PASA*, 32, 9
Fromang S., Papaloizou J., Lesur G., & Heinemann T., 2007, *A&A*, 476, 1123
Goodman, J., & Xu, G., 1994, *ApJ*, 432, 213
Guilet, J., Müller, E. & Janka, H. T., 2015, *MNRAS*, 447, 3992
Guilet, J. & Müller, E., 2015, *MNRAS*, 450, 2153
Guilet, J., Bauswein, A., Just, O., & Janka, H. T., 2017, submitted to *MNRAS*, arxiv:1610.08532
Inserra C., Smartt S. J., Jerkstrand A., *et al.* 2013, *ApJ*, 770, 128
Kasen D. & Bildsten L., 2010, *ApJ*, 717, 245
Masada Y., Sano T., & Shibata K., 2007, *ApJ*, 655, 447
Masada Y., Takiwaki T., Kotake K., & Sano T., 2012, *ApJ*, 759, 110
Menou K., Balbus S. A., & Spruit H. C., 2004, *ApJ*, 607, 564
Metzger B. D., Giannios D., Thompson T. A. *et al.*., 2011, *MNRAS*, 413, 2031
Mösta P., Ott C. D., Radice D., Roberts L. F., Schnetter E., & Haas R., 2015, *Nature*, 528, 376
Obergaulinger M., Cerdá-Durán P., Müller E., & Aloy M. A., 2009, *A&A*, 498, 241
Obergaulinger M. & Aloy M. A., 2017, *MNRAS letters*, 469, L43
Rembiasz, T., Obergaulinger, M., Cerdá-Durán, P. *et al.*, 2016, *MNRAS*, 456, 3782
Rembiasz, T., Guilet, J., Obergaulinger, M., Cerdá-Durán, P. *et al.*, 2016, *MNRAS*, 460, 3316
Rembiasz, T., Obergaulinger, M., Cerdá-Durán *et al.*, 2016, arXiv:1611.05858
Sawai H. & Yamada S., 2014, *ApJL*, 784, L10

Supernova 1987A:30 years later - Cosmic Rays and Nuclei from Supernovae and their aftermaths
Proceedings IAU Symposium No. 331, 2017
A. Marcowith, M. Renaud, G. Dubner, A. Ray & A. Bykov, eds.

doi:10.1017/S1743921317004513

Magnetically assisted explosions of weakly magnetized stars

Hidetomo Sawai[1,2] and Shoichi Yamada[2]

[1]Research Organization for Information Science & Technology, Minatojima-Minamimachi 1-5-2, Chuo, Kobe, Hyogo, Japan
email: hsawai@rist.or.jp

[2]Scinence & Engineering, Waseda University, Okubo 3-4-1, Shinjuku, Tokyo, Japan

Abstract. We carried out high resolution simulations of weakly-magnetized core-collapse supernovae in two-dimensional axisymmetry in order to see the influence of the magnetic field and rotation on the explosion. We found that the magnetic field amplified by magnetorotational instability (MRI) has a great positive impact on the explosion by enhancing the neutrino heating, provided that the progenitor has large angular momentum close to the highest value found in stellar evolution calculations. We also found that even for progenitors neither involving strong magnetic flux nor large angular momentum, the magnetic field is greatly amplified by the convection aand rotation, and this leads to the boost of the explosion again by enhancing the neutrino heating.

Keywords. supernovae: general, MHD, instabilities, methods: numerical

---

1. Introduction

The explosion mechanism of core-collapse supernovae has not yet been fully understood. Although the neutrino heating is likely the most promising mechanism, state-of-the-art simulations led to less energetic explosions than what are observed in reality, and in some cases even resulted in no explosion (e.g., the article by E. O'Connor in this proceedings).

The impacts of magnetic field and rotation on core-collapse supernovae have been studied as possible agents to drive explosion other than neutrino heating. Previous studies showed that an energetic explosion occurs when a pre-collapse iron core involves a strong magnetic field, $B_{\rm pre} \sim 10^{12}$–10^{13} G, and rapid rotation, $\Omega_{\rm pre} \sim 1$ rad s^{-1} (e.g., Yamada & Sawai 2004).

The strengths of the magnetic fields of pre-collapse cores are highly uncertain. However, it is likely that the above value of $B_{\rm pre}$ is close to the extreme limit, since it corresponds to the magnetic flux comparable to those of magnetar candidates, the most highly magnetized stars ever known.

The rotation of $\Omega_{\rm pre} \sim 1$ rad s^{-1} may also be near the possible highest value. In stellar evolution calculations of Woosley & Heger(2006), such a high speed rotation is obtained for main sequence stars with the surface rotational velocity more than a few 100 km s^{-1}. Among the 216 O-type stars in 30 Dor analyzed by Ramírez-Agudelo *et al.* (2013) only the 25% of the sample have the projected surface velocity exceeding 200 km s^{-1}.

In this article, we will see the dynamics of "weakly" magnetized core-collapse explosions based on magnetohydrodynamic (MHD) simulations, where $B_{\rm pre}$ is one order of magnitude weaker than the above value. Numerical results are shown and discussed separately for fast-rotating progenitors (based on Sawai & Yamada 2014, 2016) and for slowly-rotating progenitors (based on a work we are currently undertaking).

2. Numerical Method

We carry out two-dimensional simulations to follow the dynamics of $15M_\odot$ star (Woosley & Weaver 1995) for the central part within the radius of 500, 1000, or 4000 km covered with the polar coordinates by using a time-explicit Eulerian MHD code, Yamazakura (Sawai *et al.* 2013a). There the following MHD equations are solved with the assumptions of axisymmetry (and equatorial symmetry for the fast rotation case):

$$\frac{\partial \rho}{\partial t} + \nabla \cdot (\rho \boldsymbol{v}) = 0, \tag{2.1}$$

$$\frac{\partial}{\partial t}(\rho \boldsymbol{v}) + \nabla \cdot \left(\rho \boldsymbol{v}\boldsymbol{v} - \frac{\boldsymbol{B}\boldsymbol{B}}{4\pi}\right) = -\nabla\left(p + \frac{B^2}{8\pi}\right) - \rho \nabla \Phi, \tag{2.2}$$

$$\frac{\partial}{\partial t}\left(e + \frac{\rho v^2}{2} + \frac{B^2}{8\pi}\right) + \nabla \cdot \left[\left(e + p + \frac{\rho v^2}{2} + \frac{B^2}{4\pi}\right)\boldsymbol{v} - \frac{(\boldsymbol{v}\cdot\boldsymbol{B})\boldsymbol{B}}{4\pi}\right] = -\rho(\nabla\Phi)\cdot\boldsymbol{v} + Q_E^{\rm abs} + Q_E^{\rm em}, \tag{2.3}$$

$$\frac{\partial \boldsymbol{B}}{\partial t} = \nabla \times (\boldsymbol{v} \times \boldsymbol{B}), \tag{2.4}$$

$$\frac{\partial n_e}{\partial t} + \nabla \cdot (n_e \boldsymbol{v}) = Q_N^{\rm abs} + Q_N^{\rm em}, \tag{2.5}$$

Here, $Q_E^{\rm abs}$ and $Q_E^{\rm em}$ are, respectively, the changes of energy density due to neutrino/anti-neutrino absorptions and emissions, which are given by simplified formulae of neutrino emission and absorption (Janka 2001). $Q_N^{\rm abs}$ and $Q_N^{\rm em}$ are the similar notations for the changes of electron number density. The electron fraction is required for the evaluations of these Q valuables, which is obtained by Eq. (2.5) after bounce, and by the prescription suggested by Liebendörfer(2005) before bounce. The gravitational potential, Φ, is obtained by the Newtonian mono-pole formula. A tabulated nuclear equation of state produced by Shen *et al.* 1998a,b is utilized.

We assume that the pre-collapse core is rotating with the initial angular velocity profile of

$$\Omega(r) = \Omega_{\rm pre}\frac{r_0^2}{r_0^2 + r^2}, \tag{2.6}$$

where r is the distance from the center of the core and $r_0 = 1000$ km. The angular velocity is set as $\Omega_{\rm pre} = 2.73$ and 0.1π rad s^{-1} for rapid and slow rotation cases, respectively.

The pre-collapse magnetic field is assumed to be a dipole-like configuration produced by electric currents of a 2D-Gaussian-like distribution (see Sawai & Yamada 2016 for the detailed formula). The strength of magnetic field at the center of the core is in the range of 5×10^{10}–2×10^{11} G.

We first run low resolution simulations for the numerical domain of $r < 4000$ km from the onset of the collapse to several 100 ms after bounce. The core is covered with

$N_r \times N_\theta = 720 \times 60$ (120 for full-sphere simulations) numerical grids, where the spatial resolution is 400 m near the center. Then, using the data obtained by the low resolution simulations as the initial and boundary conditions, we perform high resolution simulations starting from a few millisecond after bounce with smaller numerical domains $r_{\rm in} < r < r_{\rm out}$, where $(r_{\rm in}/{\rm km}, r_{\rm out}/{\rm km})$ are (50, 500) and (30,1000) for fast rotation simulations and slow rotation simulations, respectively. The number and the minimum size of numerical grids in the simulation of the highest numerical cost are, respectively, $N_r \times N_\theta = 9250 \times 6400$ and 12.5 m.

3. Fast Rotation Case

When the core of a progenitor star rotates rapidly with $\Omega_{\rm pre} \sim \pi$ rad s^{-1}, the rotation is the best agency to amplify the magnetic field. While winding of the magnetic field lines is one process of amplification, the so-called magnetorotational instability (MRI) is much more efficient (Balbus & Hawley 1991, Akiyama *et al.* 2003). Since the fastest-growing wavelength of the MRI is proportional to the magnetic field, a higher resolution is necessary for a weaker seed magnetic field. In the current case, the fastest-growing wavelength is a few 100 m near the surface of proto-neutron star. It is found that the magnetic field is exponentially amplified in some high resolution simulations. We infer this is due to the MRI by the fact that the growing time scale is comparable to one obtained by the linear analysis.

The impact of MRI is apparent by comparing low and high resolution simulations. As shown in Fig. 1, the high resolution simulations in which the MRI is resolved lead to a faster shock surface propagation compared to the low resolution simulations (and even a jet-like mass ejection in the high resolution model with the relatively strong magnetic field, $B_{\rm pre} = 2 \times 10^{11}$ G). We also found that the higher the resolution (equally the better the MRI resolved), the larger the explosion energy.

Interestingly, although the shock surface propagation is accelerated in the high resolution models shown here, magnetic pressure is not so large except for the vicinity of the rotation axis in the relatively strong magnetic field model (the upper right panel of Fig. 1), and it is even negligible in the most region in the weak magnetic field model (the lower right panel). This suggests that the explosion is not boosted by magnetic force.

We found that the boost of the explosion stems from an enhancement of neutrino heating due to indirect effects of the MRI. In the simulations in which the MRI is well resolved, the angular momentum is efficiently transferred outwards from the rotation axis as the growth of MRI proceeds. Then the region right behind the shock front gains more angular momentum and expands due to the extra centrifugal force. Since the neutrino heating surpasses the cooling more easily at a larger radius, the expansion leads to the enlargement of the gain region. Indeed, Fig. 2 shows that the gain region is larger in the higher resolution simulation. We also evaluated the ratio of the advection timescale, $\tau_{\rm a}$, during which matter traverses the gain region, to the heating timescale, $\tau_{\rm h}$, within which matter gains enough energy to overcome gravity. We found that the fraction increases more rapidly for a simulation of a better MRI resolution, indicating an enhancement of the neutrino heating.

4. Slow Rotation Case

When the rotation of the core is one order-of-magnitude slower than the former fast rotation case, the MRI is no longer an efficient amplification mechanism of the magnetic field, since its timescale, approximately equal to the rotation period, is now long compared

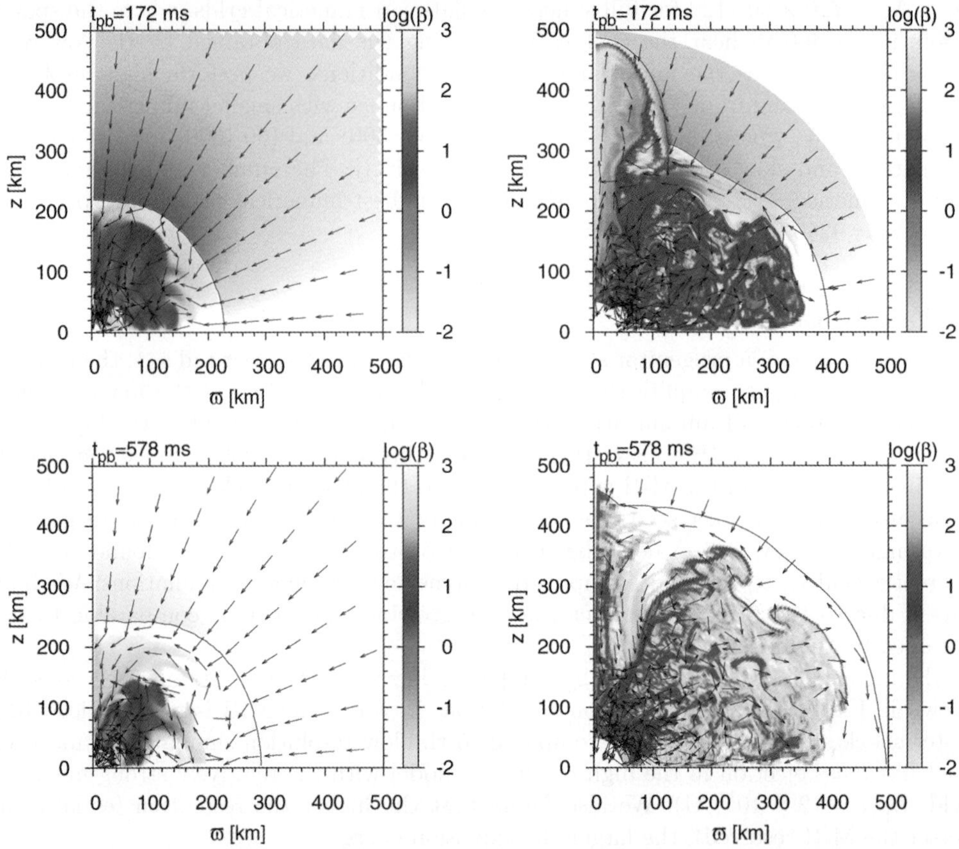

Figure 1. Color maps of the plasma beta with velocity vectors presented by arrows in the fast rotation case for models with $B_{\rm pre} = 2 \times 10^{11}$ G at $t_{\rm pb} = 172$ ms (upper panels) and $B_{\rm pre} = 5 \times 10^{10}$ G at $t_{\rm pb} = 578$ ms (lower panels). The left panels represent the results of low spatial resolution simulations with 720×60 grids while the right panels that of high spatial resolution simulations with 9250×6400 grids.

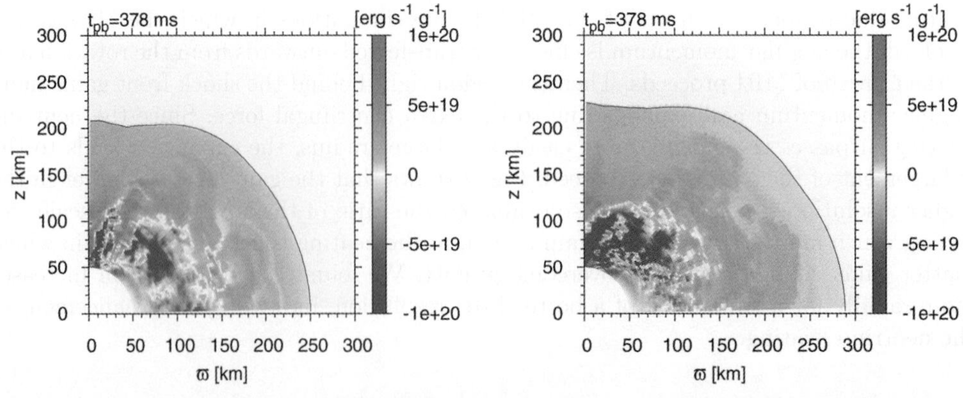

Figure 2. Color maps of the net heating rate per unit mass at 378 ms after bounce in the fast rotation case with $B_{\rm pre} = 5 \times 10^{10}$ G, for a low resolution simulation with 1160×800 grids (left) and for that of a high spatial resolution simulation with 9250×1600 grids (right). The volume of the (orange-colored) gain region is larger in the high resolution simulation.

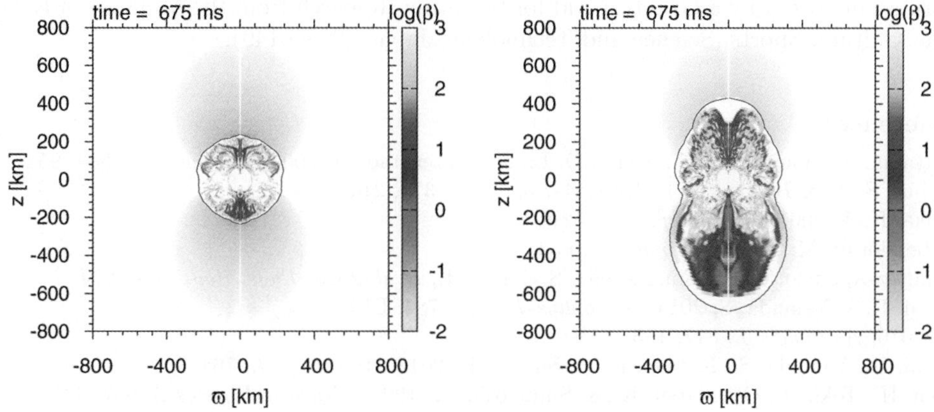

Figure 3. Color maps of the plasma beta in the slow rotation case with $B_{\rm pre} = 1 \times 10^{11}$ G for a low resolution simulation with 1754×1600 grids (left), and for that of a high spatial resolution simulation with 3504×3200 grids (right). The explosion is only found in the latter case.

to the dynamical timescale around the surface of the proto-neutron star. Nevertheless, we found an exponential growth of magnetic field in this case too for simulations with high spatial resolutions. The amplification starts right after the shock generation and its timescale is comparable to the dynamical timescale, suggesting that the amplification is due to the buoyant convection caused by the shock propagation.

We found that the amplified magnetic field boosts the explosion again by increasing the neutrino heating, but in a different way from one shown in the previous section. Although the rotation is one order-of-magnitude slower than the previous case, it still works to substantially amplify the toroidal magnetic field in a very limited layer close to the rotation axis, where the rotational period reaches down to milliseconds. The amplified toroidal magnetic field weakly pushes the matter outwards, the shock surface expands accordingly, and by the same reason as before the neutrino heating is enhanced.

The right panel of Fig. 3 shows such an explosion. Comparison with the left panel prove the essential role of the magnetic field on the dynamics. Without the great amplification of the magnetic field, the matter is not lifted to a larger radius, the neutrino heating is inefficient, and the shock surface stalls at a small radius (the left panel).

5. Conclusion

We have seen the influences of magnetic field and rotation on core-collapse supernovae in the context of relatively weak magnetic fields. We found that they indirectly boost the explosion by enhancing the neutrino heating in either case where the progenitor rotates rapidly or slowly. In both cases, the enhancement of the neutrino heating occurs due to the expansion of the shock surface. In the former case, the expansion is caused by the angular momentum transfer accompanied by the MRI, while in the latter case, it is caused by the magnetic pressure amplified by convection and rotational winding of the magnetic field lines near the rotation axis. These results found by two-dimensional axisymmetric MHD simulations must be confirmed in three dimensional simulations, which may be feasible in the era of exascale computing.

Acknowledgement

Numerical computations in this work were performed on Cray XC30 at the Center for Computational Astrophysics, the National Astronomical Observatory of Japan. This work is supported by a Grant-in-Aid for Scientific Research from the Ministry of Education, Culture, Sports, Science and Technology, Japan (26800149).

References

Akiyama, S., Wheeler, J. C., Meier, D. L., & Lichtenstadt, I. 2003, *Astrophys J.*, 584, 954
Balbus, S. A. & Hawley, J. F. 1991, *Astrophys J.*, 376, 214
Janka, H.-T. 2001, *Phys. Rep.*, 368, 527
Liebendörfer, M. 2005, *Astrophys J.*, 633, 1042
Ramírez-Agudelo, O. H., Simón-Díaz, S., Sana, H., *et al.* 2013, *Phys. Rep.*, 560, A29
Sawai, H. & Yamada, S. 2014, *Astrophys J. Lett.*, 784, L10
—. 2016, *Astrophys J.*, 817, 153
Sawai, H., Yamada, S., Kotake, K., & Suzuki, H. 2013, *Astrophys J.*, 764, 10
Shen, H., Toki, H., Oyamatsu, K., & Sumiyoshi, K. 1998a, *Nuclear Physics A*, 637, 435
—. 1998b, Progress of Theoretical Physics, 100, 1013
Woosley, S. E. & Heger, A. 2006, *Astrophys J.*, 637, 914
Woosley, S. E. & Weaver, T. A. 1995, *Astrophys J. Sup.*, 101, 181
Yamada, S. & Sawai, H. 2004, *Astrophys J.*, 608, 907

Supernova 1987A:30 years later - Cosmic Rays and Nuclei from Supernovae and their aftermaths
Proceedings IAU Symposium No. 331, 2017
A. Marcowith, M. Renaud, G. Dubner, A. Ray & A. Bykov, eds.

doi:10.1017/S1743921317004781

A minority view on the majority: A personal meeting summary on the explosion mechanism of supernovae

Noam Soker

Department of Physics, Technion, Haifa, Israel
email: soker@physics.technion.ac.il

Abstract. I present my minority view that the majority (or even all) of core collapse supernovae (CCSNe) are driven by jets rather than by neutrinos, and that the majority of type Ia supernovae (SN Ia) reach their explosion via the core degenerate scenario. New simulations presented at the meeting did not achieve an explosion of CCSNe. I critically examine other arguments that where presented in support of the neutrino-driven model, and present counter arguments that support the jet-driven explosion mechanism. The jets operate via a negative jet feedback mechanism (JFM). The negative feedback mechanism explains the explosion energy being several times the binding energy of the core in most CCSNe. We do not know yet what mechanism explodes massive stars and we do not know yet what evolutionary route leads white dwarfs to explode as SN Ia, and so we must be open to different ideas and critically examine old notions.

Keywords. supernovae: general, ISM: supernova remnants, stars: jets

---

1. Introduction

I summarize the talks on the subject of exploding mechanisms of core collapse supernovae (CCSNe) that were given during the meeting *SN 1987A, 30 years later*, that took place in La Reunion Island, France, 20-24 February 2017. This is not a review article on the subject, hence I do not list many relevant references, and I limit myself to the talks that were given at the meeting, and only to those in relation to the explosion mechanism.†

The delayed neutrino mechanism to explode CCSNe, although the most popular model in scientific meetings (e.g., Müller 2017), in the literature, and in text books, seems to fail (e.g., Papish *et al.* 2015; Kushnir 2015). In the present meeting this was clearly demonstrated by Evan O'Connor, who presented new high quality 3D simulations. In section 2 I critically summarize the talks that dealt with the delayed neutrino mechanism to explode CCSNe. I conclude that the claims for a successful neutrino-driven explosion are far from being justified. In section 3 I present the alternative jet feedback mechanism (JFM). In section 4 I discuss the role of binary interaction in CCSNe and in pre-explosion outbursts.

The distinction between most popular in the literature and most popular among stars should be made also with the scenarios of type Ia supernovae (SN Ia). Too many papers mention only the single degenerate and the double degenerate scenarios, ignoring the other three scenarios that have been proposed over the years, the WD-WD collision scenario, the double detonation scenario, and the core degenerate scenario. Several speakers at the meeting made this unjustified omission (You-Hua Chu; Esha Kundu; Anne Decourchelle). In some cases this omission brings the papers to wrong conclusions,

† The link to talks is https://iaus331.lupm.in2p3.fr/programme/scientific-programme/

like that the presence of circum-stellar matter (CSM) around a SN Ia necessarily implies the single degenerate scenario. In those cases with CSM, e.g., SN PTF11kx, the core degenerate scenario does better than the single degenerate scenario. Thus, although the single degenerate and double degenerate are the two most popular scenarios in the literature, my view is that the most popular scenario among exploding white dwarfs is the core degenerate scenario (e.g., Tsebrenko & Soker 2015).
★**Summary of section.** When studying the explosion mechanism of massive stars and the evolution of SN Ia progenitors, we must distinguish between the most popular scenarios in the literature and the scenarios that might be the most popular among the exploding stars. My view is that the two groups of scenarios do not overlap.

2. A critical examination of the neutrino-driven mechanism

2.1. *No/yes/no/yes/no/yes/no/yes/no/yes explosion*

Chris Fryer argued in his talk that most groups now produce explosions with a convective engine. He cited for example the simulations of Lentz *et al.* (2015). However, in the 3D simulation of Lentz *et al.* (2015) the diagnostic energy rises very slowly, and it is not clear at all what value it will reach at the end. He mainly cited his results from 1999 (Fryer 1999), when 3D simulations of CCSNe were not available yet. As even the supporters of the the delayed neutrino mechanism agree, this mechanism cannot explain ab explosion (kinetic) energy greater than about 2×10^{51}. Alexander Heger in his talk adopted the view that these, and only these, super-energetic CCSNe are driven by jets. I will argue later that all CCSNe are driven by jets.

The really new results about the neutrino-driven mechanism were presented by Evan O'Connor. He used the FLASH hydrodynamical code in 3D, and obtained no explosion. I note that Roberts *et al.* (2016) did obtain an explosion. The differences are that in his new results O'Connor used a different code than Roberts *et al.* (2016), as well as a different progenitor and a different equation of state. To that we add the finding of many groups that some stellar models explode and some do not explode. In most cases where an explosion does take place, the explosion (kinetic) energy is very low, $\ll 10^{51}$ erg (see discussion by Papish *et al.*2015).
★ **Summary of subsection.** No consistent and persistent explosion has been achieved by the neutrino-driven mechanism. It seems that the delayed neutrino mechanism has a generic problem (Papish *et al.*2015).

2.2. *Distribution of metals in Cassiopeia A*

Brian Grefenstette presented the distribution of ^{44}Ti in the Cassiopeia A supernova remnant (SNR; Grefenstette *et al.* 2017; also talk by Roland Diehl, and a recent paper by Lee *et al.* 2017). I present this image in the upper panel of Fig. 1. Annop Wongwathanarat, Michael Gabler, and Hans-Thomas Janka (all from the same group) presented their numerical results based on a neutrino-driven explosion, and claimed to explain the protrusions in Cassiopeia A. In the lower panel I present results from Wongwathanarat *et al.* (2015). I argue that the numerical results do not explain the observations of the north-east jet. (1) The (north-east) jet in Cassiopeia A is Si-rich, and does not seem to contain iron (which is the product of nickel). (2) The instability-fingers formed by the nickel in the simulations do not move much faster than the main shell of the supernova (helium-rich material). Such fingers will not form protrusions out of the main shell of the SNR. There are instabilities (as also expected in the JFM), but they are not the entire story.

Takuma Ikeda in his talk presented the detection of titanium at the tip of the jet in Cassiopeia A. But this, he claimed, is ^{48}Ti and not ^{44}Ti. He further suggested that it

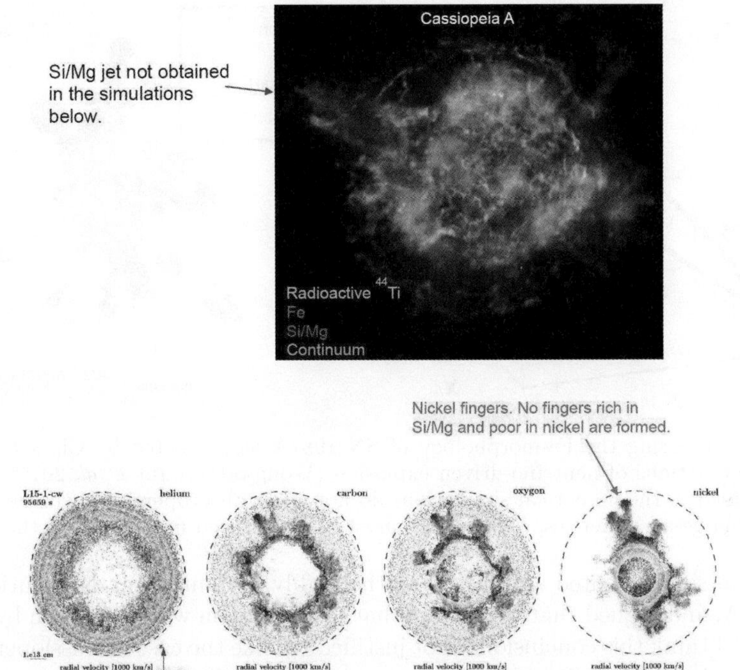

Figure 1. Comparing the observation of Cassiopeia A (Grefenstette *et al.* 2017) with numerical simulations based on the neutrino-driven explosion mechanism (Wongwathanarat *et al.* 2015). The simulations do not explain the properties of the jet.

results from incomplete Si-burning. I suggest that this must be a different shock episode than that responsible for the production of ^{56}Ni and ^{44}Ti. In the neutrino-driven explosion scenario all fingers are formed at about the same time. Later I will discuss why the JFM, where different jets are launched at different times, can better account for the properties of the jet in Cassiopeia A. A support to the jet-driven explosion scenario comes from the simulations that Salvador Orlando presented at the meeting. Those simulations produce the structure of Cassiopeia A when the two opposite jets are introduced as a different component from the smaller scale instabilities.

★ **Summary of subsection.** The instabilities that are found in 3D numerical simulations of the neutrino-driven explosion mechanism cannot explain all the properties of Cassiopeia A. Jets must be included as well. As instabilities are expected also in the JFM, the partial match between these simulations and some properties of Cassiopeia A are insufficient to support the neutrino-driven explosion mechanism.

2.3. *The morphology of SN 1987A*

Both the explosion of SN 1987A and its CSM are non-spherical. The asymmetry reveals itself in several ways. Vinay Kashyap, for example, presented X-ray maps at two times. In 2000 the eastern part of the ring was brighter in X-ray than the western part, while in 2015 it was the opposite. The uneven brightness of the inner ring in radio was presented by Giovanna Zanardo. Then there is the asymmetrical distribution of the ejecta itself, e.g., as seen by dust distribution (talk by Phil Cigan). Highly asymmetrical distributions of metals are observed in other SNRs (e.g., the talk by Ivo Seitenzahl who presented such a distribution for the SNR 1E 0102.2-7219 in the SMC).

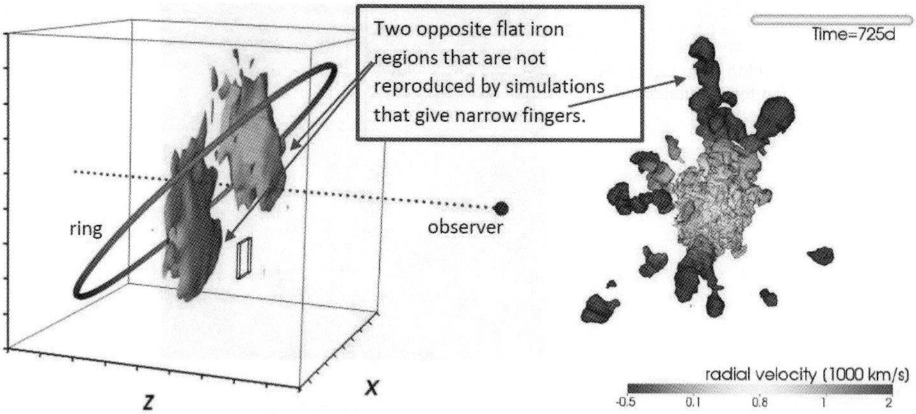

Figure 2. Comparing the Fe-morphology of SN 1987A as presented by Claes Fransson with numerical simulations of neutrino-driven explosion (Wongwathanarat *et al.* 2015). The narrow fingers obtained in the numerical simulations do not cover all properties of the Fe-morphology. I argue that large asymmetries, such as jets, must be introduced in addition to the instabilities.

Claes Fransson presented and discussed the highly asymmetrical distribution of ejecta in SN 1987A, and argued that this shows that the explosion was not driven by jets. I now explain why I think this conclusion is not justified. Unlike the case in Cassiopeia A where I argued in section 2.2 that instabilities cannot explain the Si-rich jet, here instabilities can in principle account for the distribution of the iron found in SN 1987A. Claes Fransson compared his Fe-morphology (see Larsson *et al.* 2016) with the 3D simulations conducted by Wongwathanarat *et al.* (2015). The match is far from perfect. The instabilities in the simulations form narrow Ni-rich (later turn to Fe) fingers, while the iron distribution in SN 1987A is concentrated in two opposite wide regions (the two iron regions are somewhat tilted to each other, i.e., not exactly opposite). Since this is an important point, I emphasize it in Fig. 2.

I argue that this shows that on top of the instabilities there should be a global asymmetry (that I attribute to jets). Such a global asymmetry is supported by the very interesting talk that Mikako Matsuura gave on ALMA observations of molecules. The molecules form a large torus around the center, that is tilted both to the inner ring of the CSM and to the iron morphology.

Let me elaborate on why the Fe-morphology of SN 1987A does not contradict the JFM. Consider the CCSN remnant (CCSNR) W49B. The concentration of iron along a stripe (upper-left panel of Fig. 3) brought Lopez *et al.* (2013) to argue that this CCSN was driven by jets, and that the jets symmetry axis is along the dense iron stripe. We (Bear & Soker 2017) compared the morphology of SNR W49B with the morphologies of many planetary nebulae. We found morphological features that are shared by some PNe and by the CCSNR W49B, such as a barrel-shaped main body and an 'H' shape dense gas, and used these to deduce that the jets that shaped SNR W49B were launched along the symmetry axis of the 'barrel', i.e., perpendicular to the dense iron stripe. We further speculated that this CCSNR has two opposite lobes (or ears), that are too faint to be observed. All these features are marked on Fig. 3.

The point to make here is that according to our suggestion (Bear & Soker 2017), the dense iron stripe is near (or in) the equatorial plane, and not along the jets' axis. Claes Fransson presented a dense ejecta stripe more or less in the plane of the inner ring of SN 1987A. This might resemble the dense iron stripe in the SNR W49B, and hence does not rule out jets perpendicular to the plane of the ring and perpendicular to the Fe-rich

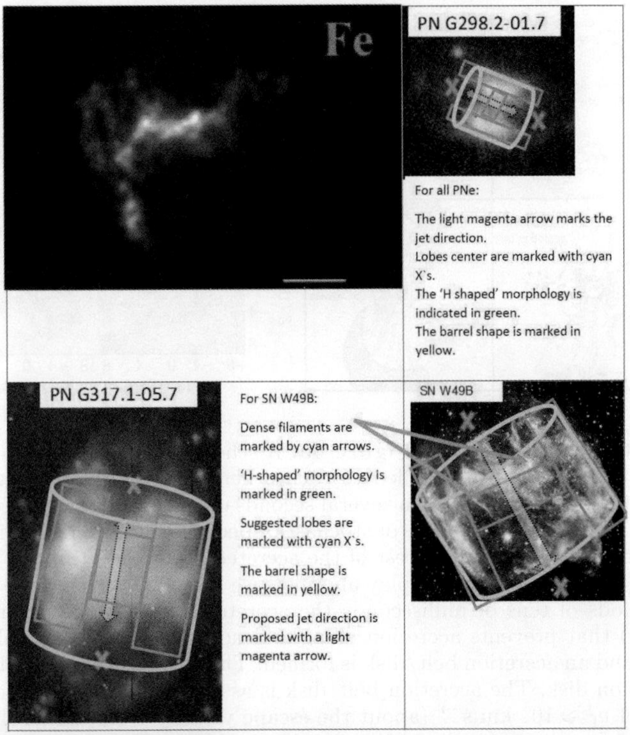

Figure 3. Comparing the morphologies of the CCSNR W49B (lower right; credit the Chandra website http://chandra.harvard.edu/photo/2013/w49b/), its dense iron regions (upper left; from Lopez *et al.* 2013) and two planetary nebulae taken from Balick (1987) and Górny *et al.* (1999). This comparison suggests that the dense iron regions might be located in the equatorial plane, as I suggest here for SN 1987A. Figure is based on Bear & Soker (2017).

stripe. Furthermore, jets launched at different episodes during the explosion might have different directions if the jittering jets scenario takes place (see next section).

★ **Summary of subsection.** The iron-morphology of SN 1987A cannot be reproduced only by instabilities that are obtained in simulations of neutrino-driven explosions. A global asymmetry must take place during the explosion. I attribute this asymmetry to jets. In that case the iron can be concentrated along a stripe in the equatorial plane, and not necessarily along the jets. In any case, according to the jittering jets mechanism jets launched at different episodes during the explosion might have different directions.

3. The jet feedback mechanism (JFM)

3.1. *The two scenarios*

An alternative mechanism to account for all CCSNe is the jet feedback mechanism (JFM; for a review see Soker 2016). Although this mechanism is much less popular in the scientific literature, I argue that it is the most popular explosion mechanism among the exploding massive stars themselves. As long as consistent explosions have not been achieved by the neutrino-driven mechanism, the community should be open to all proposed explosion mechanisms. My goal in this contribution is to bring arguments that show that the JFM has a merit.

Jet-driven CCSNe have been discussed in the literature for a long time, but mainly for specific cases where the pre-collapse core is rapidly rotating. Several speakers in the

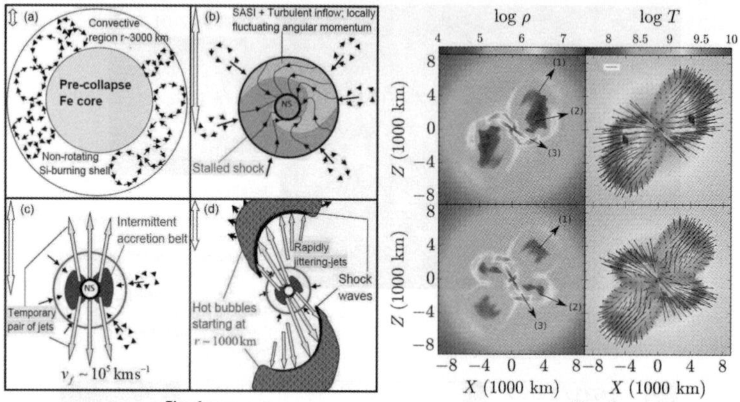

Fig. 4a Fig. 4b

Figure 4. The jittering jets scenario. **Figure 4a.** A schematic presentation of the jittering jets scenario (left). The panels are not scaled to each other; the two-sided arrow in each panel is approximately 500 km. The panels span several seconds of evolution. (a) The convective vortices in the Si-burning shell serve as a source of stochastic specific angular momentum $\mathbf{j}_a(t)$. (b) After the formation of the neutron star the rest of the accreted mass flows through the stalled shock. The spiral modes of the SASI come to play and increase the variations of $\mathbf{j}_a(t)$ in the post-shock zone. (c) For periods of tens of milliseconds the accreted gas near the neutron star possesses large enough $\mathbf{j}_a(t)$ that prevents accretion along and near the temporary poles of the angular momentum axis, and an accretion belt/disk is formed. The belt can spread in the radial direction to form an accretion disk. The accretion belt/disk is assumed to launch two opposite jets with initial velocities of $v_f \approx 10^5$ km s^{-1} (about the escape velocity from the newly formed neutron star). (d) The jittering jets penetrate through the gas close to the center, but then they are shocked and inflate bubbles. The bubbles expand and explode the core and then the star.
Figure 4b. Results of 3D simulations of the jittering jets scenario (right). Presented is the flow structure in the $y = 0$ plane at the end of the third jets-launching episode at time $t = 0.15$ s, and for two runs with different initial conditions (upper and lower rows, respectively). Left-hand panels: density, with a colour coding in logarithmic scale and units of g cm^{-3}. The three arrows depict the direction of jets' launching in the three episodes as numbered. The jets are inserted by hand, as the simulations do not include the formation of the neutron star and the jets. Right-hand panels: temperature in log scale in units of K, and velocity map. Velocity is proportional to the arrow length, with inset showing an arrow for 30,000 km s^{-1}. For more details see Papish & Soker (2014).

meeting presented this view (Niloufar Afsariardchi; Alexander Heger; Hidetomo Sawai; Kuntal Misra). I differ in that I argue that *all* CCSNe are driven by jets, and by considering the JFM. In the JFM there is no need to revive the stalled shock.

In principle, there are two scenarios within the JFM (Soker 2017b). The first one is the jittering jets scenario that I proposed in 2010 (Soker 2010). In this scenario even a collapsing core with very slow pre-collapse rotation launches jets. The accretion belt or accretion disk that launches the jets is intermittent, e.g., with a varying direction of the jets' axis. (An accretion belt stands for an accretion flow that has sub-Keplerian specific angular momentum, and so the inflow is concentrated toward the equatorial plane but it does not form an accretion disk.) Instabilities (e.g., the standing accretion shock instability, SASI) lead to stochastic accretion with fluctuating specific angular momentum of the accreted gas $\mathbf{j}_a(t)$ (both magnitude and direction). At high magnitudes of $\mathbf{j}_a(t)$ an intermittent accretion disk/belt is formed even when the pre-collapse core is slowly rotating. According to this scenario no failed CCSNe exist. The jittering jets scenario is schematically summarized in Fig. 4a.

If it will turn out that slowly-rotating cores cannot launch jets (and hence end with a failed CCSN), the second scenario, termed fixed axis scenario, will have to be considered.

In the fixed axis scenario the core must be spun-up by a binary interaction before collapse starts. According to the fixed axis scenario all CCSNe are descendants of strongly interacting binary systems, most likely through a common envelope evolution. I note that even in the jittering jets scenario many of the progenitors suffer a strong binary interaction, and hence behave like the CCSNe under the fixed axis scenario.

The main challenge of the JFM is the accretion of gas with high enough specific angular momentum ($\mathbf{j}_a$) to form at least an accretion belt. I think that our understanding of the specific angular momentum of the accreted gas is poor. The coupling of the SASI with rotation is very complicated (talk by Remi Kazeroni; Kazeroni *et al.* 2017). It is quite possible that the amplitude of $\mathbf{j}_a$ is large, and an intermittent accretion belt is formed in all cases. Such a belt can launch jets (Schreier & Soker 2016). The exact structure of the accretion flow in the case of a rapidly rotating core is not settled either (e.g., Gilkis 2016).

The JFM nicely connects regular CCSNe to super-energetic CCSNe and to gamma ray bursts (Gilkis *et al.* 2016). Super-energetic CCSNe occur when the JFM is inefficient. This is the case when the jets do not jitter and are narrow, and hence do not expel core mass from the equatorial plane. As more mass is accreted from the equatorial plane the jets carry more and more energy, leading to a super-energetic CCSN.

One very significant comment is in place here, as it is ignored too many times. Magnetars are mentioned as a powering engine of super-energetic CCSNe (in the meeting: Jerome Guilet; Stefano Valenti; Antonio de Ugarte Postigo; Yves Gallant). However, the formation of a magnetar must be accompanied by very energetic jets (Soker 2017a). Therefore, whenever magnetars are mentioned, jets must be mentioned alongside.

The JFM has a prediction on the influence of jets on the shape of the descendant SNR. After the jets explode the core, accretion stops and the jets are turned off. However, this takes some time, such that the accretion continues for a short time after explosion. For that, the jets that are launched at the very end do not encounter the dense core material and the inner envelope layers, as they have been exploded already. These two opposite jets launched at the end, can leave marks on the expanding ejecta. One such signature can be two opposite protrusions in the SNR that are called ears.

In Grichener & Soker (2017) we analysed CCSNRs that have two opposite 'ears' protruding from their main shell (some of these CCSNRs were presented at different talks in the meeting, e.g., SNR W44 by Elise Egron, Alberto Pellizzoni, Sara Loru, Yasuo Fukui, and Herman Lee, and Vela SNR by Iurii Sushch). Shu Masuda presented the high energy ($2-100$Gev) γ-ray emission from the CCSNR W44 (Uchiyama *et al.* 2012). The emission comes from two regions outside the SNR and along the ears, and can be attributed to the cosmic rays that escape from W44. They probably escape along the ears.

The ears can be formed in the CSM before the explosion, during the explosion by the last jets to be launched by the explosion process, or after the explosion, e.g., by jets from the neutron star remnant. We (Grichener & Soker 2017) discussed why it is reasonable to assume that the ears in most CCSNRs were formed by jets blown during the explosion. We estimated that the typical (kinetic) energy of the jets that inflated the ears in CCSNRs is about $5-15$ per cents of the explosion energy. This is compatible with the expectation of the JFM, whether in the jittering jets scenario or the fixed axis scenario. We further estimated that about 40 per cents of all CCSNe have ears. The image of W44 with marks of quantities used to calculate the energy of the jets is presented in Fig. 5.

★ **Summary of subsection.** The explosion JFM of massive stars might be operating in one of two scenarios, the jittering jets scenario or the fixed axis scenario. The negative feedback mechanism accounts for the explosion energy being several times the binding

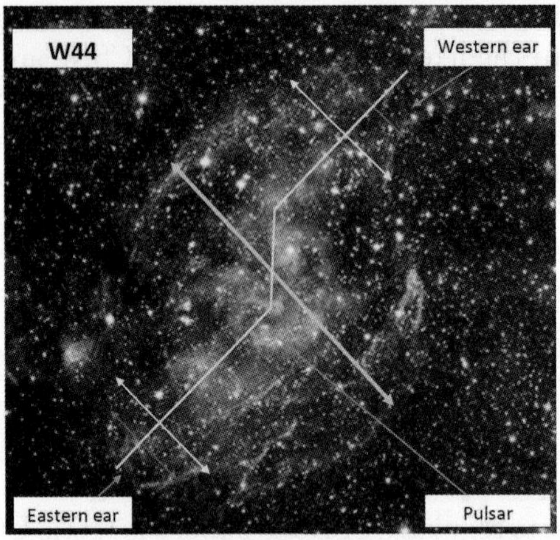

Figure 5. The SNR W44. Composite image taken from the Chandra gallery. The cyan represents X-ray (based on Shelton *et al.*2004), while the red, blue and green represent infra-red (based on NASA/JPL-Caltech). We (Grichener & Soker 2017) added three beige thick lines to schematically define the S-shape of this SNR.

energy of the core. In cases where the JFM is highly inefficient, the jets continue to be launched for a long time without removing mass from the equatorial plane, and hence the explosion leads to a super-energetic CCSN. The last jets to be launched might form two opposite protrusions, called ears, on the SNR. In cases when magnetars are formed (they are required by other models), jets carry a large amount of energy, even more than the magnetar. Therefore, jets must be mentioned whenever a magnetar is discussed for super-energetic CCSNe.

3.2. *Accounting for Cassiopeia A and SN 1987A*

The case of the jittering jets scenario and the ^{44}Ti distribution in Cassiopeia A was discussed by Gilkis *et al.* (2016). Grefenstette *et al.* (2014) argue against fast-rotating progenitors, as well as a jet-like explosion of Cassiopeia A. They suggest that the ^{44}Ti nonuniform distribution is the result of a multimodal explosion, such as expected from instabilities. As I discussed in section 2.2 the instabilities cannot reproduce the jets. The point is that the jittering jets scenario has the property of multimodal explosion, as several pairs of opposite jets are launched in different directions. The explosion has no symmetry axis or symmetry plane. The jets explode the core and the star. The last bipolar jets to be launched might propagate freely in the inner region of the star and leave an imprint on the outer ejecta. Gilkis *et al.* (2016) speculated that the jet structure in Cassiopeia A and its counter protrusion are the result of a last jet-launching episode, as expected in the jittering jets scenario.

In SN 1987A, the long axis of the iron distribution, the symmetry axis of the ring, and the symmetry axis of the molecular torus (see section 2.3), have different directions. Such a point-symmetric structure might be accounted for with the jittering jets scenario of the JFM. This is a topic of a future study with 3D simulations. At this point I only present results from two simulations of jittering jets. These runs, taken from Papish & Soker (2014), are presented in Fig. 4b. The 3D simulations were run for only 0.15 seconds and for only 3 jets-launching episodes; about 10 such episodes are expected until explosion. Nonetheless, these simulations show the complicated structure of the shocked regions in the core, where nucleosynthesis takes place. Note that different regions are shocked at different times, something that might explain large asymmetries in metals distribution in the ejecta.

★ **Summary of subsection.** The JFM, that includes jittering jets in addition to instabilities, seems to better explain the ejecta distribution in Cassiopeia A and in SN 1987A, than the neutrino-driven mechanism.

4. Roles of binary stars

Strong binary interaction can spin-up the stellar envelope and the core. Substantial core spin-up occurs when the secondary star spirals-in to the center of the CCSN progenitor. Because pre-collapse angular momentum plays a crucial role in the JFM, a binary interaction that influences the pre-explosion evolution has also implications on the explosion itself. I turn to very briefly mention two aspects of pre-explosion binary interaction.

4.1. *Pre-explosion outburst*

Several speakers at the meeting mentioned the high mass loss rate that some CCSNe experience before explosion (Sanskriti Das; Firoza Sutaria; Christina Thöne; Stefano Valenti; Yukari Ohtani). There is no indication that SN 1987A experienced a pre-explosion outburst. Poonam Chandra mentioned that the CSM that is formed in pre-explosion outbursts is highly non-spherical, and most likely has a bipolar structure. This is exactly what I expect if bright pre-explosion outbursts involve a binary interaction.

In Mcley & Soker (2014) we suggested that pre-explosion outbursts result from strong binary interaction triggered by the expansion of the primary star. The binary companion accretes mass from the extended envelope and releases a huge amount of energy through radiation and jets. The jets form the bipolar CSM into which the ejecta expand days to years after explosion. Such jets could power the pre-explosion outbursts of SN 2009ip (Soker & Kashi 2013).

Robert P. Kirshner in his talk gave a nice historic description of SN 1987A, and emphasized the clumps in the inner ring. These clumps result from instabilities in the mass loss process, as also seen in rings in some planetary nebulae, and can be reproduced by jets that compress gas into the equatorial plane (Akashi *et al.* 2015). The shape itself, of three rings, must result from a binary interaction. The asymmetrical explosion of 1987A was attributed to a binary interaction already in 1989.

Michael Bietenholz presented nice radio observations of SN 1986J, and concluded that the CSM is highly asymmetrical and/or jets were launched after the explosion. This adds to the importance of bipolar structure in CCSNRs that might result both from jets before and during the explosion.

★**Summary of subsection.** Jets might play a role in many CCSNe before the explosion. Jets launched by a companion to the CCSN progenitor might power strong pre-explosion outbursts and shape the CSM into a bipolar structure.

4.2. *Removing mass*

Type IIb CCSNe are exploding stars with a very light but extended hydrogen envelope. They amount to about $10 - 12\%$ of all CCSNe. Niharika Sravan presented her nice results of stellar evolution calculations, according to which single stars cannot be the progenitor of Type IIb CCSNe (also talk by Charlie Kilpatrick), while binary systems with a companion avoiding a common envelope evolution can account for Type IIb CCNSe, but only to about 1% of all CCSNe. I suggest here that grazing envelope evolution can substantially increase the number of binary systems that can lead to Type IIb CCSNe. In

the grazing envelope evolution the companion star grazes the envelope of the progenitor primary star, and removes most of its envelope by launching jets (Shiber *et al.* 2017).
★**Summary of subsection.** I speculate that the progenitor stars of a large fraction of type IIb CCSNe lost most of their hydrogen-rich envelope in a grazing envelope evolution.

I thank my collaborators that, despite being in the minority, continued to work with me on the literature-unpopular JFM: Amit Kashi, Oded Papish, Danny Tsebrenko, Avishai Gilkis, Ealeal Bear, and Aldana Grichener.

References

Akashi, M., Sabach, E., Yogev, O., & Soker, N. 2015, *MNRAS*, 453, 2115
Balick, B. 1987, *AJ*, 94, 671
Bear, E. & Soker, N. 2017, *MNRAS*, in press (arXiv:1611.07327)
Fryer, C. L. 1999, *ApJ*, 522, 413
Gilkis, A. 2016, arXiv:1608.05320
Gilkis, A., Soker, N., & Papish, O. 2016, *ApJ*, 826, 178
Górny, S. K., Schwarz, H. E., Corradi, R. L. M., & Van Winckel, H. 1999, *A&AS*, 136, 145
Grefenstette, B. W., Fryer, C. L., Harrison, F. A., *et al.* 2017, *ApJ*, 834, 19
Grefenstette, B. W., Harrison, F. A., Boggs, S. E., *et al.* 2014, *Nature*, 506, 339
Grichener, A. & Soker, N. 2017, *MNRAS*, in pres (arXiv:1610.09647)
Kazeroni, R., Guilet, J., & Foglizzo, T. 2017, arXiv:1701.07029
Kushnir, D. 2015, arXiv:1506.02655
Larsson, J., Fransson, C., Spyromilio, J., *et al.* 2016, *ApJ*, 833, 147
Lee, Y.-H., Koo, B.-C., Moon, D.-S., Burton, M. G., & Lee, J.-J. 2017, arXiv:1703.03551
Lentz, E. J., Bruenn, S. W., Hix, W. R., *et al.* 2015, *ApJL*, 807, L31
Lopez, L. A., Ramirez-Ruiz, E., Castro, D., & Pearson, S. 2013, *ApJ*, 764, 50
Mcley, L. & Soker, N. 2014, *MNRAS*, 445, 2492
Müller, B. 2017, arXiv:1702.06940
Papish, O., Nordhaus, J., & Soker, N. 2015, *MNRAS*, 448, 2362
Papish, O. & Soker, N. 2014, *MNRAS*, 443, 664
Roberts, L. F., Ott, C. D., Haas, R., *et al.* 2016, *ApJ*, 831, 98
Schreier, R. & Soker, N. 2016, *Research in Astronomy and Astrophysics*, 16, 001
Shelton, R., Kuntz, K., & Petre, R. 2004, *ApJ*, 611, 906
Shiber, S., Kashi, A., & Soker, N. 2017, *MNRAS*, 465, L54
Soker, N. 2010, *MNRAS*, 401, 2793
Soker, N. 2016, *New Astron. Rev.*, 75, 1
Soker, N. 2017a, arXiv:1612.01912
Soker, N. 2017b, arXiv:1702.03451
Soker, N. & Kashi, A. 2013, *ApJ*, 764, L6
Tsebrenko, D. & Soker, N. 2015, *MNRAS*, 447, 2568
Uchiyama, Y., Funk, S., Katagiri, H., *et al.* 2012, *ApJ*, 749, L35
Wongwathanarat, A., Müller, E., & Janka, H.-T. 2015, *A&A*, 577, A48

Supernova 1987A:30 years later - Cosmic Rays and Nuclei from Supernovae and their aftermaths
Proceedings IAU Symposium No. 331, 2017
A. Marcowith, M. Renaud, G. Dubner, A. Ray & A. Bykov, eds.

doi:10.1017/S1743921317004744

The infancy of supernova remnants: evolving a supernova into its remnant in 3D

Michael Gabler[1], Hans-Thomas Janka[1] and Annop Wongwathanarat[2]

[1]Max-Planck-Institut für Astrophysik, Karl-Schwarzschild-Str. 1, 85741 Garching, Germany
[2]RIKEN, Astrophysical Big Bang Laboratory, 2-1 Hirosawa, Wako, Saitama 351-0198, Japan

Abstract. Recently, first neutrino-driven supernova explosions have been obtained in 3D, self-consistent, first-principle simulations, these models are still not always exploding robustly and, in general, the explosions are not sufficiently energetic. To constrain the explosion mechanism, and the related uncertainties, it is thus very helpful to consider observational constraints: pulsar kicks, progenitor association and supernova remnants (SNR). Recent observations of asymmetries in the supernova ejecta of Cas A are very promising, to compare to long-term simulations of the explosion. In addition 3D observations of SN87A are becoming more constraining on the geometry of the ejected material during the explosion. In this talk I will discuss our efforts to model the late time evolution of a 3D supernova explosion, where we include the effects of beta decay, which inflates the structures rich in ^{56}Ni. The structures we find in the simulations depend on the quantities plotted.

Keywords. hydrodynamics, instabilities, methods: numerical, (stars:) supernovae: individual (1987A), (stars:) supernovae: general, (ISM:) supernova remnants

1. Introduction

Supernovae are amongst the most energetic events in our universe. Huge efforts have been undertaken to improve our knowledge about these phenomena by performing sophisticated simulations and by ever more detailed observations. Our current understanding of, in particular, core-collapse supernovae (CCSNe) shows us that 3D effects are very important to model them theoretically, but also to interpret the observations. First long-term multi-dimensional supernova simulations became possible at end of the 1980s and in the early 1990s (Arnett *et al.* (1989), Fryxell *et al.* (1991), Mueller *et al.* (1991), Hachisu *et al.* (1990), Hachisu *et al.* (1992), Hachisu *et al.* (1994), Yamada & Sato (1990), Herant & Benz (1991), Herant & Benz (1992), Herant & Woosley (1994), Iwamoto *et al.* (1997), Nagataki *et al.* (1998)). Later the complexity of the simulations was increased, by among other things no longer initiating the explosions with thermal bombs (Kifonidis *et al.* (2000), Kifonidis *et al.* (2003), Hungerford *et al.* (2003), Hungerford *et al.* (2005), Kifonidis *et al.* (2006), Joggerst *et al.* (2009), Joggerst *et al.* (2010), Joggerst *et al.* (2010), Couch *et al.* (2009), Gawryszczak *et al.* (2010), Couch *et al.* (2011), Ono *et al.* (2013), Ellinger *et al.* (2012), Ellinger *et al.* (2013)). In a series of papers, we also investigated how neutrino-driven explosions evolve through the progenitor up to its surface and how multi-dimensional effects and instabilities create asymmetries in the ejecta (Hammer *et al.* (2010), Wongwathanarat *et al.* (2010), Wongwathanarat *et al.* (2013), Wongwathanarat *et al.* (2015), Wongwathanarat *et al.* (2016)).

Other groups followed a different approach to investigate the CCSN phenomenon. They modeled particular supernova remnants such as SN 1987A or Cas A (Li *et al.* (1993), Blondin & Lundqvist (1993), Blondin *et al.* (2001), Tanaka & Washimi (2002),

Podsiadlowski *et al.* (2007), Dewey *et al.* (2012), Potter *et al.* (2014), Orlando *et al.* (2015), Orlando *et al.* (2016)). These studies focused on the late phases of the explosion/early phases of the remnant and the interaction with the circumstellar material. The simulations were started with spherically symmetric or artificial initial conditions after the shock break-out.

The theoretical models were stimulated by the ever more detailed observations revealing truly 3D structures and asymmetries (Nagataki *et al.* (1998), Fesen (2001), Hwang *et al.* (2004), Fesen *et al.* (2006), Milisavljevic & Fesen (2013), Grefenstette *et al.* (2014), Milisavljevic & Fesen (2015), Boggs *et al.* (2015), Grefenstette *et al.* (2017)).

In the current work we continue our previous study (Wongwathanarat *et al.* (2015)) and simulate the evolution from a 3D supernova explosion starting at bounce up to the early remnant stage including nuclear beta decay of ^{56}Ni consistently.

2. Numerics

For our computations we use the 3D, explicit finite-volume hydrodynamic code PROMETHEUS (Fryxell *et al.* (1991), Mueller *et al.* (1991)). It is based on the piecewise parabolic reconstruction method (PPM; Colella & Woodward (1984)), employs a Riemann solver for realistic gases (Colella & Glaz (1985)) and treats multi-fluid systems with the consistent multi-fluid advection scheme (CMA; Plewa & Müller (1999)). To avoid artefacts at the polar axis of a spherical polar grid and to allow for larger time steps due to a less restrictive CFL condition close to the polar axis, we use a 'Yin-Yang' grid (Kageyama & Sato (2004), Wongwathanarat *et al.* (2010)). The code has already been applied to calculate the 3D propagation of the shock and the ejecta during a neutrino-driven supernova explosion up to the shock break out (Wongwathanarat *et al.* (2013), Wongwathanarat *et al.* (2015)) and to investigate light curves for different progenitors compared to SN 1987A (Utrobin *et al.* (2015)) as well as to study the production of ^{44}Ti and ^{56}Ni in Cas A (Wongwathanarat *et al.* (2016)).

Here we extend the code version PROMETHEUS-HOTB by including the effects of additional heating due to beta decay. ^{56}Ni decays with a half-life time of $\tau_{1/2}^{Ni} = 6.07\,$d to ^{56}Co, which in turn decays to the stable ^{56}Fe with $\tau_{1/2}^{Co} = 77.23\,$d. The energy releases in γ photons and e$^+$ per decay are $Q_{Ni} = 1.72\,$MeV and $Q_{Co} = 3.735\,$MeV, respectively (Junde *et al.* (2011)). For the current study we assume that the total energy is deposited locally. This overestimates the energy trapped in the medium, because depending on the model, the ejecta become transparent to the photons released during the beta decay after about $100 - 150\,$d. However, until this time about 2/3 of the radioactive material will have decayed and our approximation only mildly overestimates the energy input due to radioactive decay. Practically, the additional energy is implemented as a source term for the specific energy per unit mass ε and the energy increase in a time interval Δt is given by

$$\Delta\varepsilon = \sum_i \frac{Q_i X_i}{m_i} \left(1 - e^{-\frac{\Delta t \ln 2}{\tau_{1/2}^i}} \right), \tag{2.1}$$

where X_i and m_i are the mass fraction and atomic mass of the respective element $i = \{Ni, Co\}$. To reduce numerical costs and numerical diffusion we use a radially moving mesh with $v_{\rm grid} \sim v_r$, which results in a quasi-Lagrangian grid, because $v_r >> \{v_\theta, v_\varphi\}$.

model	model in Wongwathanarat+2015	type	mass[$M_\odot$]	radius[10^6 km]	t_{map} [ks]	$t_{outbreak}$[ks]
W15	W15-2-cw	RSG	15	339	5.8	85
L15	L15-1-cw	RSG	15	434	5.0	96
N20	N20-4-cw	BSG	20	33.8	1.4	5.6
B15	B15-1-pw	BSG	15	39.0	3.2	7.3

Table 1. Progenitor models used in this work.

3. Models

Continuing the work of Wongwathanarat *et al.* (2015) we consider two red (RSG) and two blue super giants (BSG) progenitors (see Table 1) and evolve them for up to two years including the beta decay of ^{56}Ni to ^{56}Fe.

The evolution of the models until shock breakout in Wongwathanarat *et al.* (2015) can be summarized as follows: Initial asymmetries arise due to hydrodynamical instabilities behind the stalled shock during the first second of post bounce evolution. Once the shock expands the initial asymmetric structures move outwards and fragment due to Rayleigh-Taylor (RT) instabilities at the C+O/He and He/H shell interfaces of the progenitor. How strong the initial, big plumes fragment depends sensitively on the structure of the progenitor. In particular, the density decrease at each interface and the flatness of the density profiles within the following shell determine how much time the instabilities have to grow before the reverse shock hits the inner ejecta. At this point the plumes are decelerated and get compressed. Fragmentation and mixing occurs by the growth of RT instabilities in the region between reverse shock and shell interfaces. It is therefore suppressed when the reverse shock from the He/H interface decelerates the inner ejecta before they reach the RT unstable region at the base of the H envelope. Depending on the progenitor structure, the different models have different characteristics at shock breakout. The two RSG (W15 and L15) are qualitatively similar: there are small scale RT-fingers grouped into larger structures that arose from the initial asymmetries during the shock revival phase. In the BSG N20 the initial asymmetries were decelerated and compressed due to a faster interaction with the reverse shock, before the inner ejecta could reach the RT unstable region at the He/H interface. Thus, the model is quite spherical even at late times. Model B15 has different morphology than the other two groups. There are a few very elongated fingers in the direction of the initial asymmetries. Those almost did not interact with the reverse shock. However, the latter slowed down and compressed the material in the central region such that the fast moving fingers are much more extended than in the other models. For a more detailed description we refer to Wongwathanarat *et al.* (2015).

4. Results

Right after the shock breakout, the Ni-rich ejecta of almost all models still have not reached the homologous expansion. For a short time, the velocity differences of different structures lead to a further elongation of the Ni-fingers for models W15, L15 and in particular B15. The Ni-rich ejecta of the more spherical model N20 become slightly asymmetric. Starting at about one day, the radioactive decay of ^{56}Ni provides a significant energy source that heats up the Ni-clumps. This increase of the internal energy leads to an inflation of the corresponding structures. Small fragmented fingers start to expand and can even merge with neighbouring fingers. Thus, they create larger structures in the direction of the initial asymmetries from the onset of the explosion. In Figure 1, we plot the surfaces where the mass fraction of ^{56}Ni+^{56}Co+^{56}Fe is larger than 3%. These

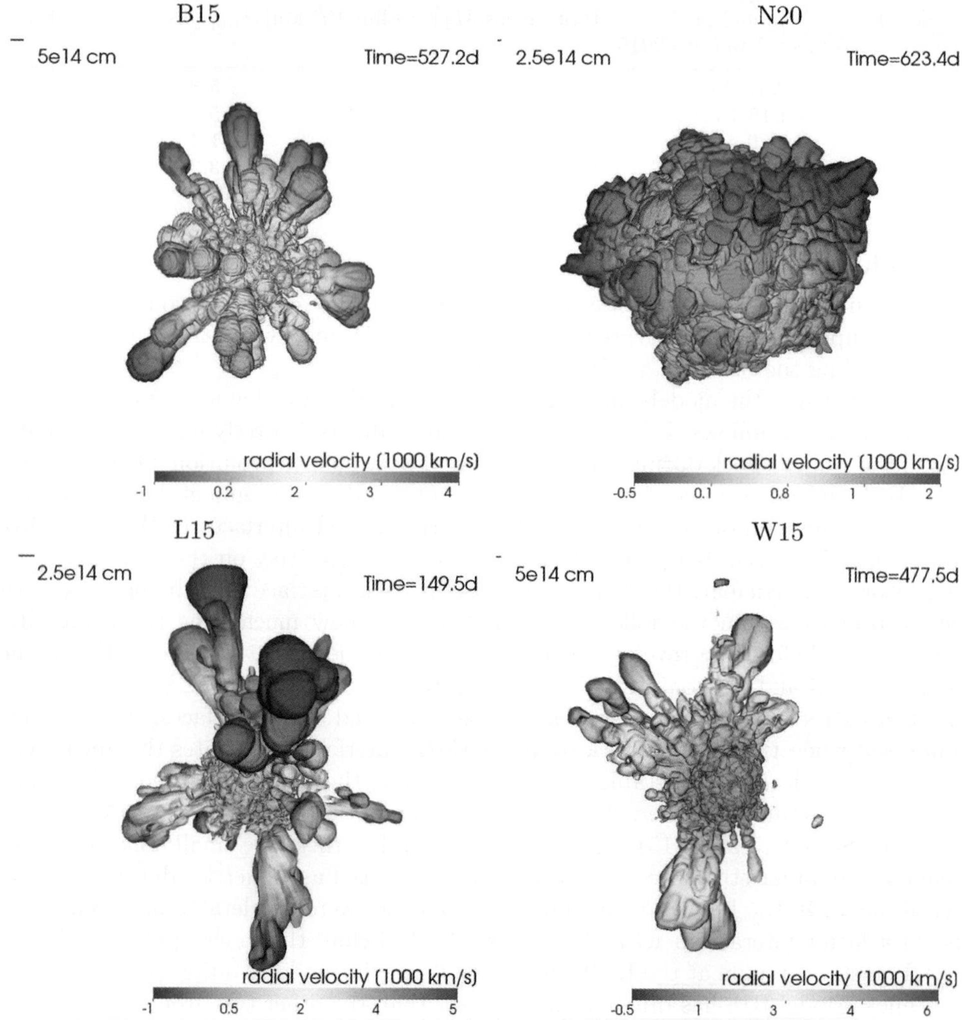

Figure 1. Isosurfaces where the elements of the decay chain of ^{56}Ni sum up to 3%. The color coding is the radial velocity.

surfaces contain most of the final ^{56}Fe and are thus representative for the most extended structures of the ejecta.

At late times, the small asymmetries in model N20 after the shock breakout, are almost completely washed out and the overall shape of the ejecta is rather close to spherical symmetry. In the other models the reverse shock slows down most of the central material which therefore forms an almost spherical shell (blue regions in panels for B15, L15, and W15 in Figure 1), while the fingers become more elongated and expand with modest extra boost by the ^{56}Ni decay heating. In the models L15 and W15 we see three main structures consisting of a few Ni-fingers. In addition, L15 has some smaller structures with only one or a few smaller fingers. Model B15 has more fingers than the two RSG, but the basic morphology elements are highly asymmetric and comparable in all three models.

Due to the expansion of the Ni-bubbles or Ni-fingers their density decreases with respect to the surrounding and matter gets accumulated at the walls of the bubbles/fingers.

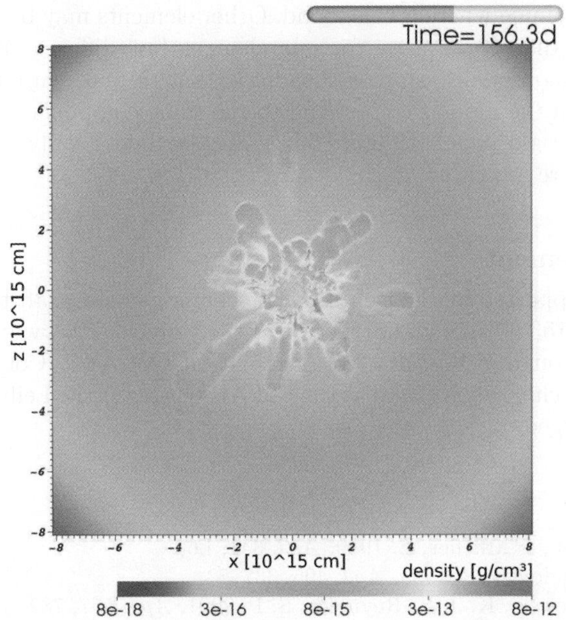

Figure 2. Crossectional slice through the young remnant for model B15, showing the mass density color coded. The walls of the Ni-rich domains have clearly higher densities than the interior of these structures.

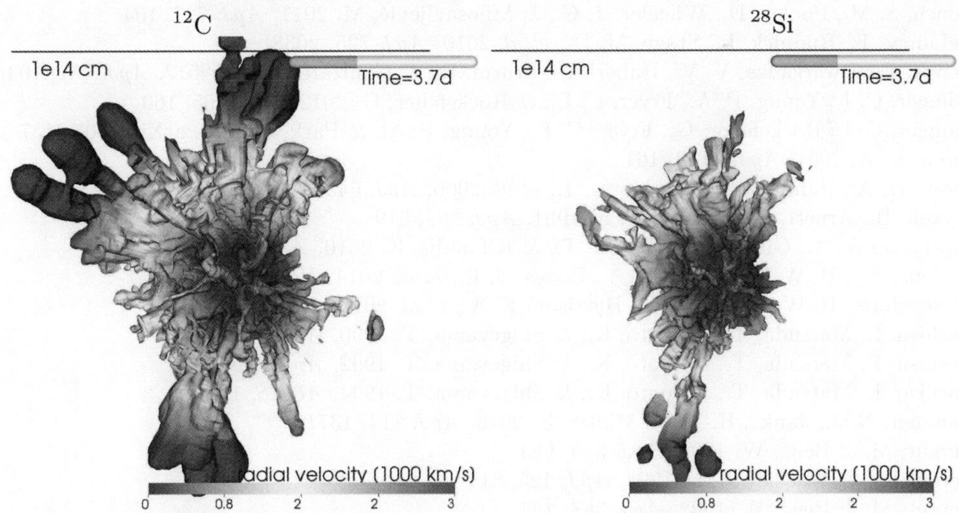

Figure 3. Surfaces enclosing 25% of ^{12}C (left) and ^{28}Si (right) with the highest respective densities for model W15.

This effect is displayed in Figure 2, where we plot a density slice of model B15 at around 150 d. Clearly, there are density enhancements at the borders of the Ni-rich structures. One also can identify a central volume which is less dense than the walls of the bubbles. ('Nickel bubble',Woosley (1988))

Observations are not only limited to Fe, for which we show most of the plots. To show the differences to other elements we plot the surfaces containing each 25% of mass with the highest densities of ^{12}C and ^{28}Si in Figure 3. As expected the heavier element ^{28}Si is located inside the volume where ^{12}C is found. Other elements may be distributed slightly differently. Similar differences arise for one element but different threshold densities. When comparing to observations, it is thus important to make sure that the structures seen and the ones in the models correspond to the same component of ejected material. For example, in the right panel of Figure 3 it looks like there is only one dominant finger, while in the left there are three.

5. Acknowledgements

This work was supported by the Deutsche Forschungsgemeinschaft through the Cluster of Excellence EXC 153 'Origin and Structure of the Universe' (www.universe-cluster.de), and by the EU through ERC-AdG No. 341157-COCO2CASA. Computer time at the Rechenzentrum Garching (RZG) and at the C2PAP-cluster at the Leibnitz-Rechenzentrum (LRZ) is acknowledged.

References

Arnett, D., Fryxell, B., & Mueller, E. 1989, *ApJ*, 341, L63
Blondin, J. M. & Lundqvist, P. 1993, *ApJ*, 405, 337
Blondin, J. M., Borkowski, K. J., & Reynolds, S. P. 2001, *ApJ*, 557, 782
Boggs, S. E., Harrison, F. A., Miyasaka, H., *et al.* 2015, *Science*, 348, 670
Colella, P. & Glaz, H. M. 1985, *Journal of Computational Physics*, 59, 264
Colella, P. & Woodward, P. R. 1984, *Journal of Computational Physics*, 54, 174
Couch, S. M., Wheeler, J. C., & Milosavljević, M. 2009, *ApJ*, 696, 953
Couch, S. M., Pooley, D., Wheeler, J. C., & Milosavljević, M. 2011, *ApJ*, 727, 104
DeLaney, T., Rudnick, L., Stage, M. D., *et al.* 2010, *ApJ*, 725, 2038
Dewey, D., Dwarkadas, V. V., Haberl, F., Sturm, R., & Canizares, C. R. 2012, *ApJ*, 752, 103
Ellinger, C. I., Young, P. A., Fryer, C. L., & Rockefeller, G. 2012, *ApJ*, 755, 160
Ellinger, C. I., Rockefeller, G., Fryer, C. L., Young, P. A., & Park, S. 2013, arXiv:1305.4137
Fesen, R. A. 2001, *ApJS*, 133, 161
Fesen, R. A., Hammell, M. C., Morse, J., *et al.* 2006, *ApJ*, 645, 283
Fryxell, B., Arnett, D., & Mueller, E. 1991, *ApJ*, 367, 619
Gawryszczak, A., Guzman, J., Plewa, T., & Kifonidis, K. 2010, *A&A*, 521, A38
Grefenstette, B. W., Harrison, F. A., Boggs, S. E., *et al.* 2014, *Nature*, 506, 339
Grefenstette, B. W., Fryer, C. L., Harrison, F. A., *et al.* 2017, *ApJ*, 834, 19
Hachisu, I., Matsuda, T., Nomoto, K., & Shigeyama, T. 1990, *ApJ*, 358, L57
Hachisu, I., Matsuda, T., Nomoto, K., & Shigeyama, T. 1992, *ApJ*, 390, 230
Hachisu, I., Matsuda, T., Nomoto, K., & Shigeyama, T. 1994, *A&AS*, 104,
Hammer, N. J., Janka, H.-T., & Müller, E. 2010, *ApJ*, 714, 1371
Herant, M. & Benz, W. 1991, *ApJ*, 370, L81
Herant, M. & Woosley, S. E. 1994, *ApJ*, 425, 814
Herant, M. & Benz, W. 1992, *ApJ*, 387, 294
Hungerford, A. L., Fryer, C. L., & Warren, M. S. 2003, *ApJ*, 594, 390
Hwang, U., Laming, J. M., Badenes, C., *et al.* 2004, *ApJ*, 615, L117
Hungerford, A. L., Fryer, C. L., & Rockefeller, G. 2005, *ApJ*, 635, 487
Iwamoto, K., Young, T. R., Nakasato, N., *et al.* 1997, *ApJ*, 477, 865
Joggerst, C. C., Woosley, S. E., & Heger, A. 2009, *ApJ*, 693, 1780
Joggerst, C. C., Almgren, A., Bell, J., *et al.* 2010, *ApJ*, 709, 11
Joggerst, C. C., Almgren, A., & Woosley, S. E. 2010, *ApJ*, 723, 353
Junde, H., Su, H., & Dong, Y. 2011, *Nuclear Data Sheets*, 112, 1513
Kageyama, A. & Sato, T. 2004, *Geochemistry, Geophysics, Geosystems*, 5, Q09005
Kifonidis, K., Plewa, T., Janka, H.-T., & Müller, E. 2000, *ApJ*, 531, L123
Kifonidis, K., Plewa, T., Janka, H.-T., & Müller, E. 2003, *A&A*, 408, 621

Kifonidis, K., Plewa, T., Scheck, L., Janka, H.-T., & Müller, E. 2006, *A&A*, 453, 661
Li, H., McCray, R., & Sunyaev, R. A. 1993, *ApJ*, 419, 824
Milisavljevic, D., & Fesen, R. A. 2013, *ApJ*, 772, 134
Milisavljevic, D., & Fesen, R. A. 2015, *Science*, 347, 526
Mueller, E., Fryxell, B., & Arnett, D. 1991, *A&A*, 251, 505
Nagataki, S., Hashimoto, M.-a., Sato, K., Yamada, S., & Mochizuki, Y. S. 1998, *ApJ*, 492, L45
Ono, M., Nagataki, S., Ito, H., *et al.* 2013, *ApJ*, 773, 161
Orlando, S., Miceli, M., Pumo, M. L., & Bocchino, F. 2015, *ApJ*, 810, 168
Orlando, S., Miceli, M., Pumo, M. L., & Bocchino, F. 2016, *ApJ*, 822, 22
Plewa, T. & Müller, E. 1999, *A&A*, 342, 179
Podsiadlowski, P., Morris, T. S., & Ivanova, N. 2007, Supernova 1987A: 20 Years After: Supernovae and Gamma-Ray Bursters, 937, 125
Potter, T. M., Staveley-Smith, L., Reville, B., *et al.* 2014, *ApJ*, 794, 174
Tanaka, T. & Washimi, H. 2002, *Science*, 296, 321
Utrobin, V. P., Wongwathanarat, A., Janka, H.-T., & Müller, E. 2015, *A&A*, 581, A40
Wongwathanarat, A., Janka, H.-T., & Müller, E. 2010, *ApJ*, 725, L106
Wongwathanarat, A., Hammer, N. J., & Müller, E. 2010, *A&A*, 514, A48
Wongwathanarat, A., Janka, H.-T., & Müller, E. 2013, *A&A*, 552, A126
Wongwathanarat, A., Müller, E., & Janka, H.-T. 2015, *A&A*, 577, A48
Wongwathanarat, A., Janka, H.-T., Mueller, E., Pllumbi, E., & Wanajo, S. 2016, arXiv:1610.05643
Woosley, S. E. 1988, *ApJ*, 330, 218
Yamada, S. & Sato, K. 1990, *ApJ*, 358, L9

Supernova 1987A:30 years later - Cosmic Rays and Nuclei from Supernovae and their aftermaths
Proceedings IAU Symposium No. 331, 2017
A. Marcowith, M. Renaud, G. Dubner, A. Ray & A. Bykov, eds.

doi:10.1017/S1743921317004549

Spatial distribution of radionuclides in 3D models of SN 1987A and Cas A

Hans-Thomas Janka[1], Michael Gabler[2] and Annop Wongwathanarat[1,3]

[1]Max Planck Institute for Astrophysics, Postfach 1317, D-85741 Garching, Germany
email: thj@mpa-garching.mpg.de

[2]Max Planck Institute for Astrophysics, Postfach 1317, D-85741 Garching, Germany
email: miga@mpa-garching.mpg.de

[3]RIKEN, Astrophysical Big Bang Laboratory, 2-1 Hirosawa, Wako, Saitama 351-0198, Japan
email: annop.wongwathanarat@riken.jp

Abstract. Fostered by the possibilities of multi-dimensional computational modeling, in particular the advent of three-dimensional (3D) simulations, our understanding of the neutrino-driven explosion mechanism of core-collapse supernovae (SNe) has experienced remarkable progress over the past decade. First self-consistent, first-principle models have shown successful explosions in 3D, and even failed cases may be cured by moderate changes of the microphysics inside the neutron star (NS), better grid resolution, or more detailed progenitor conditions at the onset of core collapse, in particular large-scale perturbations in the convective Si and O burning shells. 3D simulations have also achieved to follow neutrino-driven explosions continuously from the initiation of the blast wave, through the shock breakout from the progenitor surface, into the radioactively powered evolution of the SN, and towards the free expansion phase of the emerging remnant. Here we present results from such simulations, which form the basis for direct comparisons with observations of SNe and SN remnants in order to derive constraints on the still disputed explosion mechanism. It is shown that predictions based on hydrodynamic instabilities and mixing processes associated with neutrino-driven explosions yield good agreement with measured NS kicks, light-curve properties of SN 1987A and asymmetries of iron and ^{44}Ti distributions observed in SN 1987A and Cassiopeia A.

Keywords. supernovae: general, supernovae: individual (Cassiopeia A, SN1987A), neutrinos, hydrodynamics, instabilities, nuclear reactions, nucleosynthesis, abundances

1. Introduction

Since the seminal works by Colgate & White (1966) and Arnett (1967) more than 50 years ago, the computational modeling of supernova (SN) explosions has experienced enormous progress with respect to the numerical methods, input physics, computational accuracy, and dimensionality. Nevertheless, despite many generations of successively improved simulations, initially performed in spherical symmetry (1D), then since the 1990's also in two dimensions (2D), and in recent years also in full 3D, the physical processes that cause the explosion are not established yet (for recent reviews, see Janka (2012), Burrows (2013), Foglizzo *et al.* (2015), Janka *et al.* (2016), Müller (2016), Janka (2017b)). For the far majority of normal core-collapse SNe the delayed neutrino-driven mechanism (Wilson (1985), Bethe & Wilson (1985)) is widely considered as most likely explanation, because it taps the vast reservoir of energy that is radiated by the nascent neutron star (NS) in neutrinos and outranges the explosion energy of SNe by a factor of several hundred.

The neutrino-driven mechanism also satisfies a number of fundamental requirements that any viable scenario should fulfill (for a detailed discussion, see Janka (2017b)). First, the mechanism is *not "robust"*, because it should allow for the formation of stellar-mass black holes (BHs), whose abundent existence has been confirmed by the recent measurements of gravitional waves from binary BH mergers (Abbott *et al.* (2016)). Second, it must be *inefficient*, because it should explain why SN explosion energies are so much lower than the gigantic amount of gravitational binding energy that is available during NS or BH formation. Third, it should be *self-regulated*, because the energy transferred to the ejecta does not largely exceed the binding energy of the progenitor shells outside of the degenerate core, i.e., it is as low as several 10^{49} erg to about 10^{50} erg near the low-mass side of SN progenitors and may be $\sim(1-2)\times 10^{51}$ erg for energetic explosions of stars around $20\,M_\odot$. This clearly separates such normal SNe from the significantly more powerful but much rarer hypernovae (with a rate of less than roughly one out of thousand core-collapse events), whose energies and explosion properties point to another mechanism, probably invoking the formation of BHs or magnetars and of jet-driven outflows caused by extreme amplification of magnetic fields during the collapse of rapidly rotating progenitors (see, e.g., Woosley & Bloom (2006)). The latter are the final outcome of very special and uncommon single and binary star evolution scenarios of massive stars (e.g., Levan *et al.* (2016)).

Although first 3D simulations with energy-dependent neutrino transport have meanwhile obtained successful explosions by neutrino heating (Takiwaki *et al.* (2014), Melson *et al.* (2015a), Melson *et al.* (2015b), Lentz *et al.* (2015), Roberts *et al.* (2016), Müller (2016)), the viability of this theoretical scenario is still not generally accepted (e.g., Soker (2017a), Soker (2017b), Kushnir & Katz (2015), Blum & Kushnir (2016)). Doubts are either motivated by refering to those computational models where the numerical setups still failed to produce explosions, or they are justified by pointing to remaining shortcomings of the current calculations, for example the lack of a clear demonstration by modern simulations that neutrino-driven explosions can yield energies around 10^{51} erg or more (e.g., Soker (2017a) and references therein). Such missing pieces in the puzzle are very likely to fall into place once longer and better resolved 3D simulations are performed, larger sets of progenitors are investigated, more consistent initial conditions that account for large-scale perturbations in the convective silicon and oxygen burning shells are applied (Couch & Ott (2013), Couch *et al.* (2015), Müller & Janka (2015), Müller (2016)), and further improvements of the microphysics are used for the description of dense NS matter and of neutrino interactions in the correlated nuclear medium (see the sensitivity test by Melson *et al.* (2015b)).

Moreover, convincing alternatives to the neutrino-heating mechanism, which could elucidate the processes that initiate and power most core-collapse SNe, do not exist. Considering thermonuclear burning in the stellar carbon shell during gravitational collapse as the main energy source of the SN blast wave (Kushnir & Katz (2015)) demands a radical change of the chemical structure of progenitor stars in conflict with the results of stellar evolution calculations. This approach to overrule the common notion of stellar evolution appears as an unnecessary and overmotivated act of desperation. Similarly unsatisfactory (and not particularly elegant) is a renunciation of self-consistency in the explosion modeling by introducing, in an ad hoc manner, components whose physical origin remains unexplained, for example "jittering jets" (Soker (2017a) and references therein). Magnetic fields, which are often readily invoked to bridge gaps of reasoning, cannot be made responsible for the production of (jittering) jets in most or even in all SNe as claimed

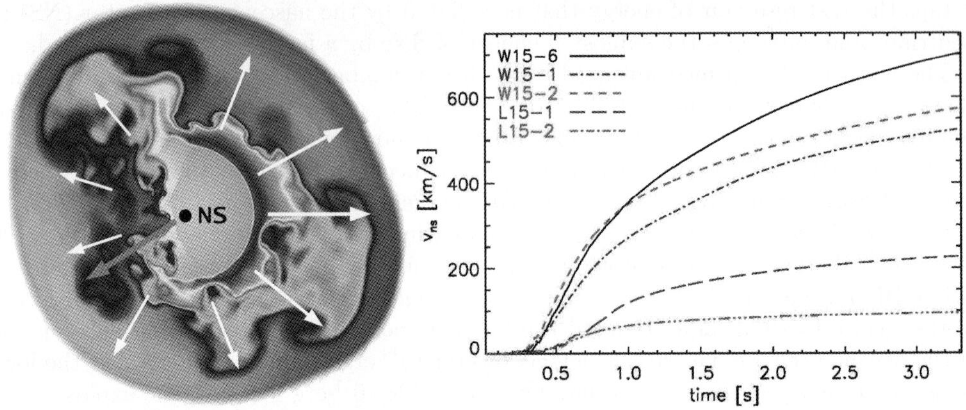

Figure 1. *Left:* Mass ejection asymmetry in a cross-sectional plane for one of the 3D simulations of neutrino-driven explosions (at 1.3 s after core bounce) published by Wongwathanarat *et al.* (2013). The NS kick direction (marked by the red arrow) is opposite to the hemisphere of the stronger explosion, where relatively more mass from silicon to the iron group is expelled. The entropy per nucleon is color coded (blue, green, yellow, red signal growing values), and the outer contour is the deformed SN shock. *Right:* NS star kick velocities as functions of time for a subset of the 3D simulations of neutrino-driven explosions in Wongwathanarat *et al.* (2013).

by Soker (2017a). Many stellar collapse simulations including the effects of magnetic fields (e.g., Moiseenko & Bisnovatyi-Kogan (2007), Burrows *et al.* (2007), Mösta *et al.* (2014), Mösta *et al.* (2015), Obergaulinger & Aloy (2017)) have demonstrated that very rapid rotation of the progenitor cores is needed to obtain magnetohydrodynamic jets by field amplification. The thus required angular momentum in stellar cores is in conflict with the moderate spin rates of white dwarfs (Kawaler (2015)) and young neutron stars (Heger *et al.* (2000), Heger *et al.* (2005)), with predictions from stellar evolution calculations including angular momentum transport through magnetic fields for progenitors of normal core-collapse SNe (Heger *et al.* (2005)), and with astroseismological measurements, which find core rotation rates that are considerably *slower* than predicted by the stellar evolution models (e.g., Beck *et al.* (2012), Eggenberger *et al.* (2012), Eggenberger *et al.* (2016)). Angular momentum separation in collapsed stellar cores by spiral modes of the standing accretion shock instability (SASI), which can develop even in nonrotating stars (e.g. Kazeroni *et al.* (2017)), has been shown to potentially lead to considerable growth of the initial B-fields (Endeve *et al.* (2012)). However, the emerging field configuration is highly turbulent and does not possess the ordered structure on large scales that is necessary for driving jet outflows. This problem will not disappear by future 3D simulations with increased grid refinement, because better resolution of the turbulent flow will foster a growth of the fields on *small scales* while it is unlikely to boost a large-scale field that could enable jet formation.

Despite these direct and indirect arguments in favor of neutrino-driven explosions, this theoretical scenario requires further consolidation by demonstrating that corresponding explosion models are able to account for the observational properties of SNe and of their remnants. In the following, a number of such aspects will be briefly summarized that provide support for the neutrino-driven mechanism.

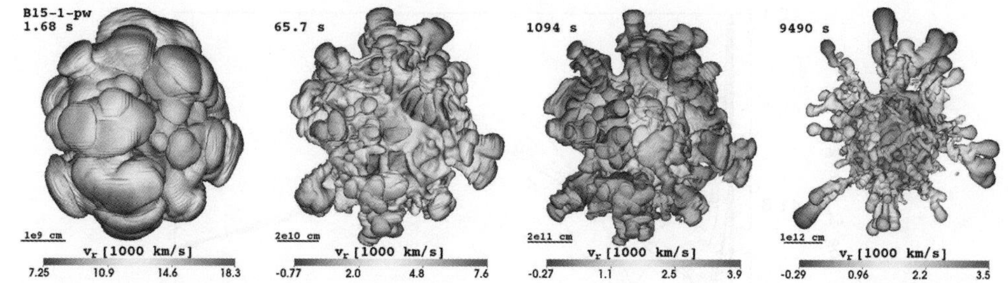

Figure 2. Fragmentation and growth of initial explosion asymmetries by secondary Rayleigh–Taylor instabilities at the C-O/He and He/H composition interfaces for a $15\,M_\odot$ BSG star. The images show isosurfaces for a mass-fraction of 3% iron-group elements at the time when the SN shock crosses the C-O/He and He/H interfaces (*left* and *second*, respectively), before the reverse shock from the He/H interface hits the iron ejecta (*third*), and at shock breakout (*right*). (Images from Wongwathanarat *et al.* (2015))

2. Neutrino-driven explosions: 3D models versus observations

In the following we present some selected results from recent 3D calculations of neutrino-driven explosions that were performed with the PROMETHEUS-HOTB code for different red (RSG) and blue supergiants (BSG) of $15\,M_\odot$ and $20\,M_\odot$ by Wongwathanarat *et al.* (2013), Wongwathanarat *et al.* (2015), Wongwathanarat *et al.* (2016). The blast waves were artifically initiated with a parametric neutrino "engine" such that explosion energies in the ballpark of quite energetic SNe like SN 1987A and Cassiopeia A were obtained. Although the modeling is not based on a fully self-consistent first-principle approach, the simulations still capture the essential physics of neutrino-driven explosions, which start by the energy deposition of neutrinos around the newly formed NS and succeed by the crucial support of hydrodynamic instabilities (convective overturn and SASI), creating large-scale asymmetries in the outgoing SN shock and the innermost ejecta already during the very first second of the explosion. The subsequent SN evolution is then followed continuously from core bounce to shock breakout at the progenitor surface and beyond (see the contribution in this volume by M. Gabler *et al.*).

Neutron-star kicks. The asymmetric onset of the explosion causes momentum transfer to the newly formed NS mainly by the long-time gravitational interaction between the compact remnant and the anisotropically expelled matter (Scheck *et al.* (2006), Wongwathanarat *et al.* (2013)). This leads to NS kicks opposite to the direction of the stronger explosion (consistent with linear momentum conservation in the disrupted star; Fig. 1, left). The kick velocities $v_{\rm ns}$ cover the full range of measured space velocities of young NSs up to nearly $1000\,{\rm km\,s^{-1}}$ (Fig. 1, right):

$$v_{\rm ns} = 211\,\frac{\rm km}{\rm s}\,\zeta\left(\frac{\alpha_{\rm ej}}{0.1}\right)\left(\frac{E_{\rm exp}}{10^{51}\,{\rm erg}}\right)\left(\frac{M}{1.5\,M_\odot}\right)^{-1}, \tag{2.1}$$

(Janka (2017a)), where $\alpha_{\rm ej}$ is the momentum-asymmetry of the innermost ejecta, $E_{\rm exp}$ the explosion energy, M the NS mass, and ζ is a numerical factor of order unity. $\alpha_{\rm ej}$ is determined by stochastic effects during the onset of the asymmetric explosion. While it is found to vary usually between 0 and $\sim$0.3 in existing 2D and 3D SN models, higher values are well possible, and kick velocities in excess of $1000\,{\rm km\,s^{-1}}$ seem in reach for extreme explosions with high energies and large asphericities.

Light curve of SN 1987A. The initial ejecta asymmetries, which are caused by the neutrino-driven mechanism and its associated hydrodynamic instabilities, lead to a global asphericity of the explosion. They also trigger efficient growth of secondary Rayleigh-

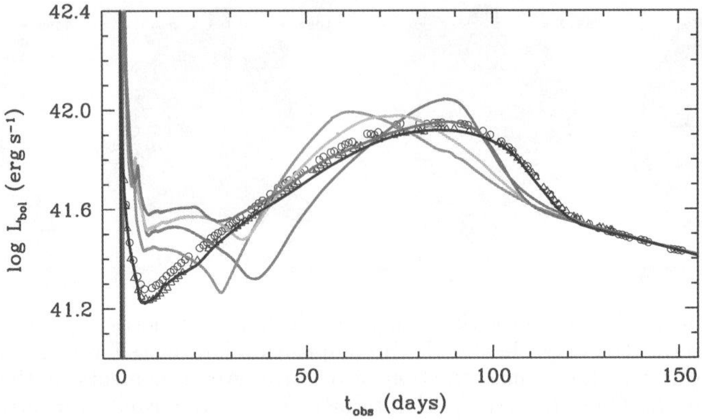

Figure 3. Bolometric light curves for SN calculations of four BSG progenitors (solid colored lines) compared to the observed light curve of SN 1987A (symbols; see Utrobin *et al.* (2015)). The neutrino-driven explosions from shock formation until beyond shock breakout were simulated in 3D, while the long-time radiation-hydrodynamical light-curve calculations were carried out in 1D, starting from spherically averaged data of the 3D explosion models. The blue line corresponds to the model shown in Fig. 2, where strong outward mixing of radioactive nickel and inward mixing of hydrogen allow for a good reproduction of the light-curve peak of SN 1987A. The imperfect match around the luminosity minimum and during the decline phase from the peak is a consequence of an overestimated radius and underestimated ejecta mass by the progenitor model. The black line corresponds to a specifically tailored stellar model.

Taylor (RT) instablities at the C-O/He and He/H shell interfaces of the progenitor by crossing density and pressure gradients after the passage of the outgoing shock (for a detailed discussion, see Wongwathanarat *et al.* (2015)).

These RT instabilities induce the fragmentation of the initial large-scale structures and counteract the deceleration of the inner ejecta by the reverse shock that moves into the metal core. As a consequence, effective outward mixing of radioactive ^{56}Ni (in extended finger-like plumes with velocities up to $\sim$3500 km s^{-1}, see Fig. 2) and inward mixing of hydrogen (to velocities close to zero) can take place. 3D simulations of neutrino-driven explosions are thus able to account for the radial mixing and high nickel velocities that are needed to explain the shape of the light curve of SN 1987A (Fig. 3), the early observations of gamma-rays and X-rays from radioactive decays, and a variety of spectral features of this SN (Arnett *et al.* (1989) and references therein).

Silicon and iron ejecta in SN 1987A. Based on detailed spectral and imaging observations with HST/STIS and VLT/SINFONI, Larsson *et al.* (2016) produced 3D maps of various chemical components of SN 1987A including silicon and iron. The morphology of these tracers of the innermost ejecta is shown in the left panel of Fig. 4. In the right panel of this figure we display the iron distribution (Si+Fe only differs insignificantly) in a 3D neutrino-driven explosion model for a 15 $M_\odot$ star with properties (energy, ejecta geometry) resembling those of SN 1987A. We only display the dense, inner regions of the iron material up to radial velocities of 2500 km s^{-1}, guided by the space velocities of the centers of the brightest regions in the left panel. Moreover, following the observational image we also omit all mass with velocities between -450 km s^{-1} and $+450$ km s^{-1} along the line of sight. Assuming that 15–30% of high-velocity Fe and Si, of which some are in extended fingers similar to those in Fig. 4, remain invisible is justified for three reasons: First, although such high-velocity material is not seen in the observations, it *must* be present in SN 1987A to explain its light curve and spectral properties (see above). Second, the fingers and plumes of high-velocity Si and Fe possess an average mass-density

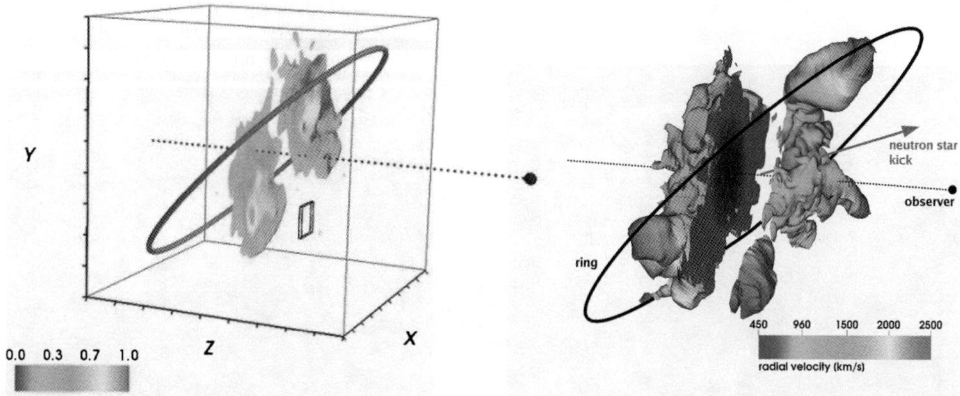

Figure 4. *Left:* 3D isosurfaces for [SiI]+[FeII] in SN 1987A, combining the levels for 30%, 50%, and 70% of the maximal intensity as indicated by the color bar from Larsson *et al.* (2016). The ring corresponds to the position of the reverse shock at the inner edge of the equatorial ring, and the dotted line marks the viewing direction with the observer location indicated by the filled circle. Tick marks on the axes are spaced in intervals of 1000 km s^{-1}. *Right:* Similar visualization of the iron distribution in a 3D neutrino-driven explosion simulation with an energy close to that of SN 1987A. The explosion has been evolved to 600 days including heating by ^{56}Ni decay. The image shows an isosurface of constant mass-density of ^{56}Fe with the radial velocity color coded. Including silicon yields a very similar morphology but higher velocities. As in the observation, the central region within $\pm$450 km s^{-1} along the line of sight has been removed. The observer direction (dashed line with bullet for observer position) was chosen for optimal similarity with the observations; the corresponding ring plane and the NS kick direction of the SN model are also indicated.

that is at least a factor of 10 lower than the density of the central bulk of iron. Third, the observed [SiI]+[FeII] emission is powered by positrons from ^{44}Ti decay (Larsson *et al.* (2016)), but titanium in the extended plumes is less abundant (at least a factor of 2) relative to iron than in the more concentrated, lower-velocity bulk. Therefore, the excitation of Si and Fe by ^{44}Ti-decay positrons must be expected to be less efficient. The resemblance of the two images in Fig. 4 is assuring, in particular since the 3D simulation was *not tailored* to match SN 1987A.

^{44}Ti and iron in Cassiopeia A. Using NuSTAR observations Grefenstette *et al.* (2014) and Grefenstette *et al.* (2017) mapped the 3D distribution of ^{44}Ti in Cas A, which is a perfect tracer of the explosion geometry of this SN. They found a clumpy, non-uniform distribution of titanium around the centre of expansion with most of the high-emission ^{44}Ti knots being concentrated in the hemisphere opposite to the NS kick direction. Without any fine tuning, one of the 3D neutrino-driven explosion simulations of Wongwathanarat *et al.* (2015) is able to account for the main properties of the observed titanium and iron distributions (Figs. 5 and 6; Wongwathanarat *et al.* (2016)). The theoretical model can explain the total yields of these nuclei estimated for Cas A, their 3D geometry in relation to the NS kick magnitude and direction, their radial and velocity distributions, and the relative variations of the ^{44}Ti/iron ratio suggested by the observations. The morphology of this explosion seems to be compatible with a neutrino-driven explosion that produced three large-scale plumes of neutrino-heated, high-entropy ejecta that are essentially located in one plane (Wongwathanarat *et al.* (2016)), which could be associated with the "tilted thick disk" identified in Cas A (see Grefenstette *et al.* (2017)).

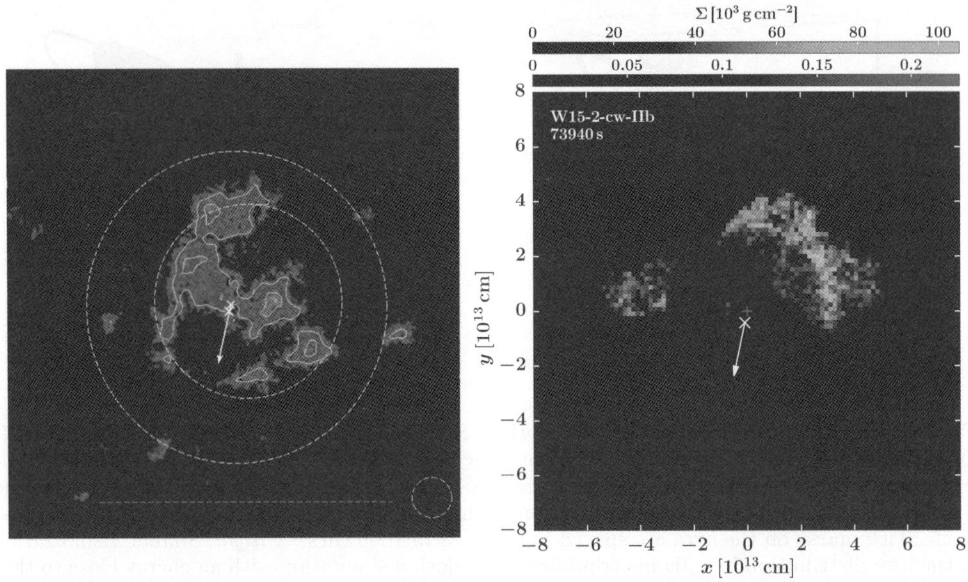

Figure 5. *Left:* ^{44}Ti distribution in Cas A with forward and reverse shock locations marked by dashed circles (reprinted by permission from Macmillan Publishers Ltd: Nature, Grefenstette *et al.* (2014), ©2014). *Right:* Distribution of ^{44}Ti (blue) and ^{56}Fe (decay product of ^{56}Ni) in a 3D simulation of a neutrino-driven explosion (Wongwathanarat *et al.* (2016)). The reverse shock is assumed to have moved inward through half of the iron, which is therefore visible and displayed only in the shock-heated outer shell. In both images the geometrical center of the expansion and the location and kick direction of the NS are indicated.

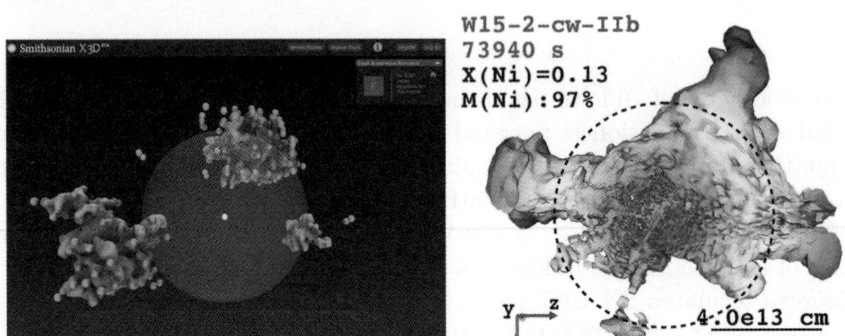

Figure 6. *Left:* Iron K-shell emission around the reverse-shock sphere of Cas A as measured by the Chandra X-ray observatory (DeLaney *et al.* (2010), Hwang & Laming (2012)). The NS is marked by a white bullet. (Image copied from 3D visualization available at `http://3d.si.edu/explorer?modelid=45`.) *Right:* Distribution of ^{56}Fe in a 3D simulation of a neutrino-driven explosion (Wongwathanarat *et al.* (2016)). The NS kick direction (red arrow) and the approximate location of the reverse shock (dashed circle, assuming 50% of the iron to be shock heated) are indicated.

3. Conclusions

Neutrino-driven explosion simulations in 3D can explain basic observational properties of the distribution of the innermost ejecta (iron-group material, radioactive ^{44}Ti) observed in young SN remnants such as SN 1987A and Cas A. Invoking new features in an ad hoc manner (e.g. "jittering jets", Soker (2017a)) instead of relying on self-consistently developing hydrodynamic instabilities that collaborate with the neutrino-heating mechanism, is not necessary.

For SN 1987A neutrino-driven explosions predict the existence of a NS as a compact remnant. In view of a diagnosed explosion energy of $(1.3 - 1.5) \times 10^{51}$ erg, massive fallback is highly unlikely for the progenitors (with He-core masses between 4 and $6\,M_\odot$) considered for SN 1987A. Because of a maximum NS mass of at least $2\,M_\odot$ (Antoniadis *et al.* (2013), Demorest *et al.* (2010)) the collapse of the compact remnant of a neutrino-driven explosion to a BH can be ruled out. If the existence of a BH in SN 1987A can be proven, radical revisions of our understanding of the SN mechanism would therefore be inescapable (e.g., Blum & Kushnir (2016)).

Acknowledgments. This work was supported by the Deutsche Forschungsgemeinschaft through the Cluster of Excellence EXC 153 "Origin and Structure of the Universe", by the European Research Council through grant ERC-AdG No. 341157-COCO2CASA. The computations were performed on Hydra of the Max Planck Computing and Data Facility.

References

Abbott, B. P., *et al.* 2016, *Phys. Rev. Lett.*, 116, 061102
Antoniadis, J., *et al.* 2013, *Science*, 340, 448
Arnett, D. 1967, *Canadian Journal of Physics*, 45, 1621
Arnett, W. D., Bahcall J. N., Kirshner R. P. & Woosley, S. E. 1989, *Ann. Rev. Astron. Astrophys.*, 27, 629
Beck, P. G., *et al.* 2012, *Nature*, 481, 55
Bethe, H. & Wilson, J. R. 1985, *Astrophys. J.*, 295, 14
Blum, K. & Kushnir, D. 2016, *Astrophys. J.*, 828, 31
Burrows, A. 2013, *Rev. Mod. Phys.*, 85, 245
Burrows, A., Dessart L., Livne E., Ott, C. D., & Murphy, J. 2007, *Astrophys. J.*, 664, 416
Colgate, S. A. & White, R. H. 1966, *Astrophys. J.*, 143, 626
Couch, S. M. & Ott, C. D. 2013, *Astrophys. J. Lett.*, 778, L7
Couch, S. M., Chatzopoulos, E., Arnett, W. D., & Timmes, F. X. 2015, *Astrophys. J. Lett.*, 808, L21
DeLaney, T., *et al.* 2010, *Astrophys. J.*, 725, 2038
Demorest, P. B., Pennucci, T., Ransom, S. M., Roberts M. S. E., & Hessels, J. W. T.. 2010, *Nature*, 467, 1081
Eggenberger, P., Montalbán J. & Miglio, A. 2012, *Astron. Astrophys.*, 544, L4
Eggenberger, P., *et al.* 2016, *Astronomische Nachrichten*, 337, 832
Endeve, E., *et al.* 2012, *Astrophys. J.*, 751, 26
Foglizzo, T., *et al.* 2015, *Publ. Astron. Soc. Australia*, 32, e009
Grefenstette, B. W., *et al.* 2014, *Nature*, 506, 339
Grefenstette, B. W., *et al.* 2017, *Astrophys. J.*, 834, 19
Heger, A., Langer, N., & Woosley, S. E. 2000, *Astrophys. J.*, 528, 368
Heger, A., Woosley, S. E., & Spruit, H. C. 2005, *Astrophys. J.*, 626, 350
Hwang, U. & Laming, J. M. 2012, *Astrophys. J.*, 746, 130
Janka, H.-T. 2012, *Ann. Rev. Nucl. Part. Sci.*, 62, 407
Janka, H.-T. 2017a, *Astrophys. J.*, 837, 84
Janka, H.-T. 2017b, *e-print arXiv:1702.08825*
Janka, H.-T., Melson, T., & Summa, A. 2016, *Ann. Rev. Nucl. Part. Sci*, 66, 341
Kawaler, S. D. 2015, *ASP Conference Series*, 493, p. 65; *e-print arXiv:1410.6934*
Kazeroni, R., Guilet, J., & Foglizzo, T. 2017, *eprint arXiv:1701.07029*
Kushnir, D. & Katz, B. 2015, *Astrophys. J.*, 811, 97

Larsson, J., *et al.* 2016, *Astrophys. J.*, 833, 147

Lentz, E. J., *et al.* 2015, *Astrophys. J. Lett.*, 807, L31

Levan, A., Crowther P., de Grijs, R., Langer, N., Xu, D., & Yoon, S.-C. 2016, *Space Science Reviews*, 202, 33

Melson, T., Janka, H.-T., & Marek, A. 2015, *Astrophys. J. Lett.*, 801, L24

Melson, T., Janka, H.-T., Bollig, R., Hanke, F., Marek, A., & Müller, B. 2015, *Astrophys. J. Lett.*, 808, L42

Mösta, P., *et al.* 2014, *Astrophys. J. Lett.*, 785, L29

Mösta, P., Ott, C. D., Radice D., Roberts, L. F., Schnetter, E., & Haas, R. 2015, *Nature*, 528, 376

Moiseenko, S. G. & Bisnovatyi-Kogan, G. S. 2007, *Astrophys. and Space Science*, 311, 191

Müller, B. 2016, *Publ. Astron. Soc. Australia*, 33, e048

Müller, B. & Janka, H.-T. 2015, *Monthly Not. R. Astron. Soc.*, 448, 2141

Obergaulinger, M. & Aloy, M.Á. 2017, *eprint arXiv:1703.09893*

Roberts, L. F., Ott, C. D., Haas, R., O'Connor, E. P., Diener, P., & Schnetter, E. 2016, *Astrophys. J.*, 831, 89

Scheck, L., Kifonidis, K., Janka, H.-T., & Müller, E. 2006, *Astron. Astrophys.*, 457, 963

Soker, N. 2017a, *e-print arXiv:1702.03451*

Soker, N. 2017b, *e-print arXiv:1703.03673*

Takiwaki, T., Kotake, K., & Suwa, Y. 2014, *Astrophys. J.*, 786, 83

Utrobin, V. P., Wongwathanarat, A., Janka, H.-T., & Müller, E. 2015, *Astron. Astrophys.*, 581, A40

Wilson, J. R. 1985, in: J.M. Centrella, J.M. LeBlanc & R.L. Bowers (eds.), *Numerical Astrophysics* (Boston: Jones and Bartlett Publ.), p. 422

Wongwathanarat, A., Janka, H.-T., & Müller, E. 2013, *Astron. Astrophys.*, 552, A126

Wongwathanarat, A., Müller, E., & Janka, H.-T. 2015, *Astron. Astrophys.*, 577, A48

Wongwathanarat, A., Janka, H.-T., Müller, E., Pllumni, E., & Wanajo, S. 2016, *e-print arXiv:1610.05643*

Woosley, S. E. & Bloom, J. S. 2006, *Ann. Rev. Astron. Astrophys.*, 44, 507

Supernova 1987A:30 years later - Cosmic Rays and Nuclei from Supernovae and their aftermaths
Proceedings IAU Symposium No. 331, 2017
A. Marcowith, M. Renaud, G. Dubner, A. Ray & A. Bykov, eds.

doi:10.1017/S1743921317004343

Gamma-ray line measurements from supernova explosions

Roland Diehl

Max Planck Institut für extraterrestrische Physik
Giessenbachstr. 1, D-85748 Garching, Germany
email: rod@mpe.mpg.de

Abstract. Gamma ray lines are expected to be emitted as part of the afterglow of supernova explosions, because radioactive decay of freshly synthesised nuclei occurs. Significant radioactive gamma ray line emission is expected from ^{56}Ni and ^{44}Ti decay on time scales of the initial explosion (^{56}Ni, τ ~days) and the young supernova remnant (^{44}Ti,τ ~90 years). Less specific, and rather informative for the supernova population as a whole, are lessons from longer lived isotopes such as ^{26}Al and ^{60}Fe. From isotopes of elements heavier than iron group elements, any interesting gamma-ray line emission is too faint to be observable. Measurements with space-based gamma-ray telescopes have obtained interesting gamma ray line emissions from two core collapse events, Cas A and SN1987A, and one thermonuclear event, SN2014J. We discuss INTEGRAL data from all above isotopes, including all line and continuum signatures from these two objects, and the surveys for more supernovae, that have been performed by gamma ray spectrometry. Our objective here is to illustrate what can be learned from gamma-ray line emission properties about the explosions and their astrophysics.

Keywords. gamma rays: observations, (stars:) supernovae: general, (stars:) supernovae: individual (Cas A, SN1987A, SN2014J), (ISM:) supernova remnants, Galaxy: kinematics and dynamics

1. Introduction

Supernova explosions have been associated with cosmic matter in hot and dense environments where nuclear reactions can change its composition. The light emitted by a supernova is typically found to rise in brightness within few days and then last for months; these characteristics are plausibly connected with the production of radioactive elements in the explosion itself, from such nuclear fusion reactions. The release of this radioactive energy into the expanding explosion thus is a plausible driver of observed supernova light curves†.

The ^{56}Ni isotope is nature's most-stable n/p-symmetric atomic nucleus, and has a prominent and specific role in the context of supernovae, with its decay first to ^{56}Co $\tau = 8.8$ days, or $T_{1/2} = 6.1$ days, and thereafter from ^{56}Co to stable ^{56}Fe $\tau = 111.7$ days, or $T_{1/2} = 77.4$ days. In particular, the near-exponential decline of supernova brightness with time and its similarity to the ^{56}Co decay time was often taken as supportive – although in view of the rapidly-expanding supernova and complex radiation-deposit and -transport and -escape physics, this nowadays appears more as a coincidence.

The production of significant amounts of ^{56}Ni in all 'normally-bright' supernovae has been a consensus, supported by successful modeling of supernova brightness behaviour

† The explosion blast and even its recombination effect after explosion has ionised the matter, these would only result in brief luminosity lasting typically hours to days at most

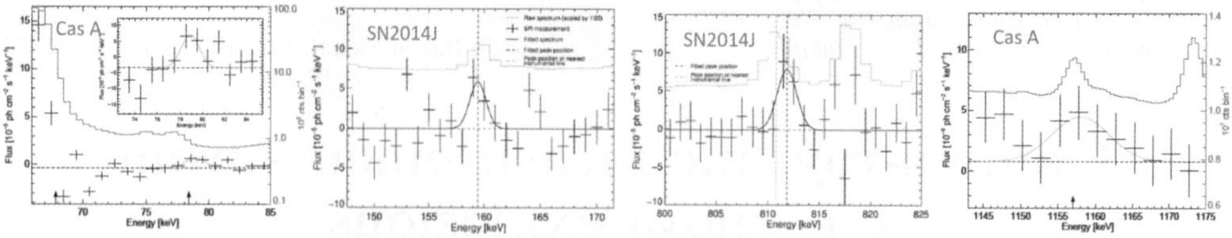

Figure 1. Examples of gamma-ray line detections from supernovae with the SPI spectrometer on INTEGRAL. Shown are *(from left to right, data points)*: Cas A ^{44}Ti line 78 keV (Siegert *et al.* 2015), SN2014J ^{56}Ni lines 158 keV and 812 keV (Diehl *et al.* 2014), and Cas A ^{44}Ti line 1157 keV (Siegert *et al.* 2015), and raw data *(histograms)*, scaled-down in intensity, to illustrate instrumental-background spectra and their line features.

based on ^{56}Ni as energy source and radiation transport physics alone. Typically, core-collapse supernovae are though to produce on the order of 0.05 M$_\odot$ of ^{56}Ni, while thermonuclear supernovae produce about ten times as much.

It has always been desirable to observe ^{56}Ni decay directly though characteristic gamma rays, thus avoiding complex radiation transport as part of the interpretation of supernova brightness. But in the absence of sufficiently sensitive telescopes and/or sufficiently nearby supernovae, this only became a reality rather recently (see Fig. 1), decades after much detail already had been disclosed from analysis of spectra and light curves in other spectral bands, from infrared to optical, UV and X-rays.

High resolution gamma ray spectroscopy began from first high-resolution detector balloons and satellites using solid state detectors with Ge as a main ingredient towards resolution $E/\Delta E$ of about 600 in the 1970ies (Mahoney *et al.* 1982; Leventhal *et al.* 1978). The evolution of missions and instrument since then now culminates with SPI on INTEGRAL (Vedrenne *et al.* 2003; Diehl 2013). The competing interests of astronomers for costly space missions make unlikely a successing next-generation mission in coming decades. Although the explosive nature leads to kinematic broadening of gamma-ray lines of several tens of keV, observing lines with keV-resolution instruments can be meaningful (see below and Fig. 2).

In this paper, we address ^{56}Ni gamma rays, and, by analogy, also discuss the astrophysical messages inherent to such type of gamma ray spectroscopy and other line emission from supernovae and their remnants.

2. Gamma rays and the physics of supernova explosions

Thermonuclear supernova explosions are believed to originate from ignition of Carbon burning of degenerate matter inside a white dwarf at densities exceeding 10^{12}g cm^{-3}. Nuclear reactions herein approach nuclear statistical equilibrium at least for a fraction of the white dwarf material, thus rearranging nucleons towards nuclei with a maximum of binding energy. The successful empirical description of the nuclear-burning conditions and their evolution in the W7 model standard (Nomoto *et al.* 1984) predicts a total ^{56}Ni mass around 0.5 M$_\odot$ (see also Iwamoto *et al.* 1999, as another standard reference). Also many other model variants for thermonuclear supernovae have predicted ^{56}Ni production within factors of a few of this (see reviews in Röpke *et al.* 2011; Hillebrandt *et al.* 2013), and ^{56}Ni radioactive energy dominates the energy input as the supernova obtains it brightness maximum, typically 20 days after the onset of the explosion (see discussion of 3D models and other radioactivities in Seitenzahl *et al.* 2013).

The main diagnostic power of gamma-ray observations lies in the brightness evolution of the characteristic gamma-ray lines from the ^{56}Ni decay chain, which includes lines from

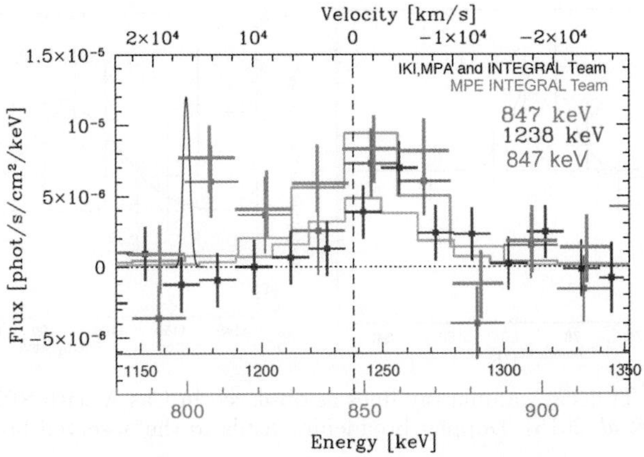

Figure 2. The ^{56}Co decay gamma ray lines as observed in SN2014J. Shown are superpositions of the line profiles of the 1238 (blue) and 847 keV lines (red, green) from this decay; the 847 keV analysis with two different analysis methods shows a consistent result in spectral detail. Doppler broadening leads to the observed broad lines, and appears consistently for both lines of ^{56}Co decay. The Gaussian black line profile shows how a line as narrow as instrumental spectral resolution would appear.

the first decay to ^{56}Co as well as to the second and final product ^{56}Fe, and the 511 keV line from positrons produced in the β^+-decay. The expanding supernova is initially dense enough to absorb all such gamma rays, thus converting energy to lower-energy radiation and internal energy of the expanding gas. As the supernova expands, more and more of the gamma rays will escape, thus becoming observable as their energy deposit in the supernova fades away. The brightness evolution thus reflects both the distribution of ^{56}Ni within the supernova and the absorbing overlying material. Measurements of the gamma ray brightness evolution therefore can *calibrate* the measurements of re-radiated radioactive energy, as they constrain the energy source. Moreover, the spatial distribution of ^{56}Ni may become observable rather directly, e.g. distinguishing inside-out explosions with ^{56}Ni enclosed in a large mass of white dwarf material from explosion scenarios where nuclear burning occurs partly or predominantly in outer regions of the exploding object.

Also core collapse of a massive star leads to high-density matter suitable for nuclear rearrangements and even nuclear statistical equilibrium (Nomoto *et al.* 2006). But in this case, the infalling and exploding gas dynamics is less constrained, thus may evolve over the time of collapse and onset of the explosion, and vary for different progenitor masses. As a result, the amount of ^{56}Ni expected from such explosions is much less and also more uncertain and variable, as equilibrium conditions only occur in a relatively small region of the available parameter space (Thielemann *et al.* 2011; Magkotsios *et al.* 2010). Moreover, the larger amount of overlying material in the outer shells above the collapsing iron core results in modest expectations of observable ^{56}Ni gamma ray emission, compared to the case of thermonuclear explosions. But another radioactive nucleus with a half life of about 60 years becomes interesting in this case, depositing radioactive energy mainly through its positrons from β^+-decay in the late supernova. At such late times the supernova is fully transparent to ^{44}Ti decay gamma rays, and these therefore allow to observe kinematics and spatial distribution of ^{44}Ti as a representative product of nuclear processing deep in the original supernova (Thielemann *et al.* 2011; Magkotsios *et al.* 2010).

Other line emission in the nuclear energy range may be expected at somewhat later times, as particle acceleration in young supernova remnant reaches a peak efficiency. This is expected within the first few hundred years of the remnant evolution, as ambient gas

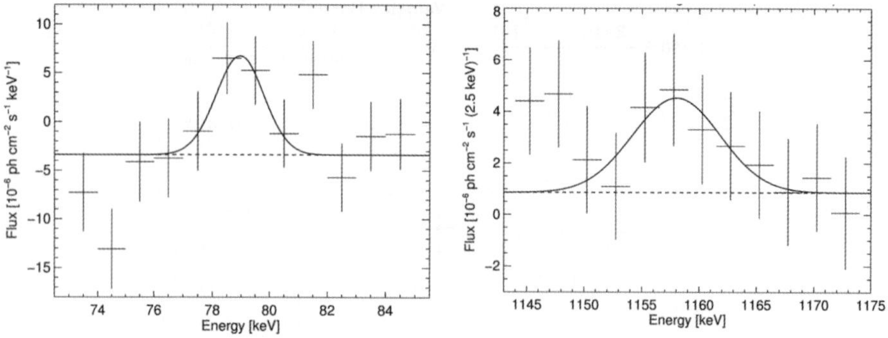

Figure 3. The ^{44}Ti decay gamma ray lines as observed in Cas A with SPI on INTEGRAL (Siegert *et al.* 2015). Doppler broadening leads to the observed broad lines.

is swept up by the still rapidly-expanding blast wave. Accelerated nuclei will result in collisionally-excited nuclei which de-excite with emission of characteristic nuclear lines (Ramaty *et al.* 1979).

Late emission of nuclear lines from long-lived radioactivities of ^{26}Al and ^{60}Fe from supernova ejecta provides another diagnostic about the cosmic importance of supernovae: The spatial distribution of ejecta at late times, as it mixes with ambient interstellar gas, determines the time scales of recycling of newly-produced nuclei from one generation of supernovae towards a next generation of stars that forms from enriched cold interstellar gas.

3. Core collapse events: SN1987 and Cas A

SN1987A was the first supernova where gamma ray lines were reported: Measured with the Solar Maximum Missions gamma-ray spectrometer (Matz *et al.* 1988), this nicely confirmed the ^{56}Ni origin of supernova light. This result moreover provided strong indications for explosion asymmetry, because the lines from ^{56}Co decay appeared much earlier than expected (see above). The GRIS balloon mission with a Ge detector provided the first measurement of the 847 keV ^{56}Co line profile (Tueller *et al.* 1990), indicating the redshift that seemed plausible from the early appearance. Later, INTEGRAL reported detection of ^{44}Ti emission from SN1987A (Grebenev *et al.* 2012), which was also measured with better spectroscopic quality by NuSTAR more recently (Boggs *et al.* 2015). Thus, SN1987A is not only the best-observed core collapse supernova that shapes what we believe today about this type of explosion (McCray & Fransson 2016), it also is a prominent example of the success of multi-wavelenth science.

The second supernova of type core collapse, which has been studied in many astronomical windows, is Cas A (Vink 2004). This supernova exploded already more than 350 years ago, so that ^{56}Ni amounts can only inferred indirectly here, from light echoes. The young Cas A supernova remnant shows several remarkable properties, such as fast-moving knots, and an apparent overturn of the original layering of material of its progenitor star.

^{44}Ti emission from Cas A has been detected first with COMPTEL. A diversity of efforts followed to confirm and improve upon this exciting possibility to measure ^{44}Ti radioactiivity from a supernova. Results were reported from OSSE on CGRO (The *et al.* 1996), RXTE (Rothschild *et al.* 1999), Beppo-Sax (Vink *et al.* 2001), and both INTEGRAL instruments IBIS and SPI (Renaud *et al.* 2006a; Martin *et al.* 2009; Siegert *et al.* 2015), culminating with NuSTAR's observations where NuSTAR's X-ray mirror allowed to spatially resolve this emission (Grefenstette *et al.* 2014). This first image of ^{44}Ti gamma-ray

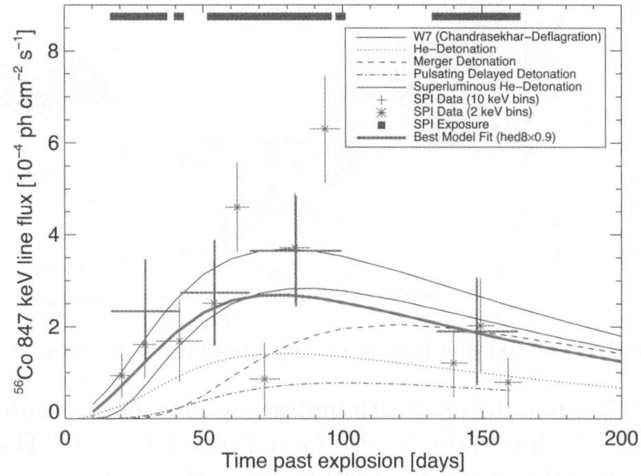

Figure 4. The ^{56}Co decay gamma ray line brightness evolution of SN2014J, as seen with SPI on INTEGRAL (Diehl *et al.* 2015).

line emission (Grefenstette *et al.* 2014) provides a key lesson on supernova nucleosynthesis and morphology, with interesting deviations from a spherical explosion (Grefenstette *et al.* 2017) (see also Grefenstette *et al.*, Janka *et al.*, Fryer *et al.*, and Wongwhatanarath, these proceedings). INTEGRAL/SPI data are being accumulated to refine spectroscopic results, as only SPI is capable to measure both the hard X-ray and gamma ray lines from the ^{44}Ti decay chain (Siegert *et al.* 2015). Preliminary latest analysis (Weinberger *et al.*, in preparation) provides interesting hints of a composite line structure, compatible with and supporting the latest spatially-resolved spectroscpic results from NuSTAR (Grefenstette *et al.* 2017). More data will be collected in the coming years of the continuing INTEGRAL mission.

4. Thermonuclear supernovae: SN2014J

The first gamma-ray line results from a thermonuclear supernova had been long awaited, and were realised with SN2014J and INTEGRAL's instruments (Churazov *et al.* 2014; Diehl *et al.* 2015). This provides an independent measurement of the ^{56}Ni mass ejected (Diehl *et al.* 2015). As a surprise, early line emission from the shortlived ^{56}Ni decay could be seen as well (Diehl *et al.* 2014). This indication of ^{56}Ni near the surface, but also the irregular rise of the ^{56}Co emission line brightness (Fig. 4) (Diehl *et al.* 2015), all provide strong indications that 1-dimensional models of a spherical explosion are too restrictive a model, and do not adequate represent reality.

5. Supernova explosions throughout the Galaxy

5.1. *Environments of supernova explosions*

Supernovae of type Ia may occur also in regions remote from the birth sites of their progenitors, while core collapse supernovae result from massive stars that are born in clusters and evolve within millions of years, so exploding near their formation sites. Therefore, ^{44}Ti emitting young supernova remnants were the objective of searches with gamma-ray surveys of the Compton (Dupraz *et al.* 1997; Iyudin *et al.* 1999) and INTEGRAL (Renaud *et al.* 2006b; Tsygankov *et al.* 2016) Observatories. The lessons from both surveys are (The *et al.* 2006; Dufour & Kaspi 2013) that typical core-collapse supernovae are not

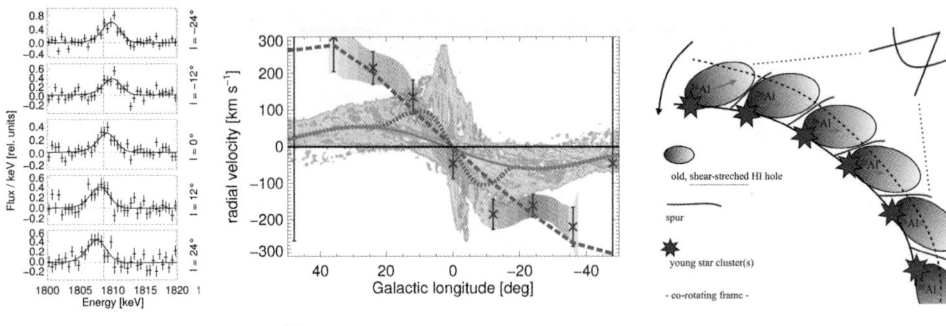

Figure 5. The ^{26}Al line lessons on superbubbles as recycling regions.

ejecting much of ^{44}Ti, while the ones with indications of strong deviations from spherical symmetry sometimes do. Interestingly, also from Tycho a hint of ^{44}Ti emission has been reported (Troja *et al.* 2014; Wang & Li 2014), while this event is commonly associated with a thermonuclear explosion.

The ^{26}Al map (Diehl *et al.* 1995) provides yet another survey of nucleosynthesis ejecta from supernovae, as its origins had been associated with massive star origins predominantly (Prantzos & Diehl 1996). Active regions as showing up in ^{26}Al have been studied in more detail. Interesting for the topic of this paper is the Orion-Eridanus region (Diehl *et al.* 2003), where several massive star groups are located on the near side of the Orion molecular clouds, and a large interstellar cavity, the Eridanus bubble, extends from those towards the Sun. In theoretical simulations and their comparisons with X- and gamma-ray emission of the region, it seems that supernova explosions drive the evolution and continuum emission characteristics of such superbubbles (Krause *et al.* 2014).

5.2. *How supernovae shape properties of the Galaxy*

The superbubble creation and evolution that has become apparent in the Orion-Eridanus example seems to even operate galaxy-wide: The ^{26}Al line shape measurements with SPI/INTEGRAL have revealed systematic Doppler shifts from large-scale galactic rotation (Fig. 5 left) (Kretschmer *et al.* 2013), that exceeds expectations from other tracers of galactic rotation by about 200 km s^{-1} (Fig. 5 center). This suggests that ^{26}Al ejection from massive-star groups systematically occurs into asymmetric superbubbles, that are formed as those massive-star groups have moved to the leading edge of spiral arms during the evolution time of the massive stars before their ^{26}Al ejection (Fig. 5 right) (Krause *et al.* 2015). This means that SN ejecta occur as a mixture of cold cavity walls and hot bubble interiors. As superbubbles break up in the later phase of their evolution, ejecta thus may escape in champagne flows or winds, and thus take considerable time to cool and condense as part of sites of new star formation: Recycling of SN ejecta appears to be a more complex process, that may not be captured in a single recycling time scale, and is not *instantaneous*, strictly speaking.

References

Boggs, S. E., Harrison, F. A., Miyasaka, H., *et al.* 2015, *Science*, 348, 670
Churazov, E., Sunyaev, R., Isern, J., *et al.* 2014, *Nature*, 512, 406
Diehl, R. 2013, *Reports on Progress in Physics*, 76, 026301
Diehl, R., Dupraz, C., Bennett, K., *et al.* 1995, *Astron.& Astroph.*, 298, 445
Diehl, R., Kretschmer, K., Plüschke, S., *et al.* 2003, *Astronomische Nachrichten Supplement*, 324, 18
Diehl, R., Siegert, T., Hillebrandt, W., *et al.* 2014, *Science*, 345, 1162

Diehl, R., Siegert, T., Hillebrandt, W., *et al.* 2015, *Astron.& Astroph.*, 574, A72
Dufour, F. & Kaspi, V. M. 2013, *Astroph.J.*, 775, 52
Dupraz, C., Bloemen, H., Bennett, K., *et al.* 1997, *Astron.& Astroph.*, 324, 683
Grebenev, S. A., Lutovinov, A. A., Tsygankov, S., & Winkler, C. 2012, *Nature*, (tbd), (accepted for publication)
Grefenstette, B. W., Fryer, C. L., Harrison, F. A., *et al.* 2017, *Astroph.J.*, 834, 19
Grefenstette, B. W., Harrison, F. A., Boggs, S. E., *et al.* 2014, *Nature*, 506, 339
Hillebrandt, W., Kromer, M., Röpke, F. K., & Ruiter, A. J. 2013, *Frontiers of Physics*, 8, 116
Iwamoto, K., Brachwitz, F., Nomoto, K., *et al.* 1999, *Astroph.J.Supp.*, 125, 439
Iyudin, A. F., Schönfelder, V., Bennett, K., *et al.* 1999, *Astrophysical Letters Communications*, 38, 383
Krause, M., Diehl, R., Böhringer, H., Freyberg, M., & Lubos, D. 2014, *Astron.& Astroph.*, 566, A94
Krause, M. G. H., Diehl, R., Bagetakos, Y., *et al.* 2015, *Astron.& Astroph.*, 578, A113
Kretschmer, K., Diehl, R., Krause, M., *et al.* 2013, *Astron.& Astroph.*, 559, A99
Leventhal, M., MacCallum, C. J., & Stang, P. D. 1978, *Astroph.J.Lett.*, 225, L11
Magkotsios, G., Timmes, F. X., Hungerford, A. L., *et al.* 2010, *Astroph.J.Supp.*, 191, 66
Mahoney, W. A., Ling, J. C., Jacobson, A. S., & Lingenfelter, R. E. 1982, *Astroph.J.*, 262, 742
Martin, P., Knödlseder, J., Vink, J., Decourchelle, A., & Renaud, M. 2009, *Astron.& Astroph.*, 502, 131
Matz, S. M., Share, G. H., Leising, M. D., Chupp, E. L., & Vestrand, W. T. 1988, *Nature*, 331, 416
McCray, R. & Fransson, C. 2016, *Ann.Rev.Astron.& Astroph.*, 54, 19
Nomoto, K., Thielemann, F.-K., & Yokoi, K. 1984, *Astroph.J.*, 286, 644
Nomoto, K., Tominaga, N., Umeda, H., Kobayashi, C., & Maeda, K. 2006, *Nuclear Physics A*, 777, 424
Prantzos, N. & Diehl, R. 1996, *Phys. Rep.*, 267, 1
Ramaty, R., Kozlovsky, B., & Lingenfelter, R. E. 1979, *Astroph.J.Supp.*, 40, 487
Renaud, M., Vink, J., Decourchelle, A., *et al.* 2006a, *Astroph.J.Lett.*, 647, L41
Renaud, M., Vink, J., Decourchelle, A., *et al.* 2006b, *New Astronomy Review*, 50, 540
Röpke, F. K., Seitenzahl, I. R., Benitez, S., *et al.* 2011, *Progress in Particle and Nuclear Physics*, 66, 309
Rothschild, R. E., Lingenfelter, R. E., Blanco, P. R., *et al.* 1999, *Nuclear Physics B Proceedings Supplements*, 69, 68
Seitenzahl, I. R., Ciaraldi-Schoolmann, F., Röpke, F. K., *et al.* 2013, *Mon.Not.Royal Astr. Soc.*, 429, 1156
Siegert, T., Diehl, R., Krause, M. G. H., & Greiner, J. 2015, *Astron.& Astroph.*, 579, A124
The, L.-S., Clayton, D. D., Diehl, R., *et al.* 2006, *Astron.& Astroph.*, 450, 1037
The, L.-S., Leising, M. D., Kurfess, J. D., *et al.* 1996, *Astron.& Astroph. Suppl.*, 120, C357+
Thielemann, F.-K., Hirschi, R., Liebendörfer, M., & Diehl, R. 2011, in Lecture Notes in Physics, Berlin Springer Verlag, Vol. 812, Lecture Notes in Physics, Berlin Springer Verlag, ed. R. Diehl, D. H. Hartmann, & N. Prantzos, 153–232
Troja, E., Segreto, A., La Parola, V., *et al.* 2014, *Astroph.J.Lett.*, 797, L6
Tsygankov, S. S., Krivonos, R. A., Lutovinov, A. A., *et al.* 2016, *Mon.Not.Royal Astr. Soc.*, 458, 3411
Tueller, J., Barthelmy, S., Gehrels, N., *et al.* 1990, *Astroph.J.Lett.*, 351, L41
Vedrenne, G., Roques, J.-P., Schönfelder, V., *et al.* 2003, *Astron.& Astroph.*, 411, L63
Vink, J. 2004, *New Astron. Rev.*, 48, 61
Vink, J., Laming, J. M., Kaastra, J. S., *et al.* 2001, *Astroph.J.Lett.*, 560, L79
Wang, W. & Li, Z. 2014, *Astroph.J.*, 789, 123

Supernova 1987A:30 years later - Cosmic Rays and Nuclei from Supernovae and their aftermaths
Proceedings IAU Symposium No. 331, 2017
A. Marcowith, M. Renaud, G. Dubner, A. Ray & A. Bykov, eds.

doi:10.1017/S1743921317006342

Supernova remnants dynamics

Anne Decourchelle

Laboratoire AIM-Paris-Saclay, CEA/DSM/Irfu - CNRS - Universite Paris Diderot, CEA-Saclay, 91191 Gif sur Yvette, France,
email: anne.decourchelle@cea.fr

Abstract. Supernova remnants are the site of a number of physical processes (shock-heating, non-equilibrium ionization, hydrodynamic instabilities, particle acceleration, magnetic field amplification). Their related emission processes provide us with a large set of observational data. Supernova remnants result from the interaction of high-velocity material ejected by the supernova explosion with the medium surrounding the progenitor star. This interaction gives rise to a double-shock structure that lasts for hundreds of years, with a forward shock and a reverse shock compressing and heating to tens million of degrees the surrounding medium and the ejecta, respectively. It is mostly in this phase that young supernova remnants provide information on their explosion mechanism through spectro-imaging observations of the ejected nucleosynthesis products and their dynamics, notably in the X-ray domain. I will review these observations and their implications for our current understanding of the dynamics of supernova remnants. I will conclude on the prospects with future facilities.

Keywords. supernova remnants, kinematics and dynamics, hydrodynamics, cosmic rays, radiation mechanisms: nonthermal, radiation mechanisms: thermal

1. Introduction

The morphology of supernova remnants (SNRs) is driven by the initial explosion mechanism of the supernova and their subsequent interaction with the circumstellar environment/ambient interstellar medium. SNRs are traced by photons emitted through different radiation mechanisms arising from various physical processes and particle populations. As a consequence, observations of SNRs display various morphologies and features depending on their initial conditions, its subsequent evolution and their current evolutionary stage, as illustrated in X-rays in Fig. 1. They tell us a lot about the composition of the emitting plasma, its kinematics, local and global properties of the remnants, and they keep track of their evolution.

The main physical process is the interaction of high-velocity ejected material from the supernova (ejecta) with its surrounding medium, leading to the formation of a high Mach number shock propagating outwards into the ambient medium (hereafter, named forward shock). As introduced by McKee (1974), the deceleration of the ejected material by the ambient material gives rise to a shock propagating inwards into the ejecta (hereafter, named reverse shock). Both shocks are responsible for compressing and heating to high temperatures the ambient medium and ejecta, respectively. They are expected to produce X-ray bremsstrahlung and X-ray emission lines. The contribution of the shocked ejecta is strongly dominant compared to the shocked ambient medium in this phase for a number of factors (range of density and temperature, composition, non-equilibrium conditions,..). X-ray detections of the shocked ambient medium are indeed very limited in young SNRs (for example, Miceli *et al.* 2012). However, fast forward shocks have the ability to accelerate efficiently particles (Blandford & Eichler 1987). The electrons, accelerated to TeV energies

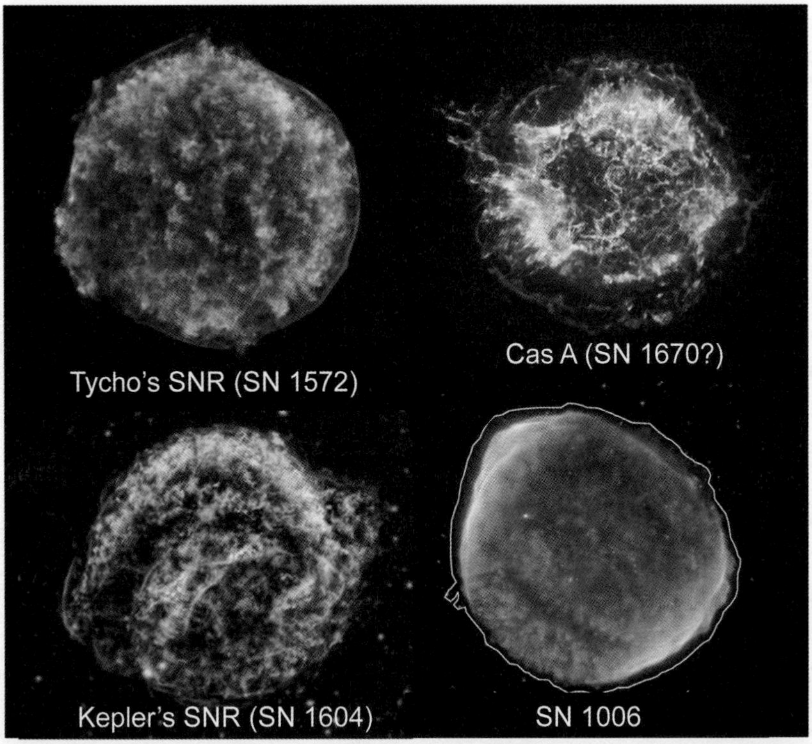

Figure 1. A gallery of young supernova remnants of different types and environments showing various X-ray morphologies observed with *Chandra* for Tycho's (Credit: NASA/CXC/Rutgers/J.Warren & J.Hughes *et al.*), Kepler's (Credit: NASA/CXC/NCSU/M.Burkey *et al.*; Optical: DSS) and Cas A's (NASA/CXC/SAO) SNRs and with *XMM-Newton* for SN1006 remnant (from Li *et al.* 2015). Note that the scale is different for each remnant.

at the forward shock, emit X-ray synchrotron emission that follows closely the forward shock rim. They can be observed as narrow filaments delineating the extension of young remnants, as seen in Fig. 1 for Tycho's, Kepler's, Cas A and SN1006 remnants. At the unstable interface between the ejecta and the ambient medium develop Rayleigh-Taylor instabilities that shape the observed fluffy X-ray ejecta emission. The Rayleigh-Taylor fingers emission from the ejecta comes relatively close to the forward shock. In between, the shocked ambient medium is invisible.

2. Hydrodynamics of young supernova remnants

In this section, I will review the hydrodynamic profile of the interaction region, bounded by both forward and reverse shocks, and which depends on the initial density profile of both the ejected material from the supernova and the ambient medium.

Chevalier (1982) has shown that in the case of expanding ejecta with a power-law density profile interacting with a stationary power-law density profile ambient medium, self-similar solutions for the structure of the interaction region can be obtained. They depend on the power law index of both density profiles. The observations of SN 1987A gave for the first time the opportunity to quantitatively challenge existing theories of core collapse supernovae and to provide constraints to the models. From the light curve constraints, Arnett (1988) derived the evolution of the ejecta density profile from core

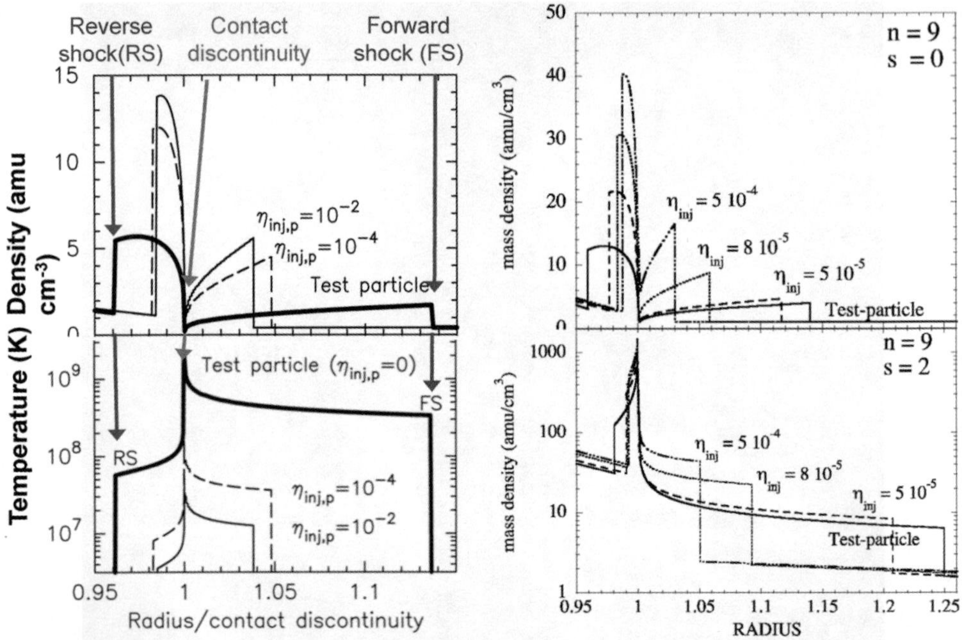

Figure 2. Left : Density and temperature structure of the interaction region for different values of the fraction of accelerated ions (from Decourchelle, Ballet, & Ellison 2000). Right : Density structure with (bottom) and without stellar wind (top) for different values of the fraction of accelerated ions.

collapse to its further evolution up to 1.85 days (see Fig. 4 in Arnett 1988). After core collapse, the shock took about 50 minutes to propagate from the core to the stellar surface. The ejecta density structure has then a steep dependence with radius in coherence with the surface layers of an hydrostatic object. As a result of the non-homologous expansion, the shock propagating through this structure accelerates and produces a power-law density profile, which is effective about 10 hours after the core collapse of SN1987A. In the phase where the self-similar solutions of Chevalier (1982) describe the evolution of the remnant, the behavior depends upon a combination of kinetic energy of the explosion per unit mass ejected. Once the reverse shock has propagated out of the power-law density profile into the inner flatter core of the ejecta, the Chevalier self-similar solutions do not apply anymore. It is by following the further evolution of the remnant that both parameters could be separately determined.

The density structure of the interaction region has also been investigated in the case of an exponential density profile in the ejecta, that approximates a set of Type Ia explosion models (Dwarkadas & Chevalier 1998). Compared to power-law profiles, one of the main results is a much flatter temperature profile in the ejecta, not favored by X-ray observations of young type Ia remnant like Tycho's SNR that indicate significant temperature variations in the ejecta.

These previous hydrodynamic models were not considering the back-reaction of particle acceleration at the shocks. However, shocks in SNRs are thought to provide a significant fraction of cosmic rays up to energies close to the knee at about 3 10^{15} eV. If a significant fraction of ions is accelerated at the shocks, it modifies the compressibility of the shocked gas and the shock properties with an overall compression ratio larger than 4 (Axford 1981). The overall interaction structure is modified as shown by Chevalier (1983), who investigated the effects of cosmic-ray pressure on the dynamics of self-similar solutions with a prescribed back-reaction of the accelerated particles. Decourchelle, Ballet & Ellison (2000) coupled these solutions to a model of ion diffusive acceleration at the shock (Berezhko & Ellison 1999) to investigate and quantify the impact on the interaction structure as a function of the fraction $\eta_{\mathrm{inj,p}}$ of protons injected into the acceleration process. When the injection $\eta_{\mathrm{inj,p}}$ is large, the compression ratio is larger than 4, the post-shock temperature is decreased and the shocked region shrinks very significantly, compared to test-particle conditions. As a result, the shock gets closer to the contact discontinuity (see Fig. 2). This is what is observed in young remnants like Tycho's and Kepler's SNR (see Fig. 1), where the forward shock is close to the contact discontinuity. This proves that ion acceleration is very efficient at the forward shock in these remnants. The distance between the forward shock and the contact discontinuity can thus provide an estimate of the efficiency of ion acceleration at the shock (see for Tycho's SNR, Warren *et al.* 2005).

The structure of the interaction region is also dependent on the nature of the ambient medium (mass density $\rho \propto r^{-s}$), notably if there is a circumstellar wind ($s = 2$), or a uniform interstellar medium ($s = 0$). A comparison between these two cases is shown in Fig. 2 (right panel) showing important differences in the structure in terms of extent and density profile of the shocked regions. The shocked ejecta are much compressed in presence of a stellar wind and displays a stronger density gradient, while the shocked ejecta are more extended than in the case of a uniform medium.

The above models are one dimensionals. There are however a few reasons why the evolution of young SNRs will not follow spherical symmetry. The first physical one is the development of Rayleigh-Taylor instabilities leading to large distortions of the shocked ejecta close to the contact discontinuity (Chevalier, Blondin & Emmering 1992), which distort in response the reverse shock and let unaffected the forward shock. Including efficient prescribed back-reaction from accelerated particle leads to possible distortion of the forward shock as well, as a result of its proximity to the contact discontinuity (Blondin & Ellison 2001). Three-dimensional simulations of young SNRs including non-linear particle acceleration have been developed by Ferrand *et al.* (2010) coupling a semi-analytical kinetic model of shock acceleration (Blasi 2004) with a 3D hydrodynamic code *RAMSES* (Teyssier 2002). Coupling 3D calculations of the out-of-equilibrium thermal X-ray emission, as well as the broad range non-thermal emission, enables synthetic emission maps of young SNRs to be confronted to the characteristic features of the observations (Ferrand, Decourchelle & Safi-Harb 2012, 2014).

3. Morphological and spectral properties of a sample of SNRs

The evolution, morphology, and spectral properties of young SNRs retain information on the physics of the explosion and on its environment. The quantities of synthesized elements depend on the SN type and the physics of the explosion. Their distribution, as well as the structure of the various physical quantities in the remnant, are markers of the initial event and its initial and subsequent conditions. The aim is to use this information to set constraints on the physics of supernovae.

For thermonuclear supernovae (SN Ia), one of the main issues concerns the nature of the companion in the binary system and the scenario leading to the explosion, either through a single degenerate system with a normal star as a companion, or through a double degenerate scenario with a white dwarf as a companion. A central question concerns the ignition of the burning and the propagation of the burning front. From the ratio between intermediate elements and iron in SN Ia, constraints could be set on the

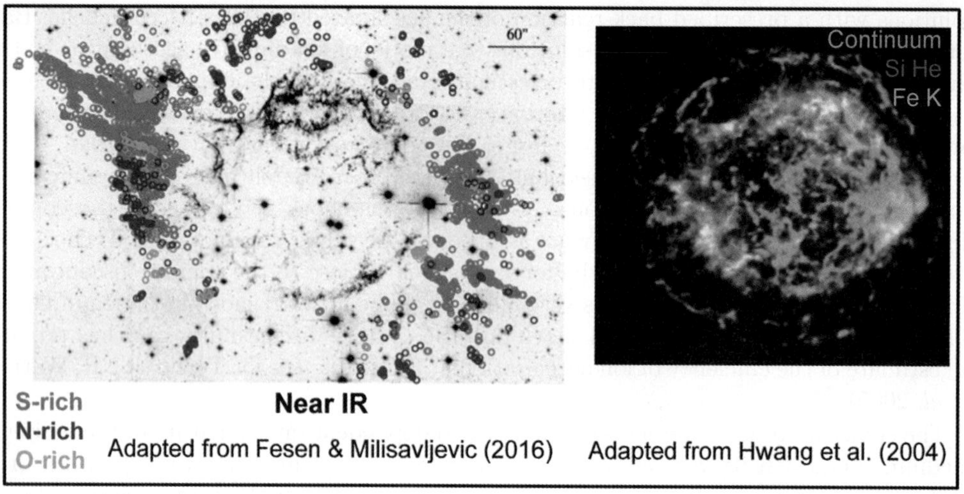

Figure 3. Cas A: Near IR image showing S-rich, N-rich et O-rich ejecta knots (left) and X-ray Chandra composite image (right, Credit: NASA/CXC/GSFC/U.Hwang *et al.*)

accretion rate from the companion star, and on the physics of propagating flame fronts (slow to fast deflagration, transition to detonation).

For core collapse supernovae, the issues deal with in particular the explosion mechanism (and related progenitors) and the value of the mass-cut to form the compact object. Elements lighter than silicon, that are produced mainly during hydrostatic evolution and spread away during the explosion, depend on the progenitor. The intermediate elements, enhanced through explosive oxygen burning, inform about the explosion energy and amount of matter accreted onto the core before the explosion. The iron-group nuclei provide detailed information on the explosion and mass-cut.

The morphology of a sample of 24 SNRs in the Milky Way and Large Magellanic Cloud has been investigated by Lopez *et al.* (2011), using statistical tools to analyze Chandra X-ray images. As a result, the authors found a clear trend in the morphology: type Ia SNRs are more symmetrical than core collapse SNRs. A systematic study of the energy of the iron K line in a sample of 23 SNRs observed with the X-ray satellite *Suzaku* was undertaken by Yamaguchi *et al.* (2014). It revealed a clear separation between type Ia SNRs with a lower centroid energy (low ionisation state) and core collapse SNRs with a higher centroid energy and higher Fe ionization state. This is a signature of different circumstellar media in both types of SNRs: Type Ia are compatible with an unmodified uniform ambient medium while core collapse supernovae explode in a circumstellar medium shaped by the progenitor wind. The overall X-ray spectra can in addition set constraints to the plausible theoretical models of both type Ia (Yamauchi *et al.* 2014) and core collapse supernovae (Patnaude *et al.* 2015).

4. A close-up view of Cassiopeia A and Tycho

Cassiopeia A (Cas A) is the remnant of a young core collapse supernova, on which has focused a large number of observations in the full electromagnetic range. Cas A evolves in a circumstellar medium resulting from the mass loss of its progenitor. As a result, the forward shock is in average significantly ahead of the contact discontinuity and of ejecta material, as expected from the hydrodynamical evolution of a remnant in a stellar wind (see Fig. 2, bottom right panel, case $s = 2$). However, X-ray observations with *Chandra* (Hughes *et al.* 2000) show a highly non-uniform distribution of elements in the shocked ejecta, reaching in some places the blast wave. They reveal a spatial inversion of a significant portion of the supernova core with the iron emission ahead of the silicon emission. The million second observation of Cas A with *Chandra* clearly reveals the bipolar structure, already observed in the optical (Fesen 2001 et references therein), of the Si-rich ejecta in the northeast and southwest directions (Hwang *et al.* 2004). Its origin has been debated with regard to the explosion mechanism of core collapse supernovae, but with insufficient energy associated with the jets to drive the explosion (Laming *et al.* 2006). Bulk X-ray Doppler velocities were derived from *XMM-Newton* (Willingale *et al.* 2002) and *Chandra* (Hwang *et al.* 2001) observations with a line-of-sight velocity scale around 2000-3000 km/s. They confirm a highly asymmetric explosion in an inhomogeneous environment with in particular two large clumps of iron expanding faster than silicon and sulphur. These measurements provide a deep insight into the explosion mechanism and expansion of the ejecta and are complementary to velocity patterns in the optical for much denser ejecta. A comparison of X-ray and optical emission, performed by Patnaude & Fesen (2014), shows the absence of significant large scale correlation but points to a number of correlated small scale features, which can be explained if the ejecta are highly inhomogeneous and display a range of densities. In the optical and near infrared, there are a very large number of detected knots spanning a large range of velocities and elements. The highest velocity metal-rich knots are located outside the outer blast wave of Cas A. They are interacting with the local unshocked circumstellar medium and gradually get disrupted. Deep HST WFC3/IR images of Cas A detected roughly 3400 outlying sulfur emitting knots with estimated velocities close to 15 000 km/s (Fesen & Milisavlevic 2016). The observed northeast and southwest jet structures are estimated to contain about 10% of the canonical 10^{51} ergs kinetic energy of the supernova. Their composition and velocities differ from any observed areas of Cas A. These findings are consistent with the dynamical scenario put forward by Burrows (2005), where jets from the proto-neutron star, produced after the neutrino-driven supernova explosion, pushed the turbulent inhomogeneous supernova ejecta. Further information from the innermost ejecta arose recently from NuSTAR observations of Cas A, and of its ^{44}Ti line emission, resulting from radioactive decays (Grefenstette *et al.* 2017).

Tycho's SNR is the remnant of a thermonuclear type Ia supernova, whose explosion was observed by the Danish astronomer Tycho Brahe in 1572. It is considered as a prototype of type Ia remnant. Its morphology is globally regular, but displays small scale inhomogeneities and large scale features. Constraints on the explosion mechanism came through its global X-ray spectrum (Decourchelle *et al.* 2001), whose spectral features are best explained by a delayed detonation models (Badenes *et al.* 2006). A fantastic observation was that of the optical light echo of the supernova event scattering on ambient interstellar clouds, whose spectrum was shown to be perfectly consistent with spectra of extragalactic standard type Ia supernovae (Krause *et al.* 2008). A thorough investigation of the spatial distribution of the shocked elements using *XMM-Newton* has put into evidence a radial stratification of the ejecta composition (see Fig. 4, left panel) with Fe located inner to the Ca and Si, as well as anisotropies in the distribution of Ca-rich and Fe-rich ejecta (Miceli *et al.* 2015). The region with high iron equivalent width shows indications of Ti line emission, in addition to Cr. The iron-peak nuclei seem to be spatially correlated in the ejecta, in agreement with the expectations from type Ia models. However, observations with *Suzaku* revealed that the outer southeast iron knot exhibits no emission from Cr, Mn or Ni, pointing at a peak temperature of roughly 5.4×10^9 K and a neutron excess of less than 2×10^{-3} (Yamaguchi *et al.* 2017). This

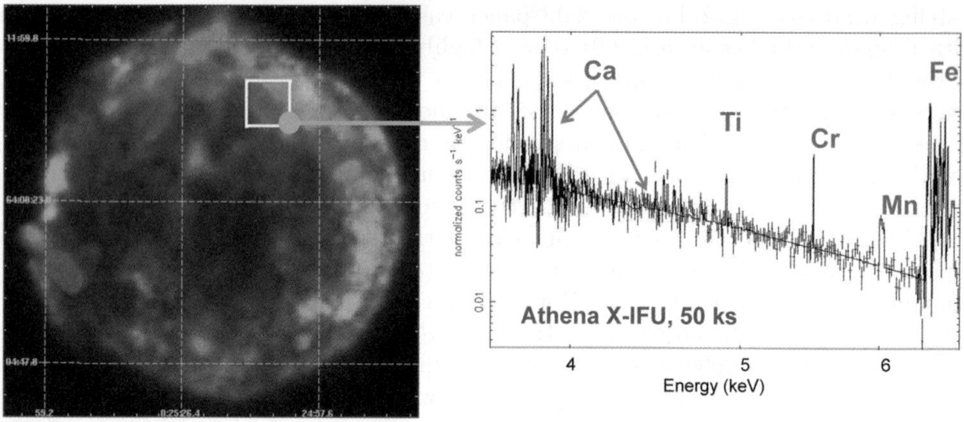

Figure 4. Left: *XMM-Newton* EPIC color-composite image of Tycho's SNR in the 4.4-6.1 keV energy band (green) and of the Ca (red) and Iron (blue) equivalent width (see Miceli *et al.* 2015). Right: *Athena* X-IFU 50 ks simulations of the spectrum of a region of 1 arcmin^2 region in Tycho's SNR (Courtesy of Marco Miceli).

rules out an origin of this knot in the deep, dense core of a Chandrasekhar-mass white dwarf and favors either incomplete Si burning or an α-rich freeze-out regime, probably close to the boundary of the remnant.

X-ray observations, in complement notably to optical and infrared observations, are keys to investigate the composition, kinematics and properties of the ejecta in young SNRs. X-rays give specifically access to the hot plasma of the shocked ejecta and provide diagnostics on the emitting conditions (ionization state of the elements, temperatures, densities). The progress in the field of young SNRs has been considerable since the launch of *Chandra* and *XMM-Newton* X-ray observatories, thanks to their capability of performing spatially resolved spectroscopy. However, to progress in the understanding of the explosion mechanism at play in supernovae, a key observational requirement is to improve on the spectral resolution of current X-ray spectro-imagers, which hampers our scientific investigation of the 3D structure and properties of the shocked ejecta. This is what will be possible with the ESA *Athena* X-ray observatory.

5. A new X-ray mission for spatially-resolved high-resolution spectroscopy : Athena

Athena is the ESA L2 mission, selected in June 2014, and dedicated to the hot and energetic Universe. The launch is planned for 2028 with an Ariane 6 rocket on an L2 orbit. It consists of a telescope, based on Silicon Pore Optics, of 2 m^2 at 1 keV and 5 arcsec HEW for a focal length of 12 m (Willingale *et al.* 2013) and of two instruments:
- the X-IFU (X-ray Integral Field Unit) designed to provide a spectral resolution of 2.5 eV, a field of view of 5 arcmin (Barret *et al.* 2013).
- the WFI (Wide Field Imager), designed to provide an energy resolution of 125 eV and a large field of view of 40 arcmin, with a high countrate capability (Rau *et al.* 2013).

The expected impact of Athena in the study and understanding of SNRs is fantastic, covering a number of key issues in the field (Decourchelle *et al.* 2013).

Athena will be in particular revolutionary for investigating young SNRs. The X-IFU instrument, with its high energy resolution, combined with a good spatial resolution, and

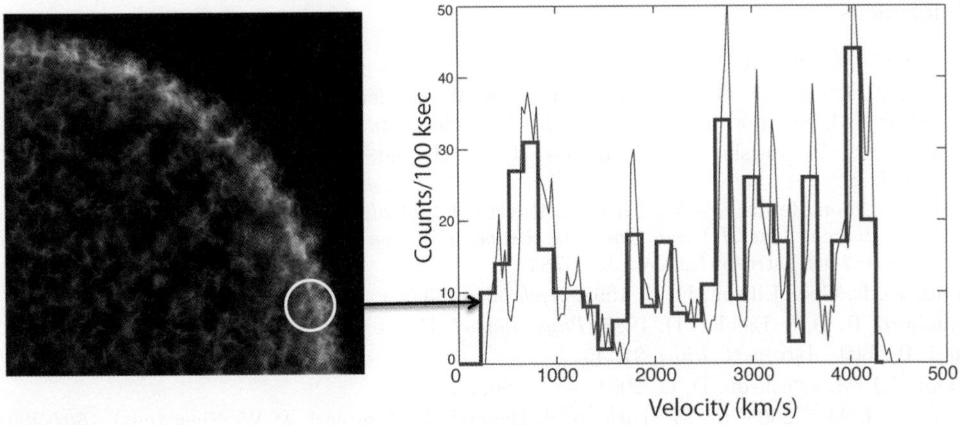

Figure 5. Left: Numerical 3D simulations of thermal X-ray emission in Tycho-like SNR (Ferrand, Decourchelle, Safi-Harb 2014). Right: Expected *Athena* X-IFU line-of-sight velocity structure in the silicon K line of a region in the simulated ejecta (circle in left panel). In solid line, the profile derived from hydrodynamical simulations, in bold line *Athena* X-IFU simulations (Courtesy: A. Decourchelle, G. Ferrand & R. Smith).

large effective area, will provide the first detailed 3D mapping of the hot ejected material by deciphering the various ejecta components in the line of sight and will give access to emission lines from rare elements, like Titanium, Chromium and Manganese. Figure 4 (right panel) shows the X-IFU expected spectrum from a iron-rich region in Tycho's SNR. The faint Ti K line is detected at 6 sigma in a region of 1 arcmin^2.

Each observed region of the remnant includes a range of velocities in the line of sight that depends, as presented previously, upon the hydrodynamic structure of the interaction region, as well as local instabilities and initial stratification/inhomogeneities of the supernova material. The objective is to measure the velocity of the ejecta structures in the line of sight as a function of position in the remnant, along with the elemental abundances, temperatures, and ionization states. This will allow us to determine the geometry, the structure and properties of the ejecta, offering an insider view into the supernova explosion, revealing the underlying explosion mechanism, the initial conditions in the progenitor, as well as structures in the circumstellar and interstellar media. Figure 5 illustrates the ability of *Athena* X-IFU to recover the line-of-sight velocity structure of the bright Si K line in the ejecta.

6. Conclusion

Multi-wavelength observations of young galactic SNRs have provided a wealth of information on the nucleosynthesis yields, the dynamics of the ejecta and its asymmetries/inhomogeneities, the wide range of conditions in the ejecta from dense optical knots to diffuse X-ray plasma and the properties of the shocks. As a result, the observational constraints are tightened on the physics of the explosion and the properties of the progenitor. However, we lack a detailed 3D mapping of the distribution and conditions of the different elements in the hot ejecta that encompasses a large fraction of the remnant, and that is intimately linked to the supernova explosion mechanism and to the progenitor properties. The need is to perform spectroscopic diagnostics at the relevant spatial scales, including rare elements, to yield bulk and turbulent velocities, temperatures, ionization states and abundances of the various components of the shocked ejecta. *Athena* L2 ESA mission, to be launched in 2028, is awaited to fulfill these objectives.

References

Arnett, W. D. 1988, *ApJ*, 331, 377

Axford, W. I. 1981, in: G. Setti, G. Spada, & A. W. Wolfendale (eds.), *Origin of Cosmic Rays*, Proc. IAU Symposium No. 94 (Dordrecht: Reidel), p. 339

Badenes, C., Borkowski, K. J., Hughes, J. P., Hwang, U., & Bravo, E. 2006, *ApJ*, 645, 1373

Barret, D., den Herder, J.-W., Piro, L., Ravera, L., *et al.* 2013, *Supporting paper for the science theme "The Hot and Energetic Universe" to be implemented by the Athena+ X-ray observatory*, astro-ph/arXiv:1308.6784

Berezhko, E. G. & Ellison, D. C. 1999, *ApJ*, 526, 385

Blandford, R. D. & Eichler, D. 1987, *Phys. Report*, 154, 1

Blasi, P. 2004, *Astropart. Phys*, 21, 45

Blondin, J. M. & Ellison, D. C. 2001, *ApJ*, 560, 244

Burrows, J. M., 2005, in: M. Turatto, S. Benetti, L. Zampieri, & W. Shea (eds) *1604-2004: Supernovae as Cosmological Lighthouses*, Proc. ASP Conf. Ser. No. 342 (San Francisco: Astronomy Society of the Pacific), p184

Chevalier, R. A. 1982, *ApJ*, 258, 790

Chevalier, R. A. 1983, *ApJ*, 258, 790

Chevalier, R. A., Blondin, J. M., & Emmering, R. T. 1992, *ApJ*, 392, 118

Decourchelle, A., Ballet, J., & Ellison, D. C. 2000, *ApJ*, 365, L218

Decourchelle, A., *et al.* 2001, *ApJ*, 543, L57

Decourchelle, A., Costantini, E., Badenes, C., Ballet, J., & Bamba, A. *et al.* 2013, *Supporting paper for the science theme "The Hot and Energetic Universe" to be implemented by the Athena+ X-ray observatory*, astro-ph/arXiv:1306.2335

Dwarkadas, V. V. & Chevalier, R. A. 1998, *ApJ*, 497, 807

Ferrand, J., Decourchelle, A., Ballet, J., Teyssier, R., & Frachetti, F. 2010, *A&A*, 509, L10

Ferrand, J., Decourchelle, A., & Safi-Harb, S. 2012, *A&A*, 760, 34

Ferrand, J., Decourchelle, A., & Safi-Harb, S. 2014, *A&A*, 789, 49

Fesen, R. A. 2001, *ApJS*, 133, 161

Fesen, R. A. & Milisavljevic, D. 2016, *ApJ*, 818, 17

Grefenstette,J. P. *et al.* 2017, *ApJ*, 834, 19

Hughes,J. P., Rakowski,C. E., Burrows, D. N. & Slane, P. O. 2000, *ApJ*, 528, L109

Hwang, U., Szymkowiak, A. E., Petre, R., & Holt, S. S. 2001, *ApJ*, 560, L179

Hwang, U., *et al.* 2004, *ApJ*, 615, L117

Li, J.-T., Decourchelle A., Miceli, M., Vink, J., & Bocchino, F. 2015, *MNRAS*, 453, 3953

Krause, O., *et al.* 2008, *Nature*, 456, 617

Laming, J. M., Hwang, U., Radics, B., Lekli, G., & Takacs, E. 2006, *ApJ*, 644, L260

Lopez, L. A., Ramirez-Ruiz, E., Huppenkothen, D., Badenes, C., & Pooley, D. A. 2011, *ApJ*, 732, 114

Miceli, M., Bocchino, F., Decourchelle, A., Maurin, G., Vink, J., *et al.* 2012, *A&A*, 546, A66

Miceli, M., Sciortino, S., Troja, E., & Orlando, S., 2015, *ApJ*, 805, 120

McKee, C. F. 1974, *ApJ*, 188, 335

Patnaude, D. J. & Fesen, R. A. 2014, *ApJ*, 789, 138

Patnaude, D. J. *et al.* 2015, *ApJ*, 803, 101

Rau, A., Meidinger, N., Nandra, K., Porro, M., Barret, D., et al. 2013, *Supporting paper for the science theme "The Hot and Energetic Universe" to be implemented by the Athena+ X-ray observatory*, astro-ph/arXiv:1308.6785

Warren, J. S., Hughes, J. P., Badenes, C., Ghavamian, P., McKee, C. F. *et al.* 2005, *ApJ*, 634, 376

Willingale, R., Bleeker, J. A. M., van der Heyden, K. J., Kaastra, J. S., & Vink, J. 2002, *A&A*, 381, 1039

Willingale, R., Pareschi, G., Christensen, F., & den Herder, J.-W. 2013, *Supporting paper for the science theme "The Hot and Energetic Universe" to be implemented by the Athena+ X-ray observatory*, astro-ph/arXiv:1307.1709

Yamaguchi, H., Badenes, C., Petre, R., Nakano, T., Castro, D. *et al.* 2014, *ApJ*, 785, L27

Yamaguchi, H., Hughes, J. P., Badenes, C., Bravo, E., Seitenzahl, I. V., *et al.* 2017, *ApJ*, 834, 124

Supernova 1987A:30 years later - Cosmic Rays and Nuclei from Supernovae and their aftermaths
Proceedings IAU Symposium No. 331, 2017
A. Marcowith, M. Renaud, G. Dubner, A. Ray & A. Bykov, eds.

doi:10.1017/S1743921317004902

MHD Simulation of Supernova Remnants

D. Wu[1,2], M. F. Zhang[1,3], S. S. Shan[1,3] and W. W. Tian[1,4]

[1]Key Laboratory of Optical Astronomy, National Astronomical Observatories, Chinese Academy of Sciences, Beijing 100012, China
email: wudan@bao.ac.cn

[2]College of Information Science and Technology, Beijing Normal University, Beijing 100875, China

[3]School of Astronomy and Space Science, University of Chinese Academy of Sciences, 19A Yuquan Road, Shijingshan District, Beijing 100049, China

[4]Department of Physics & Astronomy, University of Calgary, Calgary, Alberta T2N 1N4, Canada

Abstract. We present some Magnetohydrodynamic (MHD) simulations for supernova remnants (SNRs), which hints the thicknesses of SNRs' shells are likely related to the density and density distribution of surrounding interstellar medium (ISM). The simulations show clear formation and evolution of reverse shocks. In addition, we find that stellar winds can blow a bubble around the progenitor stars then lead to weaker radio emission at the center of SNRs.

Keywords. Methods: numerical, ISM: supernova remnants, MHD

1. Introduction

Supernova shock interacts with the ISM then generate SNRs. There are many unsolved puzzles about SNRs, and the key to unlock these is to figure out the evolution of SNRs. Full of randomness, the evolution of SNRs is not only dependent on the progenitor, but also influenced by the surrounding ISM. But, MHD simulations can help us understand such a complicated process.

The shells of different SNRs usually have different thicknesses, which are determined by many conditions, such as the ejecta mass, explosion kinetic energy and the density of ISM. And they can be also influenced by some random parameters, such as surrounding molecular clouds (MCs) and even the shapes of MCs (Koo & Moon 1997). In the simulations, we adjust these initial conditions to study ISM' contributions to the thickness. To get more reasonable initial conditions, we take the stellar wind of progenitor star, magnetic field and the nearby MCs into consideration. Assuming the radio flux density of SNRs totally originates from synchrotron emission, we convert the simulation results to radio flux density maps and smooth them in terms of the resolution of real observation, so that we can compare the simulations with observations easily.

2. Simulation model and results

Ignoring gravitation and dissipative effects, we can get an ideal conservation equation set, including mass, energy, momentum and magnetic conservations. On the basis of the set, we perform the simulation on PLUTO Mignone *et al.* (2007), a code used to study MHD. We assume a supernova with an explosion kinetic energy of 2×10^{51} erg and an ejecta mass of 15 $M_{\odot}$. The hydrogen atom number densities of ISM are set to 2 cm^{-3} and 20 cm^{-3}. The initial radius is set to 1 pc, based on which the initial velocity and

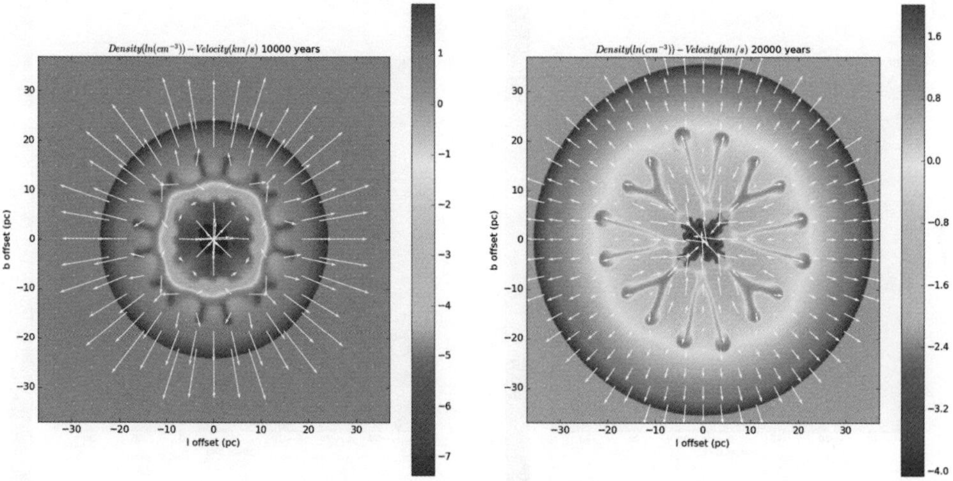

Figure 1. The density-velocity maps. The top tow images show the simulated results after 10000, 20000 years with a initial ISM density of 2 cm^{-3}. White arrows stand for velocities.

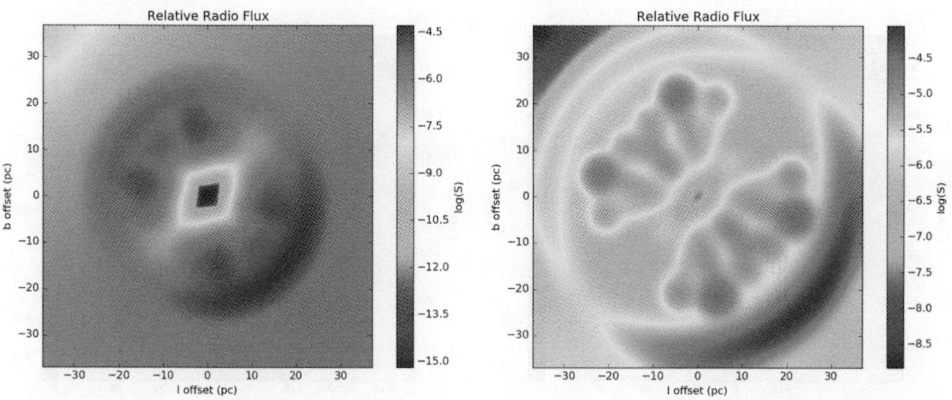

Figure 2. Relative radio flux maps. The images shows the corresponding relative radio flux in Fig. 1.

density can be estimated by the work of Leahy & Williams (2017). To simulate a one-side SNR, we add a magnetic gradient increasing from upper left to lower right. The magnetic field in the explosion center is set to 9 μG and the magnetic field gradient is 0.1 μG/pc. The relative radio flux is calculated by $i(\nu) = C\rho B_{\perp}^{\beta+1}\nu^{-\beta}$ (Orlando *et al.* 2007). In the simulation, β is set to 0.25. If we use a telescope with a resolution of $4'$ observe our simulated SNRs at about 4 kpc, the observed images should be the simulation results smoothed by a gaussian function with a standard deviation of 14 referring to the number of pixels.

Fig. 1 shows the simulated density-velocity maps at different ages. Fig. 2 shows the corresponding relative radio flux maps. In Fig. 3 and Fig. 4 , we adopt a power law density distribution $N(\rho) = N_0\rho^{-\alpha}$, in which N is the pixel numbers at different densities α is set to 2.4, i.e., low-density regions are more than high-density regions. Fig. 3 shows simulated density-velocity maps at 20 cm^{-3}, while Fig. 4 shows the relative radio flux maps at 2 cm^{-3}.

To understand the influence of stellar winds, we use another parameter set, in which the explosion kinetic energy is 0.8×10^{51} erg, the density of ISM is about 20 cm^{-3}. The

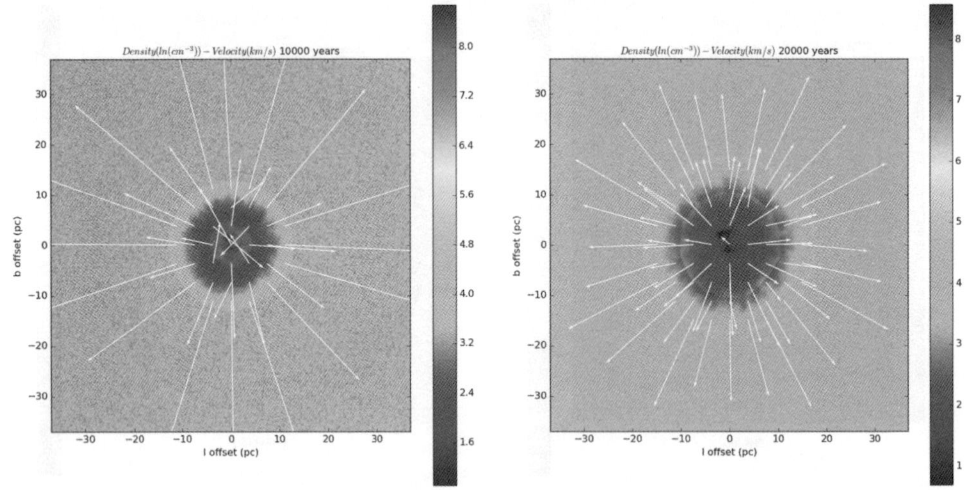

Figure 3. The density-velocity maps. The two images show the simulated results after 10000, 20000 years with a initial ISM density about 20 cm^{-3}. White arrows stand for velocities.

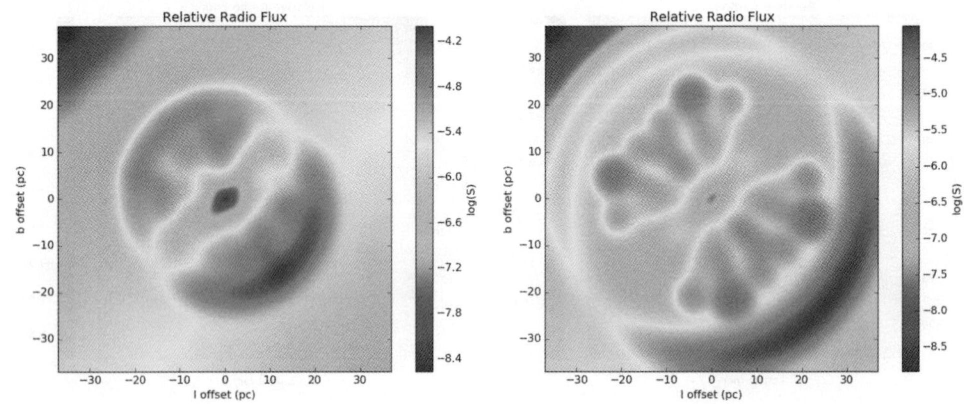

Figure 4. Relative radio flux maps. The two images show the radio flux map after 10000, 20000 years with a initial ISM density about 2 cm^{-3}.

density distribution is set based on the background CO data from GRS (Jackson *et al.* 2006). The stellar winds model is the standard model in Lamers & Cassinelli (1999). We simulate a progenitor star blowing stellar winds in one million years and use the final density distribution as the initial density. To study more details, we get simulation results after 30000 years and do not smooth them. In Fig. 5, we show the results with and without stellar winds after 30000 years.

3. Discussion

In Fig. 1, we can see the velocity changes direction towards the explosion center at 10000 years, which indicates the formation of reverse shocks. In Fig. 2, there are two shells and obvious inside structures at 2 cm^{-3} maps. If the flux density of darker shell is lower than the sensitivity of a telescope, it is not visible, then we can only see one shell. In addition, these inside structures will make the shells thicker. Comparing it with Fig. 1, it is reasonable to guess reverse shocks and instabilities lead to these inside structures.

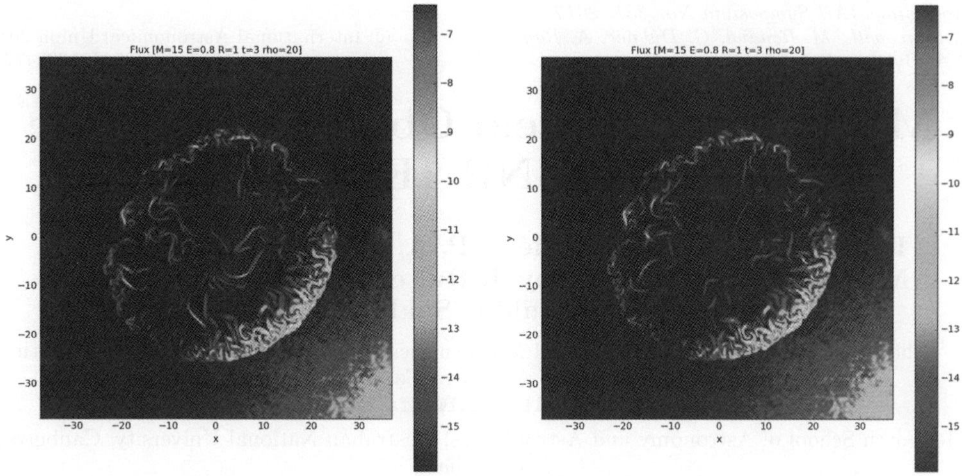

Figure 5. Relative radio flux maps without smoothing. The two pictures are both the flux at same ISM background. The right one takes the stellar winds into consideration.

In Fig. 3, reverse shocks become secondary forward shocks after they converge to the explosion center and form two layers with the main forward shocks. However, we ignore the dissipative effects in the simulation, if not, the secondary forward shocks will be much weaker. The inner layer, generated by reverse shocks, is low-density and high-temperature, so it is possible to radiate at X-ray. If it is real, then the phenomenon can be used to explain that the X-ray sizes of SNRs are usually less than radio sizes. The inner structures are easier to be distinguished from outer layers in Fig. 4. Comparing Fig. 4 with Fig. 2, we think the thickness of shells is also related to the density distribution.

Fig. 5 indicates the stellar winds of progenitors can only make the central small regions weaker and more uniform at radio wavelength during the evolution of SNRs. So the radio images of SNRs usually have a weak center and thick shells. In Fig. 5, we can see the diagonally asymmetric structure like Fig. 2 and Fig. 4. This is entirely dependent on the magnetic field gradient. Although strong magnetic field can obstruct the shocks and make the ISM denser, the magnetic field here is so weak that such an effect can be ignored.

We will apply the simulations to some observed SNRs, and further test our conclusion.

References

Denis, A. Leahy & Jacqueline, Williams 2017, *AJ*, 153, 239

Jackson, J. M., Rathborne, J. M., Shah, R. Y., Simon, R., Bania, T. M., Clemens, D. P., Chambers, E. T., Johnson, A. M., Dormody, M., Lavoie, R., & Heyer, M. H. 2006 *ApJS*, 163, 145

Jun, B.-I. & Norman, M. L. 1996, *ApJ*, 465, 800

Koo, B.-C. & Moon, D.-S. 1997, *ApJ*, 475, 194

Lamers, H. J. G. L. M. & Cassinelli, J. P. 1999, *Introduction to Stellar Winds*

Mignone, A., Bodo, G., Massaglia, S., Matsakos, T., Tesileanu, O., Zanni, C., & Ferrari, A. 2007, *ApJS*, 170, 228

Orlando, S., Bocchino, F., Reale, F., Peres, G., & Petruk, O. 2007, *A&A*, 470, 927

Supernova 1987A:30 years later - Cosmic Rays and Nuclei from Supernovae and their aftermaths
Proceedings IAU Symposium No. 331, 2017
A. Marcowith, M. Renaud, G. Dubner, A. Ray & A. Bykov, eds.

doi:10.1017/S174392131700477X

MUSE Integral Field Observations of the Oxygen-rich SNR 1E 0102.2-7219

Ivo R. Seitenzahl[1,2,3], Frédéric P. A. Vogt[4], Jason P. Terry[5], Michael A. Dopita[2], Ashley J. Ruiter[2,3], Parviz Ghavamian[6] and Tuguldur Sukhbold[7]

[1]School of Physical, Environmental and Mathematical Sciences, University of New South Wales, Australian Defence Force Academy, Canberra, ACT 2600, Australia.
email: i.seitenzahl@adfa.edu.au

[2]Research School of Astronomy and Astrophysics, Australian National University, Canberra, Australia.

[3]ARC Centre for All-sky Astrophysics (CAASTRO).

[4]European Southern Observatory, Av. Alonso de Córdova 3107, 763 0355 Vitacura, Santiago, Chile.

[5]Department of Physics and Astronomy, University of Georgia, USA.

[6]Department of Physics, Astronomy and Geosciences, Towson University, Towson, MD 21252, USA.

[7]Department of Astronomy and Center for Cosmology & Astro-Particle Physics, The Ohio State University, Columbus, OH 43210, USA.

Abstract. We have observed the oxygen-rich SNR 1E 0102.2-7219 with the integral field spectrograph WiFeS at Siding Spring Observatory and discovered sulfur-rich ejecta for the first time. Follow-up deep DDT observations with MUSE on the VLT (8100 s on source) reaching down to a noise level of $\sim 5 \times 10^{-20}$ ergs^{-1}cm^{-2}Å^{-1}spaxel^{-1} have led to the additional discovery of fast-moving hydrogen as well as argon-rich and chlorine-rich material. The detection of fast-moving hydrogen knots challenges the interpretation that the progenitor of 1E 0102 was a compact core of a Wolf-Rayet star that had shed its *entire* envelope. In addition to the detection of hydrogen and the products of oxygen-burning, this unprecedented sharp ($0.2''$ spaxel size at $\sim 0.7''$ seeing) and deep MUSE view of an oxygen-rich SNR in the Magellanic Clouds reveals further exciting discoveries, including [Fe XIV]λ5303 and [Fe XI]λ7892 emission, which we associate with the forward shock. We present this exciting data set and discuss some of its implications for the explosion mechanism and nucleosynthesis of the associated supernova.

Keywords. supernova remnants

1. Introduction

1E 0102.2-7219 (1E 0102 for short) belongs to the class of *oxygen-rich* young supernova remnants (O-rich YSNRs) showing remarkably little-to-no hydrogen emission. The prominent oxygen forbidden-line emission observed in the filaments and knots (see Figs. 1, 2) is understood to arise from shocked ejecta from a core-collapse supernova that had been stripped of (most of) its hydrogen envelope prior to explosion. Unlike most SNRs, where the optical emission comes from interstellar medium ionized by the forward shock, O-rich YSNRs are special in that one can study the actual ejecta composition of the supernova, providing unique and direct insights into explosion mechanisms and nucleosynthesis conditions in core-collapse supernovae. Only a handful of such O-rich

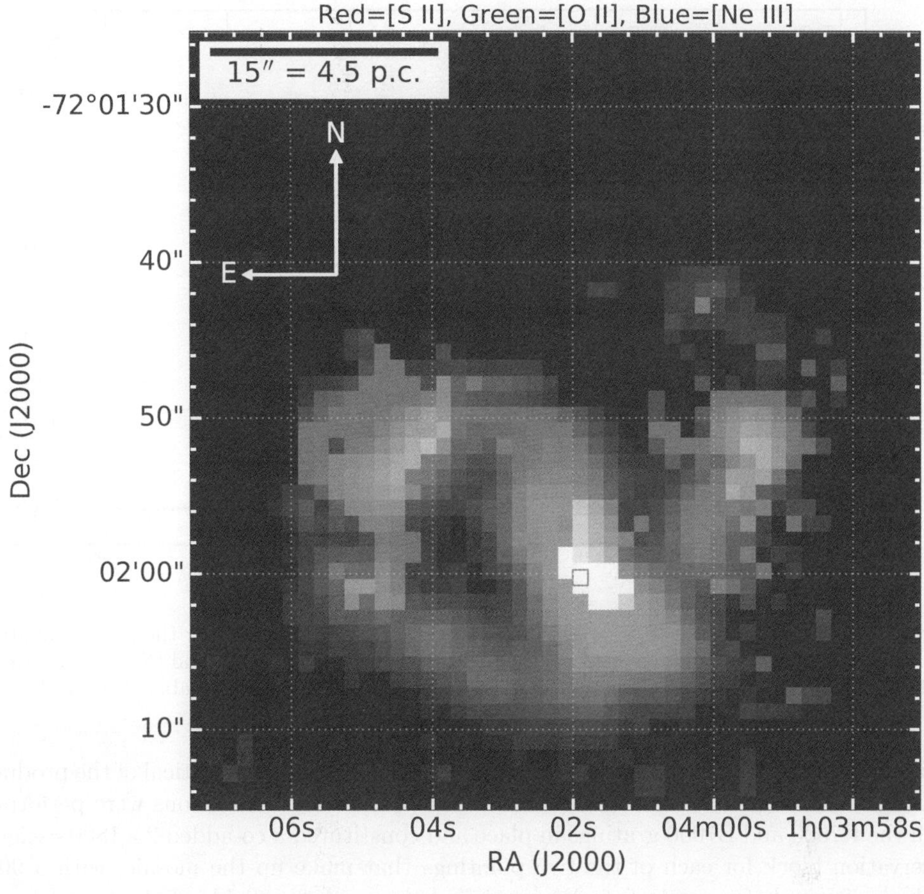

Figure 1. Three color (RGB) image of 1E 0102.2-7219 showing the regions where fast-moving [S II]$\lambda\lambda$6716,6731 (red), [O II]$\lambda\lambda$3726,3729 (green), and [Ne III]$\lambda\lambda$3869,3967 (blue) ejecta is detected with WiFeS.

YSNRs are known: the best studied examples include Cas A in our Galaxy and 1E 0102 in the Small Magellanic Cloud (SMC).

Since the discovery of the high-velocity oxygen-rich supernova ejecta in 1E 0102 (Dopita *et al.* 1981), subsequent searches (including HST FOS spectroscopy and WFC3 imaging) for the expected products of oxygen burning, in particular sulfur, have returned negative results (e.g., Lasker & Golimowski 1991, Blair *et al.* 2000). These searches have only detected carbon, oxygen, neon, and magnesium, but no products of oxygen-burning, i.e., no sulfur, silicon, calcium, or argon. The angular size of 1E 0102 (50″ in diameter) and the large Doppler shifts associated with its fast-moving ejecta ($v \gtrsim 1000\,\mathrm{km\,s^{-1}}$) have been strong limitations for previous long-slit spectroscopic and narrow-band imaging searches, which have either failed to map locations of interest and/or were blind to fast-moving ejecta altogether.

2. WiFeS observations

In our observations (P.I.: Seitenzahl) on the nights of Aug. 7 and Aug. 8, 2016 with the WiFeS Integral Field Spectrograph on the 2.3 m telescope at Siding Springs Observatory, we have detected fast-moving (mostly, but not exclusively blue-shifted) [S II]$\lambda\lambda$6716, 6731

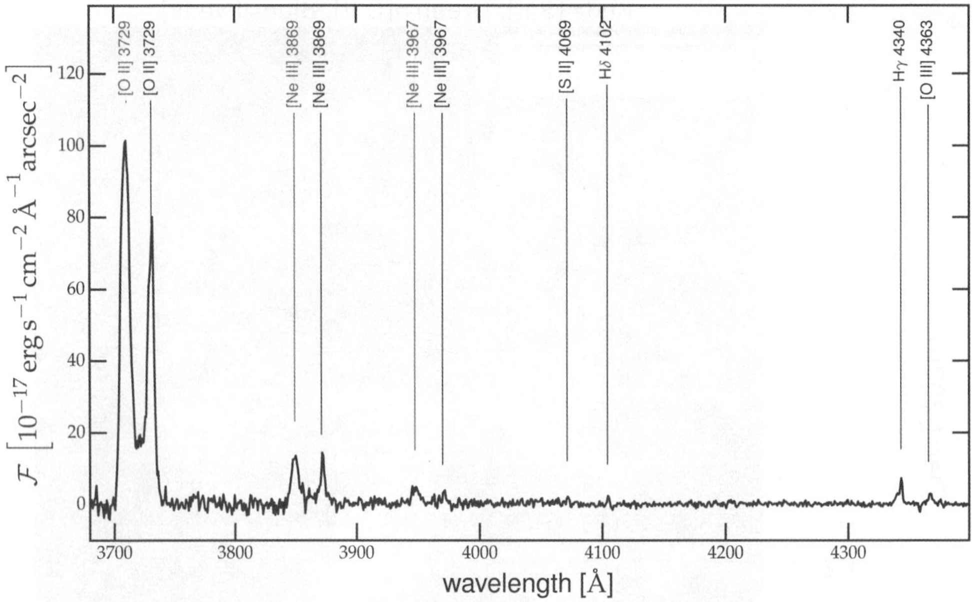

Figure 2. Blue end of the WiFeS spectrum of the bright knot indicated by the purple square in Fig. 1. The broad, fast-moving ejecta component is labeled in blue ([O II] and [Ne III] inaccessible with MUSE), the narrow line emission at the local velocity of 1E 0102 is labeled in black.

emission (see Fig. 1). These data represent the first detection in the optical of the products of oxygen-burning in the ejecta of 1E 0102. The first set of observations were performed with the R7000 and B7000 gratings in place and constituted a co-added 2×1800 s science observation block for each of the two pointings that make up the mosaic, with a 900 s dark sky frame before and after. We re-visited the same two fields of the target for an additional 2×1800 s per science field on the nights of Oct. 28 and Oct. 29, 2016, this time with the B3000 in the blue arm to include [O II]λλ3726,3729 and [Ne III]λλ3869,3967 (see Fig. 2). All data were reduced and flux-calibrated with the PyWiFeS pipeline (Childress *et al.* 2014).

3. MUSE observations

We observed 1E 0102 with the Multi-Unit Spectroscopic Explorer (MUSE) in Service Mode at the Very Large Telescope on the night of October 7, 2016 via Director's Discretionary Time (DDT) program 297.D-5058 (P.I.: F.P.A. Vogt). We obtained a total of 9×900 s exposures on-source, with appropriate 180 s dark sky exposures away from the SMC in between the science exposures. The data frames were individually reduced and then combined into a single data cube with the ESO Reflex workflow (Freudling *et al.* 2013), for further details see Vogt *et al.* 2017b. Compared to the $1''$ by $1''$ spaxel size and $\sim 1.5''$ seeing of the WiFeS cube, the $0.2''$ by $0.2''$ spaxel size and $\sim 0.7''$ seeing of the MUSE cube offers a much sharper view of 1E 0102 (see Fig. 3). In addition to gaining in spatial resolution, the MUSE observations have increased the sensitivity by a factor of approximately 100, improving from the WiFeS noise level of a few times $10^{-18}\,\mathrm{ergs^{-1}cm^{-2}\AA^{-1}spaxel^{-1}}$ to a noise level of a few times $10^{-20}\,\mathrm{ergs^{-1}cm^{-2}\AA^{-1}spaxel^{-1}}$.

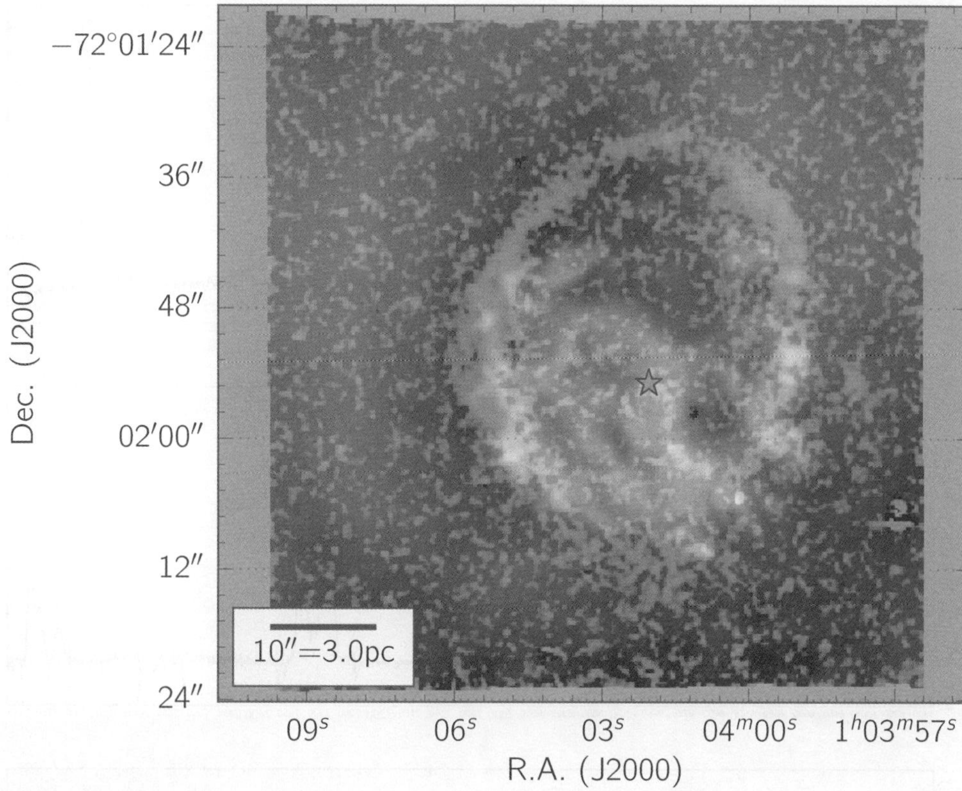

Figure 3. MUSE view of 1E 0102, reproduced from Vogt *et al.* 2017b. Shown in magenta is [Fe XIV] (5300 – 5308 Å slice through the cube) and in green [O III] (4900 – 5100 Å slice through the cube). The center of expansion derived by Finkelstein *et al.* (2006) is marked with a star symbol. The clumpy O-rich ejecta is seen in bright green mostly inside the magenta [Fe XIV] ring. The more diffuse [O III] emission outside the magenta [Fe XIV] ring is attributed to the photoionization precursor ahead of the forward shock.

4. Summary

In this contribution, we have presented optical integral field spectroscopy observations of 1E 0102.2-7219 performed with the WiFeS and MUSE instruments. The combined data cubes extend from 3500 Å to 9350 Å. In addition to mapping out the (previously known) O- and Ne-rich ejecta in unprecedented detail, we have discovered fast-moving [S II], [S III], [Ar III], Hα, Hβ, and possibly [Cl II], which we attribute to be components of the supernova ejecta. S, Ar, and Cl are main nucleosynthesis products of O-burning, which we detect here as ejecta in the optical in 1E 0102 for the first time. The discovery of Balmer line emission in the ejecta indicates that the progenitor star must not have been entirely stripped of its hydrogen envelope when it exploded. We have also discovered high ionization coronal lines of Fe ([Fe X], [Fe XI], and [Fe XIV]) and the photoionization precursor emission, which can be used to derive the forward shock parameters (see Vogt *et al.* 2017b).

5. Acknowledgements

This research has made use of BRUTUS, a Python module to process data cubes from integral field spectrographs hosted at `http://fpavogt.github.io/brutus/`. For

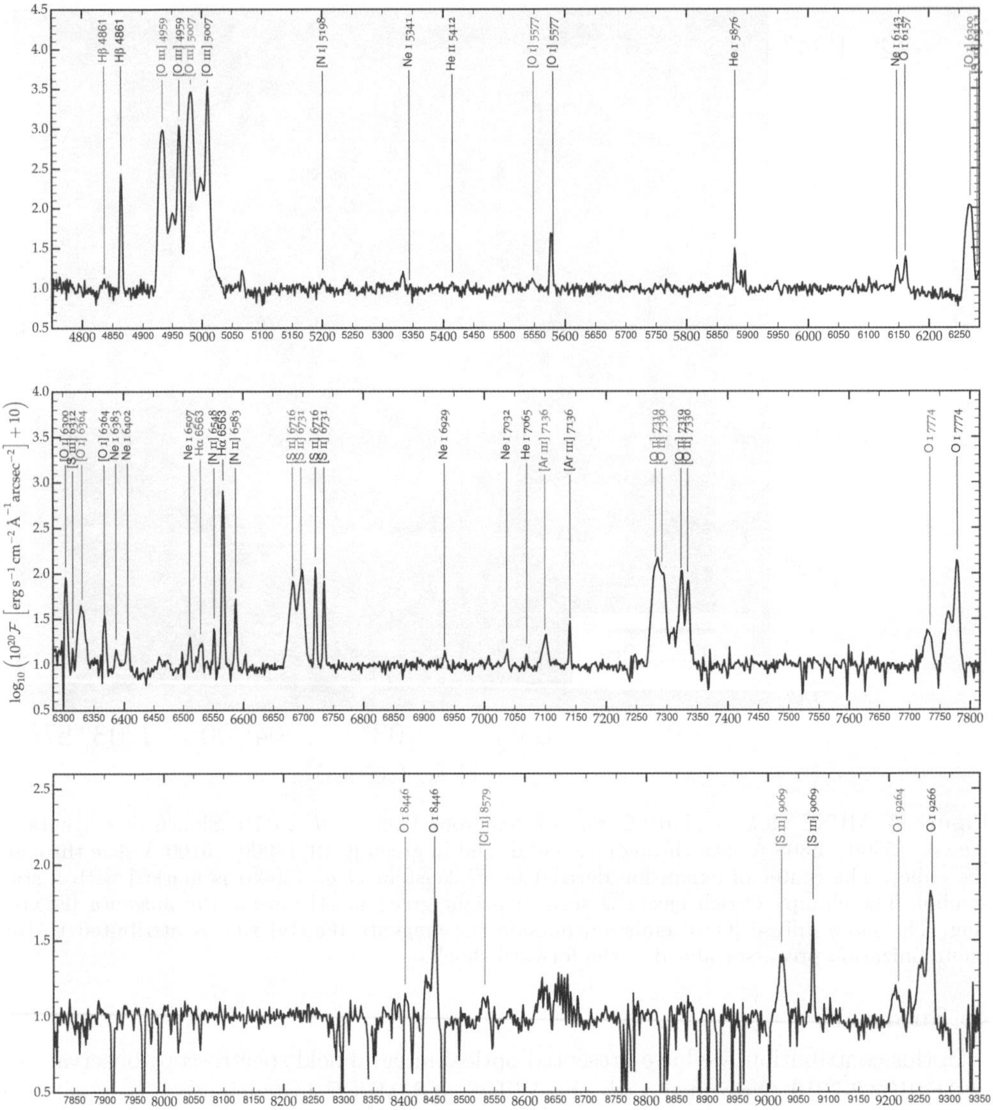

Figure 4. MUSE spectrum of the same S-bright knot as for the WiFeS spectrum in Figs. 1 and 2. The broad, fast-moving ejecta component is labeled in blue, the narrow line emission at the local velocity of 1E 0102 is labeled in black. Note in particular the clear detections of blue-shifted [S II]$\lambda\lambda$6716,6731, [S III]λ9069, [Ar III]λ7136, and the (marginal) detection of [Cl II]λ8579. Weak lines of Hα and Hβ are seen at the same Doppler shift.

this analysis, BRUTUS relied on STATSMODEL (Seabold & Perktold 2010), MATPLOTLIB (Hunter 2007), ASTROPY, a community-developed core Python package for Astronomy (Astropy Collaboration 2013), APLPY, an open-source plotting package for Python hosted at `http://aplpy.github.com`, and MONTAGE, funded by the National Science Foundation under Grant Number ACI-1440620 and previously funded by the National Aeronautics and Space Administration's Earth Science Technology Office, Computation Technologies Project, under Cooperative Agreement Number NCC5-626 between NASA and the California Institute of Technology. This research has also made use of the ALADIN interactive sky atlas (Bonnarel *et al.* 2000), of SAOIMAGE DS9 (Joye & Mandel 2003) developed by Smithsonian Astrophysical Observatory, of NASA's Astrophysics Data System, and of the NASA/IPAC Extragalactic Database (NED; Helou *et al.* 1991), which is operated by the Jet Propulsion Laboratory, California Institute of Technology, under contract with the National Aeronautics and Space Administration. IRS was supported in part by the Australian Research Council Laureate Grant FL0992131 and the Future Fellowship Grant FT160100028. AJR acknowledges funding through the ARC Centre of Excellence for All-sky Astrophysics (CAASTRO) through project number CE110001020. PG thanks the Stromlo Distinguished Visitor Programme. FPAV and IRS thank the CAASTRO AI travel grant for generous support. Based in part on observations made with ESO Telescopes at the La Silla Paranal Observatory under programme ID 297.D-5058[A] and the ANU 2.3 m telescope at Siding Spring Observatory.

References

Astropy Collaboration, Robitaille, T. P., Tollerud, E. J., *et al.* 2013, *A& A*, 558, A33
Blair, W. P., Morse, J. A., Raymond, J. C., *et al.* 2000, *ApJ*, 537, 667
Bonnarel, F., Fernique, P., Bienaymé, O., *et al.* 2000, *A& A Supplement Series*, 143, 33
Childress, M. J., Vogt, F. P. A., Nielsen, J., & Sharp, R. G. 2014, *ApSS*, 349, 617
Dopita, M. A., Tuohy, I. R., & Mathewson, D. S. 1981, *ApJ Lett.*, 248, L105
Finkelstein, S. L., Morse, J. A., Green, *et al.* 2006, *ApJ*, 641, 919
Freudling, W., Romaniello, M., Bramich, D. M., *et al.* 2013, *A& A*, 559, A96
Hunter, J. D. 2007, *Computing in Science and Engineering*, 9, 90
Joye, W. A. & Mandel, E. 2003, in *Astronomical Data Analysis Software and Systems XII ASP Conference Series*, Vol. 295, 489
Helou, G., Madore, B. F., Schmitz, M., *et al.* 1991, *Databases and On-line Data in Astronomy*, 171, 89
Lasker, B. M. & Golimowski, D. A. 1991, *ApJ*, 371, 568
Seabold, S. & Perktold, J. 2010, in *Proc. of the 9th Python in Science Conference*, 57–61
Vogt, F. & Dopita, M. A. 2010, *ApJ*, 721, 597
Vogt, F. P. A., Seitenzahl, I. R., Dopita, M. A., & Ruiter, A. J. 2017a, *PASP*, 129, 058012
Vogt, F. P. A., Seitenzahl, I. R., Dopita, M. A., & Ghavamian, P. 2017b, accepted for publication in *A& A Lett.*, arXiv:1705.04322

Discussion

KIRSHNER: It's great you finally found [S II] – in a way we always knew it had to be there. You should try to put a limit on [Ca II].

Supernova 1987A:30 years later - Cosmic Rays and Nuclei from Supernovae and their aftermaths
Proceedings IAU Symposium No. 331, 2017
A. Marcowith, M. Renaud, G. Dubner, A. Ray & A. Bykov, eds.

doi:10.1017/S1743921317004847

High-resolution imaging of SNR IC443 and W44 with the Sardinia Radio Telescope

E. Egron[1], A. Pellizzoni[1], M. N. Iacolina[1], S. Loru[1,2], M. Marongiu[1], S. Righini[3], M. Cardillo[4], A. Giuliani[5], S. Mulas[2], G. Murtas[2] and D. Simeone[2]

[1]INAF, Osservatorio Astronomico di Cagliari, Via della Scienza 5, 09047 Selargius, Italy

[2]Dip. di Fisica, Università degli Studi di Cagliari, SP Monserrato-Sestu, KM 0.7, 09042 Monserrato, Italy

[3]INAF, Istituto di Radio Astronomia di Bologna, Via P. Gobetti 101, 40129 Bologna, Italy

[4]INAF, Osservatorio Astrofisico di Arcetri, Largo E. Fermi 5, 50125 Firenze, Italy

[5]INAF, Istituto di Astrofisica Spaziale e Fisica cosmica di Milano, via E. Bassini 15, 20133 Milano, Italy

email: `egron@oa-cagliari.inaf.it`

Abstract. We present single-dish imaging of the well-known Supernova Remnants (SNRs) IC443 and W44 at 1.5 GHz and 7 GHz with the recently commissioned 64-m diameter Sardinia Radio Telescope (SRT). Our images were obtained through on-the-fly mapping techniques, providing antenna beam oversampling, automatic baseline subtraction and radio-frequency interference removal. It results in high-quality maps of the SNRs at 7 GHz, which are usually lacking and not easily achievable through interferometry at this frequency due to the very large SNR structures. SRT continuum maps of our targets are consistent with VLA maps carried out at lower frequencies (at 324 MHz and 1.4 GHz), providing a view of the complex filamentary morphology. New estimates of the total flux density are given within 3% and 5% error at 1.5 GHz and 7 GHz respectively, in addition to flux measurements in different regions of the SNRs.

Keywords. ISM: supernova remnants, ISM: individual (IC443, W44), radio continuum: ISM

1. Introduction

Although radio emission is the most common identifier of Supernova Remnants (SNRs) and a prime probe for refining models, high-resolution images at frequencies above 5 GHz are scarse even for bright and well-known SNRs such as IC443 and W44. These SNRs belong to the remnant "mixed morphology" class (Rho & Petre 1998), characterised by a highly filamentary radio shell (synchrotron emission) and a centrally concentrated X-ray emission. Providing good angular resolution and sensitivity maps at high radio frequencies could help to better constrain the multi-wavelength scenario since local changes in the radio spectrum trace the actual energy distribution of the different electron populations responsible for both radio and part of gamma-ray emissions.

In the frameworks of the Astronomical Validation (AV) and Early Science Program (ESP) with the 64-m single-dish Sardinia Radio Telescope (SRT), we produced, for the first time, single-dish deep imaging at 7 GHz of IC443 and W44 in addition to imaging them at 1.5 GHz. We obtained accurate flux density measurements through on-the-fly (OTF) mapping procedures and using state-of-the-art imaging techniques.

2. Observations with SRT

Located on the Sardinia island in Italy, SRT is designed to observe in the $0.3 - 115$ GHz frequency range. At present, three receivers are available for observers: a 7-feed K-band receiver (18 – 26.5 GHz), a mono-feed C-band receiver (5.7 – 7.7 GHz), and a coaxial dual-feed L-P band (1.3 – 1.8 GHz; 305 – 410 MHz) receiver (Bolli *et al.* 2015).

A first series of observations of IC443 and W44 was performed during the AV of SRT, from May 27 to December 10 2014 (~12 hrs of effective time on targets). This phase was devoted to testing the performance of the telescope and the acquisition systems (Prandoni *et al.* 2017). The targets were observed at 7.24 GHz with a bandwidth of 680 MHz, using the Total-Power (TP) backend. Subsequently, a program dedicated to the observations of SNRs was approved during the Early Science phase of SRT, for a total exposure time on targets of $\sim$ 20 hrs in C and L-bands. The observations were performed in "shared-risk mode" at 7.0 GHz (bandwidth=1,200 MHz), and 1.55 GHz (bandwidth=460 MHz) between February 14 and March 24 2016, operating simultaneously the TP and SARDARA (SArdinia Roach2-based Digital Architecture for Radio Astronomy; Melis *et al.* in preparation) backends in piggy-back mode.

We carried out the mapping of IC443 and W44 through On-the-Fly (OTF) scans. This technique implies that the data acquisition is continuously ongoing while the antenna performs constant-speed scans across the sky (typically a few degrees/min), alternatively producing maps along the Right Ascension (RA) and Declination (Dec) directions. In particular, the sampling time was set to 40 ms during AV activities and 20 ms during the ESP, which implies the acquisition of $> 10 - 20$ samples/beam for each scan passage (largely oversampling the beam w.r.t. Nyquist sampling) and an accurate evaluation of flux errors. In addition, beam oversampling allows us to efficiently reject outlier measurements ascribed to radio frequency interference (RFI). The length of the scans is chosen according to the size of the source (typically of the order of $0.5 - 1°$ for SNRs). In order to properly reconstruct the morphology of the observed source and its associated flux, the scan-dependent baseline (background flux) must be correctly subtracted. Ideally, each scan should be free from significant source contribution (and RFI contamination) for 40 – 60% of its length/duration, in order to properly identify and subtract the baseline component. This requirement is not trivially satisfied for targets located in crowded regions of the Galactic Plane, as for the case of W44.

Data analysis was performed using the SRT Single-Dish Imager (SDI), which is a tool designed to perform continuum and spectro-polarimetric imaging, optimized for OTF scan mapping, and suitable for all SRT receivers/backends (Egron *et al.* 2016). The core of our prodecure is to fully exploit the availability of a significant number of measurements per beam (and then per pixel, typically chosen to be about 1/4 of the HPBW), in order to have a straightforward evaluation of statistical errors (through standard deviation of the measurements in each pixel), efficient RFI outliers removal and accurate background baseline subtraction.

3. IC443

We produced the maps of IC443 and measured the integrated flux density of the source by defining a centroid of the diffuse emission and considering a radius of 0.5° (centroid coord: $06^h 16^m 58.00^s + 22°31'36.0''$). Integrated flux densities of the maps obtained by simultaneous observations using the TP and SARDARA (operating in piggy-back mode) during the ESP are fully consistent at 7 GHz and also compatible with AV observations performed more than one year earlier within 1σ. Averaging these results, we obtain a

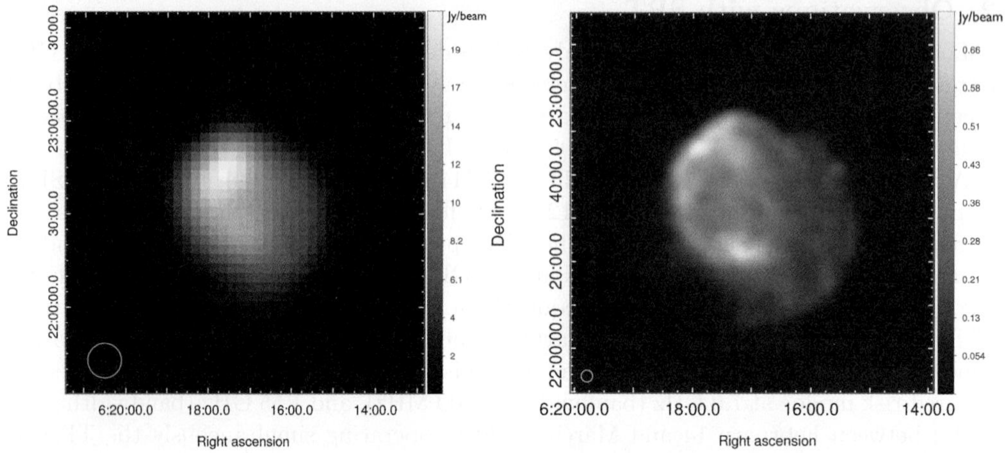

Figure 1. SRT continuum map of SNR IC443 obtained at 1.55 GHz with the SARDARA backend (left) and at 7.2 GHz with the TP backend (right). The green circles on the bottom left indicate the beam size at the observed frequencies. Pixel sizes are 3.0′ and 0.6′ for 1.55 GHz and 7.2 GHz maps, respectively, which corresponds to 1/4 HPBW.

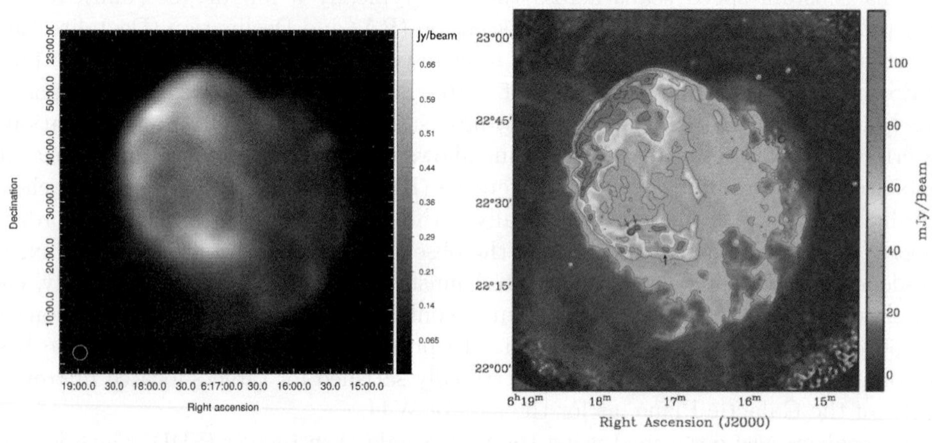

Figure 2. Comparison of the continuum maps of IC443 obtained with SRT at 7 GHz (left) with that obtained combining VLA and Arecibo at 1.4 GHz (right; Lee *et al.* 2008).

continuum flux of 66.8 ± 2.9 Jy at 7 GHz. The integrated flux measured with SARDARA at 1.55 GHz is 133.7 ± 4.0 Jy. The total averaged maps of IC443 obtained at both frequencies are presented in Figure 1.

Our measurements are consistent with continuum flux density measurements reported in the literature at 1.4 GHz ($S_{1.4\mathrm{GHz}}$ = 130 ± 13 Jy originally obtained by Green 1986 then scaled accordingly to Baars *et al.* 1977 flux density scales) and at 6.6 GHz ($S_{6.6\mathrm{GHz}}$ = 70 ± 15 Jy by Dickel 1971), within 1σ. It is worth noting that typical continuum flux errors for IC443 in the literature are of the order of $\sim 10-15\%$, while we provided more accurate measurements. This is mostly due to our oversampled maps in which, for each pixel, tens of OTF baseline-subtracted scans are available, providing straightforward error measurements through standard deviation estimates.

We then compared the SRT map with high-resolution maps of IC443 obtained by combining interferometric and large-area single-dish observations. VLA and Arecibo lower frequency data at 1.4 GHz were combined together to achieve an extremely good

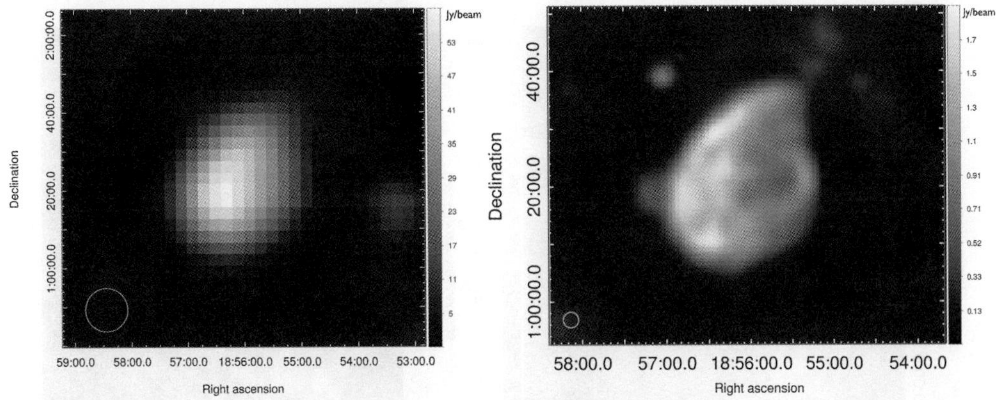

Figure 3. SRT continuum map of SNR W44 obtained at 1.55 GHz with the SARDARA backend (left) and at 7.2 GHz with the TP backend (right). Pixel sizes are as in Fig. 1.

sensitivity and angular resolution of about 40″ (Lee *et al.* 2008). The main features of the morphology of IC443 revealed with SRT at 7 GHz are consistent with those in the above observations, as testified by Figure 2, and with the image obtained at 330 MHz (not shown in this paper) with the VLA (Castelletti *et al.* 2011).

IC443 consists in two nearly concentric shells, which present a complex structure with the presence of filaments (not resolved, but detected in the SRT image), and a clear difference in the radio continuum intensity. The mean surface brightness associated with the brightest rims of IC443 is ~ 0.6 Jy/beam at 7 GHz (measured with SRT), which is about five times larger than that corresponding to the western halo.

4. W44

The OTF maps of W44 obtained with SRT at 1.55 and 7 GHz are presented in Figure 3. The integrated flux density of W44 was measured by considering a radius of 0.32° around $18^h 56^m 05.00^s + 01°21'36.0''$. We obtained 214.4 ± 6.4 Jy at 1.55 GHz and 93.7 ± 4.0 Jy at 7 GHz. The flux obtained at 7.0 GHz with SRT is in agreement with Hollinger & Hobbs (1966): $S_{8.3\mathrm{GHz}} = 95 \pm 23$ Jy. The continuum fluxes presented in the literature at ~ 1.4 GHz provide a wide scatter, as shown in Table 1. In particular, measurements at 1.4 GHz by Giacani *et al.* (1997) ($S_{1.4\mathrm{GHz}} = 210 \pm 20$ Jy) and Castelletti *et al.* (2007) ($S_{1.4\mathrm{GHz}} = 300 \pm 7$ Jy) are inconsistent. The value we measured with SRT at 1.55 GHz is comparable within 1σ with most other L-band flux measurements obtained in the '60s and with Giacani *et al.* (1997). Note also the small error bars on our value.

We then compared our maps with high-resolution VLA images of the remnant at 1,465 MHz (Jones *et al.* 1993) and 324 MHz (Castelletti *et al.* 2007), which were obtained using interferometric multiple-configurations (see Figure 4 for a comparison with the 324 MHz map). The radio emission of W44 is characterized by an asymmetric limb-brightened shell. The brightest filaments are also detected with SRT (although not resolved), and they most likely result from radiative shocks driven into clouds or sheets of dense gas (Jones *et al.* 1993). The brightest emission occurs along the eastern boundary at $\sim 18^h 56^m 50.00^s + 01°17'00.0''$. It results from the interaction between W44 and dense molecular clouds observed in this region (Seta *et al.* 2004). Using Spitzer observations at 24 μm and 8 μm, Castelletti *et al.* (2007) identified a circular HII region centered at $18^h 56^m 47.9^s + 01°17'54.0''$ and named G034.7-00.6 with the IRAS point source 18544+0112, which is a young stellar object located on its border. This feature

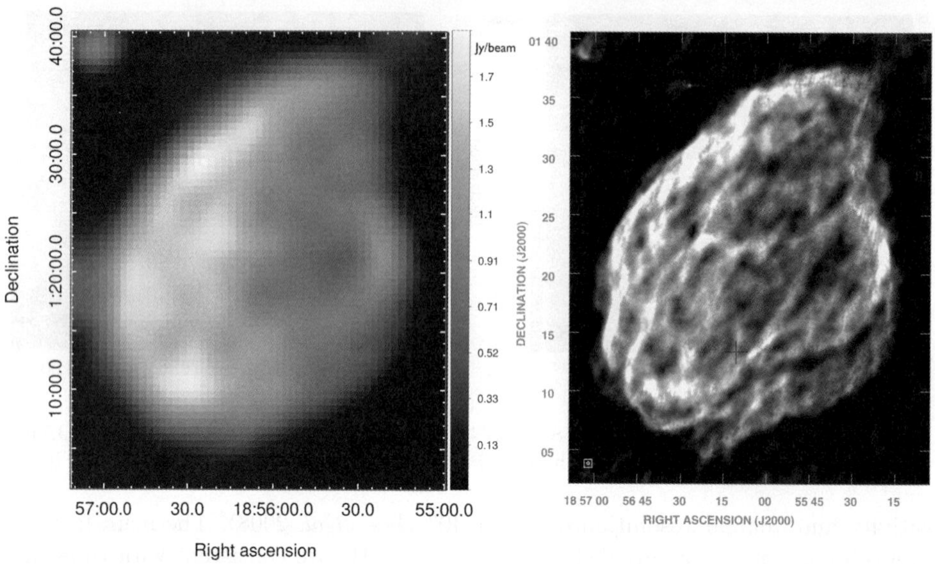

Figure 4. Comparison of the continuum maps of SNR W44 obtained with SRT at 7.0 GHz (left) with that obtained with the VLA at 324 MHz (right) by Castelletti *et al.* (2007). The beam size in the SRT map is represented by the point-like source (known as a HII region G035.040-00.510) in the top left corner.

Table 1. Comparison of the integrated flux densities of W44 measured in the 1.39 to 1.55 GHz frequency range, including the new SRT value at 1.55 GHz. Tabulated values have been reported from the literature as they were originally published.

Frequency (GHz)	Flux density (Jy)	References
1.39	173 ± 35	Westerhout (1958)
1.40	188 ± 23	Pauliny-Toth *et al.* (1966)
1.40	173 ± 26	Kellermann *et al.* (1969)
1.41	236 ± 47	Scheuer (1963)
1.41	236 ± 35	Beard & Kerr (1969)
1.41	275 ± 27	Altenhoff *et al.* (1970)
1.42	180 ± 36	Leslie (1960)
1.44	210 ± 20	Giacani *et al.* (1997)
1.44	300 ± 7	Castelletti *et al.* (2007)
1.55	214 ± 6	Our work

appears at about 0.7 Jy/beam in the SRT 7 GHz image at $18^h 57^m 10.00^s + 01°18'00.0''$. To the west, a short bright arc is visible at $18^h 55^m 20.00^s + 01°22'00.0''$. It corresponds to the SNR shock colliding with a molecular cloud located in this region, which is consistent with bright optical filaments (Giacani *et al.* 1997) and IR observations. The map of W44 includes a classified HII region G035.040-00.510, as listed for example in the WISE Catalog of Galactic HII regions, at $18^h 57^m 04.00^s + 01°38'45.0''$, which is visible in the SRT and VLA images at low frequencies and other wavelengths (IR with Spitzer). The corresponding flux density measured with SRT is 0.94 ± 0.05 Jy at 7.2 GHz.

5. Summary

We obtained accurate images of IC443 and W44 probing the capabilities of the Sardinia Radio Telescope by operating single-dish OTF maps. Our results show reliable performances of the instrumentation over a two-year-long time span (gain stability <5%) and provide self-consistency checks in measurements performed by simultaneous piggy-back observations with two different backends. We compared the values of the integrated flux densities obtained with SRT at 1.55 GHz and 7 GHz with those presented in the literature. SRT flux estimates are given with a very precise error of 3% and 5% at 1.5 GHz and 7 GHz, respectively

Interferometry can "pass the baton" to single-dish techniques to image large structures ($\sim 1^\circ$ or more) at high frequencies, since synthesis imaging becomes difficult in this context. In fact, the most prominent features highlighted in the SRT maps at 7 GHz are coherent with the high-resolution, low-frequency interferometric images of IC443 and W44. Thus, good mapping quality of large structures can be maintained even at high-frequencies if deep and oversampled OTF single-dish mapping is provided.

This work represents a first scientific milestone for SRT, testifying its capabilities and performances in single-dish imaging, and exploiting different receivers and backends.

6. Work in progress

By coupling the SRT maps at 1.5 and 7 GHz, we obtain spatially-resolved spectral-indices maps of IC443 and W44 (see Pellizzoni *et al.*, these proceedings). Accurate spectral imaging at such high-radio frequencies are necessary in order to understand the physical properties and better constrain the multi-wavelength scenario (leptonic versus hadronic models). Moreover, observations of both targets performed at 21.4 GHz with SRT K-band 7-feed receiver allow us to investigate the complex morphology of these SNRs at high frequency (see Loru *et al.*, these proceedings). We will then perform spectral imaging by coupling 7 and 21 GHz maps.

A new multi-feed S-band receiver (3 – 4.5 GHz) is under construction at SRT. At the moment, we performed OTF maps of IC443 and W44 using the central feed as a first test. The high quality of the maps is very promising. SRT represents a great instrument to investigate the morphology and spectrum of SNRs at 1.5, 4, 7 and 21 GHz and search for possible spectral break.

The next steps are the analysis of spectral lines thanks to the SARDARA backend, the production of polarization maps, and performing maps in Q-band (33 – 50 GHz) as soon as the receiver will be installed at SRT.

References

Baars, J. W. M., Genzel, R., Pauliny-Toth, I. I. K. *et al.* 1977, *A&A*, 61,99
Bolli, P., Orlati, A., Stringhetti. L, *et al.* 2015, *Journal of Astronomical Instrumentation*, 4, 1550008
Castelletti, G., Dubner, G., Brogan, C., & Kassim, N. E. 2007, *A&A*, 471, 537
Castelletti, G., Dubner, G., Clarke, T., & Kassim, N. E. 2011, *A&A*, 534, A21
Dickel, J. R. 1971, *PASP*, 83,343
Egron, E., Pellizzoni, A., & Iacolina, M. N. 2016, *INAF - Osservatorio Astronomico di Cagliari. Internal Report*, N.59
Green, D. A. 2014, *Bull. Astr. Soc. of India*, 42, 47
Hollinger, J. P. & Hobbs, R. W. 1966, *Science*, 153, 1633
Jones, L. R., Smith, A., & Angelini, L. 1993, *MNRAS*, 265, 631
Lee, J. J., Koo, B. C., Yun, M. S., *et al.* 2008, *ApJ*, 135, 796
Prandoni, I., Murgia, M., Tarchi, A., *et al.* 2017, *A&A* Accepted for publication
Rho, J. & Petre, R. 1998, *ApJL*, 503, L167
Seta, M., Hasegawa, T., & Sakamoto, S. 2004, *ApJ*, 127, 1098

Supernova 1987A:30 years later - Cosmic Rays and Nuclei from Supernovae and their aftermaths
Proceedings IAU Symposium No. 331, 2017
A. Marcowith, M. Renaud, G. Dubner, A. Ray & A. Bykov, eds.

doi:10.1017/S1743921317005221

High-resolution spectral imaging of SNR W44 and IC443 at 22 GHz with the Sardinia Radio Telescope

S. Loru[1,3], A. Pellizzoni[1], E. Egron[1], N. Iacolina[1], S. Righini[2], M. Marongiu[1], S. Mulas[3], G. Murtas[3], D. Simeone[3], M. Pilia[1], M. Bachetti[1], A. Trois[1], R. Ricci[2], A. Melis[1] and R. Concu[1]

[1]INAF-Osservatorio Astronomico di Cagliari
Via della Scienza, 5 - 09047 Selargius (CA), Italy

[2]INAF- Istituto di Radioastronomia
Via P. Gobetti, 101 40129 Bologna, Italy

[3]Dipartimento di Fisica-Università di Cagliari
S.P. Monserrato-Sestu Km 0,700 I-09042 Monserrato (CA), Italy
email: saraloru@oa-cagliari.inaf.it

Abstract. In the framework of the Astronomical Validation and Early Science activities of the Sardinia Radio Telescope (SRT, www.srt.inaf.it), we performed 22 GHz imaging observations of SNR W44 and IC443. Thanks to the single-dish imaging performances of SRT and innovative ad hoc imaging techniques, we obtained maps that provide a detailed view of the structure of the remnants. We are planning to exploit the high-frequency radio data of SNRs to better characterize the spatially-resolved spectra and search for possible spectral steepening or breaks in selected SNR regions, assessing the high-energy tail of the region-dependent electron distribution.

Keywords. (ISM:) supernova remnants, ISM: individual (IC443, W44), ISM: radio continuum

1. Introduction

Galactic Supernova Remnants (SNRs) are commonly identified as extended radio sources with non-thermal spectra. To date, a multi-wavelength coverage (from radio to ultra-high energies) has been carried out for part of these SNRs. The nature of the high-energy emission is still debated. The co-spatial study of the radio and gamma-ray emissions may provide crucial information for disentangling the two basic possible scenarios represented by leptonic (Inverse Compton and bremsstrahlung emission) and hadronic models (π^0 mesons decay emission). Radio emission is expected up to $20 - 50$ GHz, but even for the most interesting and bright objects, high-resolution images at frequencies above 5 GHz in the confused regions of the Galactic Plane are lacking and not easily achievable (mostly due to the low flux densities in this range). In fact, interferometric imaging synthesis of large structures ($> 1°$) at high-frequencies becomes unfeasible. Radio continuum observations performed with single-dish telescopes allow us to obtain accurate flux density measurements of large SNR structures at high frequency ($\gtrsim 5$ GHz). Most of the features and details highlighted in high-resolution low-frequency interferometric images of SNRs (Castelletti *et al.* 2007 and 2011; Gao *et al.* 2011), can be also observed and studied in high-frequency single-dish images thanks to the very stable gaussian beam of SRT and to innovative ad hoc imaging techniques based on "On The Fly" (OTF) scans (Prandoni *et al.*, 2017). These characteristics are crucial to investigate both integrated and spatially-resolved spectral indexes of SNRs.

2. Data analysis and results

In the framework of Early Science activities (Egron *et al.* 2017), we performed observations at 21.4 GHz with the SRT K-band 7-feed receiver. By using the "Best Space Coverage configuration" (Bolli *et al.* 2015), we provided 7 simultaneous equally-spaced (OTF) scans, covering the whole area scanned by the multi-feed receiver (Fig.1). In this way, we obtained SNR maps characterized by about 14 scans per beam. This oversampling is fundamental to efficiently reject outlying measurements (due to radio-frequency interferences) and to optimize final image accuracy.

Thanks to its simple morphology and the availability of extensive flux density measurements up to high radio frequency (useful to cross-check calibrations), SNR 3C10 represents an ideal target for testing our data analysis tool (SRT Single Dish Imager; Egron *et al.*, 2016) on SRT K-band data. In Fig.2 we show a multi-feed image of 3C10 at 21.4 GHz with an angular resolution of $0.87'$, obtained by SRT on February 24 2016 (in the frameworks of the Early Science Program). This image is produced by merging the individual maps associated to each feed. We obtained a preliminary integrated flux density of ~ 8 Jy for 3C10 consistent with previous measurements at high radio frequency (observations at 14 to 18 GHz; N. Hurley-Walker *et al.* 2009).

After testing it on 3C10, we applied our K-band data analysis procedure on the larger and complex SNR W44 (Fig.3). This map will allow us to study for the first time the morphology and the flux density of W44 at high radio-frequencies. We used integrated flux density measurements obtained from SRT data at 21.4 GHz, 7.0 GHz and 1.5 GHz (see Egron *et al.* these proceedings) in order to provide a first radio continuum spectrum of W44 (Fig.3-6). Our preliminary result indicates a possible spectral curvature/break between around 5-10 GHz. This break in the spectrum of W44 would be consistent with a steepening of the primary particle spectrum at 10 GeV as suggested by gamma-ray observations of SNR W44 (Giuliani *et al.* 2011; Ackermann *et al.* 2013; Cardillo *et al.* 2014).

Bremsstrahlung and IC bumps observed in gamma-rays band are bounded to synchrotron spectral slope and cut-off in the radio domain. For this reason, a careful characterization of lepton contributions through high-frequency radio-observed parameters is necessary to disentangle hadron contributions in the high-energy spectra and better constrain SNRs as cosmic rays emitters. However, the assumption of a single electron population for the whole SNR is too simplistic. Indeed, a spread in the spectra slope distribution, through different SNR regions, has been highlighted by preliminary spatially-resolved spectral measurements obtained for SNR W44 and IC443 by using SRT data at 1.5 and 7.0 GHz (Pellizzoni *et al.*, these proceedings). Spatially-resolved high-frequency spectra measurements could be important to better disentangle different region-dependent physical processes in these SNRs.

3. Perspective for future SRT data analysis at high-frequencies

We plan to apply the same calibration procedure adopted for 3C10 and W44 to K-band images of SNR IC443, in order to look for a possible spectral steepening, as observed for W44. In Fig.7 we show a preliminary uncalibrated map of SNR IC443 obtained from SRT observations performed on March 23 2016. Furthermore, coupling K-band maps with L-band (1.5 GHz) and C-band (7.0 GHz) maps, we could obtain more accurate spectral index maps to search for possible spectral steepening or breaks in selected SNR regions, assessing the high energy tail of the region-dependent electron distribution.

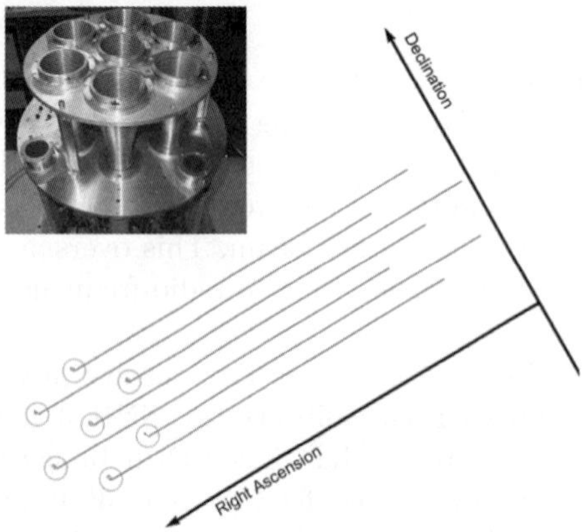

Figure 1. Schematic view of the "Best Space Coverage configuration" and SRT K-band 7-feed receiver (Upper left).

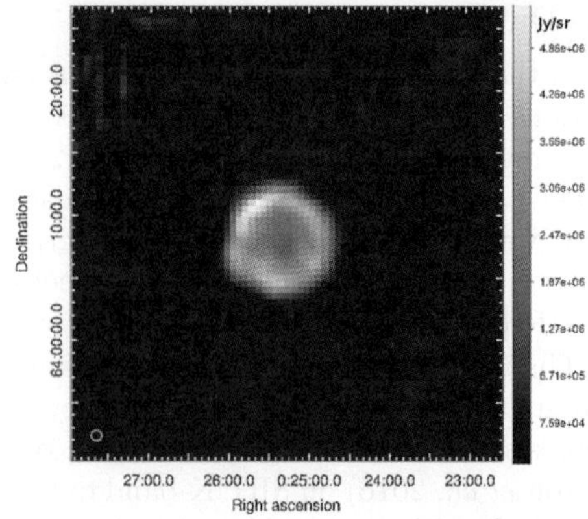

Figure 2. SRT radio continuum map of 3C10 at 21.4 GHz. The beam size is 0.87' FWHM (green circle).

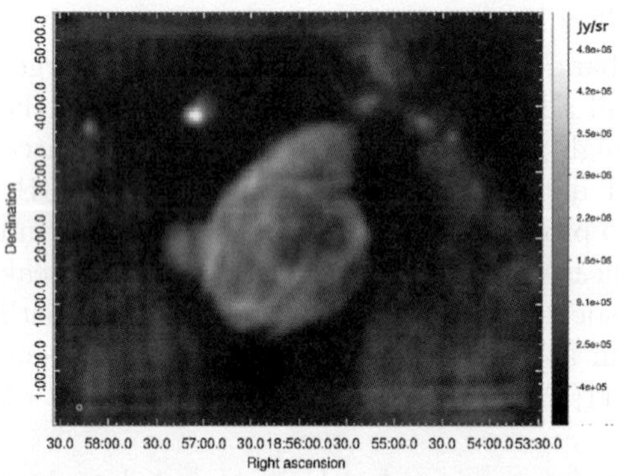

Figure 3. SRT radio continuum map of W44 at 21.4 GHz. The beam size is 0.87' FWHM (green circle on the bottom left). The Galactic plane and other nearby diffuse and point-like sources are clearly visible in this map.

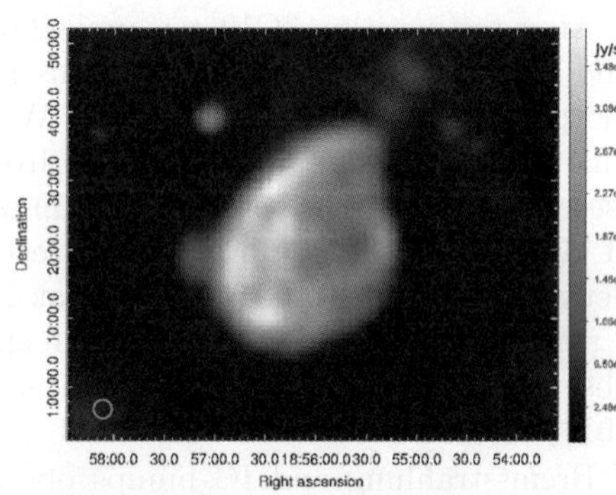

Figure 4. SRT radio continuum map of W44 at 7.0 GHz. The beam size is 2.71' FWHM (green circle on the bottom left).

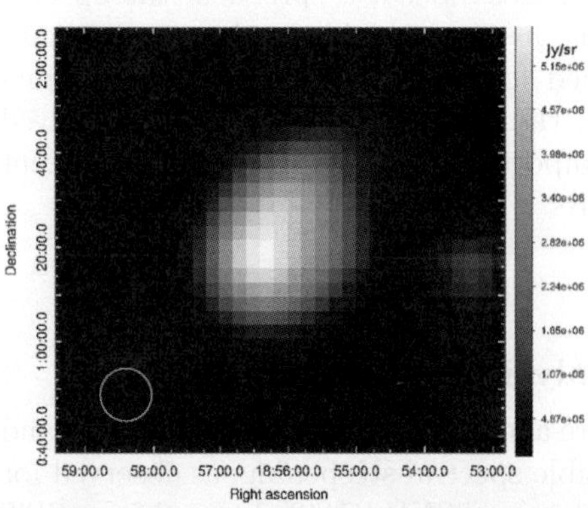

Figure 5. SRT radio continuum map of W44 at 1.5 GHz. The beam size is 11.1' FWHM (green circle on the bottom left).

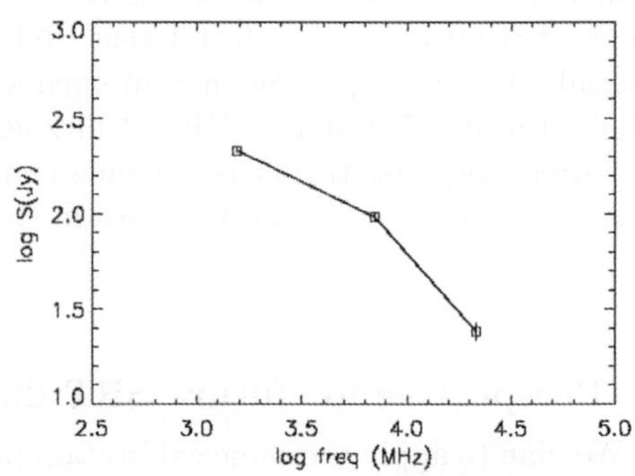

Figure 6. Integrated radio continuum spectrum of W44 obtained by using the flux density measurements related to the maps at 21.4, 7.0 and 1.5 GHz shown in Fig. 3, 4 and 5.

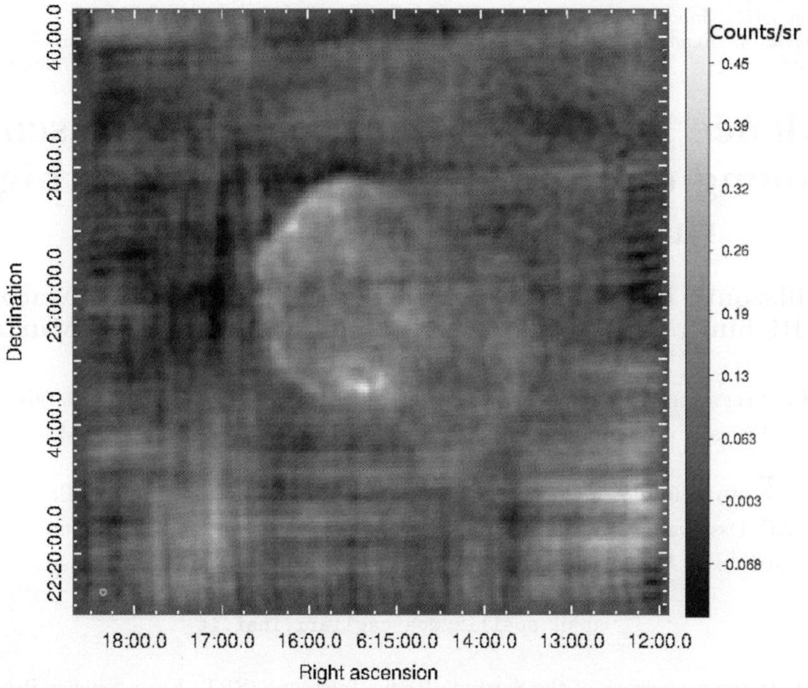

Figure 7. Preliminary SRT radio continuum map of IC443 (uncalibrated) at 21.4 GHz. The beam size is 0.87' FWHM (green circle on the bottom left).

References

Ackermann, M. *et al.* 2013, *Science*, 339, 807

Bolli, P., *et al.*, 2015, *Journal of Astronomical Instrumentation*, 4, 1550008-880

Castelletti, G., Dubner, G., Brogan, C., & Kassim, N. E. 2007, *A&A*, 471, 537

Castelletti, G., Dubner, G., Clarke, T., & Kassim, N. E. 2011, *A&A*, 534, A21

Cardillo, M., Tavani, M., Giuliani, A., Yoshiike, S., Sano, H., Fukuda, T., Fukui, Y., Castelletti, G., & Dubner, G. 2014, *A&A*, 565, A74

Egron, E. *et al.*, 2017, preprint (arXiv:1705.06886)

Egron, E., Pellizzoni, A., Iacolina, M. N., Loru, S., Righini, S., Trois, A., SRT Astrophysical Validation Team, 2016, INAF-Internal Report N.59

Hurley-Walker, N. *et al.* 2009, *MNRAS*, 396, 365

Gao, X. Y., Han, J. L., Reich, W., Reich, P., Sun, X. H., & Xiao, L., 2011, *A&A*, 529, A159

Giuliani *et al.* 2011, *ApJ*, 742, L30

Prandoni *et al.* 2017, preprint (arXiv:1603.06134v1)

Supernova 1987A:30 years later - Cosmic Rays and Nuclei from Supernovae and their aftermaths
Proceedings IAU Symposium No. 331, 2017
A. Marcowith, M. Renaud, G. Dubner, A. Ray & A. Bykov, eds.

doi:10.1017/S1743921317004811

Evidence for a wide electron spectra scatter among different SNR regions from high radio-frequency observations

A. Pellizzoni[1], E. Egron[1], M. N. Iacolina[1], S. Loru[1,2], M. Marongiu[1], S. Righini[3], M. Cardillo[4], A. Giuliani[5], S. Mulas[2], G. Murtas[2] and D. Simeone[2]

[1]INAF, Osservatorio Astronomico di Cagliari, Via della Scienza 5, 09047 Selargius, Italy

[2]Dip. di Fisica, Università degli Studi di Cagliari, SP Monserrato-Sestu, KM 0.7, 09042 Monserrato, Italy

[3]INAF, Istituto di Radio Astronomia di Bologna, Via P. Gobetti 101, 40129 Bologna, Italy

[4]INAF, Osservatorio Astrofisico di Arcetri, Largo E. Fermi 5, 50125 Firenze, Italy

[5]INAF, Istituto di Astrofisica Spaziale e Fisica cosmica di Milano, via E. Bassini 15, 20133 Milano, Italy

email: `apellizz@oa-cagliari.inaf.it`

Abstract. In the framework of the Sardinia Radio Telescope (SRT) Early Science Program, we obtained single-dish high-resolution imaging of the Supernova Remnants IC443 and W44 at 7 GHz. By coupling them with SRT 1.5 GHz maps, we provided spatially-resolved spectral measurements that are highlighting a spread in spectral slope distribution. The observed features range from flat or slightly inverted spectra corresponding to bright radio limbs and filaments, to relatively steep spectra in fainter radio regions. Different theoretical possibilities explaining the above challenging findings are discussed. In particular, we exclude that the observed region-dependent wide spread in spectral slope distribution could be related to absorption processes. Our high-frequency results can be directly related to distinct electron populations in the SNRs including secondary hadronic electrons and resulting from different shocks conditions and/or undergoing different cooling processes. Integrated fluxes associated with the whole SNRs obtained by SRT in comparison with previous results in the literature support the evidence for a slight spectral steepening above 1 GHz for both sources, which could be related to primary electrons or more likely secondary hadronic electrons cut-offs.

Keywords. ISM: supernova remnants – ISM: individual objects: W44, IC443 – radio continuum: ISM

---

1. Introduction

Despite the growing interest in high-energy GeV−TeV emission and its connection with radio emission, multi-wavelength data of SNRs are sparse and often missing. Synchrotron radio emission is expected at least up to $20-50$ GHz, and is produced by electrons with energies in the GeV range and magnetic fields of $> 10-100$ μG. However, for the most interesting and bright objects, high-resolution images at frequencies >5 GHz in the confused regions of the Galactic Plane are lacking and not easily achievable through interferometry due to the very large SNR structures.

Integrated fluxes are typically available in the literature for up to $5-10$ GHz, while SNR spatially-resolved fluxes are largely unexplored in this frequency range and above. In particular, a better characterisation of radio spectra (synchrotron spectral indices and breaks) is necessary for constraining the high-energy spectra in the frame of IC/ bremmstrahlung leptonic models and probe them versus hadronic models. For example, the gamma-ray fluxes and spectra are strongly dependent on the IC photon target parameters and spectral breaks in the electron distribution, and in most cases there are no unambiguous model solutions without detailed measurements of the latter, since the former is intrinsically uncertain (Cardillo *et al.*, 2014; Ackermann *et al.*, 2013; Giuliani *et al.*, 2011). Furthermore, a careful verification (through single-dish imaging, possibly in combination with interferometric data) of co-spatiality of the radio high-frequency and gamma-ray emissions could provide a crucial test of the hypothesis of hadron vs. electron emission. In fact, recent constraints on cosmic ray emission from SNRs and related models (e.g. Ackermann *et al.*, 2013) are based on integrated radio fluxes only (no spatially resolved spectra), implying the simplistic assumption of a single electron population for the whole SNR.

SNR IC443 and W44 represent ideal targets for better testing the above models, thanks to their interesting complex morphology and availability of extensive multi-wavelength data, from radio to gamma rays.

We focused on the need for spectral imaging of IC443 and W44 with good angular resolution and sensitivity at high-radio frequencies obtained by the recently commissioned Sardinia Radio Telescope (SRT; *www.srt.inaf.it*).

2. Integrated and spatially-resolved radio spectra of SNR IC443 and W44

Single-dish radio mapping with SRT of IC443 and W44 at L (1.55 GHz, HPBW 11.1') and C-band (7 GHz, HPBW 2.71') is described by Egron *et al.* (these proceedings).

Comprehensive spatially-resolved spectra were determined by combining C-band and L-band maps produced with the same parameters and topology (same projection centre and pixel size). In particular, spectral index maps were obtained by convolving data at low resolution (accordingly to L-band resolution). The corresponding images are reported in Fig. 1 and 2 for IC443 and W44, respectively, together with the surface brightness maps of the sources in C-band. Since on-the-fly single-dish imaging techniques allow us precise measurements of the density flux for each direction in the sky (i.e. for each pixel map), the spectral index estimate is directly obtained fitting the two L-band and C-band measurements for each pixel. Single-dish flux measurements are straightforward with respect to the more complex interferometric data processing, and related errors are typically lower (conservatively, about 3% in L-band). On the other hand, the available angular resolution for low-frequency single-dish imaging is much worse than that from interferometric images. Our first low-resolution result for single-dish spectral mapping of SNRs gives us an estimate of region-dependent spectral indices and a direct correlation with the corresponding brightness of the largest sub-regions targets (small structures are missing in our spectral index maps). More accurate (i.e. higher angular resolution) images could be provided for example through combination of C-band and K-band (22 GHz) images, the latter having about 40" HPBW (see Loru *et al.*, these proceedings).

For both sources we observed a significant spread in the distribution of the spatially-resolved spectra ranging from a flat (for W44) or slightly inverted (up to -0.5 for IC443) spectral index to a relatively steep spectrum (α>0.7) compared to a mean value of ~0.5 (where the flux density is $S \propto \nu^{-\alpha}$). Relative errors among adjacent pixels are <1%. Our data highlight a correlation between the brightest flux density regions and flat spectral indices for IC443 and W44. These regions are mainly located on the edge and on the brightest filaments of the SNRs. They match with the bulk of the gamma-ray emission

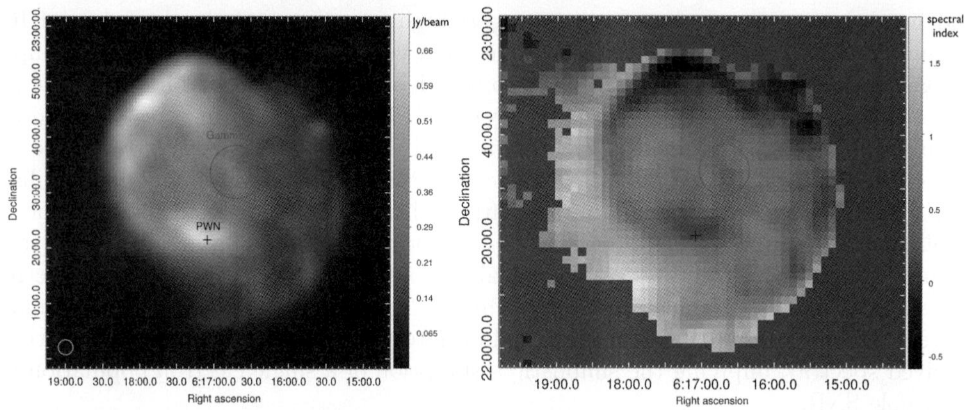

Figure 1. Continuum map of SNR IC443 at 7 GHz (left) and a preliminary spectral index map obtained by using 1.55 and 7 GHz data (right). The plus simbol indicates the position of the PWN powered by the neutron star CXOU J061705.3+222127, whereas the circle indicates the bulk of the gamma-ray emission seen with VERITAS (see text).

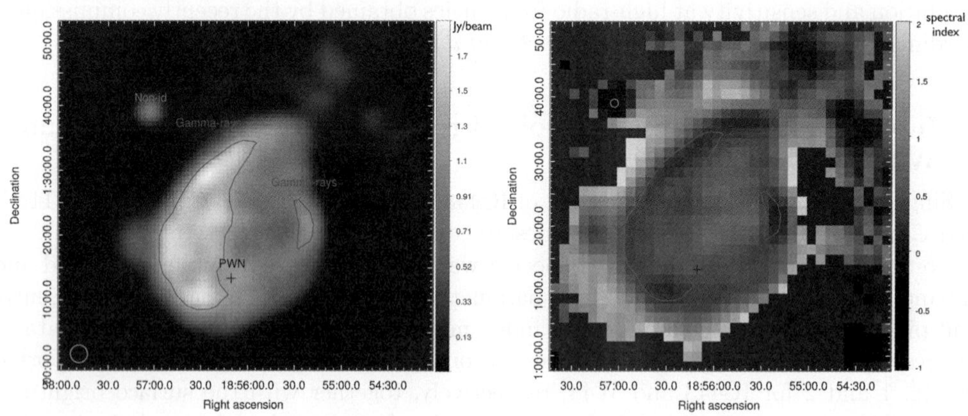

Figure 2. Continuum map of SNR W44 at 7 GHz (left) and a preliminary spectral index map obtained by using 1.55 and 7 GHz data (right). The cross and the circle indicate the position of the PWN powered by the pulsar PSR B1853+01 and a bright point-like source, respectively. The highlighted areas correspond to the gamma-ray emissions detected with AGILE and Fermi-LAT (see text).

in the case of W44. The approximate location of the gamma-ray emission detected with VERITAS and Fermi-LAT for IC443 (Humensky *et al.*, 2015), and with AGILE (Giuliani *et al.*, 2011) and Fermi-LAT in the case of W44 (Abdo *et al.*, 2010), was indicated in the Fig. 1 and 2.

Our measurements for IC443 provided integrated flux densities of $S_{1.55\mathrm{GHz}} = 133.7 \pm 4.0$ Jy and $S_{7\mathrm{GHz}} = 66.8 \pm 2.9$ Jy. For W44 we obtained $S_{1.55\mathrm{GHz}} = 214.4 \pm 6.4$ Jy and $S_{7\mathrm{GHz}} = 93.7 \pm 4.0$ Jy (see Egron *et al.*, these proceedings). It is worth noting that typical continuum flux errors for these sources in the literature are of the order of $\sim 10 - 15\%$, while we provided more accurate measurements. This is mostly due to our oversampled maps in which, for each pixel, tens of OTF baseline-subtracted scans are available, providing straightforward error measurements through standard deviation estimates.

From a weighted fit of the SRT data, our spectral index estimate for IC443 in the interval 1.55 – 7.2 GHz is $\alpha = 0.46 \pm 0.03$. This result is in perfect agreement with a spectral index estimate of $\alpha = 0.47 \pm 0.06$ obtained through a weighted fit of all previous

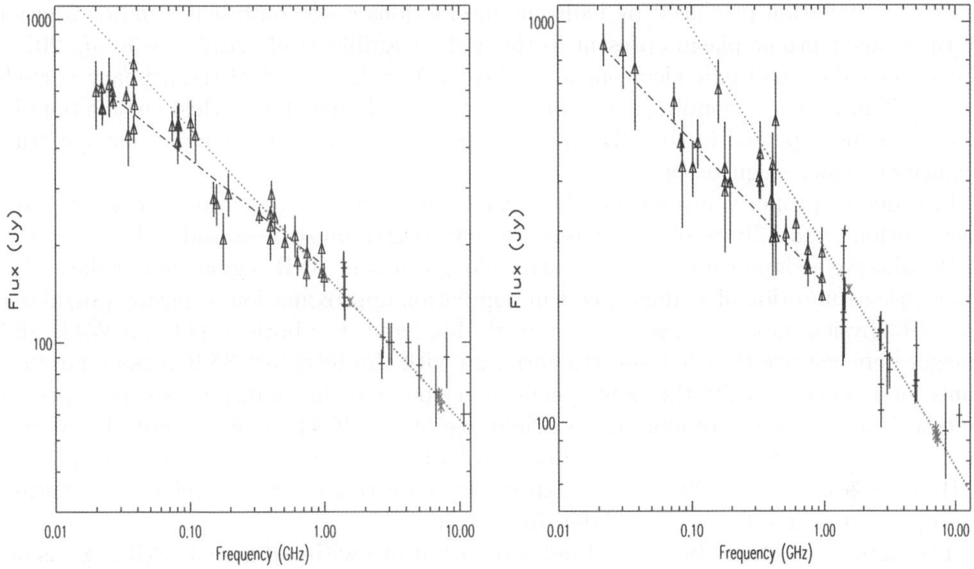

(a) IC443 (b) W44

Figure 3. Integrated radio continuum spectra of IC443 (a) and W44 (b) obtained from the flux density values (black points) listed in Castelletti *et al.*, 2011 and Castelletti *et al.*, 2007, and SRT measurements (asterisks). The spectral indices derived from SRT data at high frequencies (1.5 − 7.2 GHz) are steeper (dotted line) than that obtained from lower frequency measurements (0.02 − 1 GHz; triangles, dot-dashed line fits). These possible spectral turn-offs around ∼1 GHz were not evident from the large spread in previous measurements, in particular at high frequencies.

radio data available in the literature considering the frequency range 1.39 − 8.0 GHz (see Table 2 in Castelletti *et al.*, 2011 for a direct comparison). Focusing the same analysis on the literature data at low frequencies (0.02 – 1.0 GHz), we obtain that the IC443 spectrum flattens to $\alpha = 0.33 \pm 0.01$. Thus, a slight steepening of the IC443 spectrum ($\Delta\alpha \sim 0.1$) around ∼1 GHz is suggested considering previous literature data (though not statistically significant: $\sim 2\sigma$ level) and is now confirmed at $> 4\sigma$ confidence level by SRT observations (see Fig. 3).

W44 displays a similar spectral behaviour as compared to IC443. In fact, the high-frequency spectral index estimated by SRT for W44 is $\alpha = 0.55 \pm 0.03$ (1.55 − 7.2 GHz). Looking at low-frequency flux measurements reported in Table 2 of Castelletti *et al.* (2007), the resulting spectral index is significantly softer (see Fig. 3). A weighted fit to the data in the 0.02 – 1.0 GHz range gives a spectral index $\alpha = 0.36 \pm 0.02$ associated with a very significant goodness-of-fit (reduced $\chi^2 = 1.06$). Thus, even for W44, SRT data suggest a spectral steepenening at high frequencies, which could be related to the approach to a possible break in electron energy distribution (as averaged for the whole SNR extent). It is possibly associated with cooling processes and/or intrinsic particle distribution features at the shock.

A typical primary particle spectrum (both hadronic and leptonic) is expected to have a high energy cut-off which depends on the SNR age and other physical parameters (i.e. ambient density and magnetic field). In fact, gamma-ray observations pointed out a steepening of the primary particle spectrum at energies of ∼ 10 GeV for W44 and ∼ 100 GeV for IC443 (Cardillo *et al.*, 2014; Ackermann *et al.*, 2013; Giuliani *et al.*, 2011) that implies synchrotron break/cut-off at frequencies >10 GHz. However, a secondary electron population produced by hadronic interactions could represent a major fraction of the whole leptonic plasma present in the SNR (Cardillo *et al.*, 2016; Lee *et al.*, 2015). These secondary hadronic electrons are expected to take $\sim$ 10% of the primary particle energy. Thus, a corresponding synchrotron spectrum change due to this particle population could be expected in the GHz range, in addition to the primary particle spectrum change at higher frequencies.

In order to properly investigate the actual parameters of region-dependent electron distributions, spatially-resolved, high-frequency spectra must be considered.

We observe a significant spectral scatter along different SNR regions in our data. The simplistic assumption of a single-electron population approximation, typically postulated in SNR physical modelling, seems manifestly hindered. For both IC443 and W44, SRT images demonstrate that flat spectra correlate with the brightest SNR regions near the limbs or filaments, while the fainter central regions and halos display steeper spectra. Spectral flattening corresponding to bright regions of IC443 (e.g. toward the eastern boundary) was also evident from the analysis of VLA data at lower frequencies of 74$-$330 MHz (Castelletti *et al.*, 2011), although in this case this effect was related to thermal absorption, unlikely working at higher frequencies.

The significant spread in spectral index distributions within different SNR regions observed by SRT could in principle be related to several and possibly concurring processes.

Thermal absorption (free-free) was invoked to explain (A) the low-frequency cut-offs (<50 MHz) observed in the integrated SNR spectrum of IC443, and (B) the apparent spread in spectral index distributions seen by the VLA at <1 GHz that is possibly correlated with the nonuniform optical depth along the SNR. The reported average free-free continuum optical depth derived for IC443 by Castelletti *et al.* (2011) is negligible for frequencies > 50 MHz, but in principle absorption processes could be effective at higher frequencies if strong local enhancements of the optical depth are present. In fact, extrapolating the reported free-free optical depth local peaks for IC443 ($\tau_{74} \sim 0.3$ at 74 MHz) to $\nu > 1$ GHz, we obtain absorption coefficients $\exp(-\tau_{74}(\nu/74_{\rm MHz})^{-2.1}) \sim 1$. Thus, we can exclude that free-free thermal absorption could be responsible for the observed spectral index, region-dependent scatter at high frequencies.

The observed spread in the spectral index distribution of both IC443 and W44 could instead be related to an intrinsic variety in the primary and secondary electron spectra (spectral slopes and breaks) produced in shocks that are located in different SNR/PWN environments, -i.e. several region-dependent electron populations are present. Naive shock acceleration theory predicts a single particle spectrum that is a power law in momentum with spectral index $\sim$2, with little fluctuation depending on the actual shock parameters (e.g. Sturner *et al.*, 1997 and reference therein). The resulting synchrotron spectral slope is then expected to be $\sim$0.5, a value that is compatible with our integrated spectra above $\sim$1 GHz. On the other hand, standard models fail to predict flat or inverted spectra as seen in bright SNR selected regions. Even in the ultrarelativistic regime and assuming a high shock compression factor ($\sim$4), electron spectra slopes are expected to be above $\sim$1.5 (Ellison *et al.*, 1996; Ellison *et al.*, 1996), thus implying a synchrotron spectral index >0.2 that is incompatible with our results.

In order to overtake this incongruency, we note that a spectral flattening effect related to the contribution of secondary hadronic electrons could be present in addition to the primary electron canonical distribution (Cardillo *et al.*, 2016). In fact, the correlation among bright flat-spectrum SNR regions and gamma-ray emission could represent a possible signature of enhanced hadronic emission (and then significant secondary hadronic electrons injection). For W44, gamma-ray emission seems to be associated with the bright radio rims and filaments of the SNR (Cardillo *et al.*, 2014; Giuliani *et al.*, 2011; Abdo *et al.*, 2010). Instead, the IC443 gamma-ray emission seems to be anticorrelated with SNR limbs, although it is still associated with a relatively bright radio filament (with a flatter spectrum w.r.t. average central region spectra) close to a molecular cloud enhancing IC/bremsstrahlung emission (Humensky *et al.*, 2015). Thus, the observed spread in spectral slopes could in principle reflect a region-dependent amount of secondary electron production, however this hypothesis cannot be clearly proven by present multi-wavelength data.

On the other hand, spectral steepening could also be related to strongly-enhanced, region-dependent cooling. However, this hypothesis is disfavoured by the evidence for a correlation among bright flat-spectrum radio regions and gamma-ray emission (i.e. the electron cooling to gamma-ray energies does not significantly affect radio spectra). Furthermore, models of temporal evolution of nonthermal particle and photon spectra at different stages of shell-type SNR lifetime indicate that no significant steepening of the spectral index due to synchrotron cooling is expected from a particle gas drifting away from the shock region on a time-scale of $10^4 - 10^5$ yrs (Sturner *et al.*, 1997), unless a significant local enhancement of the magnetic field can be envisaged. However, both primary and secondary electron distribution cut-offs could approach the synchrotron radio-emitting range on these time-scales, thus providing a change in the synchrotron spectral index at high frequencies. In fact, as reported above, a slight steepening of the integrated spectra of both IC443 and W44 at frequencies >1 GHz is observed. Region-dependent spectral slopes could reflect the presence of different electron distribution cut-off energies. In order to assess spectral curvature and cut-offs, high-resolution spectral imaging at frequencies >10 GHz is required.

3. Summary

In the context of the early science observations with SRT, we obtained single-dish, high-resolution maps of SNR IC443 and W44 at 7 GHz, which provide accurate continuum flux density measurements (see Egron *et al.*, these proceedings). By coupling SRT 7 GHz maps with SRT 1.5 GHz maps, we obtained (low resolution) spatially-resolved spectral measurements that are suggesting a spread in spectral slope distribution, ranging from flat spectra corresponding to bright regions, to relatively steep spectra ($\alpha \sim 0.7$) in fainter radio structures of the SNRs.

We exclude that the observed region-dependent wide spread in spectral slope distribution could be associated to absorption processes. Our high-frequency findings can be directly related to distinct electron populations in the SNRs including secondary hadronic electrons and resulting from different shocks conditions and/or undergoing different cooling processes.

Integrated fluxes associated with the whole SNRs obtained by SRT in comparison with previous results in the literature support the evidence for a possible slight spectral steepening above ~1 GHz for both sources (needing confirmation), which could be related to primary electrons or more likely secondary hadronic electrons cut-offs.

Disentanglement among different theoretical possibilities for explaining the above findings could be provided through the analysis of further high-frequency/high-resolution imaging data. In particular, spatially-resolved, spectral curvature/break measurements could be obtained by coupling $1.5 - 7$ GHz maps with higher-frequency observations, as for instance at $20 - 22$ GHz using SRT's K-band multi-feed receiver (see Loru *et al.*, these proceedings).

References

Abdo, A. A. *et al.* 2010, *Science* 327, 1103
Ackermann, M. *et al.* 2013, *Science* 339, 807
Cardillo, M. *et al.* 2014, *A&A* 565, A74
Cardillo, M., Amato, E., & Blasi, P. 2016, *A&A* 595, 58
Castelletti, G., Dubner, G., Clarke, T., & Kassim, N. E. 2007, *A&A* 471, 537
Castelletti, G., Dubner, G., Clarke, T., & Kassim, N. E. 2011, *A&A* 534, A21
Ellison, D. C., Baring, M. G., & Jones, F. C. 1995, *ApJ* 453, 873
Ellison, D. C., Baring, M. G., & Jones, F. C. 1997, *ApJ* 473, 1029
Giuliani, A. *et al.* 2011, *ApJ* 742, L30
Humensky, B., the VERITAS Collaboration 2015, *ICRC* 34, 875
Lee, S-H., Patnaude, D. J., Raymond, J. C., Nagataki, S., Slane, P. O., & Ellison, D. C. 2015, *ApJ* 806, 71
Sturner, S. J., Skibo, J. G., Dermer, C. D., & Mattox, J. R. 1997, *ApJ* 490, 619

Supernova 1987A:30 years later - Cosmic Rays and Nuclei from Supernovae and their aftermaths
Proceedings IAU Symposium No. 331, 2017
A. Marcowith, M. Renaud, G. Dubner, A. Ray & A. Bykov, eds.

doi:10.1017/S1743921317004689

Low frequency radio counterparts of HESS J1731−347 a.k.a SNR G353.6−0.7

A. J. Nayana and Poonam Chandra

National centre for Radio Astrophysics
Tata Institute of Fundamental Research
PO Box 3, Pune 411007
email: nayana@ncra.tifr.res.in
email: poonam@ncra.tifr.res.in

Abstract. HESS J1731−347 a.k.a. SNR G353.6−0.7 is one of the five known very high energy (VHE, Energy > 0.1 TeV) shell-type supernova remnants. We carried out Giant Metrewave Radio Telescope (GMRT) observations of this TeV SNR in 1390, 610 and 325 MHz bands. We detected the 325 and 610 MHz radio counterparts of the SNR G353.6−0.7 (Nayana *et al.* 2017). We also determined the spectral indices of individual filaments and our values are consistent with the non-thermal radio emission. We compared the radio morphology with that of VHE emission. The peak in radio emission corresponds to the faintest feature in the VHE emission. We explain this anti-correlated emission in a possible leptonic origin of the VHE γ-rays.

Keywords. acceleration of particles–radiation mechanisms: non-thermal–cosmic rays–ISM: individual objects:SNR G353.6−0.7–ISM: supernova remnants–radio continuum: ISM.

1. Introduction

The origin of Galactic cosmic rays remains a fundamental question since its discovery. Supernova remnant (SNR) shocks are considered as one of the possible sites of particle acceleration to cosmic ray energies (Blasi 2013). A very high energy (VHE; Energy > 100 GeV) shell like γ-ray source associated with a SNR provides a unique opportunity to study particle acceleration in SNR shocks. However, there are only five SNRs with spatially resolved shell like VHE emission (Acero *et al.* 2015). These are RX J1713.7−3946 (Aharonian *et al.* 2004), RX J0852.0−4622 (Aharonian *et al.* 2007), RCW 86 (Aharonian *et al.* 2009), SN 1006 (Acero *et al.* 2010) and HESS J1731−347 (Abramowski *et al.* 2011). HESS J1731−347 was discovered in the HESS Galactic plane survey as an unidentified VHE γ-ray source (Aharonian *et al.* 2008). A faint shell-like radio source SNR G353.6−0.7 with a diameter of ≈ 30' was identified in spatial coincidence with this VHE source by Tian *et al.*(2008) in the 1.4 GHz southern Galactic plane survey (SGPS) map (Haverkorn *et al.* 2006) and the 843 MHz Molonglo Galactic plane survey (MOST) map (Green *et al.* 1999). A deeper γ-ray observation of HESS J1731−347 revealed shell like morphology (Abramowski *et al.* 2011) and it became the fifth member of the VHE shell-SNR group. X-ray emission was detected from HESS J1731−347 in the archival *ROSAT* data (Aharonian *et al.* 2008; Tian *et al.* 2008). The non-thermal X-ray emission attests the presence of electron population of TeV energies in the SNR (Tian *et al.* 2010; Abramowski *et al.* 2011; Bamba *et al.* 2012). A central compact object (CCO) was detected in the geometric centre of the SNR using archival *XMM-Newton* data (Halpern & Gotthelf 2010).

We carried out the 325, 610 and 1390 MHz observations of the SNR G353.6−0.7 with the Giant Metrewave Radio Telescope (GMRT) and detected the SNR at 325 and 610 MHz. we determined the spectral indices of bright individual filaments. We also compared the radio morphology with that of the VHE emission. Our results are published in (Nayana *et al.* 2017) and here we summarize the results presented in (Nayana *et al.* 2017).

2. GMRT observations and data analysis

We observed SNR G353.6−0.7 in the 325, 610 and 1390 MHz bands with the GMRT. The GMRT primary beams at 325, 610 and 1390 MHz are 85.2, 43 and 26.2 arcmin respectively. Since the SNR is 30' in diameter, a single pointing was done at 325 and 610 MHz whereas four pointings were done at 1390 MHz to cover the full extent of the SNR. Flux density calibrator was observed in the beginning and end of each observing run at all frequencies to calibrate antenna gains. Phase calibrator was observed in regular intervals through out the observing run to calibrate for phase variations due to ionospheric fluctuations. The phase calibrator was used for bandpass calibration as well. The observing bandwidth was 33 MHz split into 256 channels at all frequencies.

The data were analysed using Astronomical Image Processing System (AIPS). Initial flagging and calibration was done using standard AIPS tasks. Calibration was done for single channel and the solutions were applied to all the channels. Bandpass calibration was done using the phase calibrator. Few channels were averaged in the target source data so as to reduce the effect of bandwidth smearing. Initially high resolution images were made using visibility data from baselines greater than 1 km, i.e. excluding the GMRT central square data. Few rounds of phase only self-calibration and 2 rounds of amplitude and phase self-calibration was performed. All the clean components were subtracted from the uv data and low resolution image was made using small uv range i.e only the central square data. This procedure was followed to remove the compact sources in the field and image the extended structure of the SNR with better sensitivity.

The SNR was detected in the 325 and 610 MHz bands. We did not detect the SNR at 1390 MHz due to the sensitivty limitation and the lack of short baselines.

3. Results and discussion

3.1. *Morphology of the SNR*

We detected SNR G353.6−0.7 at 325 and 610 MHz frequencies (Fig. 1) with an angular extent of 30 arcmin (Nayana *et al.* 2017). The shell structure of the SNR with filaments is clearly seen in the 325 MHz GMRT map whereas the SNR is partially detected in the 610 MHz map (Fig. 1, top panels). The brightest filament in the 325 and 610 MHz map is the south-east filament. The faintest emission is from the northern region of the SNR which is not detected in the 610 MHz map due to sensitivity limitation. The brightest filaments seen in 325 and 610 MHz GMRT maps are also seen in 1.4 GHz SGPS map and 843 MHz MOST map (Fig. 1, bottom panels). A compact source is seen inside the SNR in the 1.4 GHz SGPS map and 843 MHz MOST map (Fig. 1, bottom panels) which is not seen in the 325 and 610 MHz GMRT maps (Fig. 1, top panels). This is due to the point source subtraction done before making low resolution images in our analysis. The compact source seen towards the west of the SNR is a H_{II} region (Condon *et al.* 1998). The extended wing like emission seen further west is likely to originate in H_{II} regions since these regions show bright emission in 60 μm IRAS map (Neugebauer *et al.* 1984) and 22 μm WISE map (Wright *et al.* 2010). The integrated flux density of the SNR at 325 MHz is 1.84 ± 0.15 Jy. Assuming a typical spectral index of −0.5 for SNRs (Jones *et al.* 1998), we obtained a flux density of 0.88 Jy at 1420 MHz. Tian *et al.*(2008) reported the

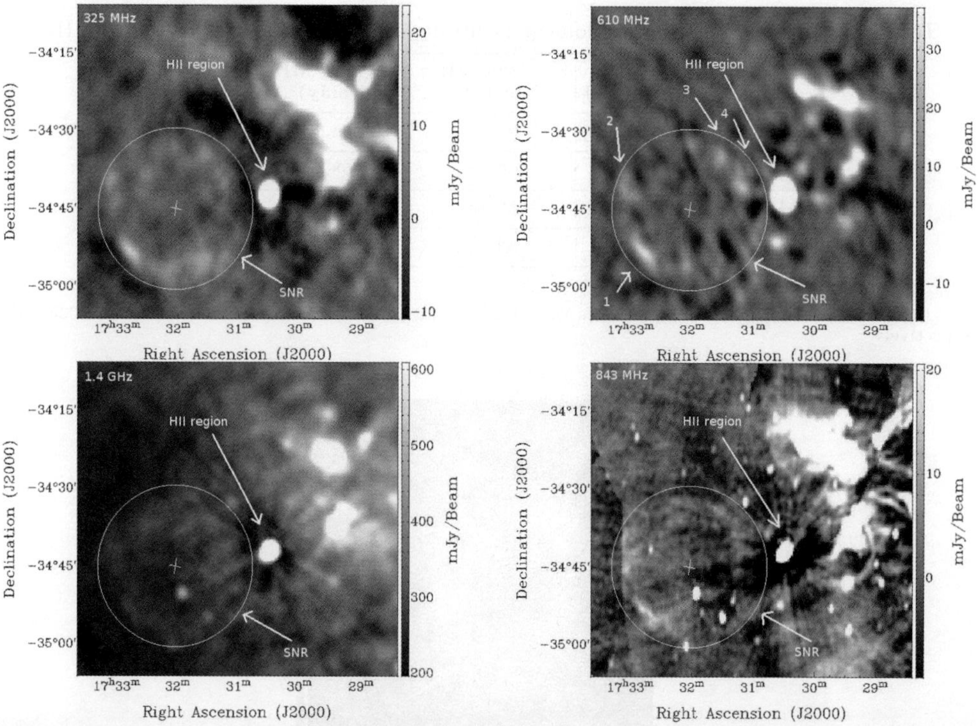

Figure 1. *Top left*: The GMRT low resolution map of the field containing SNR G353.6−0.7 at 325 MHz. Map resolution is 135×97 arcsec^2 and the rms is 4.8 mJy beam^{-1}. *Top right*: The GMRT low resolution map of the field containing SNR G353.6−0.7 at 610 MHz. Map resolution is 150×105 arcsec^2 and the rms is 4 mJy beam^{-1}. Here the areas marked by 1, 2, 3, 4 are the filaments for which the spectral indices were estimated. Compact sources have been removed from both the 325 and 610 MHz GMRT maps. For comparison, we show the SGPS map (Haverkorn *et al.* 2006) of the field containing SNR G353.6−0.7 at 1420 MHz with a resolution of 100×100 arcsec^2 and an rms of 10 mJy beam^{-1} (*bottom left*), as well as the MOST map (Green *et al.* 1999) of the field containing SNR G353.6−0.7 at 843 MHz with a resolution of 49.71×43.00 arcsec^2 and an rms of 3 mJy beam^{-1} (*bottom right*). The circles in all four images are of the same size and centred at the CCO position (marked with a cross) and indicate the area enclosing SNR G353.6−0.7. The colour scheme used in the maps is cubehelix (Green 2011). Figure reproduced from (Nayana *et al.* 2017).

flux density of SNR G353.6−0.7 as 2.2 ± 0.9 Jy at 1420 MHz from combined Australia Telescope Compact Array and Parkes telescope data. Our flux density estimation at 1420 MHz match within 1.5σ with that reported by Tian *et al.* (2008).

3.2. *Spectral Index*

Spectral indices were determined for the four prominent filaments detected in both 325 and 610 MHz maps (Fig. 1, top panel). The spectral index values of filament 1,2,3 and 4 are −0.70 ± 0.19, −1.11 ± 0.22, −0.50 ± 0.30 and −0.15 ± 0.32 respectively. The spectral indices are broadly consistent with the non-thermal synchrotron emission. The flux densities of each filament at 325 and 610 MHz and their physical dimensions in angular units are listed in Table1. The spectral index of filament 2 = −1.11±0.22 seems to be steeper than the typical synchrotron spectral index (α=−0.5). This steepening is consistent with the synchrotron cooling and could be indicative of a higher magnetic field strength in that region of the SNR.

Table 1. Physical dimensions of four filaments and flux densities at 325 and 610 MHz.

Filament	Length (arcmin)	Width (arcmin)	S_{int}(325 MHz) (mJy)	S_{int}(610 MHz) (mJy)	Spectral Index
Filament 1	11	3	106.22	68.67	−0.70 ±0.19
Filament 2	6	2.8	121.66	58.94	−1.11 ±0.22
Filament 3	5.4	3.7	56.90	41.55	−0.50 ±0.30
Filament 4	5.9	3.4	42.99	39.45	−0.15 ±0.32

Notes:
[1]S_{int}(610 MHz) and S_{int}(325 MHz) denoted the integrated flux density of the filaments at 610 and 325 MHz respectively.

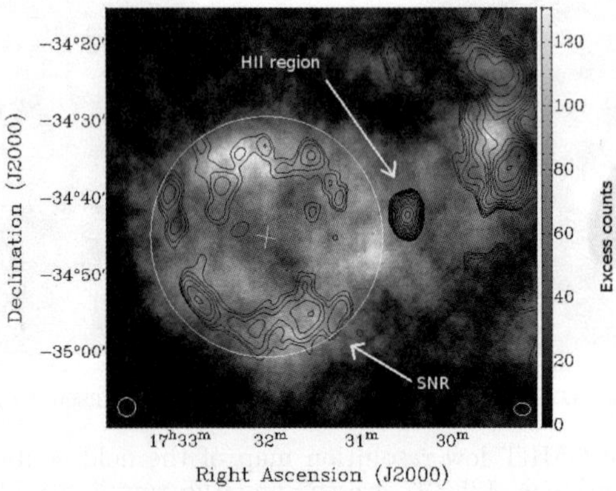

Figure 2. TeV γ-ray excess map (Abramowski *et al.* 2011) (resolution 144×144 arcsec^2; shown in the bottom left corner) overlaid with the GMRT 325 MHz map (resolution 135×97 arcsec^2; shown in the bottom right corner) contours. The scale of the γ-ray excess map is excess counts per smoothing gaussian width (σ = 0.04 deg). The contours are $\pm\sqrt{2}^n\sigma$ mJy beam^{-1} ($n = -2, 2, 3, 4, 5, 6$; σ=1.5 mJy beam^{-1}). The circle indicates the area enclosing SNR G353.6−0.7 centred at the CCO position (marked with a cross). The colour scheme used in the map is cubehelix (Green 2011). Figure reproduced from (Nayana *et al.* 2017).

3.3. *Comparison with the VHE emission*

We compared the low frequency radio morphology revealed in the GMRT maps with the VHE emission in the HESS map (Fig. 2). The shell like morphology of the SNR is clearly seen in both radio and VHE emission. The spatial extent of radio emission fairly overlaps with that of VHE emission except towards the west of the SNR. Here, the VHE emission is bright but the radio emission is not detected above the sensitivity limit of the map. The emission in radio and VHE does not seem to be correlated. The brightest filaments in the radio are the south-east and the eastern filaments (Fig. 2) which are faintest in the VHE emission. The peak in VHE emission (northern filament) corresponds to the faintest filament in radio. The variation in radio brightness can be due to the variation of magnetic field strength or due to the variation of ambient interstellar medium (ISM) density. A density gradient in the ambient ISM can happen if there is an encounter of SNR shock with a nearby molecular cloud. So far no evidences for such an interaction is reported in the case of SNR G353.6−0.7. The ^{12}CO spectrum derived from the vicinity of SNR G353.6−0.7 does not show any kinematic signatures of a possible interaction of the SNR shock with a molecular cloud (Abramowski *et al.* 2011). Thus the azimuthal variation of radio brightness is most likely due to the variation of the magnetic field strength. The spectral indices of different filaments of the SNR G353.6−0.7 indicate the azimuthal variation of magnetic field (Table 1). Here the spectral index of filament 2 is −1.11 which is roughly 0.5 steeper compared to the typical SNR spectral index of −0.5, consistent with synchrotron cooling, indicating higher magnetic field in this filament. Thus the magnetic field strength is relatively high in the region of the south-east and eastern filaments. Since radio emissivity is proportional to $B^{3/2}$, these filaments are bright in radio. If the VHE γ-rays are produced due to inverse compton scattering of photons, there are fewer IC electrons in this region due to synchrotron cooling and hence the VHE emission is faint here. The magnetic field strength is low towards the northern region of the SNR, and this causes the effect of synchrotron cooling to be weak and the IC emission becomes dominant. Towards the west, magnetic field strength is likely to be similar to or less than that of the northern region, and the synchrotron emission is below the sensitivity limits where VHE emission is bright. Thus we explain the anti-correlated emission seen in SNR G353.6−0.7 in radio and VHE γ-rays in a possible leptonic scenario.

4. Summary

We summarize the results presented in Nayana *et al.* (2017). We observed HESS J1731−347 also known as SNR G353.6−0.7 with the GMRT at 325, 610 and 1390 MHz and detected the radio morphology of the SNR G353.6−0.7 at 325 and 610 MHz. The complete shell morphology of the SNR is seen at 325 MHz and the SNR is partially seen at 610 MHz. The integrated flux density of the SNR G353.6−0.7 at 325 MHz is 1.84 ± 0.15 Jy. We also determined the spectral indices of four bright filaments of the SNR G353.6−0.7. The spectral index values range from −0.15 to −1.11, consistent with the non-thermal emission. We compared the low frequency radio morphology with VHE γ-ray emission. The brightest feature in radio correspond to the faintest feature in the VHE emission. we explain this anti-correlated emission in a leptonic scenario of γ-ray emission.

References

Acero, F., Aharonian, F., Akhperjanian, A. G., *et al.* 2010, *A&A*, 516, A62

Acero, F., Lemoine-Goumard, M., Renaud, M., *et al.* 2015, *A&A*, 580, A74

Aharonian, F. A., Akhperjanian, A. G., Aye, K.-M., *et al.* 2004, *Nature*, 432, 75

Aharonian, F., Akhperjanian, A. G., Bazer-Bachi, A. R., *et al.* 2007, *ApJ*, 661, 236

Aharonian, F., Akhperjanian, A. G., Barres de Almeida, U., *et al.* 2008, *A&A*, 477, 353

Aharonian, F., Akhperjanian, A. G., de Almeida, U. B., *et al.* 2009, *ApJ*, 692, 1500

Bamba, A., Pühlhofer, G., Acero, F., *et al.* 2012, *ApJ*, 756, 149

Blasi, P. 2013, *A&AR*, 21, 70

Condon, J. J., Cotton, W. D., Greisen, E. W., *et al.* 1998, *AJ*, 115, 1693

Green, A. J., Cram, L. E., Large, M. I., & Ye, T. 1999, *ApJS*, 122, 207

Green, D. A. 2011, *Bulletin of the Astronomical Society of India*, 39, 289

Halpern, J. P. & Gotthelf, E. V. 2010, *ApJ*, 710, 941

Haverkorn, M., Gaensler, B. M., McClure-Griffiths, N. M., Dickey, J. M., & Green, A. J. 2006, *ApJS*, 167, 230

H. E. S.. S. Collaboration, Abramowski, A., Acero, F., *et al.* 2011, *A&A*, 531, A81

Jones, T. W., Rudnick, L., Jun, B.-I., *et al.* 1998, *PASP*, 110, 125

Nayana, A. J., Chandra, P., Roy, S., *et al.* 2017, *MNRAS*, 467, 155

Neugebauer, G., Habing, H. J., van Duinen, R., *et al.* 1984, *ApJ* (Letters), 278, L1

Tian, W. W., Leahy, D. A., Haverkorn, M., & Jiang, B. 2008, *ApJ* (Letters), 679, L85

Tian, W. W., Li, Z., Leahy, D. A., *et al.* 2010, *ApJ*, 712, 790

Wright, E. L., Eisenhardt, P. R. M., Mainzer, A. K., *et al.* 2010, *AJ*, 140, 1868

Supernova 1987A:30 years later - Cosmic Rays and Nuclei from Supernovae and their aftermaths
Proceedings IAU Symposium No. 331, 2017
A. Marcowith, M. Renaud, G. Dubner, A. Ray & A. Bykov, eds.

doi:10.1017/S1743921317004434

Detecting Thermal X-Ray Emission and Proper Motions in RX J1713.7-3946

Satoru Katsuda[1], Fabio Acero[2], Nozomu Tominaga[3,4], Jean Ballet[2], Yasuo Fukui[5], Junko S. Hiraga[6], Katsuji Koyama[7], Shiu-Hang Lee[8], Koji Mori[9], Shigehiro Nagataki[10], Yutaka Ohira[11], Robert Petre[12], Hidetoshi Sano[5,13], Yoko Takeuchi[14], Toru Tamagawa[14], Naomi Tsuji[15], Hiroshi Tsunemi[16] and Yasunobu Uchiyama[15]

[1]Department of Physics, Faculty of Science & Engineering, Chuo University, 1-13-27 Kasuga, Bunkyo, Tokyo 112-8551, Japan
email: katsuda@phys.chuo-u.ac.jp

[2]Laboratoire AIM, IRFU/SAp – CEA/DRF – CNRS – Universite Paris Diderot, Bat. 709, CEA-Saclay, Gif-sur-Yvette Cedex, France

[3]Department of Physics, Faculty of Science and Engineering, Konan University, 8-9-1 Okamoto, Kobe, Hyogo 658-8501, Japan

[4]Kavli Institute for the Physics and Mathematics of the Universe (WPI), The University of Tokyo, 5-1-5 Kashiwanoha, Kashiwa, Chiba 277-8583, Japan

[5]Department of Physics, Nagoya University, Furo-cho, Chikusa-ku, Nagoya 464-8601, Japan

[6]Department of Physics, School of Science and Technology, Kwansei Gakuin University, Sanda 669-1337, Japan

[7]Department of Physics, Graduate School of Science, Kyoto University, Kitashirakawa Oiwake-cho, Sakyo-ku, Kyoto 606-8502, Japan

[8]Department of Astronomy, Kyoto University, Kitashirakawa-Oiwake-cho, Sakyo-ku, Kyoto 606-8502, Japan

[9]Department of Applied Physics and Electronic Engineering, Faculty of Engineering, University of Miyazaki, 1-1 Gakuen Kibanadai-Nishi, Miyazaki 889-2192, Japan

[10]Astrophysical Big Bang Laboratory, Riken, Wako, Saitama 351-0198, Japan

[11]Department of Physics and Mathematics, Aoyama-Gakuin University, Sagamihara, Kanagawa 252-5258, Japan

[12]Astrophysics Science Division, NASA Goddard Space Flight Center, Greenbelt, MD 20771, USA

[13]Institute for Advanced Research, Nagoya University, Furo-cho, Chikusa-Ku, Nagoya 464-8601, Japan

[14]High Energy Astrophysics Laboratory, RIKEN Nishina Center, 2-1 Hirosawa, Wako-shi, Saitama 351-0198, Japan

[15]Department of Physics, Rikkyo University, 3-34-1 Nishi-Ikebukuro, Toshima, Tokyo 171-8501, Japan

[16]Department of Earth and Space Science, Osaka University, 1-1 Machikaneyama-cho, Toyonaka, Osaka 560-0043, Japan

Abstract. We report detections of thermal X-ray line emission and proper motions in the supernova remnant (SNR) RX J1713.7-3946, the prototype of the small class of synchrotron dominated SNRs. Based on deep XMM-Newton observations, we find clear line features including Ne Lyα, Mg Heα, and Si Heα from the central portion of the remnant. The metal abundance ratios suggest that the thermal emission originates from core-collapse SN ejecta arising from a relatively low-mass ($\lesssim$20 $M_\odot$) progenitor. In addition, using XMM-Newton observations on a 13 yr time interval, we have measured expansion in the southeastern rim to be $\sim$0.75$''$ yr^{-1} or

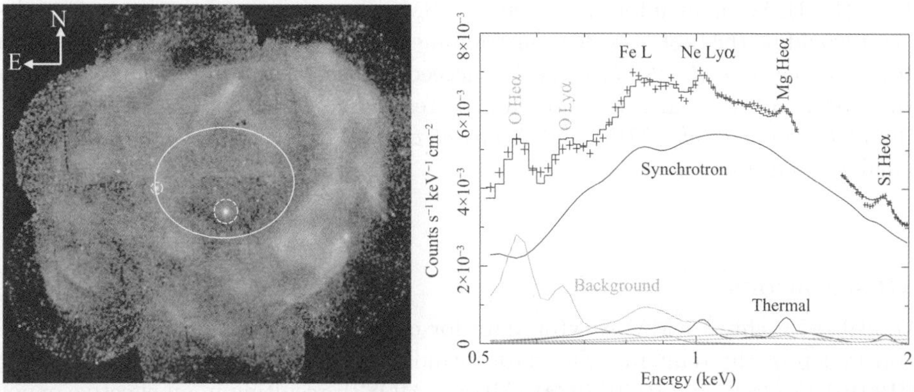

Figure 1. Left: XMM-Newton's surface brightness map of the entire RX J1713.7-3946. The large white ellipse shows where we extract spectrum shown in the right-hand panel. Two small white ellipses show regions where we excluded point-like sources from our spectral analysis. Right: Non X-ray background-subtracted XMM-Newton's MOS1+2 spectrum extracted from the ellipse in the left-hand panel. Solid curves represent the best-fit model. Note that the energy band of 1.4–1.6 keV is ignored owing to the imperfect subtraction of the instrumental Al-K line emission.

$\sim$3500 km s^{-1} at a distance of 1 kpc. Given this, we derive an upstream density to be $\sim$0.01 cm^{-3}, compatible with the lack of thermal X-rays from the shocked ambient medium. We also estimate the age of the remnant to be $\sim$1200–1600 yr, roughly consistent with the idea that RX J1713.7-3946 is the remnant of SN 393.

Keywords. shock waves, ISM: abundances, ISM: individual (RX J1713.7-3946)

1. Introduction

Synchrotron X-ray emission has been ubiquitously detected from shells of young supernova remnants (SNRs), following the first discovery in SN 1006 (Koyama *et al.* 1995). This is clear evidence that electrons are accelerated to very high energies (up to $\sim$100 TeV) at SNR shocks, cementing the long-suspected link between SNRs and cosmic rays. In addition, recent detections of TeV and GeV gamma-rays from some SNRs have also suggested acceleration of protons which are the majority of cosmic rays (McEnery *et al.* 2017).

RX J1713.7-3946, the X-ray image of which is presented in Fig. 1 left, is the brightest X-ray synchrotron and TeV gamma-ray SNR in our Galaxy, making it an exceptionally important object for the study of cosmic-ray acceleration (Koyama *et al.* 1997; Aharonian *et al.* 2007). Despite its importance, there are many unresolved problems with this SNR. For example, we do not know if the gamma-ray emission is dominated by accelerated protons (e.g., Uchiyama *et al.* 2007; Fukui *et al.* 2012; Sano *et al.* 2013) or electrons (e.g., Abdo *et al.* 2011; Lee *et al.* 2012). One of the key parameters to reveal their relative contributions is the ambient density. In most SNRs, ambient densities can be estimated by the analysis of thermal X-ray emission, whereas RX J1713.7-3946 does not allow for such estimates, because its X-ray emission is dominated by synchrotron radiation; despite extensive studies for two decades since its discovery, no thermal line emission had been detected (e.g., Tanaka *et al.* 2008). Recently, Katsuda *et al.* (2015) reported the first detection of thermal X-rays from the central region of the remnant. The thermal X-ray emission is likely to originate from shocked SN ejecta, rather than the shocked interstellar medium (ISM). Without information on the ISM, expansion measurements of the X-ray remnant would be the best hope for constraining the ambient densities. Tsuji & Uchiyama (2016) and Acero *et al.* (2017) recently succeeded in measuring proper motions in the northwestern and the southeastern filaments, respectively. Here, we concisely review the results of Katsuda *et al.* (2015) and Acero *et al.* (2017), including new information on the analysis and discussions.

2. Observations

We utilized archival XMM-Newton data for our spectral analysis that focuses on the central region of the remnant. The data include Obs.IDs 0207300201 (MOS1+MOS2), 0722190101 (MOS2), and 0740830201 (MOS1+MOS2), resulting in an effective exposure time of 375 ks for one MOS sensor. To measure proper motions in the southeastern rim, we used two data sets (Obs.IDs 0093670401 and 0743030101) with a time baseline of 13 yr. The first- and second-epoch data were taken on March 14, 2002 and March 11, 2015 with the effective exposure times of 11.7/5.7 ks (MOS/pn) and 64.1/48.0 (MOS/pn), respectively.

3. Spectral Analysis

We first generated a softness-ratio map from the XMM-Newton data, to identify best locations where thermal X-ray emission might be detectable. The idea is that softer regions should be promising sites for searching for thermal X-ray emission, because both less absorption and fainter synchrotron emission make the spectrum softer, enhancing the presence of thermal X-rays in the dominant synchrotron continuum. While Katsuda *et al.* (2015) selected the relatively small, softest area around the center of the remnant, we here enlarge the area by including its northern vicinity, so that we can improve the photon statistics. The updated region is shown as an ellipse in Fig. 1 left.

Figure 1 right exhibits the non X-ray background-subtracted XMM-Newton (MOS1+2) spectrum, from which we confirm that the line features at 1 keV (Ne Lyα), 1.35 keV (Mg Heα), and 1.85 keV (Si Heα) are clearly detected. We fit it by an absorbed thermal plus power-law component model with additional components for the sky background (OFF1 in Katsuda *et al.* (2015)). The best-fit model is shown in Fig. 1 right, and the best-fit parameters are listed in Table 1. We estimated the abundance ratios of Mg/Ne, Si/Ne, and Fe/Ne to be ~2.3, ~2.0, and ~0.16, respectively. The Mg/Ne and Si/Ne ratios are consistent with those in Katsuda *et al.* (2015), whereas the Fe/Ne ratio is slightly higher than the previous work, which is probably due to the imperfect modeling of the sky background. That Fe is deficit in this plasma strongly suggests that it originates from core-collapse SN ejecta.

Table 1. Best-fit parameters

$N_{\rm H}$ (10^{21} cm^{-2})	Γ	$kT_{\rm e}$ (keV)	$\log(n_{\rm e}t)$ (cm^{-3} s)	Ne	Mg	Si	Fe	χ^2/d.o.f.
6.0±0.1	$2.29^{+0.02}_{-0.1}$	$0.58^{+0.22}_{-0.09}$	11.7±0.1	$3.1^{+4.8}_{-0.4}$	$7.1^{+10.9}_{-0.7}$	$6.1^{+9.1}_{-1.7}$	$0.5^{+0.7}_{-0.1}$	492 / 339

Notes:
The values of elemental abundances are multiples of solar values. Other elemental abundances are fixed to the solar values according to Wilms *et al.* (2001).

4. Proper-Motion Measurements

To guide our region definition, we ran a Canny edge detection algorithm, because the exact shape and edge position of the SNR are not easy to determine by eye. The resulting edges, shown as contours in the upper panel of Fig. 2, perfectly outline the shape of the SNR edges, even in faint regions where the filament fades out. Based on the edge images shown in Fig. 2, two main structures labeled "outer region" and "inner region" (towards the inside of the SNR which is probably due to projection effects) were identified as promising regions to investigate the filament's proper motion. We used four X-ray point sources for our image alignment by using the SAS tool `eposcorr`. After alignment, the mean 2002-2015 residual offset of the four registration sources is 0.9 arcsec, which we consider as our systematical uncertainty.

To estimate the proper motion between the two epochs, we built a shock profile model based on the 2015 high statistic observation (assuming no errors), which is then compared to the 2002 profile (see Fig. 2 lower panels). The latter profile is in counts and is the sum of the MOS1 and MOS2 camera. We note that the number of counts for some bins of the 2002 radial profile falls below 25 counts per bins and we have therefore used Cash statistics (Cash 1979) rather than the χ^2 estimator. We calculate the C-value for various shifts and amplitudes (the intensity ratio between the 2002 and 2015 profiles), finding the best-fit shifts to be 0.61 ± 0.05(stat.)±0.07(sys.)$''\,\mathrm{yr}^{-1}$ and $0.75^{+0.05}_{-0.06}$(stat.)±0.07(sys.)$''\,\mathrm{yr}^{-1}$ for the inner and outer regions, respectively. These can be translate into shock speeds of 2880 ± 240(stat.)±330(sys.) $\mathrm{km\,s^{-1}}$ and 3550^{+250}_{-290}(stat.)±330(sys.) $\mathrm{km\,s^{-1}}$ at a distance of 1 kpc (Fukui *et al.* 2003).

5. Discussion & Conclusions

We have detected X-ray line emission from the synchrotron-dominated SNR RX J1713.7-3946, thanks to the long integrated XMM-Newton exposure. We measured the abundance ratios of Mg/Ne, Si/Ne, and Fe/Ne in the thermal component to be ~2.3, ~2.0, and ~0.16, respectively. The deficit of the Fe abundance immediately tells us that this emission originates from core-collapse SN ejecta rather than the swept-up ISM. The abundance pattern among Ne, Mg, and Si is consistent with those expected for relatively low-mass, $\lesssim 20\,M_\odot$, progenitor stars rather than higher-mass progenitors (e.g., Tominaga *et al.* 2007).

Single stars whose masses are $\lesssim 20\,M_\odot$ should result in SNe IIP from a theoretical point of view (Heger *et al.* 2003). Also, Wang *et al.* (1997) estimated the absolute magnitude at maximum light was probably between -12 and -14, implying a highly subluminous core-collapse SN event. For low-luminosity Type IIP events, the initial shock speeds are known to range from ~2000 $\mathrm{km\,s^{-1}}$ to ~6000 $\mathrm{km\,s^{-1}}$ (Spiro *et al.* 2014). This conflicts with the case for RX J1713.7-3946, for which the initial blastwave must be faster than ~6000 $\mathrm{km\,s^{-1}}$, i.e., the current SNR radius of 9.6 pc divided by the age of 1600 yr inferred from the potential association with SN 393 (Wang *et al.* 1997). Instead, the relatively fast blastwave speed suggests that the SN of RX J1713.7-3946 was a stripped envelope, Type Ib/c or Type IIb, event. The apparent discrepancy between the abundance pattern and the shock speed could be explained if RX J1713.7-3946 arose from a Type Ib/c (or IIb) explosion of a relatively low-mass star forming a close binary, in which the (low-mass) progenitor had experienced a binary interaction to remove its massive H envelope.

In order to estimate the age and the ambient density, we compared proper motions measured in the southeastern rim with an SNR evolutionary model in Tang & Chevalier (2017), obtaining an age of 2250 ± 170(stat.)±200(sys.) yr or 1760 ± 130(stat.)±150(sys.)

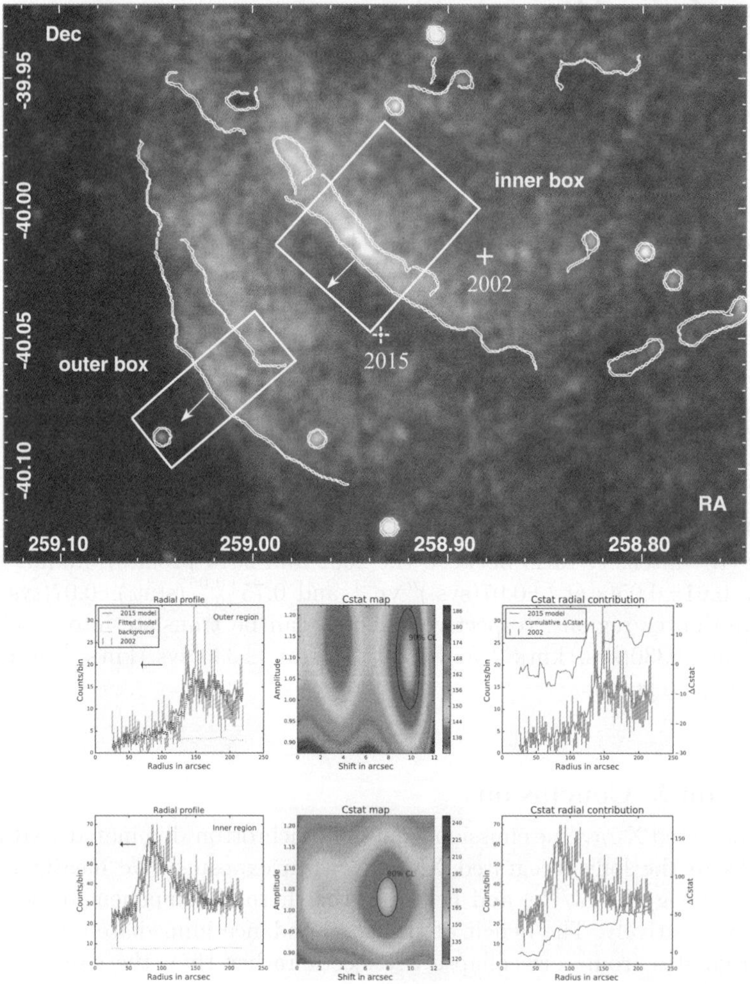

Figure 2. Upper panel: Exposure corrected and background subtracted count map in the 0.6–6 keV energy band of the 2015 XMM-Newton observation. The white contours are the results of the Canny edge detection algorithm. The white boxes represent the regions used to extract the radial profiles shown in lower panels, and the white crosses are the 2002 and 2015 pointing positions. The arrows indicate the direction of the shock motion. The middle-row panels: Radial profiles, C-stat maps and C-stat radial contribution from the outer region defined in the upper panel. The arrow in the left panel indicates the direction of the shock motion. We fit the 2002 profiles by 2015-based models with free normalizations and shifts, obtaining the best-fit models named "Fitted model". The right panel shows the cumulative statistics radial contribution (C-stat $= C_0(r) - C_{\rm bestfit}(r)$) taken at the best-fitted amplitude and where $C_0(r)$ is the Cstat radial contribution at a 0 arcsec model shift. The bottom-row panels: Same as the middle-row panels, but for the inner region defined in the upper panel.

yr in the case where the shock is evolving in a wind profile ($\rho \propto r^{-s}, s = 2$) or in a uniform medium ($s = 0$), respectively. In both cases, we derived an upstream density of the order of $0.01\,{\rm cm}^{-3}$, compatible with the lack of thermal X-rays from the shocked ISM. The ages roughly agree with that of SN 393.

We perform another age estimate, based on a simple comparison of expansion indices (m) measured for several SNRs; $m = d(\log R)/d(\log t) = (dR/dt)/(R/t) = \mu/R\,t$, where R is the angular radius of the filament whose proper motion (μ) has been measured and

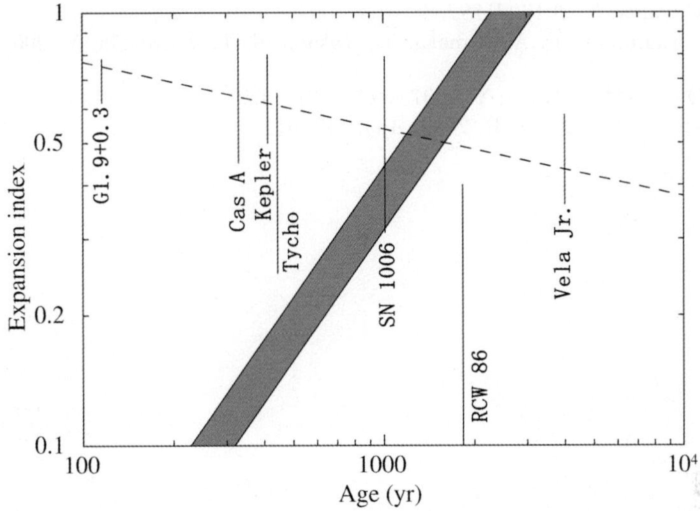

Figure 3. Expansion indices against the ages of young Galactic SNRs together with the best-fit power-law model shown as a dashed line. The red area indicates the 90% confidence interval for the m-value of RX J1713.7-3946.

t is the age of the remnant. Figure 3 plots m-values for several SNRs (Katsuda *et al.* 2016). The best-fit power-law function is shown as a dashed line. The red area indicates the 90% confidence interval for the m-value of RX J1713.7-3946, calculated from the proper motion of 0.62–0.87″ yr^{-1} and the SNR radius of 33′. This allows us to estimate an age of ∼1200–1600 yr, consistent with the idea that RX J1713.7-3946 is the remnant of SN 393.

This work is supported by the Japan Society for the Promotion of Science KAKENHI grant number 16K17673 (SK).

References

Abdo, A. A., *et al.* (Fermi Collaboration) 2011, *ApJ*, 734, 28
Acero, F., Katsuda, S., Ballet, J., & Petre, R. 2017, *A& A*, 597, 106
Aharonian, F. A., *et al.* (HESS Collaboration) 2017, *A& A*, 464, 235
Cash, W. 1979, *ApJ*, 228, 939
Fukui, Y., Moriguchi, Y., Tamura, K., *et al.* 2003, *PASJ*, 55, L61
Fukui, Y., Sano, H., Sato, J., *et al.* 2012, *ApJ*, 746, 82
Heger, A., Fryer, C. L., Woosley, S. E., Langer, N., & Hartmann, D. H. 2003, *ApJ*, 591, 288
Katsuda, S., Acero, F., & Tominaga, N., *et al.* 2015, *ApJ*, 814, 29
Katsuda, S., Tanaka, M., Morokuma, T., Fesen, R., & Milisavljevic, D. 2016, *ApJ*, 826, 108
Koyama, K., Petre, R., Gotthelf, E. V., Hwang, U., Matsuura, M., Ozaki, M., & Holt, S. S. 1995, *Nature*, 378, 255
Koyama, K., Kinugasa, K., Matsuzaki, K., Nishiuchi, M., Sugizaki, M., Torii, K., Yamauchi, S., & Aschenbach, B. 1997, *PASJ*, 49, L7
Lee, S.-H., Ellison, D., & Nagataki, S. 2012, *ApJ*, 750, 156
McEnery, J. *et al.* 2017, *these poceedings*
Sano, H., Tanaka, T., Torii, K., *et al.* 2013, *ApJ*, 778, 59
Spiro, S., Pastorello, A., Pumo, M. L., *et al.* 2014, *MNRAS*, 439, 2873
Tominaga, N., Umeda, H., & Nomoto, K. 2007, *ApJ*, 660, 516
Tang, X. & Chevalier, R. A. 2017, *MNRAS*, 465, 3793
Tanaka, T., Uchiyama, Y., Aharonian, F. A., *et al.* 2007, *ApJ*, 685, 988

Tsuji, N. & Uchiyama, Y. 2016, *PASJ*, 68, 108
Uchiyama, Y., Aharonian, F. A., Tanaka, T., Takahashi, T., & Maeda, Y. 2007, *Nature*, 449, 576
Wang, Z. R., Qu, Q.-Y., & Chen, Y. 1997, *A&A*, 318, L59
Wilms, J., Allen, A., & McCray, R. 2000, *ApJ*, 542, 914

Supernova 1987A:30 years later - Cosmic Rays and Nuclei from Supernovae and their aftermaths
Proceedings IAU Symposium No. 331, 2017
A. Marcowith, M. Renaud, G. Dubner, A. Ray & A. Bykov, eds.

doi:10.1017/S1743921317004835

Searching for Young Type Ia Supernova Remnants in M83

Chuan-Jui Li[1,2], You-Hua Chu[1,2,3] and Po-Sheng Ou[1,2]

[1]Institute of Astronomy and Astrophysics, Academia Sinica, P.O. Box 23-141, Taipei 10617, Taiwan
email: cjli@asiaa.sincia.edu.tw, yhchu@asiaa.sincia.edu.tw

[2]Graduate Institute of Astrophysics, National Taiwan University, Taipei 10617, Taiwan

[3]Department of Astronomy, University of Illinois, 1002 West Green Street, Urbana, Illinois, USA

Abstract. Young Type Ia supernova remnants (SNRs) are characterized by Balmer-dominated optical spectra, well-defined shell morphologies, $> 10^{36}$ ergs s^{-1} X-ray luminosities, and a lack of massive stars and dense interstellar gas in their vicinity. Applying these characteristics and using archival deep *HST* and *Chandra* observations of M83, we search for young Type Ia SNRs in this spiral galaxy. This is a very difficult task!

Keywords. ISM: supernova remnants

---

1. Introduction

M83 (NGC 5236), at a distance of 4.61 Mpc, is a nearby large grand-design spiral galaxy (see Figure 1). Supernova remnants (SNRs) in M83 have been identified using the criterion [S II]/Hα ratio > 0.3 in Dopita *et al.* (2010) and Blair *et al.* (2012). Using a deep, 790 ks, *Chandra* ACIS X-ray observation of M83, Long *et al.* (2014) compared X-ray point sources with SNR candidates identified from [S II]/Hα ratios or nonthermal radio sources, and confirmed 87 SNRs. Using an expanded Hubble Space Telescope (*HST*) WFC3 survey of M83, in conjunction with the deep *Chandra* observation of M83, Blair *et al.* (2014) made a targeted search of young SNRs, resulting in 63 objects.

These surveys have been successful in finding SNRs expanding into a dense medium and exhibiting forbidden lines with high [S II]/Hα ratios or young super-luminous SNRs dominated by emission from O-rich SN ejecta. However, young Balmer-dominated Type Ia SNRs have escaped detection because of their lack of forbidden lines. Adopting the characteristics of Type Ia SNRs in the Large Magellanic Cloud (LMC) reported by Ou *et al.* (2017), we started to search for young Type Ia SNRs in M83.

2. Search for Young Type Ia SNRs in M83

The young Type Ia SNRs in the LMC are characterized by (1) Balmer-dominated optical spectra, (2) well-defined shell structure, (3) X-ray luminosities of a few $\times 10^{36}$ erg s^{-1}, and (4) Population II stellar and interstellar environments.

To search for Balmer-dominated SNRs in M83, we start with the *Chandra* X-ray sources from Long *et al.* (2014), and examine the *HST* Hα images for shell structures associated with the X-ray sources. We immediately find it a very difficult task, as our initial search did not turn up any obvious candidates for Type Ia Balmer-dominated SNRs. Only three candidates are marginally identified (shown in Figure 2).

Figure 1. *HST* image of M83.

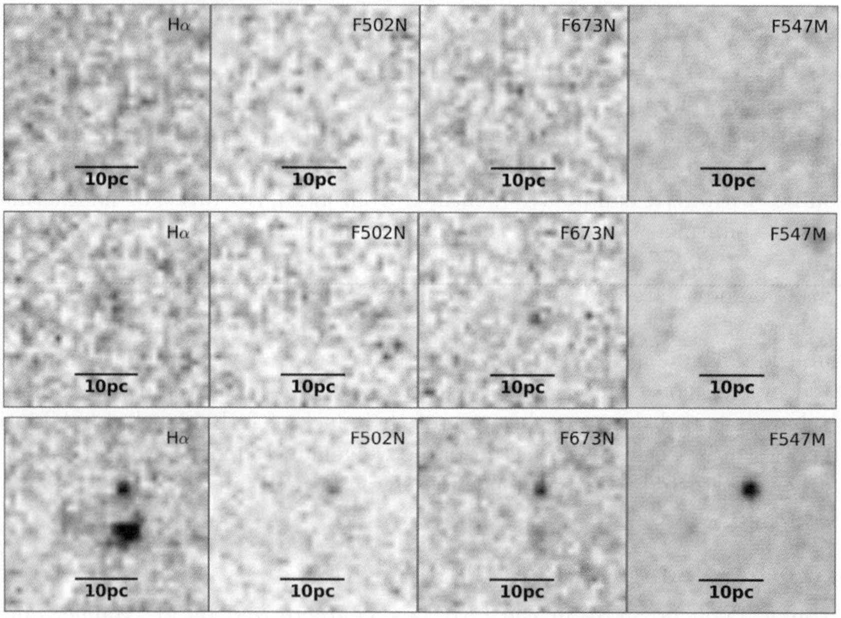

Figure 2. Images of 3 Type Ia SNR candidates (from left to right) showing Hα (F657N), [O III] (F502N), [S II] (F673N), and stellar continuum (F547M).

Are these candidates viable? Why do Balmer-dominated Type Ia SNRs appear so scarce in M83? To answer these questions, we scale their LMC counterparts and estimate their expected surface brightness if located in M83. The SNR 0519-69.0 is a 33" × 33" Balmer-dominated Type Ia SNR in the LMC at a distance of 50 kpc, reported by Smith

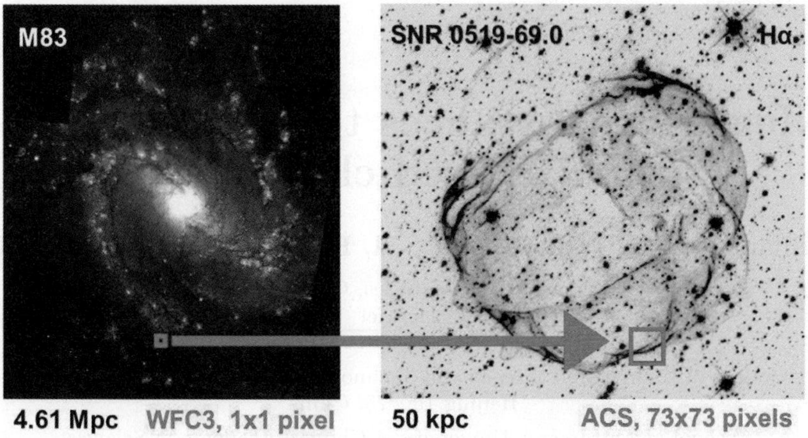

Figure 3. For *HST* images, a WFC3 pixel (0.0396" × 0.0396") in M83 corresponds to 73 × 73 ACS pixels in the LMC.

et al. (1991). For *HST* images, a WFC3 pixel (0.0396" × 0.0396") in M83 corresponds to 73 × 73 ACS pixels in the LMC (see Figure 3). The Balmer filaments in 0519-69.0 are ∼0.3" (∼6 pixels) wide; thus, only a small fraction of a 73 × 73 area centered on a filament will be filled with emission. The average surface brightness of a 73 × 73 area in 0519-69.0 will be its expected surface brightness in M83. We find the average surface brightness of 0519-69.0 to be (4-11) × 10^{-16} erg cm^{-2} s^{-1}arcsec^{-2}. The *HST* Hα image of M83 has a background of ∼3.4 × 10^{-16} erg cm^{-2} s^{-1}arcsec^{-2}. Comparisons between these figures indicate that a Balmer-dominated Type Ia SNR in M83 is expected to be detected at generally ∼2-3σ level and up to ∼5σ level for the brightest filaments.

The three objects in Figure 2 indeed are detected at ∼2-3σ levels. Their sizes, smaller than 15pc, are consistent with young SNRs. The lack of dense ionized interstellar gas and massive stars in their vicinity and their X-ray luminosities of a few × 10^{36} ergs/s are consistent with those of young Type Ia SNRs. The detection thresholds of the [S II] or [O III] images are too high to provide meaningful upper limits of these forbidden line emission. We suggest that the three objects in Figure 2 are promising candidates for Balmer-dominated Type Ia SNRs in M83.

Acknowledgements

This project is supported by MoST grant 104-2112-M001-044-MY3.

References

Dopita, M. A., Blair, W. P., Long, K. S., *et al.* 2010, *ApJ*, 710, 964
Blair, W. P., Winkler, P. F., & Long, K. S. 2012, *ApJS*, 203, 8
Blair, W. P., Chandar, R., Dopita, M. A., *et al.* 2014, *ApJ*, 788, 55
Long, K. S., Kuntz, K. D., Blair, W. P., *et al.* 2014, *ApJS*, 212, 21
Ou, P-.S., Chu, Y.-H., & Li, C.-J., 2017, poster at this meeting
Smith, R. C., Kirshner, R. P., Blair, W. P., & Winkler, P. F. 1991, *ApJ*, 375, 652

Supernova 1987A:30 years later - Cosmic Rays and Nuclei from Supernovae and their aftermaths
Proceedings IAU Symposium No. 331, 2017
A. Marcowith, M. Renaud, G. Dubner, A. Ray & A. Bykov, eds.

doi:10.1017/S1743921317004914

Measuring distances to Galactic SNRs using the red clump stars

S. S. Shan[1,3], D. Wu[1,2]†, H. Zhu[1], M. F. Zhang[1,3] and W. W. Tian[1,3]

[1]National Astronomical Observatories, Chinese Academy of Sciences
Datun Road, Chaoyang District, Beijing 100012, China
email: tww@nao.cas.cn

[2]College of Information Science and Technology, Beijing Normal University,
Beijing 100875, China

[3]School of Astronomy, University of Chinese Academy of Sciences,
Beijing 100049, China

Abstract. Reliable distances to Galactic Supernova remnants (SNRs) are essential to constrain parameters that reveal the evolutional process of SNRs. We carry out a project to measure SNRs' distances in the first quadrant of the Galaxy. In this project, red clump stars (RCS) are used as standard candle to build the optical extinction (A_V)-(D) distance relation in each direction of extinction-known SNRs. Here, G5.7-0.01, G54.1+0.3 and G78.2+2.1 are taken as typical examples. We obtain the distance of $3^{+0.4}_{-0.3}$ kpc for G5.7-0.01, the lower limit of 5.8 kpc for G54.1+0.3, the upper limit of 2 kpc for G5.7-0.01. The results are consistent with distances from kinematic measurements. Hence, we highlight the RCS method can independently trace the distance to the SNRs.

Keywords. ISM: supernova remnants, methods: data analysis

1. Introduction

Distances of SNRs play important roles in obtaining basic parameters of SNRs like the age, the physical size, the expansion velocity and the explosion energy of the progenitor supernovae. There are several approaches to measure SNRs' distances such as kinematic method, proper motion measurement, Sedov estimate, and Σ-D relation. However, only about 20% of currently known Galactic SNRs have reliable distance measurements (Green 2014).

Red clump stars (RCS) are usually low mass stars in the early stage of core He-burning and concentrate in the color-magnitude diagram (CMD). The helium cores of RCS almost have the same mass. Meanwhile, the absolute magnitudes of RCS weakly depend on metal abundance in infrared band. Therefore, RCS are good enough to be standard candles. Durant & van Kerkwijk (2006) measured the running of reddening with distance in the direction of Galactic anomalous X-ray pulsars and successfully obtained the distances of 5 pulsars. Güver *et al.* (2010) gave a distance estimate to the Low-mass X-ray binaries 4U 1608-52 through the interstellar extinction traced by RCS. Zhu *et al.* (2015) used the similar measurement to determine the distance of SNR G332.5-5.6. We closely follow this method to systematically measure the distances to SNRs in the first Galactic quadrant. Here, we discuss RCS method and the results of SNRs G5.7-0.01, G54.1+0.3 and G78.2+2.1.

† S. S. Shan and D. Wu are common first author for the paper.

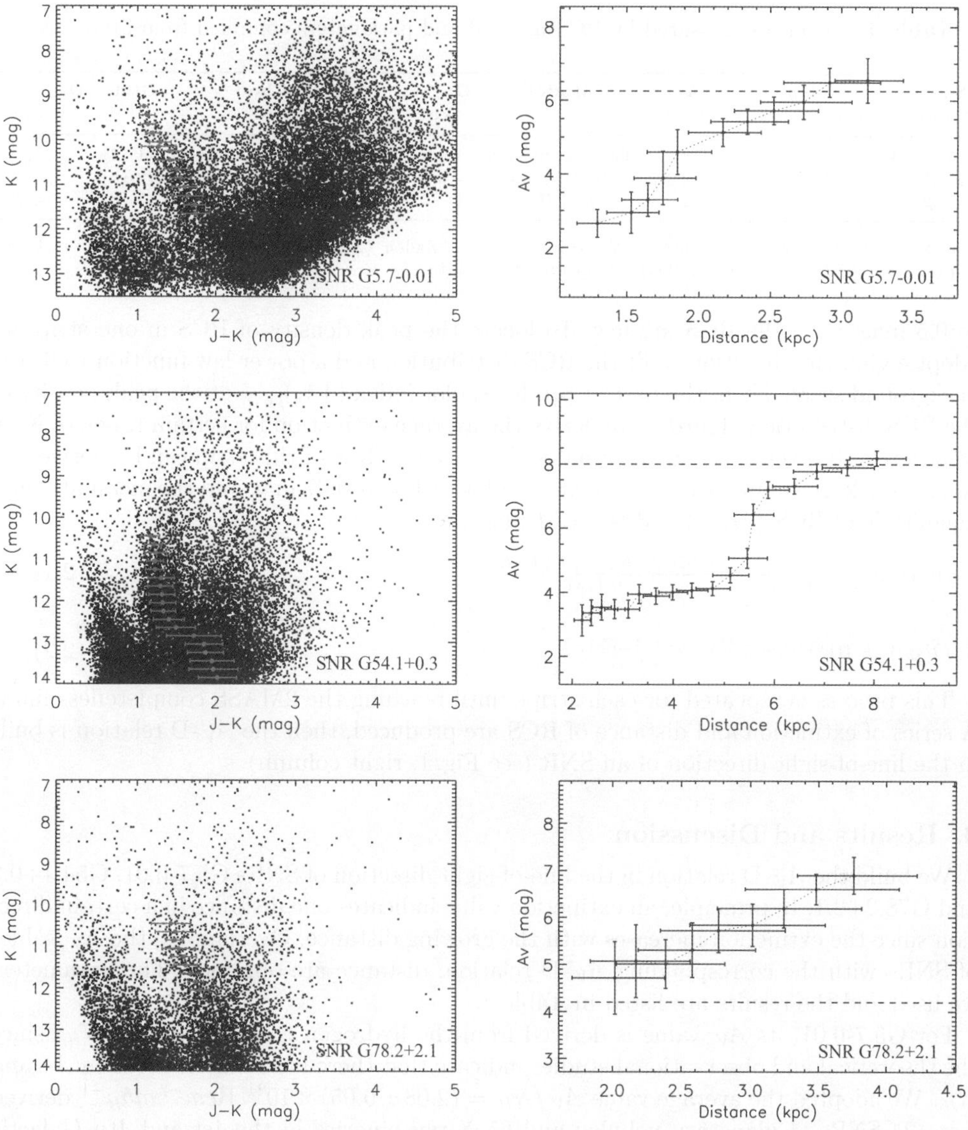

Figure 1. Left column: color-magnitude diagram for stars within 0.5 deg^2 of SNR G5.7-0.01, G54.1+0.3 and G78.2+2.1, respectively. The red dot and lines show the fitted location of the RCS density peak and its extent with 1σ. Right column: The derived A_V-D relation along the direction of G5.7-0.0, G54.1+0.3 and G78.2+2.1. The dashed line is A_v value in the direction of an SNR.

2. Method

Firstly, we select the stars in the 0.5 deg^2 area around the center of a SNR in an interval of (Galactic longitude)$\triangle l\, 1°\times\triangle b\, 0.5°$(Galactic latitude) from the 2MASS All-sky Point Source Catalog in the J and K_s (hereafter K) band (Skrutskie *et al.* 2006). Then, their magnitudes in J and K bands are used to construct the CMD (K vs. J-K). The RCS concentrate in a distinct region of the CMD (see Fig. 1, left column). The bulk of stars in the left region of the CMD are predominantly main sequence stars; while those in the right region are mainly dwarfs and RGBs. Secondly, we divide the stars in m_K into successive strips. The width of each strip is usually 0.3 mag. It wil be extended

Table 1. Distances measured by RCS method and parameters compiled from literatures.

Sourse Name	N_H ($10^{21} Hcm^{-2}$)	A_V (mag)	Model	D_{known} kpc	Method	Ref.	$D_{(thiswork)}$ kpc
G5.7-0.01	13±1		Thermal plasma	3.2	kinematic measurement	(1)(2)	$3^{+0.4}_{-0.3}$
G54.1+0.3		8.0±0.7	Spectra fits	6.2	kinematic measurement	(3)(4)	> 5.5
G78.2+2.1		4.2±0.8	H_α/H_β	1.7-2.6	kinematic measurement	(5)(6)	< 2

Reference: (1) Joubert *et al.* (2016); (2) Hewitt & Yusef-Zadeh (2009); (3) Koo *et al.* (2008); (4) Leahy *et al.*(2008); (5) Mavromatakis (2003); (6) Leahy *et al.* (2013)

to 0.5 mag when the RCS are few. To locate the peak density of RCS in one strip, we adopt a Gaussian function to fit the RCS distribution and a power law function to fit the background stars. Then, the best fit produce color index $(J\text{-}K)_{peak}$ at the peak density of the RCS distribution. Thirdly, we derive the average extinction for certain range of K by equation 2.1(Güver *et al.* 2010). Finally, distance for RCS of this strip can be derived by equation 2.2. Assuming that the intrinsic color $(J\text{-}K)_0$ is 0.67 mag and the mean absolute magnitude of RCS (M_K) band is −1.61 mag.

$$A_V(mag) = \frac{A_K}{0.1137} = \frac{0.67[(J-K)_{peak} - (J-K)_0]}{0.1137} \tag{2.1}$$

$$D(pc) = 10^{[0.2(m_K - M_K + 5 - 0.1137 A_V)]} \tag{2.2}$$

This process is repeated for each stripe until reaching the 2MASS completeness limit. A series of extinction and distance of RCS are produced, then the A_V-D relation is built in the line-of-sight direction of an SNR (see Fig. 1, right column).

3. Results and Discussion

We build the A_V-D relation in the line-of-sight direction of SNRs G5.7-0.01, G54.1+0.3 and G78.2+2.1. In principle, an extinction value indicates one distance in a certain direction since the extinction increases with the growing distance. Overlapping the A_V values of SNRs with the corresponding A_V-D relation, distance are obtained. The parameters we used and the results are listed in Table 1.

For G5.7-0.01, its A_V value is derived from the hydrogen column density (N_H) since the theoretical and observational studies indicate that there is a relation between A_V and N_H. We adopted the average value $A_V/N_H = (2.08 \pm 0.05) \times 10^{21}\ Hcm^{-2}mag^{-1}$ derived from 24 SNRs, 4 planetary nebulae and 65 X-ray binaries in the 1st and 4th Galactic quadrants, which is based on high quality data (Zhu *et al.* in prep.). In this case, the A_V value in the direction of G5.7-0.01 is 6.25±0.48 mag, and the distance to G5.7-0.01 is estimated as 3.0 kpc.

According to the A_V-D relation in the direction of G54.1+0.3, there is an extinction jump around 5.5 kpc. It may be caused by a huge molecular cloud or the increasing population of K dwarf which makes it difficult to identify RCS. In this situation, a conservative estimate should be made as 5.5 kpc for the lower-limit distance.

On the CMD built in the direction of G78.2+2.1, the density of RCS is too small to be identified when the brighter magnitude is below 10 mag. The upper-limit distance of G78.2+2.1 can be estimated as 2 kpc. Leahy *et al.* (2013) determined the distance of G78.2+2.1 from 1.7 kpc to 2.6 kpc. Combining the two results, the distance to G78.2+2.1 is range from 1.7 kpc to 2 kpc.

In summary, our results show that the three distances estimated by RCS method are consistent with those measured by kinematic method. Extinction measurement based on RCS provides a reliable and independent clue to the SNRs' distance. In the future, we will apply this method to systematically measure the distances to Galactic SNRs. With the SNRs' distances, we will further re-calibrate the Σ-D relation and calculate distance-dependent parameters of SNRs.

4. Acknowledgements

We all acknowledge supports from NSFC program (11473038, 11603039).

References

Durant, M. & van Kerkwijk, M. H. 2006, *ApJ*, 1070, 1081
Green, D. A. 2014, *Supernova Environmental Impacts*, 188, 196
Güver, T., *et al.* 2010, *ApJ*, 964, 973
Hewitt, J. W. & Yusef-Zadeh, F. 2009, *ApJ*, 694, L16
Joubert, T., Castro, D., Slane, P., & Gelfand, J. 2016, *ApJ*, 816, 63
Koo, B.-C., *et al.* 2008, *ApJ*, 673, L147
Leahy, D. A., Tian, W. W., & Wang, Q. D. 2008, *AJ*, 136, 1477
Leahy, D. A., Green, K., & Ranasinghe, S. 2013, *MNRAS*, 436, 968
Mavromatakis, F. 2003, *A& A*, 398, 153
Skrutskie, M. F., *et al.* 2006, *AJ*, 1163, 1183
Zhu, H., Tian, W. W., & Wu, D. 2015, *MNRAS*, 3470, 3474
Zhu, H., Tian, W. W., Li, A., *et al.* 2017, submitted to *MNRAS*

Supernova 1987A:30 years later - Cosmic Rays and Nuclei from Supernovae and their aftermaths
Proceedings IAU Symposium No. 331, 2017
A. Marcowith, M. Renaud, G. Dubner, A. Ray & A. Bykov, eds.

doi:10.1017/S1743921317005233

Cosmic ray astroparticle physics: current status and future perspectives

Fiorenza Donato

Department of Physics, Torino University, and INFN, Sezione di Torino
via P. Giuria 1, 10125 Torino, Italy
email: donato@to.infn.it

Abstract. The data we are receiving from galactic cosmic rays are reaching an unprecedented precision, over very wide energy ranges. Nevertheless, many problems are still open, while new ones seem to appear when data happen to be redundant. We will discuss some paths to possible progress in the theoretical modeling and experimental exploration of the galactic cosmic radiation.

Keywords. Cosmic rays, dark matter, Galaxy.

1. Introduction

A large contribution to the knowledge and understanding of the Galaxy is given by the observation of the most energetic particles, the CRs (CRs). These relativistic particles, reaching the Earth from the outer space, are either primary nuclei, arriving directly from their sources, or secondary products of the spallation processes (i.e. fragmentation by nuclear destruction) taking place during the propagation from the sources through the interstellar medium (ISM). They include nuclei from hydrogen to very high Z, and possibly in different isotopic compositions, electrons, antimatter (positrons and antiprotons measured up to now), and in some extensions also gamma rays.

The main challenges we are facing are i) the identification of the sources of CRs, ii) the mechanism through which low energy particles, the seeds being single elements, dust or grains at rest in the ISM are accelerated to such high energy to be called CRs, and iii) the understanding of their propagation through the galactic magnetic fields. Due to the wide energy range explored by detectors and the rapidly changing flux, different experimental techniques are needed in diverse energy regions. Direct measurements of the primary CRs can be performed up to 10^{14} eV. Above this indicative threshold, the low fluxes due to the steeply falling spectrum force us to exploit indirect methods, detecting for example the extensive air showers generated by the interaction of CRs in the atmosphere.

Some components of the CRs are also a peculiar means for the exploration of dark matter (DM) signals coming from the Galaxy. The most common hypothesis for particle DM consists of weakly interacting massive particles (WIMPs), supposed to be cold thermal relics after the Big Bang and to build up the galactic dark haloes. The so-called indirect DM investigation technique relies in the search for its stable annihilation products in the halo of galaxies, and in particular in the Milky Way. The indirect DM detection focuses on rare components in charged CRs (CRs), as well as in γ-rays and, more generally, in multiwavelength photons (Salati, Donato & Fornengo (2010)).

2. Galactic CRs

Galactic CRs (GCR) are supposed to be produced by nucleosynthesis processes in the stellar interior, where they are fractionated, ejected and then accelerated to CR energies and diffused throughout the whole Galaxy. They reach the Solar System, where the slowest CRs experience a spectral distortion by the solar wind called solar modulation. Charged particles are bent in the galactic magnetic fields, so there is no obvious way of tracking them back to their sources. The energy spectrum which is observed at Earth is folded with the source one through an energy dependent diffusion coefficient which shapes the effects of the galactic magnetic field in an effective way, and which can be deduced only phenomenologically. These peculiarities are at variance with γ rays, which point directly to the sources.

The most realistic propagation models are the diffusion ones, which account for spatial dependence of sources, CR densities and in principle of the ISM. The Galaxy is usually shaped as a thin gaseous disk where all the astrophysical sources are located, embedded in a thick diffusive magnetic halo. Diffusive models, besides being more realistic and closer to a physical interpretation for each component, have proven to be successful in reproducing the nuclear, antiproton and radioactive isotopes data. They also allow to treat contributions from dark matter (or other exotic) sources located in the diffusive halo.

The relevant transport equation for a charged particle wandering through the magnetic inhomogeneities of the galactic magnetic field writes in terms of the differential density $N(E, \vec{r})$ as a function of the total energy E and the position $\vec{r}$ in the Galaxy. Assuming steady-state ($\partial N/\partial t = 0$), the transport equation for a given nucleus (the subscript j is omitted) can be written in a compact form as

$$(-\vec{\nabla} \cdot (K\vec{\nabla}) + \vec{\nabla} \cdot \vec{V_C} + \Gamma_{\rm rad} + \Gamma_{\rm inel})N + \frac{\partial}{\partial E}\left(bN - c\frac{\partial N}{\partial E}\right) = \mathcal{S}\,. \tag{2.1}$$

The first bracket in the l.h.s. accounts for: i) spatial diffusion $K(\vec{r}, E)$, ii) convection with speed $\vec{V_C}(\vec{r})$, iii) the (possible, for some isotopes with half lifetime τ_0) radioactive decay rate $\Gamma_{\rm rad}(E) = 1/(\gamma\tau_0)$ (γ here is the Lorentz factor), iv) the destruction rate $\Gamma_{inel}(\vec{r}, E) = \sum_{ISM} n_{\rm ISM}(\vec{r})v\sigma_{\rm inel}(E)$ due to collisions with the interstellar medium. In this last expression $n_{\rm ISM}(\vec{r})$ is the density of the interstellar medium in the various locations of the Galaxy and in its different H and He components, and $\sigma_{\rm inel}(E)$ is the destruction (inelastic) cross section for a given nucleus. The coefficients b and c are respectively first and second order gains/losses in energy. The source term $\mathcal{S}$ includes primary sources of CRs (e.g. supernovae), secondary sources due to the fragmentation of heavier nuclei, and secondary decay-induced sources. The quasi-linear theory leads to a rigidity power law for the diffusion coefficient, which is usually assumed to have the form:

$$K(E) = K_0\beta^\eta R^\delta \tag{2.2}$$

(β is the Lorentz factor). K_0 is linked to the level of the hydromagnetic turbulence and δ to the density spectrum of these irregularities at different wavelength. η is usually set to 1, while a different value parameterizes very low energy deviations. The lack of information on the magnetic field irregularities prevents us from a precise determination of the diffusion coefficient, which is instead possible only from interpretation of CR data. The most important observable at these regards is the ratio B/C of the boron to carbon fluxes. It is considered a kind of standard candle in the CR phenomenology, being the ration of an almost secondary species to an almost primary one.

The solution to the diffusion equation (Eq.2.1) has been deeply investigated and several different techniques proposed in the literature can lead to similar fluxes at the Earth, at least for stable nuclei (Strong *et al.* (2007), Maurin *et al.* (2002))

Semi-analytical models are treated in Maurin *et al.* (2001). Fully numerical solution is the one adopted by the GALPROP code Strong & Moskalenko (1998), which is particularly useful if aiming at the production of gamma rays from charged CRs. Gamma rays do not diffuse and are therefore observed at their source, so that a full spatial treatment is required. Comparable results for the propagation of stable primary and secondary nuclei have been obtained with the similar Dragon code (Evoli *et al.* (2008), Evoli *et al.* (2017)). Recently, the PICARD code has been presented as well in Kissmann (2014). For a full treatment of galactic CRs physics we refer to Castellina & Donato (2012).

3. Antimatter in CRs

The presence of a small amount of antimatter in CRs is predicted from spallation reactions of incoming protons and helium nuclei on the ISM (contributions from CRs with higher Z being negligible). The spallation products of these inelastic scatterings account for quarks and gluons, which immediately hadronize. In particular, the hadronization process includes the production of a small amount of antiprotons, antideuterons and positrons (the latter induced by decay processes). Antiprotons and positrons in CRs have been measured in recent years with increasing accuracy and in an energy window spanning from few hundreds of MeV to hundreds of GeV. For antideuterons, only upper limits have been set.

The search for cosmic antimatter is a further test of the propagation model. The study of light antimatter, due to its tiny flux, is optimal to search for contributions with spectral peculiarities, such as non-thermal production or annihilation from dark matter (DM) particles in the galactic dark halo. The latter are usually predicted to be Weakly Interacting Massive Particles (WIMPs).

The indirect dark matter detection is based on the search for anomalous components due to the annihilation of DM pairs in the galactic halo, in addition to the standard astrophysical production (Salati, Donato & Fornengo (2010), Donato *et al.* (2004)). These contributions are potentially detectable as spectral distortions in various cosmic antimatter fluxes, as well as γ rays and ν's

$$\chi + \chi \to q\bar{q}, W^+W^-, \ldots \to \bar{p}, \bar{D}, e^+ \; \gamma \,\&\, \nu' s \; . \tag{3.1}$$

The search for exotic signals in cosmic antimatter is motivated by the very low astrophysical counterpart, acting as a background. Detection of the DM annihilation products has motivated the spectacular development of several new experimental studies.

3.1. *Antiprotons*

Antiprotons ($\bar{p}$) may be produced by spallation of high–energy primary nuclei impinging on the atoms of the ISM inside the galactic disc (Donato *et al.* 2001). They represent the background when searching for small peculiar contributions such as signals from DM annihilation.

The secondary antiproton flux has been predicted with small theoretical uncertainties (Donato *et al.* 2001) and reproduces astonishingly well the data from 200 MeV up to at 100 GeV (Donato *et al.* 2010). This can be seen in Fig. 1 where we plot, along with the solar-demodulated $\bar{p}/p$ data (data corrected by the effect of the solar wind), the curves bounding the propagation uncertainty on the $\bar{p}$ calculation. We also illustrate the uncertainty related to the production cross sections. The secondary contribution alone

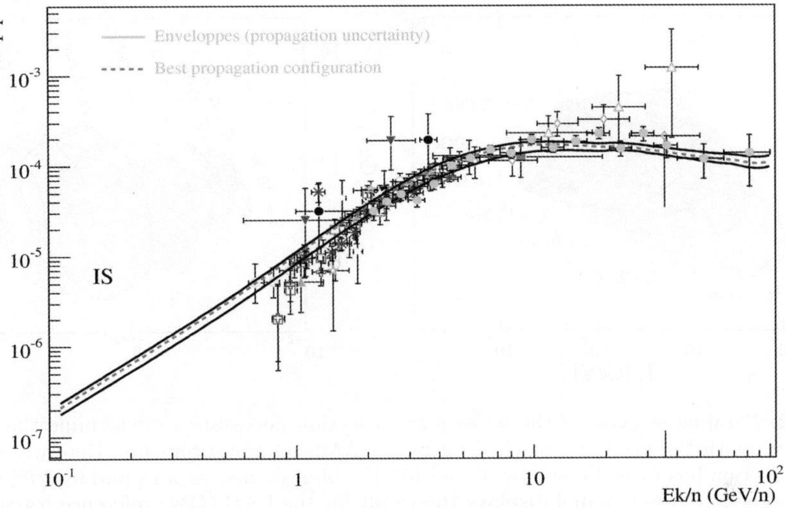

Figure 1. Propagation uncertainty envelopes of the IS $\bar{p}/p$ ratio and two parameterizations of the production cross section (see Donato *et al.* (2010)).

explains the PAMELA data on the whole energetic range. It is not necessary to invoke an additional component (such as, for instance, from DM annihilation in the galactic halo – see below) to the standard astrophysical one.

The annihilation of DM candidate particles throughout the whole Milky Way volume may generate primary antiprotons. In this case the WIMP annihilations take place all over the diffusive halo. The antiproton signal from annihilating DM particles leads to a primary component directly produced throughout the diffusive halo. The variation of the astrophysical parameters induces a much larger theoretical uncertainty on the primary than on the secondary flux: in the first case, the uncertainty reaches two orders of magnitude for energies $T_{\bar{p}} = 1$ GeV, while in the second case it never exceeds 25% (see Fig. 1 for the secondary population; for the primary DM component, see Fig. 3). The reason is in the location of the sources: the primary flux due to dark matter annihilation originates in the whole diffusive halo and is very sensitive to the halo size (which is varied between 1 and 15 kpc) and the convective velocity. It is of the utmost importance to constrain the propagation parameters in order to evaluate any possible exotic contribution to the $\bar{p}$ flux from sources in the whole halo.

In addition to propagation uncertainties, we notice that the secondary flux, produced by the spallation of CR protons and helium nuclei on the ISM, is affected by nuclear uncertainties. As shown in Fig. 1, they are dominated by uncertainties in the production cross sections, in particular the $pp, pHe, Hep, HeHe$ inclusive cross sections into $\bar{p}$+X. We have very recently determined the characteristics that a high energy experiment should have in order to measure the pp cross section, and similarly the other ones, with an accuracy so good not to exceed the errors on the AMS-02 antiproton data (Donato *et al.* 2017). In Fig. 2 we give our results as functions of the kinematical variables in both the LAB and CM reference frames. It shows the parameter space that has to be covered in order to guarantee the AMS-02 precision level on the $\bar{p}$ source term, if the $p + p \rightarrow \bar{p} + X$ cross section is determined with 3% uncertainty within the blue shaded regions and by 30% outside the contours. The plot is done for the LAB (left panel, a) and CM (right panel, b) reference frame variables. For the LAB frame we show the contours as functions of η and T, for selected values of $T_{\bar{p}}$ from 1.1 (the lowest energy) below 30%

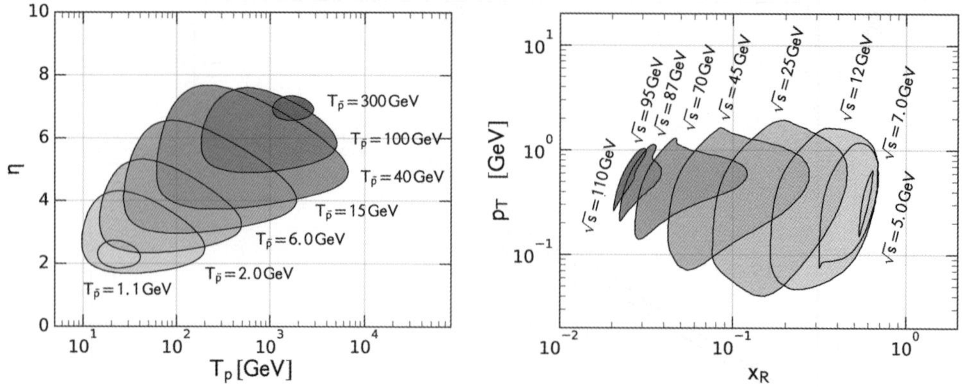

Figure 2. Parameter space of the pp to $\bar{p}$ cross section necessary to determine the antiproton source term with the accuracy reached by recent AMS-02 measurements. Here we require that the cross section has to be known by 3% within the blue shaded regions and by 30% outside the contours. The left (right) panel displays the result for the LAB (CM) reference frame variables.

uncertainty in the CR $\bar{p}$ flux to 300 GeV. As expected the contour size decreases when $T_{\bar{p}}$ approaches to 1 GeV, because there the AMS-02 uncertainty on the antiproton flux reaches 30%. A similar explanation holds for large $T_{\bar{p}}$. Antiprotons of increasing energy require the coverage of increasing η values. For example, the $\sigma_{\rm inv}(p+p \to \bar{p}+X)$ at $T_{\bar{p}}$=2 GeV is known at 3% level if data were taken with proton beams between 10 and 200 GeV and pseudorapidity from 1.8 to 4. If the whole AMS-02 energy range had to be covered with high precision, one should collect $p+p \to \bar{p}+X$ cross section data with proton beams from 10 GeV to 6 TeV, and η increasing from 2 to nearly 8. We address to Donato *et al.* 2017 for all the details.

3.2. *Antideuterons*

It was shown that the antideuteron spectra deriving from DM annihilation are expected to be much flatter than the secondary astrophysical component at low kinetic energies, $T_{\bar{d}}$ = 2-3 GeV/n, thus offering a potentially very clear indirect detection channel (Donato, Fornengo & Salati (2000), Donato, Fornengo & Maurin (2008)). Antideuterons have not been measured so far, and only upper limits exist. The dedicated GAPS experiment has been recently approved and will fly in Antarctica in few years (Aramaki *et al.* (2016)).

The secondary $\bar{D}$ flux is the sum of the six contributions corresponding to p, He and $\bar{p}$ CR fluxes impinging on H and He IS gas (other reactions are negligible). The solution to the propagation equation has the same expression as for secondary antiprotons (Donato, Fornengo & Maurin (2008)). The production of cosmic antideuterons is based on the fusion process of a $\bar{p}$ and $\bar{n}$ pair – being the pair produced from CRs collisions or from DM annihilation in the dark galactic halo. One of the simplest but powerful treatment of the fusion of two or more nucleons is based on the so–called coalescence model which, despite its simplicity, is able to reproduce remarkably well the available data on light nuclei and antinuclei production in different kinds of collisions.

We present in Fig. 3 a possible experimental scenario. The secondary $\bar{D}$ flux is plotted alongside the primary flux from DM particle (m_χ =50 GeV) annihilating in the halo. The three curves bound the propagation uncertainties which – as for the case of $\bar{D}$ from galactic DM, grossly rescaled by a factor 10^3 – span almost two orders of magnitude on the whole energetic range. The present BESS upper limit is at a level of $2{\cdot}10^{-4}$ $(\mathrm{m}^2$ s sr GeV/n$)^{-1}$. We also plot the estimated sensitivities of the gaseous antiparticle spectrometer GAPS on a long duration balloon flight (LDB) and an ultra–long duration

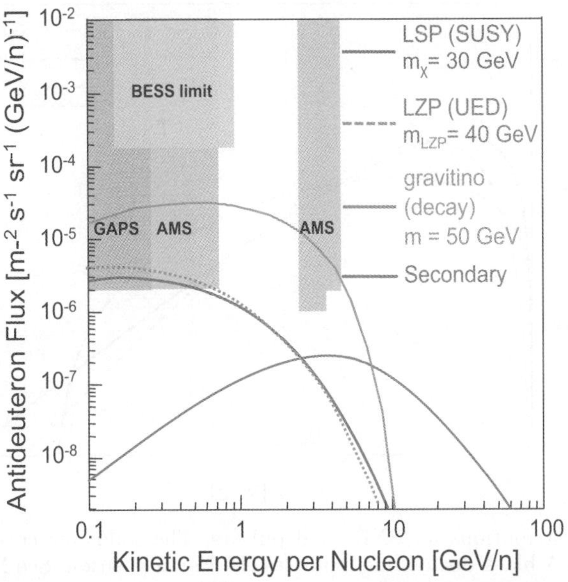

Figure 3. Predicted antideuteron flux as a function of kinetic energy per nucleon for a 30 GeV neutralino, a 40 GeV extra-dimensional Kaluza-Klein neutrino, and a 50 GeV gravitino. The antideuteron limits from BESS are shown, along with the projected sensitivities of AMS-02 after 5 years of operation and GAPS after three 35-day flights. The MED Galactic propagation scenario is assumed. These predictions use a coalescence momentum that is set to 195 MeV and the Einasto dark matter density profile. See Aramaki *et al.* 2016 for details.

balloon mission (ULDB) and of AMS–02 for three years of data taking. The perspectives to explore a part of the parameter space where DM annihilations are mostly expected (i.e. the low–energy tail) are promising. We refer to Aramaki *et al.* (2016) for details and references.

3.3. *Positrons*

Secondary positrons are produced – like antiprotons and antideuterons – by the spallation of the interstellar medium when impinging high–energy particles (Salati, Donato & Fornengo (2010)). The main production channel is the collision of protons with hydrogen atoms at rest producing charged pions $\pi^{\pm}$ which decay into muons $\mu^{\pm}$. The latter are also unstable and eventually lead to electrons and positrons. Positrons may also be produced through kaons although this channel is rare. In the case of positrons and electrons, Eq.2.1 describing the propagation of CRs throughout the DH is dominated by space diffusion and energy losses. Above a few GeV, synchrotron radiation in the galactic magnetic fields as well as inverse Compton scattering on stellar light and on CMB photons dominate.

The positron fraction $e^+/(e^+ + e^-)$ has been measured by the PAMELA satellite experiment and confirmed by the AMS-02 exeriment, which also measures absolute leptons fluxes. It increases with energies, at variance with the predictions from pure secondary production of cosmic positrons. A viable explanation of the experimental result resides in the additional contribution of astrophysical sources accelerating leptons in their sites. The cosmic fluxes of positrons and leptons are quite sensitive to the presence of sources in the near Galaxy (few kpc), whose physics can be explored also in this peculiar channel. This result is illustrated in Fig. 6, where the $e^+/(e^+ + e^-)$ has been calculated adding to the secondary production (mostly relevant for positrons) the contributions from standard astrophysical sources, such as supernova remnants and pulsars. The most recent

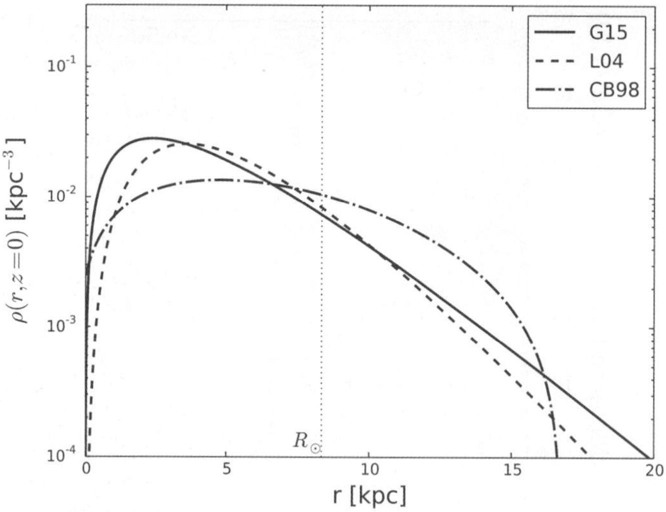

Figure 4. Radial distributions for SNRs and pulsars. The solid line corresponds to the G15 Green (2015) model. A black dotted line indicates the Earth position. See Manconi *et al.* (2017) for details.

experimental data on $e^+ + e^-$ and e^+ spectra are displayed together with predictions for fluxes from galactic sources, as well as the secondary component. We find a very good fit to AMS data, with source parameters consistent with literature results. We refer to Manconi *et al.* (2017) for all the details and the relevant references.

4. Anisotropies in the electrons

Recently, the lepton data have also been analyzed in terms of their arrival direction. Indeed, no significant anisotropy has been detected, and upper limits on the dipole term Δ have been set for the electron plus positron flux. Indeed, the dipole anisotropy in the leptons arriving at the Earth can be a remarkable observable to constrain the properties of astrophysical emitters of e^+ and e^-. In particular, as visible in Fig. 6, leptonic data can be interpreted with the emission of electrons and positrons from Supernova Remnants (SNRs), Pulsar Wind Nebulae (PWNe) and the secondary production given by the spallation reactions of primary CRs with the nuclei of the interstellar medium (ISM).

In Fig. 4 we reproduce the radial profile G15, together with few other largely used models. obtained from pulsars in the ATNF catalog (L04) and SNR observations (CB). The L04 and G15 models are similar around the Solar System, while they show a remarkable difference toward the Galactic center. At variance, the CB $\rho(r)$ is significantly flat for most of radii. Since electrons detected at the Earth have been produced by nearby (very few kpc) sources, we are mostly interested in the differences in the profiles around the Solar System. In Fig. 5 we plot our sample of *near* SNRs together with the most powerful PWNe projected on the Galactic plane. The source position is identified by the Galactic longitude (l, deg) and distance (d, kpc) to the Earth. The color scale reflects the intensity of the integrated electron flux at the Earth from $E = 50$ GeV up to $E = 5$ TeV, which is an important observable when computing the anisotropy from a single source. The most powerful PWNe Monogem, Geminga and J2043+2740, give an integrated electron flux above 10^8 (cm^2 s sr)$^{-1}$.

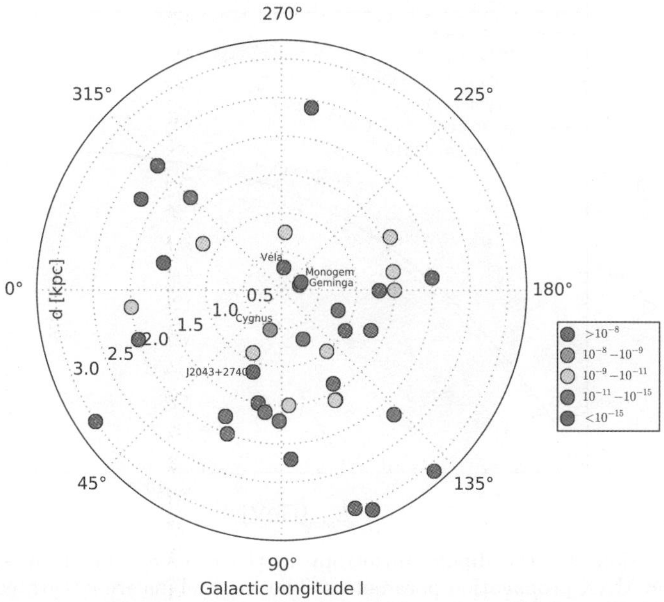

Figure 5. Sample of single SNRs and PSRs in terms of their Galactic longitude l [deg] and distance to the Earth [kpc] (located at the center of the circle). The color scale of the dots quantifies the electron flux integrated from 50 GeV to 5 TeV, in units of $(\mathrm{cm}^2\ \mathrm{s\ sr})^{-1}$.

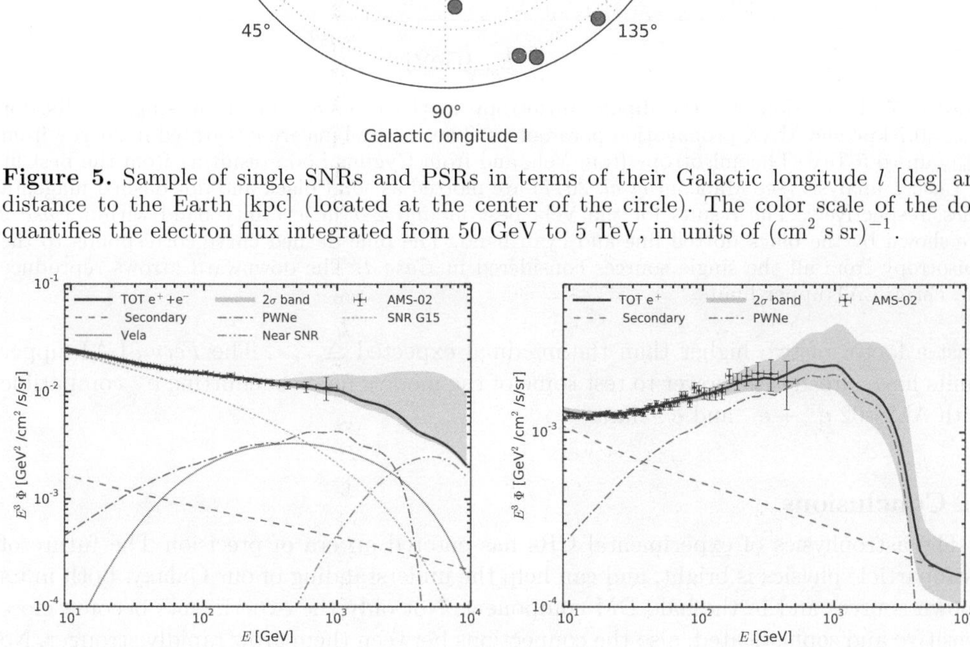

Figure 6. Fit to $e^+ + e^-$ (left panel) and e^+ (right panel) AMS-02 data with MAX propagation model and $R_{\mathrm{cut}} = 0.7$ kpc for *Case 2* (see Manconi *et al.* 2017 for details). All the components for the best fit are displayed together with the 2σ uncertainty band on the total flux. Line coding as follows: solid black, sum of all the components in the plot; red dashed, secondary e^+ and e^-; blue dash-dotted: e^+ and e^- from all ATNF PWNe; green dotted: e^- from far SNR; black dotted: e^- from Vela SNR; magenta double dash-dotted: e^- from all other near ($R \leqslant 0.7$ kpc) SNRs. The left (right) panel shows the $e^+ + e^-$ (e^+) flux.

The results for the $e^+ + e^-$ anisotropy are summarized in Fig. 7 as a function of E_{min} energy, and compared to *Fermi*-LAT upper limits. We plot $\Delta_{e^+e^-}$ for the best fit on the AMS-02 data, for the Vela and Cygnus Loop SNRs. We also plot the anisotropy for Vela with its 2-σ uncertainty band. The Vela anisotropy is an increasing function of E_{min}, at least up to few hundreds GeV, and depending on the model parameters. Its is predicted with an uncertainty of one order of magnitude or larger, and only mildly dependent on the bin of integrated energy. For E_{min} below 200 GeV, the *Fermi*-LAT upper limits lie in the Vela anisotropy band, while for higher energies the experimental data they are at

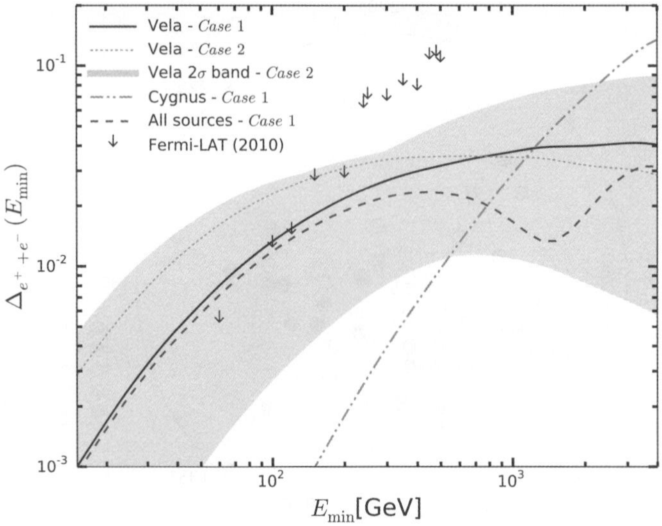

Figure 7. Predictions for the dipole anisotropy in the $e^+ + e^-$ flux from single SNRs, for $R_{\rm cut}$=0.7 kpc and MAX propagation parameters. The energy bins are integrated in energy from $E_{\rm min}$ up to 5 TeV. The anisotropy from Vela and from Cygnus Loop resulting from the best fit of *Case 1* analysis (see Manconi *et al.* 2017) are plotted as solid black and dot-dashed magenta lines, respectively. The results for the Vela best fit and 2-σ uncertainty band within *Case 2* are shown by the black dotted line and cyan band. The blue dashed curve corresponds to the anisotropy from all the single sources considered in *Case 1*. The downward arrows reproduce the *Fermi*-LAT upper limits.

least a factor of two higher than the maximal expected $\Delta_{e^+e^-}$. The *Fermi*-LAT upper limits have already the power to test some of the models for Vela emitting e^- compatible with AMS-02 $e^+ + e^-$ and e^+ data.

5. Conclusions

The astrophysics of experimental CRs has entered an era of precision The future of astroparticle physics is bright, and can help the understanding of our Galaxy, both in its known sources and in the halo DM component. Not only the experiments become more sensitive and sophisticated, also the connections between them grow rapidly stronger. No analysis is nowadays complete without the study of counterparts in other particle types and wavelengths. The experimental search for dipole anisotropies in the lepton sector encourages the growth of an astronomy of charged CRs. The forthcoming measurements will be crucial to ascertain the presence of DM particles in the halo of our Galaxy and to hint at their particle physics nature.

References

Aramaki, T. *et al.* 2016, *Phys. Rep.* 618, 1

Castellina, A. & Donato, F. 2012, *"Planets, Stars and Stellar Systems"*, *Springer* 5, 4760, arxiv:1110.2981

Donato, F., Fornengo, N., & Salati, P. 2000, *PRD*, 62, 043003

Donato, F. *et al.* 2001, *ApJ*, 563, 172

Donato, F. *et al.* 2004, *PRD*, 69, 063501

Donato, F., Fornengo, N., & Maurin, D. 2008, *PRD*, 78, 043506

Donato, F., Korsmeier, M., & Di Mauro, M. 2017, *subm. PRD*, arxiv:1704.03663

Evoli, C., Gaggero, D., Grasso, D., & Maccione, L. 2008, *JCAP* 10, 18
Evoli, C. *et al.* 2017, *JCAP*, 02, 015
Green, D. A. 2015, *MNRAS*, 454, 1517
Kissmann, R. 2014, *Astropart. Phys.*, 55, 37
Manconi, S., Donato, F., & Di Mauro, M., 2017, *JCAP* 01, 006
Maurin, D., Donato, F., Taillet, R., & Salati, P. 2001, *ApJ* 555, 585
Maurin, D. & others 2002, *astro-ph/0212111*
Salati, P., Donato, F., & Fornengo, N. 2010, *Indirect dark matter detection with cosmic antimatter. Cambridge University Press* 26, 521
Strong, A. W. & Moskalenko, I. V. 1998, *ApJ* 509, 212
Strong, A. W., Moskalenko, I. V., & Ptuskin, V. S. 2007, *Ann. Rev. Nucl. Part. Sci.* 57, 285

Supernova 1987A:30 years later - Cosmic Rays and Nuclei from Supernovae and their aftermaths
Proceedings IAU Symposium No. 331, 2017
A. Marcowith, M. Renaud, G. Dubner, A. Ray & A. Bykov, eds.

doi:10.1017/S1743921317004793

Supernova Remnant-Cosmic Ray connection: a modern view

Giovanni Morlino

Gran Sasso Science Institute, viale F. Crispi 7, 67100 L'Aquila, Italy.
email: giovanni.morlino@infn.gssi.it

Abstract. The Cosmic Ray (CR) physics has entered a new era driven by high precision measurements coming from direct detection (especially AMS-02 and PAMELA) and also from gamma-ray observations (Fermi-LAT). In this review we focus our attention on how such data impact the understanding of the supernova remnant paradigm for the origin of CRs. In particular we discuss advancement in the field concerning the three main stages of the CR life: the acceleration process, the escape from the sources and the propagation throughout the Galaxy. We show how the new data reveal a phenomenology richest than previously thought that could even challenge the current understanding of CR origin.

Keywords. acceleration of particles, radiation mechanisms: nonthermal, (ISM:) cosmic rays, supernova remnants, gamma rays: observations.

1. Introduction

Understanding the origin of Galactic cosmic rays (CR) is a puzzle consisting of three main pieces that must fit in with each other to produce a coherent picture and can be summarized as follow: 1) finding the main sources able to accelerate particles up to the highest energy we observe; 2) understanding how particles escape from their sources and are released into the interstellar medium (ISM); 3) understanding how particles propagate through the Galaxy before reaching the Earth. In this concise review I will outline some of the many recent observational findings that, to some extend, are challenging the current understanding of the Galactic CR origin. I will list those discoveries in connection to the three pillars outlined above.

The main sources of Galactic CRs are thought to be the shocks produced by the explosion of SNe, mainly because of energetic reasons, but also because the presence of accelerated particles is unequivocally inferred from the non-thermal radiation emitted by supernova remnants (SNR). The shock acceleration process is maybe the most studied mechanism related to the CR physics (see, e.g., Drury, 1983, and Blasi, 2013, for a review). In § 2 we will focus our attention mainly on the particle spectrum and the maximum energy predicted in the framework of Diffusive Shock Acceleration (DSA), showing how recent observations of SNRs in the γ-ray band are challenging the theoretical predictions.

The final spectrum released into the Galaxy by SNRs is determined by the escaping process and could be significantly different from the spectrum accelerated at the shock. Compared to the acceleration phase, the process of particle escaping has received much little attention, partially because the theoretical difficulty in describing how particles can reach their maximum energy and how they propagate in the transition region between the shock and the undisturbed ISM, where the average level of Galactic CRs dominates over the flux coming from a single source. In addition there are observational difficulties in detecting signals from this transition region with a noticeable exception: the detection of

γ-ray emission from molecular clouds (MC) locate close enough to SNRs. The interesting it such kind of systems has been renewed recently thanks to the measurements of the ionization degree in these same clouds which can provide information on the CR spectrum below the threshold of the γ-rays production, paving the way to the study of CR spectrum from MeV up to TeV energies, and possibly beyond. This argument will be discussed in § 4.

Finally, the study of the propagation mechanism through the Galaxy is receiving a lot of attention in the very recent years, thanks to the impressive amount of data coming from direct measurements of the CR flux and to information coming from the diffuse Galactic γ-ray background. The interpretation of these new data probably requires a revision of the current understanding of the propagation process and could even shed light on the role that CRs have in the evolution of the Galaxy. We will discuss such topics in § 4.

2. Acceleration phase

2.1. *Gamma-ray emission from SNRs*

It is beyond any doubt that SNRs can accelerate CRs. The question is rather which is the total amount of energy channeled into relativistic particles and which is the final spectrum injected into the ISM. γ-ray observations provide a privileged tool to answer these questions, allowing to directly infer the properties of accelerated hadrons, something that cannot be done in any other wavelength.

It is well know that the γ-ray emission can be either produced by hadronic processes (π^0 decay from hadronic collisions) or by leptonic one (inverse Compton – IC) and often it is not easy to distinguish between the two processes. On a general ground, the detection of hadronic emission would favor the high efficient acceleration scenario, while the leptonic interpretation forces one to assume a quite low acceleration efficiency. This binomial is mainly due to the different magnetic field strength required: IC scenario usually needs very low magnetic field ($\sim 10\mu$G, comparable with the ISM value) in order to simultaneously account for radio, X-ray and γ-ray emission. On the contrary, hadronic scenarios requires much larger values, of the order of few hundreds μG, that cannot result from the simple compression of interstellar magnetic field, but requires some sort of amplification. Indeed, the magnetic field amplification is thought to be a manifestation itself of an efficient acceleration process (see § 2.2).

I real cases it is not always easy to distinguish between leptonic and hadronic scenarios. The case of RX J1713 is very significative to this respect. This remnant has been considered for long time the best candidate for an efficient acceleration scenario, mainly due to its high γ-ray luminosity. The detection of γ-ray emission in the range [1-300 GeV] by the Fermi-LAT satellite have shown an unexpected hard spectrum which, at a first glance, seems to be more in agreement with a leptonic scenario. Nevertheless, a deeper analysis shows that neither the hadronic nor the leptonic scenarios, taken in their simplest form, can unequivocally explain the observations (Morlino *et al.*, 2009; Gabici & Aharonian, 2016), hence the issue remains open.

From a theoretical point of view the diffusive shock acceleration (DSA) predicts a power low in momentum $\propto p^{-4}$ (which translates into an energy spectrum $\propto E^{-2}$ for relativistic energies). A remarkable property of DSA is that this prediction turns out to be very solid: even when the complex chain of non-linear effects is taken into account, the deviation from a straight power law are very mild. On the other hand, observations shows a much more varied phenomenology. The spectra inferred from γ-ray observations

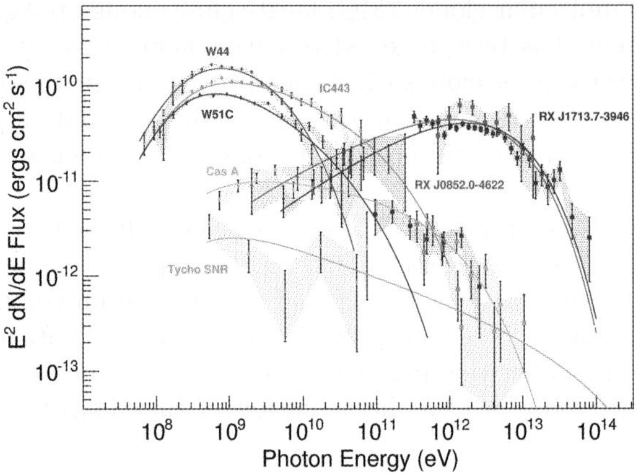

Figure 1. γ-ray spectrum detected from some SNRs. The color code characterizes the age of the remnants: very young (Tycho and Cas A, $t \lesssim 400$ yr), young (RX J0852 and RX J1713, $t \simeq 1000 \div 2000$ yr) and middle age (W51C, W44 and IC443, $t \simeq 1 \div 2 \times 10^4$ yr). Figure taken from Funk (2015).

ranges from $E^{-2.8}$ up to $E^{-1.5}$, with an average slope $\approx -2.2 \div -2.3$. Some authors have suggested the presence of a general trend of the γ-ray spectrum with the age see Funk, 2015. The idea is summarized in the Figure 1: very young SNRs (still in the free expansion phase) show a slight steep power law with index $\approx E^{-2.2} \div E^{-2.4}$ (like Tycho and Cas A); young SNRs ($t \approx 1000 - 2000$ yr) shows a quite hard spectrum more easily explained in terms of leptonic emission (but see the caveat above); middle aged SNRs show spectra of undoubtedly hadronic origin†, steeper than $\propto E^{-2.5}$. Moreover the first two categories have been detected up to tens of TeV, implying a maximum energy of primary protons up to $\lesssim 100$ TeV (maybe one order of magnitude less for Tycho and Cas A), while the middle aged SNRs have been detected up to much lower energies. Nevertheless, it is hard to say whether such a trend is intrinsic in the SNR evolution or if it is rather due to environmental effects. It has been shown, for example, that the presence of a non negligible fraction of neutral hydrogen can significantly modify the shock structure through the process of charge exchange and ionization, resulting in spectra steeper than E^{-2} (Morlino *et al.*, 2010, Morlino & Blasi, 2016). Another complication may be produced by the presence of dense clumps in the CSM (Gabici & Aharonian, 2014). In this case, when the shock crosses the clumps, hydrodynamical instabilities can amplify the magnetic field all around the clump, making difficult for particles at low energies to penetrate inside the clump compared to the most energetic ones. The final result would be a γ-ray spectrum harder then the parent proton spectrum.

Finally, steeper spectra may also result from a more subtle aspect of the non linear DSA theory. If the magnetic field is efficiently amplified, the speed of the magnetic turbulence could be enhanced up to a non negligible fraction of the shock speed, hence reducing the effective compression factor felt by accelerated particles (Morlino & Caprioli, 2012). Unfortunately such mechanism dependents on plasma conditions which are very difficult to test.

† Two middle aged SNRs, IC443 and W44, also show the characteristic pion bump around ≈ 280 MeV, un unmistakable signature of hadronic origin of γ-rays (Ackermann *et al.*, 2013).

2.2. *Maximum energy and magnetic field amplification*

The maximum energy required to explain the CR spectrum observed at Earth is around few PeV for protons and then it scales like the nuclear charge, for heavier nuclei. In the framework of DSA applied to SNRs, there is a general consensus on the fact that the maximum energy is reached before the beginning of the Sedov-Taylor (ST) stage. The reason is that particles diffusing ahead of the shock will be cached by the shock only if the shock is moving faster than the particle. The average displacement of a particle diffusing ahead of the shock is $(Dt)^{1/2}$ while the shock radius moves like $R_s(t) \propto t^\alpha$ hence particles will be recaptured only if $\alpha > 1/2$. But in the ST phase $\alpha = 2/5$, hence particles can escape, while in the free expansion phase $\alpha(\lesssim 1) > 1/2$ (the exact value of α slightly depends on the velocity profile of the ejecta and on the density profile of the circumstellar medium). This reasoning became more complicated if we try to account for the time dependence of the magnetic field amplification but, because we do expect the amplification to increase with increasing shock speed, the general feeling is that the maximum energy should stop increasing even before the end of the free expansion phase. Now, the Sedov-Taylor phase begins at $t_{\rm ST} \approx 50\,(M_{\rm ej}/M_\odot)^{5/6}\,(E_{\rm SN}/10^{51}{\rm erg})^{-1/2}\,(n_{\rm ISM}/{\rm cm}^{-3})^{-1/3}$ yr, which, for typical values of the parameters, ranges between 50 an 200 years, while the acceleration time is given by $t_{acc} \approx 8D/u_{sh}^2$. In the framework of linear DSA, the diffusion coefficient is given by $D = r_L v/(3\mathcal{F})$, where r_L is the Larmor radius of particles and v their speed, while $\mathcal{F}(k)$ is the logarithmic power spectrum of magnetic waves with wave-number k. Equating the acceleration time with $t_{\rm ST}$ we get the following estimate for the maximum energy:

$$E_{\rm max} = 5\times 10^{13}\, Z\,\mathcal{F}(k_{\rm min}) \left(\frac{B_0}{\mu{\rm G}}\right)\left(\frac{M_{\rm ej}}{M_\odot}\right)^{-\frac{1}{6}}\left(\frac{E_{\rm SN}}{10^{51}{\rm erg}}\right)^{\frac{1}{2}}\left(\frac{n_{\rm ISM}}{{\rm cm}^{-3}}\right)^{-\frac{1}{3}} {\rm eV}\,, \quad (2.1)$$

where $k_{\rm min} = 1/r_L(E_{\rm max})$ is the wave number resonant with particles at maximum energy. We notice that more realistic estimates of the maximum energy (for example accounting for the fact that the shock speed is slightly decreasing also during the ejecta-dominated phase) usually return somewhat lower values.

Eq.(2.1) depends only weakly on the environmental parameters, while depends strongly on the level of magnetic turbulence. As a consequence, the maximum energy of protons could reach few PeV only if $\mathcal{F}(k_{\rm min}) \gg 1$, namely the magnetic turbulence at the scale of $r_L(E_{\rm max})$ must be much larger than the pre-existing field, i.e. $\delta B \gg B_0$. Clearly if this condition were realized, the linear theory used to derive the diffusion coefficient would not hold anymore. Apart from that, the value of turbulence in the ISM at scales relevant here is $\delta B/B_0 \lesssim 10^{-4}$ ((Armstrong *et al.*, 1981),) hence, in absence of any mechanism able to amplify the magnetic turbulence, SNR shocks could accelerate protons only up to the irrelevant energy of a few GeV.

This puzzle has been partially solved by the idea that the same accelerated particles can amplify the magnetic field upstream through the *resonant streaming instability* while they try to diffuse far away from the shock (see Drury, 1983, or Blasi, 2013, for a review). Nevertheless the resonant instability can only produce $\delta B \lesssim B_0$, i.e. $\mathcal{F} \lesssim 1$, resulting in a maximum energy for protons of $10-100$ TeV. Hence more effort is needed to fill up the last decade of energy to reach the PeV. The solution to this conundrum probably resides in other types of instabilities that CRs can excite, the most promising one being the *non-resonant Bell instability* (Bell, 2004). This instability results from the $\mathbf{j}\times\mathbf{B_0}$ force that the current due to escaping particles produces onto the plasma and grows very rapidly for high Mach number shocks. However, the scales that get excited are very small compared with the gyration radii of accelerated particles. Hence, it is not clear if the highest energy

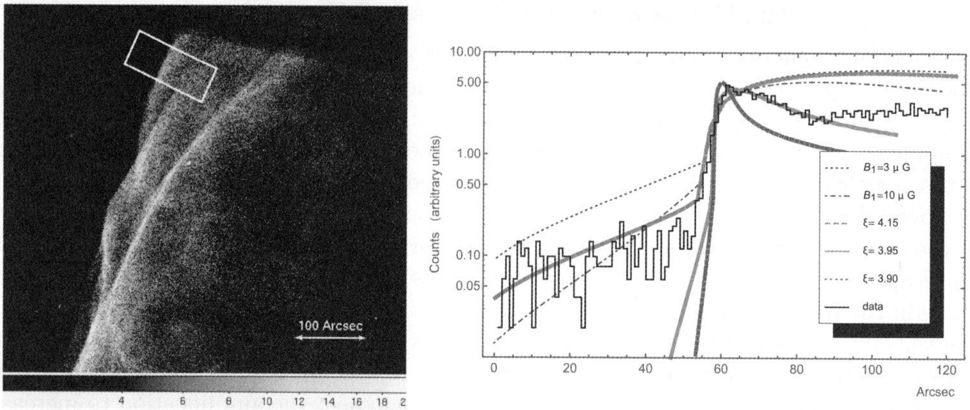

Figure 2. Left: *Chandra* observation of the Northeastern shock of SN 1006. Right: Photon counts as a function of the position corresponding to the white box in the left picture. The curves represent theoretical predictions of non-thermal X-ray profile assuming different acceleration efficiency (Morlino *et al.*, 2010). The plot shows that models without magnetic amplification also in the upstream region (thin curves) over predict the X-ray emission from this region and, simultaneously, cannot account for the bump at the shock location.

particles can be efficiently scattered. Indeed, hybrid simulations seem to confirm that the non-resonant Bell instability grows much faster than the resonant one for Mach number $\gtrsim 30$ and produces $\mathcal{F} \gg 1$ (Reville & Bell,2012; Caprioli & Spitkovsky, 2014). The same simulations also show that the instability produces a complex filamentary structure which could be able to scatter particles efficiently.

The efficiency of Bell instability is determined by the strength of the return current that balance the current of escaping CRs which, in turn, depends on the density of the circumstellar medium and on the shock speed. Indeed, Bell *et al.* (2013) showed that the conditions to reach PeV energies are rather special and can be obtained only during the first few weeks or decades following the explosion of stars occurring in dense circumstellar winds. On the contrary, type Ia SNe, occurring in a less dense ISM, should be able to produce only ~ 100 TeV protons. A similar conclusion has been reached by Cardillo *et al.* (2015). If this scenario would be confirmed, it would not be a surprise the fact that we have not detected yet a single SNR acting as a PeVatron. The chances for such a discovery will be surely enhanced with the forthcoming Cherenkov facilities (CTA).

We conclude this section recalling that one of the most remarkable finding of the last decade is that large magnetic field strength, much larger than the average galactic one, have been inferred in almost all young SNRs, through the observation of thin X-ray filaments at the forward shocks (see Vink, 2012; Ballet, 2006, for recent reviews). This provide a strong support to the presence of some amplification mechanisms. One should keep in mind, however, that the magnetic amplification can increase $E_{\rm max}$ only if it occurs both upstream and downstream of the shock otherwise particles could escape either from one side or the other. Having a magnetic amplification downstream only is quite an easy task: in fact, the shocked plasma is usually highly turbulent and hydrodynamical instabilities can trigger the amplification, converting a fraction of the turbulent motion into magnetic energy, as shown by Giacalone & Jokipii (2007). Conversely, there are no reasons, in general, to assume that the plasma where a SNR expands is highly turbulent to start with. Indeed, evidences that the amplification occurs also upstream have been collected from the X-ray observations of SNR shocks (Morlino *et al.*, 2010; Eriksen *et al.*, 2011) (see also Figure 2) confirming that CRs are the main agent to drive such a process.

3. Escaping from the sources

In the test-particle picture of DSA, accelerated particles are advected downstream of the shock and will be confined in the interior of the SNR until the shock disappears and the SNR merges into the ISM. At that point particles will be released in the ISM but they would have lost part their energy because of the adiabatic expansion of the remnant: hence the requirements in terms of maximum energy at the source would be even more severe than they already are. Therefore, effective escape from upstream, while the acceleration is still ongoing, is fundamental if high energy particles must be released in the ISM. The description of how particle escape from a SNR shock has not been completely understood yet, the reason being the uncertainties related to how particles reach the maximum energies (a careful description of the numerous problems involved can be found in Drury, 2011). Below we just describe the general framework.

Let us assume that the maximum momentum, $p_{\max,0}$, is, indeed, reached at the beginning of the ST phase as discussed in 2.2 and, afterwards, it drops with time as $p_{\max}(t) \propto (t/t_{\rm ST})^{-\beta}$, with $\beta > 0$. The energy in the escaping particles of momentum p is

$$4\pi f_{\rm esc}(p)\, pc\, p^2 dp = \xi_{\rm esc}(t)\, \frac{1}{2}\rho u_{\rm sh}^3\, 4\pi R_{\rm sh}^2\, dt \tag{3.1}$$

where $\xi_{\rm esc}(t)$ is the fraction of the incoming energy flux, $\frac{1}{2}\rho u_{\rm sh}^3 4\pi R_{\rm sh}^2$, that is converted into escaping flux. If the expansion occurs in a homogeneous medium with $R_{\rm sh} \propto t^{\alpha}$ and $V_{\rm sh} \propto t^{\alpha-1}$, therefore, since $dt/dp \propto t/p$, from Eq. (3.1) we have:

$$f_{\rm esc}(p) \propto p^{-4} t^{5\alpha-2} \xi_{\rm esc}(t). \tag{3.2}$$

It follows that in the ST phase, where $\alpha = 2/5$, the spectrum released in the ISM is $f_{\rm esc}(p) \propto p^{-4}$ only if $\xi_{\rm esc}$ keeps constant with time. It is worth stressing that this p^{-4} has nothing to do with the standard result of the DSA in the test-particle regime. Neither does it depend on the detailed evolution in time of the maximum momentum. It solely depends on having assumed that particles escape the SNR during the adiabatic phase. Notice also that in more realistic calculations $\xi_{\rm esc}$ should decreases with time, leading to a spectrum of escaping particles which is even harder than p^{-4} (Caprioli *et al.*, 2010). On the other hand, the total spectrum of particles injected into the ISM by an individual SNR is the sum of two populations: particles escaping during the acceleration and particles released after the shock dissipates. The latter component is not modified by the escaping process and should reflect closely the accelerated spectrum. An interesting byproduct of this scenario is that the intersection of the two populations may tentatively be identified with the change in spectral index at $\sim$ 300 GeV detected by PAMELA and AMS-02 (Bell, 2015). We will discuss this feature more in details in the next section.

A further complication arises when one try to account for the environment when SNR expands. While type Ia SN are thought to expand in a pretty homogeneous ISM, core-collapse SNe expand into the wind/bubble system generated by the progenitor stars. In such a situation the remnant evolution differs significantly from the classical ST solution and could results in CR spectra different from the the standard p^{-4}. In spite of such complications, it seems that when one sum the contribution of CRs from different type of SNRs, the final results approach p^{-4} (Caprioli *et al.*, 2010). Nevertheless one should take this result with caution, in that it is obtained extrapolating the effect of streaming instability calculated in the linear regime, to the non linear one, and neglecting the more complicated effect of the non-resonant instability, which should play a major role when the acceleration efficiency is high. In conclusion, the theory does not provide yet a solid answer to the question of escaping from sources. Therefore, it is of the outermost importance to look for possible observational constraints that can help to understand such a process.

MC-SNR associations. A possible way to investigate the escaping process is to look for γ-ray emission from MC-SNR associations. When MCs are located close enough to a SNR, escaping CRs can "illuminate" the cloud, producing γ-rays through interaction with the cloud gas. The resulting γ-rays can provide information on the escaping CR spectrum but also on the diffusion coefficient in the region between the SNR and the MC, which determines how fast particles escape (Gabici *et al.*, 2009).

MC-SNR systems could also be used to shed light on the lowest energy part of the CR spectrum, namely below the pion threshold of $\sim$ 280 MeV, allowing us to test models on CR acceleration and propagation over an energy interval spanning from the MeV to the TeV domain. Such impressive result can be reached looking for the ionization level of MC (see Gabici & Montmerle, 2015, for a review). In fact, even though CRs with energy below $\sim$ 280 MeV carry a subdominant fraction of the bulk energy of CRs, they nevertheless play a pivotal role in regulating the properties of the interstellar medium, by ionising and heating the gas, and thus driving interstellar chemistry in MCs, where ionizing photons cannot penetrate (Dalgarno, 2006). Moreover, being the main regulators of the ionization fraction inside MCs, low energy CRs set the level of coupling between magnetic field and gas and thus influence the process of formation of stars and planets. Hence, finding observational ways to trace them and quantify these effects is of prime interest. A remarkable result has been already achieved studying, among the others, the MC complex around the SNR W28, where the MC closest to the SNR shows a strong enhancement of the ionization rate (Vaupré *et al.*, 2014) signature of the fact that low energy CRs has reached only the closest MC, while high energy CRs also reached the more farther MCs, producing the characteristic γ-ray emission.

4. Propagation through the Galaxy

After leaving the near-source region, CRs start their journey across the Galaxy and eventually escape from the Galactic environment. In the standard picture used to describe Galactic propagation, CRs are produced by sources located in the thin Galactic disc and then diffuse in a larger volume where the presence of some magnetic turbulence is thought to scatter efficiently CRs. Beyond such *magnetic halo*, the turbulence is assumed to vanish such that particles can escape freely into the intergalactic space and the CR density reduces to ≈ 0 (Berezinski *et al.*, 1990). The presence of a magnetic halo with a thickness H much larger than the disc thickness, h, is required by the estimate of the residence time of CRs into the Galaxy which, assuming a diffusion coefficient $D(E)$ spatially constant across the whole halo, is given by the simple expression $\tau_{\rm esc} = H^2/D(E)$. Such quantity can be derived from the flux of radioactive nuclei present in the CR spectrum (like ^{10}Be, ^{26}Al and ^{36}Cl, see Yanasak *et al.* (2001)) plus the ratio of secondary over primary (stable) nuclei (among which the one measured with better accuracy is B/C). Combining those information one gets the halo size $H \approx 3 \div 5$ kpc and the diffusion coefficient $D(R) \simeq 1 \div 5 \times 10^{28} (R/{\rm GV})^{-\delta}$ cm^2 s^{-1}, with $\delta \approx 1/3$, in good agreement with the Kolmogorov theory of turbulence (Aguilar *et al.*, 2015). Those values give $\tau_{\rm esc} \approx 150$ Myr for particle's rigidity around 1 GV †.

† Notice that the residence time estimated by Yanasak *et al.* (2001) using radioactive nuclei is only $\sim$ 15 Myr. This result is obtained in the framework of the *leaky-box* model and is, to some extend, misleading in that it does not account for the spatial extention of the halo. It is known that when the decay time of unstable nuclei is smaller than the residence time, the leaky-box model gives incorrect results Berezinski *et al.*, 1990.

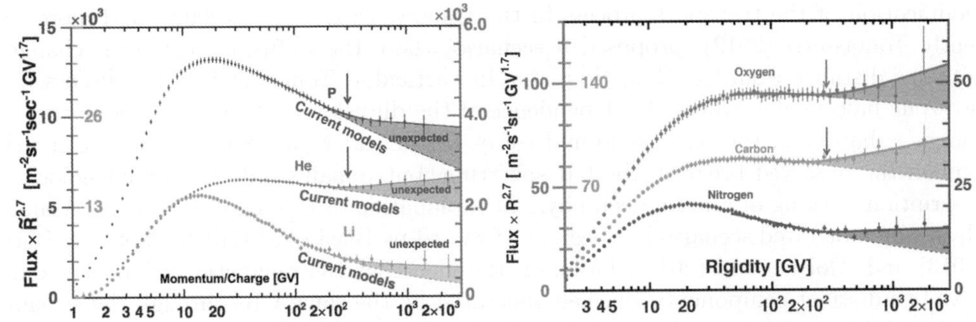

Figure 3. Spectra of proton, Helium and Lithium (left plot) and Oxygen, Carbon, Nitrogen (right plot) as a function of rigidity. The shaded regions show the discrepancy above ~ 300 GV of the best fit models to data with the basic prediction of the standard model for CR propagation. Figures from Yan (2017).

An independent proof to the existence of a magnetic halo comes from observations in the radio band which points towards the presence of diffuse synchrotron emission, revealing the presence of electrons and magnetic field above and below the galactic plane (Beuermann *et al.* ,1985). The determination of the halo size in this case is not an easy task, partially due to the yet unknown structure of Galactic magnetic field. Nevertheless, comparison with numerical models for the CR electron distribution favors a halo size of $\sim$ 10 kpc (Orlando & Strong, 2013) while a size $\lesssim 2$ kpc seems to be strongly disfavored (Di Bernardo *et al.* , 2015). It is worth mentioning that radio halos with a similar extension have been observed in others spiral galaxies observed edge-on(e.g. NGC 4631, NGC 891).

Nevertheless, the halo model, at least in its basic version, does not provide a coherent physical picture. The value of H required to explain the data is not well understood: it should be consistently determined by the spatial extension of the magnetic turbulence produced inside the Galactic disc (mainly by SN explosions), rather than used as a fit parameter. Moreover it is difficult to imagine a sharp boundary where the particle transport abruptly changes from diffusion to free streaming regime and it is also unclear if such a boundary should be the same at all energies, as naively assumed. Finally, the assumption that the diffusion coefficient should be equal everywhere in the halo is also a strong simplification. The above criticisms are strengthened by the recent discovery of several anomalies both in the CR spectrum measured at Earth and in the diffuse Galactic γ-ray emission which both suggest a more complex situation with respect to the toy model depicted above. A detailed discussion of all these aspects can be found in Amato & Blasi (2017) while here we restrict our attention on two main results.

We start considering the anomalies in the local CR spectrum. Recent measurements carried out by PAMELA and AMS-02 have revealed that the spectra of protons and Helium are characterized by a spectral hardening above ~ 300 GV (Adriani *et al.* 2011; Aguilar *et al.*, 2015a,2015b). Based on preliminary results by AMS-02, shown by Yan (2017), a similar break might be present also in the spectra of heavier nuclei as shown in Fig. 3. In standard approaches to CR transport, these findings can be accommodated imposing breaks in either the diffusion coefficient or the injection spectrum. We already discussed in the previous section a possible way to produce a break in the injected spectrum resulting from the superposition of two components: one due to particles escaping during the acceleration process and the other due to particles released after the SNR dissolution. Nevertheless, it is not clear if such a model remains viable once the integration over many SNRs is taken into account. Much attention has been payed to a possible modification of the transport process in the Galaxy. Dogiel *et al.* (2002), and more recently Tomassetti (2012), proposed a scenario where the diffusion coefficient changes with the distance from the Galactic plane. In particular, Tomassetti (2012) discussed a two-zone halo model, where the dependence of the diffusion coefficient on the particle's energy is harder in the outer halo and softer to the inner one. Such a model can well explain the observed break in the CR spectrum, but remains only a phenomenological description, lacking of a more deep physical explanation of the halo's structure. A more physically motivated scenario has been put forward by Blasi *et al.* (2012), Aloisio & Blasi (2013) and Aloisio *et al.* (2015), based on the idea that the magnetic turbulence could have two distinct components: one self-generated by the same CRs through the streaming instability and one resulting from the Kolmogorov cascade of turbulence injected at larger scale (presumably by SN explosions). In this scenario the diffusion is dominated at low energies by self-generated waves, while at larger energies, where the CR energy density becomes too low to generate a significant amount of waves, the Kolmogorov turbulence dominates and the spectral break in the CR spectrum should mark the energy where this transition occurs.

One should keep in mind, however, that direct measurements can only probe the local CRs spectrum, hence the question whether such a spectrum reflects some peculiarities of the local environment, rather than being representative of the average Galactic CR spectrum, still remains open. Relevant information on the CR distribution in the rest of the Galaxy can be obtained looking at the diffuse γ-ray emission which is thought to originate mainly from π^0 decay produced by collisions of CRs with the interstellar gas. Recent results presented by the Fermi-LAT collaboration (Acero *et al.*, 2017) and independently by Yang *et al.* (2017), show a substantial variation of the CR spectrum as a function of the distance from the Galactic Center. The spatial distribution of the CR density in the outer Galaxy appears to be weakly dependent upon the galactocentric distance, as found in previous studies as well, while the density in the central region of the Galaxy was found to exceed the value measured in the outer Galaxy. At the same time, Fermi-LAT data suggest a gradual spectral softening while moving outward from the center of the Galaxy to its outskirts, with a slope ranging from 2.6, at a distance of ~ 3 kpc, to 2.9 in the external regions. These findings represent a challenge for standard calculations of CR propagation based on assuming a uniform diffusion coefficient within the Galactic volume and may point towards a spatial-dependence of the CR transport properties, as already discussed by Evoli *et al.* (2012) and Gaggero *et al.* (2015). Indeed, this behavior could reflect, once again, the presence of self-generated turbulence which should dominate the particle scattering at low energies, the ones probed by Fermi-LAT. Recchia *et al.* (2016a), indeed, shows that the combination of self-generated turbulence plus the advection due to the wave motion at the Alfvén speed can account for both the spatial CR density and the spectral slope (see Figure 4). This result depends, to some extend, on the assumption of the Galactic magnetic field structure, which determines the effectiveness of the streaming instability.

Galactic winds. We conclude this section discussing an interesting development of CR transport models that may have a large impact on the overall Galaxy evolution. Many galaxies produces a large scale wind that may affect star formation, through the regulation of the amount of gas available (Crain *et al.*, 2007; Stinson *et al.*, 2013) and inject hot gas in the galactic halo. Signatures for the presence of such hot gas have been presented also for the Milky Way through the detection of absorption lines and continuum emission in the X-ray band (Miller & Bregman, 2015). Galactic winds are thought to be powered by either SN explosions or starburst radiation (see e.g. Chevalier & Clegg, 1985), but in our Galaxy the power injected by SNe and stars is too small to drive a large

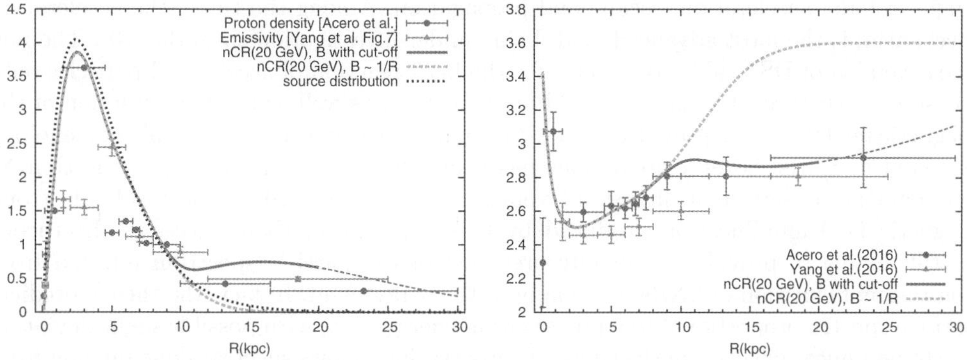

Figure 4. CR density (left) and spectral slope (right) inferred from Fermi-LAT data as a function of galactocentric distance. Data correspond to two different analysis, (Acero *et al.*, 2016) and (Yang *et al.*, 2017), as labelled. Solid-red and dashed-green curves show the CR density at $E > 20$ GeV (left) and the spectral slope (right) predicted by Recchia *et al.* (2016a) using two different assumption for the Galactic magnetic field strength. Finally, the dotted-black line on the left shows the distribution of CR sources (Green, 2015).

scale wind. On the other hand, the escape of CRs from the Galaxy inevitably leads to a gradient in the CR pressure that acts as a force on the background plasma, in the direction opposite to the gravitational pull (Ipavich *et al.*, 1975; Breitschwerdt *et al.*, 1991; Everett *et al.*, 2009). This force, added to the thermal pressure gradient, may contribute to lunch a galactic wind if the gas can be accelerated to supersonic speeds, otherwise the material can be lifted up and fall down in what is known as Galactic fountains. The existence of a Galactic wind would significantly change the transport properties of CRs. The main change is due to the presence of an advection speed, $u_{\rm adv}(z)$, which increases moving from the Galactic plane towards larger distances. As a consequence, above some distance, $s_*(E)$, given by the condition

$$\frac{s_*^2}{D(E)} \approx \frac{s_*}{u_{\rm adv}(s_*)}, \tag{4.1}$$

the advection will dominate over diffusion and particles will be unable to diffuse back to the Galaxy. In other words, this distance s^* represent an effective energy-dependent boundary which replaces the role of the halo size H. Such a model is particularly attractive because s_* is not artificially imposed, like H, but it is rather self-consistently given by the wind solution (Ptuskin *et al.*, 1997). Nevertheless, the wind model that includes only a diffusion coefficient due to self-generated turbulence, predicts a CR spectrum remarkably different from the one observed at Earth, especially at high energies where it is too steep. A better agreement with observations can be obtained assuming a pre-existing turbulence only in the near disc region (Recchia *et al.*, 2016b). In conclusions wind models seem to have a great potential to provide a comprehensive picture of the CR propagation, but at the moment such models are still immature and a lot of work needs to be done to include self consistently several relevant piece of physics like plasma cooling and heating processes, magnetic field damping and cascade of large scale turbulence. The complexity of such task will probably require time dependent simulations.

5. Conclusions

We presented a brief overview of some recent advancement in the field of CR physics, discussing separately the three main pillars that constitute the SNR paradigm for the origin of CRs: acceleration, escape and propagation. Among the three, the acceleration mechanism is the most advanced field, being under investigation since the '70s. The non linear version of DSA, able to account for the back reaction of accelerated particles onto the shock structure, the magnetic field amplification, as well as the back reaction of the magnetic field itself, has provided a robust framework to interpret the rich phenomenology observed in SNRs. Several predictions has been verified, like, for example, the thin X-ray filaments observed in almost all young SNRs, interpreted as the result of strong magnetic field amplification produced by CRs. In spite of these successes the theory is still unable to provide a firm interpretation of the particle spectrum inferred from non-thermal emission of SNRs, especially in the γ-ray band. In fact, the theory predicts particle spectra proportional to a power low in energy E^{-2}, with possible slight deviation due to non linear effects. On the contrary, observations shows spectral slope ranging from -1.5 to -2.5 for young remnants, while in the case of middle aged SNRs the spectrum can be even steeper. This rich phenomenology is probably due to environmental effects, which are difficult to model. Moreover, the possible implications of those effects on the overall Galactic CR spectrum are, at the moment, unclear.

The escaping process has never been studied in great details, due to some theoretical difficulties but also to the lack of data coming from the escaping particles. A possible way to study this mechanism is through the γ-ray emission from SNR-MC associations. Few such complex have been already detected in γ-rays, but the future Cherenkov Telescope Array is expected to find tens of those systems, paving the way to a systematic study.

Finally the propagation process has received a lot of attention in the last few years, thanks to an avalanche of new data coming from both direct measurements (AMS-02 and PAMELA) and to indirect informations inferred from the diffuse γ-ray emission detected by Fermi-LAT. Thanks to those data, the basic propagation model start to shows its limitations, forcing the community to build more refined models. In particular I think we are close to a better understand of the magnetic turbulence that determine the CR propagation properties and, connected to this, also on the structure of the magnetic halo that surrounds the Galaxy.

It is worth mentioning that others anomalies found in the CR spectrum and not discussed here, like the rising positron fraction and the anti-proton flux, are driving new ideas that are questioning the basic pillars of the CR propagation (se, e.g., Katz *et al.*, 2010; Blum *et al.*, 2013; and Lipari, 2016). Whether these ideas can provide viable alternatives to the current paradigm for the origin of CRs is a matter of debate, which shows, anyhow, that the CR physics is undergoing a new phase of growing, potentially rich of new discoveries.

References

Acero, F., Ackermann, M., Ajello, M., Albert, A., *et al.* 2017 ApJS 223, 26
Adriani, O., *et al.* 2011, *Science*, 332, 69
Aguilar, M., Ali Cavasoza, L., Ambrosi, G., *et al.* 2016, *Phys. Rev. Lett.*, 117, 231102
Aguilar, M., *et al.*, 2015a, *Phys. Rev. Lett.*, 114, 171103
Aguilar, M., *et al.*, 2015b, *Phys. Rev. Lett.*, 115, 211101
Aloisio, R. & Blasi, P. 2013, JCAP, 7, 001
Aloisio, R., Blasi, P., & Serpico, P. D. 2015, *A&A*, 583, A95
Amato, E. & Blasi, P. 2017, preprint arXiv:1704.05696
Armstrong, J. W., Cordes, J. M., & Rickett, B. J. 1981, *Nat*, 291, 561
Ballet, J. 2006, Advances in Space Research, 37, 1902
Bell, A. R., Schure, K. M., Reville, B., & Giacinti, G. (2013) *MNRAS*, 431, 415
Bell, A. R. (2015) *MNRAS*, 447, 2224

Berezinski, V. S., Bulanov, S. V., Dogiel, V. A., & Ptuskin, V. S. 1990, book edited by Ginzburg, V.L., published by North-Holland (Amsterdam)
Beuermann, K., Kanbach, G., & Berkhuisjen, E. M. 1985, *A&A*, 153, 17.
Blasi, P., Amato, E., & Serpico, P. D. 2012, *Phys. Rev. Lett.*, 109, 061101
Blasi, P. 2013, *A&ARv*, 21, 70
Blum, K., Katz, B., & Waxman, E. 2013, *Phys. Rev. Lett.*, 111, 1101
Breitschwerdt, D., McKenzie, J. F., & Voelk, H. J. 1991, *A&A*, 245, 79
Caprioli, D., Amato, E., & Blasi, P. 2010, *Astropart. Phys.*, 33,160
Caprioli, D. & Spitkovsky, A. 2014, *ApJ*, 794, 46
Cardillo, M., Amato, E., & Blasi, P. (2015) *APh*, 69, 1
Crain, R. A., Eke, V. R., Frenk, C. S., *et al.* 2007, *MNRAS*, 377, 41
Dalgarno, A. 2006, *PNAS*, 103, 12269
Di Bernardo, G., Grasso, D., Evoli, C., & Gaggero, D. 2015, *ASTRA Proceedings*, 2, 21
Dogiel, V. A., Schonfelder, V., & Strong, A. W. 2002, *ApJ Lett.*, 572, L157
Drury, L'O. C. 1983, RPPh, 46, 973
Drury, L'O.C. (2011) *MNRAS* 415, 1807
Everett, J. E., Zweibel, E. G., Benjamin, R. A., *et al.* 2008, *ApJ*, 674, 258-270
Eriksen, K., Hughes, J. P., Badenes, C., *et al.* 2011, *ApJ*, 728, L28
Evoli, C., Gaggero D., Grasso D., & Maccione L. 2012, *Phys. Rev. Lett.*, 108, 211102
Funk, S. 2015, *Ann. Rev. of Nuc. and Part. Sci.*, 65, 245.
Gabici, S., Aharonian, F. A., & Casanova, S. 2009, *MNRAS*, 396, 1629
Gabici, S. & Montmerle, T. 2015, *proceeding to the ICRC* 2015, 34, 29
Gabici, S. & Aharonian, F. 2016, *EPJ Web of Conferences*, Vol. 121, id.04001
Gabici, S. & Aharonian, F. 2014, *MNRAS*, 445, L70
Gaggero, D., Urbano A., Valli M., & Ullio P. 2015, *Phys. Rev. D*, 91, 083012
Giacalone, J. & Jokipii, J. R. 2007, *ApJ*, 663, L41
Giacinti, G., & Bell, A. R. (2015), *ICRC proceeding*, 34, 519
Green, D. A. 2015, *MNRAS*, 454, 1517
Ipavich, F. M. 1975, *ApJ*, 196, 107
Katz, B., Blum, K., Morag, J., & Waxman, E. 2010, *MNRAS*, 405, 1458
Lipari, P. 2016, *Ph. Rev. D.*, 95, 3009
Morlino, G., Amato, E., & Blasi, P. 2009, *MNRAS*, 405, 21
Morlino, G., Amato, E., Blasi, P., & Caprioli, D. 2010, *MNRAS*, 405, L21
Morlino, G. & Caprioli, D. 2012, *A& A*, 538, 81
Morlino, G., Blasi, P., Bandiera, R., Amato, E., & Caprioli, D. 2013, *ApJ*, 768, 148
Morlino, G. & Blasi, P. 2016, *A&A*, 589, 7
Miller, M. J. & Bregman, J. N. 2015, *ApJ*, 800, 14
Orlando, E. & Strong, A. 2013, *MNRAS*, 436, 2127
Ptuskin, V. S., Voelk, H. J., Zirakashvili, V. N., & Breitschwerdt, D. 1997, *A&A*, 321, 434
Recchia, S., Blasi, P., & Morlino, G. 2016, *MNRAS*, 462, 88
Recchia, S., Blasi, P., & Morlino, G. 2016, *MNRAS*, 462, 4227
Reville, B. & Bell, A. R. 2012, *MNRAS*, 419, 2433
Stinson, G. S., Brook, C., Macció, A. V., *et al.* 2013, *MNRAS*, 428, 129
Tomassetti, N. 2012, *ApJ. Lett.*, 752, L13
Vaupré, S., Hily-Blant, P., Ceccarelli, C., Dubus, G., Gabici, S., & Montmerle, T. 2014, *A&A*, 568, 50
Vink, J. 2012, *A&ARv*, 20, 49
Yan, Q. 2017, *CR nuclei measurements with AMS-02: latest results and perspectives*, Presented at XSCRC2017: Cross sections for Cosmic Rays, CERN
Yanasak, N. E., Wiedenbeck, M. E., Mewaldt, R. A., Davis, A. J., *et al.* 2001, *ApJ*, 563, 768
Yang, R., Aharonian, F., & Evoli, C. 2017, *PhRvD*, 93,123007

Supernova 1987A:30 years later - Cosmic Rays and Nuclei from Supernovae and their aftermaths
Proceedings IAU Symposium No. 331, 2017
A. Marcowith, M. Renaud, G. Dubner, A. Ray & A. Bykov, eds.

doi:10.1017/S174392131700480X

X-ray Synchrotron Polarization from Turbulent Plasmas in Supernova Remnants

Matthew G. Baring

Department of Physics and Astronomy - MS 108, Rice University,
6100 Main Street, Houston, Texas 77251-1892, USA
email: baring@rice.edu

Abstract. As supernova remnants (SNRs) age, they become efficient cosmic ray accelerators at their outer shell shocks. The current paradigm for shock acceleration theory favors turbulent field environs in the proximity of these shocks, turbulence driven by current instabilities involving energetic ions. With the imminent prospect of dedicated X-ray polarimeters becoming a reality, the possibility looms of probing turbulence on scales that couple to the super-TeV electrons that emit X-rays. This paper presents model X-ray polarization signatures from energetic electrons moving in simulated MHD turbulence of varying levels of "chaos." The emission volumes are finite slabs that represent the active regions of young SNR shells. We find that the turbulent field energy must be quite limited relative to that of the total field in order for the X-ray polarization degree to be as strong as the radio measures obtained in some remnants. Results presented are pertinent to the planned IXPE and XIPE polarimeters.

Keywords. supernova remnants, cosmic rays, radiation mechanisms: nonthermal, polarization, plasmas, turbulence.

---

1. Introduction

One of the most important contributions of the *Chandra* X-ray Observatory to supernova remnant (SNR) science has been the discovery of thin synchrotron rims in the outer shells of several remnants. This advance was enabled by *Chandra*'s exquisite imaging capability. Among these sources are SN1006 (Long *et al.* 2003) and Cassiopeia A (Vink & Laming 2003), the original *first light* target for the Observatory. The sharp rise and fall of the X-ray flux in these filamentary rims, on angular scales of a few arcseconds, implies a very short synchrotron cooling timescale. This yields the unavoidable interpretation that the embedded magnetic field is on the order of 20–70μGauss, substantially above values normally interpreted from standard invocations of field compression in MHD shocks. Thus a quandary emerged: how could such anomalous **field enhancements** come about?

In the years just prior to and around this time, the concept of **turbulent amplification** of magnetic fields mediated by cosmic ray-driven instabilities in SNR shocks was being developed, led by the papers by Lucek & Bell (2000) and Bell (2005). The anisotropy of the most energetic ions on the scales of the thickness of remnant shells seeds growth of MHD fluctuations to the levels of $(\delta B/B)^2 \sim \mathcal{M}_{\rm A} P_{\rm cr}/\rho u^2 \gtrsim 3 - 10$. Here $\mathcal{M}_{\rm A}$ is the Alfvénic Mach number, $P_{\rm cr}$ is the cosmic ray (CR) pressure, and ρu^2 is the fluid ram pressure. Basically, the CR pressure gradient does work on the field energy density. The interplay between the thermal gas, the cosmic rays and the field energy introduces inherently non-linear elements to the dynamics of the shock environs (e.g. Vladimirov, Ellison & Bykov 2006). The observation of thin rims propelled the extensive development of models of the so-called **Bell instability**, and the paradigm that SNR shells contain fields that are turbulent and enhanced has become very popular.

We are now at the dawn of a new era in X-ray imaging, namely with the advent of X-ray polarimetry. NASA announced earlier this year the selection of the Imaging X-ray Polarimetry Explorer (IXPE, see Weisskopf *et al.* 2013) as a Small Explorer mission, to be launched in four or so years. This will serve as a pathfinder for discovery of polarized sources in X-rays, with a special focus on imaging and therefore on supernova remnants. IXPE's nominal $30''$ angular resolution will be able to probe the diffusive scales of the most energetic electrons emitting synchrotron X-rays in SNRs. It will thus be able to cast light on the level and nature of turbulence in SNR rims and shells, with around 8 prime SNR targets in its advertised science program (Weisskopf *et al.* 2013).† Subsequent X-ray polarimetry initiatives such as XIPE (Soffitta *et al.* 2013) and eXTP offer additional prospects for probing the SNR environment in the not too distant future.

In preparation for such a data influx, exploration of what one might expect to see with an X-ray polarimeter is strongly motivated. A seminal study of the flux and polarization degree variations in turbulent field models of synchrotron-emitting remnants was published in Bykov *et al.* (2009), outlining how images sampling various angular scales can capture the information of field turbulence. Here we present results from an incipient simulation of synchrotron radiation from charges in prescribed MHD turbulence using a somewhat different construction, but notably with a decidedly different emphasis. Here the simulation output provides measures of the fluctuations, mean and standard deviation of the polarization Stokes parameters, and demonstrates the strong correlation between these and the variance of Kolmogorov turbulence in SNR shells.

2. Synchrotron Polarization from Electrons in Magnetic Turbulence

To assess the nature of polarized synchrotron emission in turbulence near SNR shocks, a simulation of charge motion in prescribed magnetostatic field fluctuations was developed. The construction was similar to the approach of Giacalone & Jokipii (1999) in their charge diffusion study, in that 1D turbulence was superposed upon a uniform background field $\mathbf{B}_0$, which is radially-directed along the x-axis. The MHD turbulence assumes the form

$$\delta\mathbf{B} \; = \; \sum_k \delta B_k \left\{ \cos\phi_\perp\, \hat{y} + \sin\phi_\perp\, \hat{z} \right\} \, \cos\bigl(kx + \phi_k\bigr) . \tag{2.1}$$

The slab nature is embodied in the property that the wavenumbers $\mathbf{k}$ of turbulence only possess a component $k\hat{x}$ along the direction of the unperturbed field $\mathbf{B}_0$, and the transverse field perturbations are in the $(y,\, z)$-plane, perpendicular to the plane of the sky (adopted for simplicity though not necessarily true, it is a testable assumption nonetheless). The wave phase ϕ_k and the azimuthal orientation $\phi_\perp$ were chosen randomly. The power spectrum of the fluctuations was presumed to have a 1D Kolmogorov form:

$$\frac{\langle \delta B_k \rangle^2}{B_0^2} \; = \; \sigma^2 \left(\frac{k}{k_-} \right)^{-5/3} \quad , \quad k_- \leqslant k \leqslant k_+ . \tag{2.2}$$

Here σ^2 is the wave variance at the base of the inertial range, i.e. at $k = k_- \equiv 2\pi/\lambda_{\rm stir}$, with $\lambda_{\rm stir} \sim R_{\rm shell}/3$ being the stirring scale for the cascading turbulence. Also, for the simulations herein, $k_+ = 100 k_-$ was adopted with values of k sampled randomly on a logarithmic scale over the inertial range. Representative simulated turbulent fields are illustrated in Fig. 1 via their projections onto the sky plane, which is taken to encapsulate the radial vector through the slab. The absence of any time dependence (magnetostatic assumption) in $\delta\mathbf{B}$ is a tantamount to presuming non-relativistic plasma dynamics.

† See also the IXPE web page at `https://wwwastro.msfc.nasa.gov/ixpe/`.

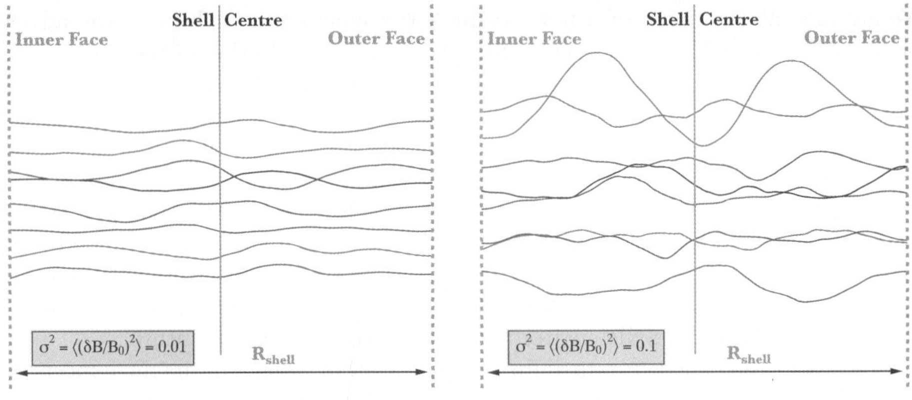

Figure 1. Schematics for the geometry of SNR shell containing turbulent fields. The depicted field lines are from the prescription of Kolmogorov slab turbulence in Eq. (2.2), with the variance $\sigma^2 = \langle(\delta B/B_0)^2\rangle$ of the turbulence at the stirring scale $R_{\rm shell}/3$ being $\sigma^2 = 10^{-2}$ for the left panel, and $\sigma^2 = 10^{-1}$ for the right panel. Only the projections of the field lines in the plane of the Figure are represented; the color coding serves only to aid distinguishability of the lines.

Yet the synchrotron-radiating electrons are indeed ultra-relativistic. They were injected at the inner boundary ($x = -10$) of the slab, distributed randomly in the (y, z) plane, and with the initial momentum vector directions selected randomly from an isotropic distribution. The Lorentz factors were selected randomly on a logarithmic scale from a power law distribution over a narrow range spanning a factor of two in Γ_e . The actual scale was a convolution of $\Gamma_e/\sqrt{B_0}$ so as to position the synchrotron spectral information neatly in the X-ray band. The electrons were then propagated into the slab, solving the Lorentz force equation in the turbulent field structure in incremental intervals along their trajectories using a fourth-order Euler ODE solver. Their paths and acceleration vectors were logged for use in the polarization algorithms.

2.1. *Synchrotron Polarization Characteristics*

To develop a general idea of the properties of synchrotron radiation polarization in SNR shells, the synchrotron emissivity is integrated over the charges' trajectories. This polarized emissivity is principally dependent on three quantities at each point along the lepton path: the magnetic field vector and the electron's Lorentz factor and pitch angle. Since the charges were injected into the slab with an isotropic angular distribution, provided the turbulence is not too large, they retain approximate isotropy throughout the slab; this is relinquished somewhat when $\delta B/B \sim 1$. The local acceleration vector determines the instantaneous pitch angle, and combined with $\mathbf{B}$, they serve as parametric input for the standard synchrotron emissivity formulae (e.g. see Rybicki & Lightman 1979). The electron Lorentz factors are distributed, and for our purposes here, a power-law distribution $n_e(\gamma) \propto \gamma^{-p}$ with $p = 3$ was presumed. This generates a synchrotron intensity spectrum $I_\nu \propto \nu^{-1}$, which approximates the steep synchrotron slopes in the keV X-ray band seen in SNRs that are interpreted as portions of exponential turnovers. Accordingly, the assignment of synchroton photon electric field vectors can be made with the usual $\|$ and $\perp$ designation and textbook probabilities. Thus

$$\Pi \equiv \frac{I_\perp - I_\|}{I_\perp + I_\|} \to \frac{p+1}{p+7/3} \to \frac{3}{4} \tag{2.3}$$

is the degree of polarization for a uniform, unturbulent field ($\sigma \to 0$).

These polarization assignments are then converted into contributions to the Stokes parameters as an electron continues its trajectory. If the projection of the magnetic field onto the *sky plane* $(x,\, y)$ lies parallel to the mean field vector $\mathbf{B}_0$ in the x-direction, then one is at liberty to assign $Q = I, U = 0$ for the $\parallel$ state, and $Q = -I, U = 0$ for the $\perp$ state. Since this field orientation does not point to an observer, the zero circular polarization sets $V = 0$. Adding components to $\mathbf{B}$ in the z-direction perpendicular to the sky plane does not alter this polarization, but it does change the synchrotron intensity, since the field vector has been modified. Adding a field component in the y-direction amounts to rotating the sky plane-projected field about the z-axis through some angle χ (the **position angle**), and this then mixes the Stokes parameters (tensor elements) so that $Q = I\cos 2\chi, U = I\sin 2\chi$ for a $\parallel$ state contribution, and $Q = -I\cos 2\chi, U = I\sin 2\chi$ for a $\perp$ state one. These are then added for each step along the electron path, sampling different local field vectors and varying pitch angles. As long as the local field components are inferior to the total sky plane-projected $\mathbf{B}$, the circularity is simply $V = 0$. The result is an integrated ensemble of Stokes Q, U and I information.

For the purposes of illustration, the emission was discretized using the cubic slab structure in a sequence of slices abutting each other along the shock face, but also in bins of equal thickness in the shock normal (x) direction. Thus, a slice constitutes a volume $-r \leqslant x \leqslant +r, -nr/N \leqslant y \leqslant +nr/N, -r \leqslant z \leqslant +r$, and each bin defines a rectangular prismatic rod of volume $-nr/N \leqslant x \leqslant +nr/N, -nr/N \leqslant y \leqslant +nr/N, -r \leqslant z \leqslant +r$ (for integer $n \leqslant N$) that is aligned normal to the sky plane. Here r is the slab half-thickness, which scales with the thickness $R_{\rm shell}$ of the SNR shell; $r = 10$ is chosen here. Also, $N = 4$ is chosen for the purposes of illustration. This approximates the sky plane *pixelation density* achievable with IXPE for typical remnants as afforded by its angular resolution of $\sim 30''$ – see various figures in Bykov *et al.* (2009) for a range of intensity imagery realizable with X-ray polarimetry observations of turbulent SNR shells.

The synchrotron Stokes parameter contributions from 10^4 electrons injected with a narrow range of Lorentz factors into slab at the $x = -10$ boundary were recorded, added, and assigned to each (x, y) pixel according to the position of each charge as it moved along its trajectory. Results for two turbulence cases are ilustrated in Figs. 2 (for $\sigma^2 = 10^{-2}$) and 3 (for $\sigma^2 = 0.1$). The four panels each plot histograms for each slice, color-coded to enhance the visual distinguishability, given the overlapping/entanglement of the histograms. The intermingling of such traces is expected because of the diffusive meandering of electrons relative to pure gyrohelixes in uniform magnetic fields. The benchmark values for a uniform background field, $\mathbf{B}_0$ are indicated with the horizontal red lines with double arrowheads. For the chosen index, the polarization degree $\Pi = \sqrt{Q^2 + U^2}/I$ realizes a value of $3/4$, and the position angle is $\chi = 0$ for the Stokes parameter convention adopted here. The introduction of turbulence reduces the polarization degree value as expected, with $\Pi(\sigma) \simeq 3/4 - 3\sigma^2/2$. The contributions of Q and U to the deviation from the uniform field value are of the same magnitude, but not identical. While Q fluctuates a little in departing from the baseline $3/4$ value, the statistical scatter in values of U is an unambiguous marker of the turbulence. The average value of U is close to zero, but the standard deviation scales as σ^2. Accordingly, collecting information about the mean and standard deviation of the position angle χ affords a second direct diagnostic on the strength of the turbulence, namely $|\chi| \sim \sigma^2$. This latter result is anticipated: the variance in polarization direction should directly scale with the variance in magnetic field directions. These results obviously are obtained in the domain when the pixel scale is of the order of the turbulence stirring scale. When the stirring scale is much smaller than the polarimeter resolution scale, the laminar structure of the field will emerge in the results, corresponding to $\sigma^2 \ll 1$ domains.

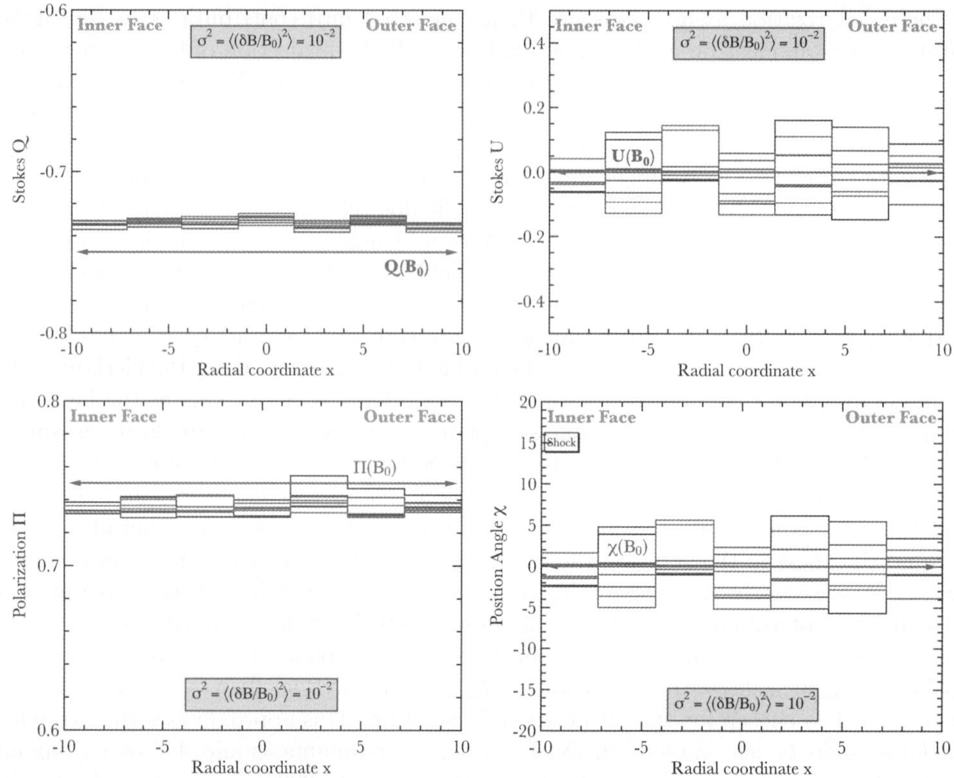

Figure 2. Synchrotron X-ray polarization quantities from power-law electrons $n_e(\gamma) \propto \gamma^{-p}$ with distribution index $p = 3$, for a sequence of slices of the shell that are oriented in the x-direction normal to the planar shock, and whose centers are laterally distributed with equidistant separations along the shock. Each section of a histogram thus represents the polarization measure associated with a rectangular prism segment of a slice, with a square cross section in the plane of Fig. 1. The color coding is to aid distinguishability of results for different slices, with color correspondence through all panels. The field turbulence had a Kolmogorov spectrum above the stirring scale $\Delta x = 20/3$, at which the variance was $\sigma^2 = 10^{-2}$. The Stokes Q and U values are depicted in the upper panels, with circularity V being zero. The polarization degree Π and position angle $\chi = 1/2 \arctan(U/Q)$ are in the bottom two panels. The values of all four quantities for a uniform field $\mathbf{B}_0$ are indicated by the red double-headed long arrows.

These illustrations provide key insights into the coupling between field turbulence and X-ray synchrotron polarization, even though the slab geometry does not precisely model the spatial non-uniformity and curvature of SNR shell environments. The polarization degrees represented in the two Figures are substantially higher than those measured in the radio for SN 1006 in Reynolds & Gilmore (1993) and Reynoso, Hughes & Moffett (2013), suggesting that the variance of field turbulence on the diffusion/gyrational scales of GeV radio-emitting electrons (much smaller than $R_{\rm shell}$) must be on the order of $\sigma^2 \lesssim 1$. Yet, the position angle imagery for the VLA data on SN 1006 indicates a coherent component to the field. The X-ray band samples much larger diffusion scales, on the order of the shell thickness. On these lengthscales, the cosmic-ray driven magnetic field amplification models predict $\delta B/B \sim 1$, for which the results here suggest not only polarization degrees $\Pi < 0.4$, but that the position angle information in the 2–8 keV window should be highly disordered, with the variance in χ providing a direct measure of the variance in the field fluctuations. Moreover, the coupling constant should be dependent on the binning or pixelation size of the data, providing constraints on the stirring scale for

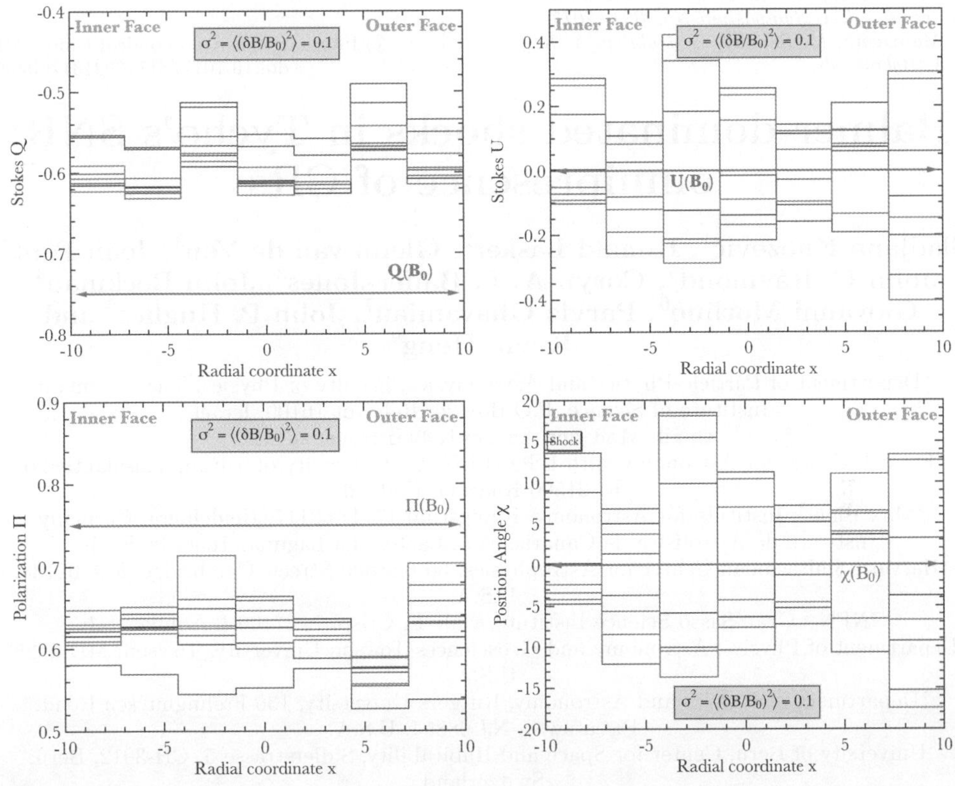

Figure 3. Synchrotron X-ray polarization quantities from power-law electrons as in Fig. 2, but here for a Kolmogorov spectrum with a higher variance of $\sigma^2 = 10^{-1}$ at the stirring scale. The enhanced depolarization and greater standard deviation in the position angle are evident.

the turbulence. We note that if IXPE measures polarization levels in excess of 30–40%, theorists will have to revise the paradigm of turbulence generation in SNR shocks.

These brief insights elicit an exciting indication of the prospects for advances in our understanding of MHD turbulence in supernova remnant shells in the coming era of imaging X-ray polarimetry.

References

Bell, A. R. 2005, *MNRAS*, 358, 181

Bykov, A. M., Uvarov, Yu. A., Bloemen, J. B. G. M., *et al.* 2009, *MNRAS*, 399, 1119

Giacalone, J. & Jokipii, J. R. 1999, *Astrophys. J.*, 520, 204

Long, K. S., Reynolds, S. P., Raymond, J. C., *et al.* 2003, *Astrophys J.*, 586, 1162

Lucek, S. G. & Bell, A. R. 2000, *MNRAS*, 314, 65

Reynolds, S. P. & Gilmore, D. M. 1993, *Astronomical J*, 106, 272

Reynoso, E. M., Hughes, J. P., & Moffett, D. A. 2013, *Astrophys. J.*, 145:104

Rybicki, G. B. & Lightman, A. P. 1979 *Radiative Processes in Astrophysics* (Wiley, New York)

Soffitta, P., Barcons, X., Bellazzini, R., *et al.* 2013, *Experimental Astronomy*, 36, 523

Vink, J. & Laming, J. M. 2003, *Astrophys. J.*,, 584, 758

Vladimirov, A., Ellison, D. C., & Bykov, A. 2006, *Astrophys. J.*, 652, 1246

Weisskopf, M. C., Baldini, L., Bellazini, R., *et al.* 2013, in *Proc. SPIE*, Vol. 8859, UV, X-Ray, and Gamma-Ray Space Instrumentation for Astronomy XVIII, 885908

Supernova 1987A:30 years later - Cosmic Rays and Nuclei from Supernovae and their aftermaths
Proceedings IAU Symposium No. 331, 2017
A. Marcowith, M. Renaud, G. Dubner, A. Ray & A. Bykov, eds.

doi:10.1017/S1743921317004550

Balmer-dominated shocks in Tycho's SNR: omnipresence of CRs

Sladjana Knežević[1], Ronald Läsker[2], Glenn van de Ven[3], Joan Font[4], John C. Raymond[5], Coryn A. L. Bailer-Jones[3], John Beckman[4], Giovanni Morlino[6], Parviz Ghavamian[7], John P. Hughes[8] and Kevin Heng[9]

[1]Department of Particle Physics and Astrophysics, Faculty of Physics, The Weizmann Institute of Science, P.O.Box 26, Rehovot 76100, Israel
email: `sladjana.knezevic@weizmann.ac.il`
[2]Finnish Centre for Astronomy with ESO (FINCA), University of Turku, Väisäläntie 20, FI-21500 Kaarina, Finland
[3]Max Planck Institute for Astronomy, Königstuhl 17, D-69117, Heidelberg, Germany
[4]Instituto de Astrofísica de Canarias, Vía Láctea, La Laguna, Tenerife, Spain
[5]Harvard-Smithsonian Center for Astrophysics, 60 Garden Street, Cambridge, MA 02138, U.S.A.
[6]INFN - Gran Sasso Science Institute, viale F. Crispi 7, 67100 L'Aquila, Italy
[7]Department of Physics, Astronomy and Geosciences Towson University, Towson, MD 21252, U.S.A.
[8]Department of Physics and Astronomy, Rutgers University, 136 Frelinghuysen Road, Piscataway, NJ 08854, U.S.A.
[9]University of Bern, Center for Space and Habitability, Sidlerstrasse 5, CH-3012, Bern, Switzerland

Abstract. We present wide-field, spatially and highly resolved spectroscopic observations of Balmer filaments in the northeastern rim of Tycho's supernova remnant in order to investigate the signal of cosmic-ray (CR) acceleration. The spectra of Balmer-dominated shocks (BDSs) have characteristic narrow (FWHM $\sim 10\,\mathrm{km\,s^{-1}}$) and broad (FWHM $\sim 1000\,\mathrm{km\,s^{-1}}$) H$\alpha$ components. CRs affect the Hα-line parameters: heating the cold neutrals in the interstellar medium results in broadening of the narrow Hα-line width beyond $20\,\mathrm{km\,s^{-1}}$, but also in reduction of the broad Hα-line width due to energy being removed from the protons in the post-shock region. For the first time we show that the width of the narrow Hα line, much larger than $20\,\mathrm{km\,s^{-1}}$, is not a resolution or geometric effect nor a spurious result of a neglected intermediate (FWHM $\sim 100\,\mathrm{km\,s^{-1}}$) component resulting from hydrogen atoms undergoing charge exchange with warm protons in the broad-neutral precursor. Moreover, we show that a narrow line width $\gg 20\,\mathrm{km\,s^{-1}}$ extends across the entire NE rim, implying CR acceleration is ubiquitous, and making it possible to relate its strength to locally varying shock conditions. Finally, we find several locations along the rim, where spectra are significantly better explained (based on Bayesian evidence) by inclusion of the intermediate component, with a width of $180\,\mathrm{km\,s^{-1}}$ on average.

Keywords. Tycho's SNR, Balmer-dominated shocks, CR precursor, broad-neutral precursor

1. Introduction

Balmer-dominated shocks (BDSs) are observed around many supernova remnants (SNRs) as bright edge-on sheets of shocked gas. BDSs are non-radiative, collisionless and relatively high velocity ($\gtrsim 200\,\mathrm{km\,s^{-1}}$) shocks interacting with dilute ($\sim 1\,\mathrm{cm^{-3}}$) partially ionized medium (Heng 2010). Spectroscopically BDSs exhibit bright Balmer lines, in particular narrow (NL) and broad (BL) Hα components (Chevalier *et al.* 1980).

Both components originate downstream of the shock in radiative decays of either pre-shock hydrogen atoms that have been excited in the shock or have undergone charge exchange (CE) reactions with the post-shock protons. The latter process produces a so-called broad-neutrals. In an idealized scenario, the pre-shock gas is not affected (e.g. temperature, ionization state) by the oncoming shock until the latter arrives. However, emission from the post-shock gas can influence the physical conditions in the pre-shock region. This phenomenon is termed a precursor. The shocked gas can be a source of photons, particles and waves that can overtake the shock and heat the pre-shock gas. Mach number, ambient density and orientation of the magnetic field define the precursor type. Its existence can affect the lines' widths, intensities and even their shapes.

Balmer filaments are very faint ($\sim 10^{-16}\,\mathrm{erg\,s^{-1}\,cm^{-2}\,arcsec^{-2}}$). This requires an instrument with a very good efficiency. Furthermore, they are usually extended ($\sim 1'$) and have complex structures, thus we need a large field-of-view (FOV) to cover them in their entirety and a high spatial resolution is desirable to analyze individual structures of the shock. Finally, to be able to resolve both components, i.e. $\sim 10\,\mathrm{km\,s^{-1}}$ NL width and $\sim 1000\,\mathrm{km\,s^{-1}}$ BL width, a high spectral resolution and a large-enough spectral coverage to include all of the BL are needed. At the moment, there is no a single instrument that satisfies all these requirements.

Tycho's SNR is one of remnants famous for its prominent Balmer filaments. They have been imaged at high spatial resolution with the Hubble Space Telescope (Lee *et al.* 2010). However, spectroscopically only a few bright, small regions in these filaments were observed, and remained spatially unresolved (Ghavamian *et al.* (2000), Lee *et al.* (2007)). In this paper, we will present spatially resolved observations of the northeastern (NE) filaments that cover and spectrally resolve the narrow component. As an additional step forward, our analysis employs Bayesian inference to reliably quantify confidence intervals and to compare models by means of the evidence for possibly multiple narrow- and intermediate-line (IL width $\sim 100\,\mathrm{km\,s^{-1}}$) components.

2. Data and Analysis

With the FOV of $3.4'\times 3.4'$ and the pixel scale of $0.2''$ of the Galactic Hα Fabry-Pérot Spectrometer (GHαFaS) on the William Herschel Telescope (WHT) we covered the entire Tycho's NE rim and spatially resolved its Balmer filaments. The GHαFaS spectral coverage of $392\,\mathrm{km\,s^{-1}}$ and spectral dispersion of $8.1\,\mathrm{km\,s^{-1}}$ (the instrument profile is Gaussian, see Blasco-Herrera *et al.* 2010) precluded simultaneous analysis of the broad component. We modified the standard procedure for GHαFaS data reduction described by Hernandez (2008) to reduce the $\approx 9.6\,\mathrm{h}$ observations. We carried out wavelength and phase calibration for each individual exposure, from which we obtained individual data-subcubes. We did not use the optical derotator so that we could use the largest possible GHαFaS FOV. Therefore, before co-adding the data-subcubes, we aligned and derotated them. After co-addition of the subcubes, we obtained a final datacube of 48 calibrated constant-wavelength channels. Following the same procedure, we created a background and a flatfield cubes. We modeled the background flux and flatfield image in individual exposures, and subsequently processed the individual background and flatfield frames in exactly the same manner as the corresponding data frames.

The seeing of $1''$ sets the lower size of the binned area that we want to analyze to 19 pixels. Following this requirement, we binned pixels of the NE filaments to get 82 Voronoi spatial-spectral bins (Cappellari & Copin 2003) with nearly equal signal-to-noise (S/N) of 10. Furthermore, we excluded bins that cover area larger than 400 pixels, so that the estimated $\approx 2\%$ residual background variations do not significantly affect our measurements.

Our goal is to find a model that we define as $M(\theta) = S(\theta) \times F + B$, where S characterize the shock emission, F flatfield, B background, and its parameters θ that best describe the data. We model the shock emission with a Gaussian profile to account for a NL, while in our small spectral window, the BL is sufficiently well approximated by a constant component. Presence of a precursor introduces either a wider NL and/or a split in the NL, or an IL (we assume that all lines have a Gaussian profile), depending on a precursor's type. Heating and momentum transfer in a cosmic-ray (CR) precursor result in a NL component wider than the maximal thermal width of $20\,\mathrm{km\,s^{-1}}$, but also in a split in the NL if we have two shocks inclined with the respect to the line-of-sight (LOS). CE in a broad-neutral (BN) precursor would produce an IL. BN precursor, created by fast BNs that overtake the shock, is expected in shocks with velocities of $2500\,\mathrm{km\,s^{-1}}$ (as in the NE rim of Tycho; Ghavamian *et al.* 2001) that propagate in a partially neutral medium, and formation of a CR precursor is expected if the acceleration process in the shock is efficient.

To account for possible NL split or IL presence, we apply several parametrized models to each data set (bin and spectrum): two single-NL models (NL and NLIL), and two double-NL models (NLNL and NLNLIL). To find which of the four models and their parameters best describe the data, we perform Bayesian analysis. Bayes' theorem defines the *posterior* probability density function (PDF) as a function of model M's parameters θ: $P(\theta|D, M) = P(D|\theta, M)P(\theta|M)/P(D|M)$, where $P(\theta|M)$ is the *prior* – parameter PDF that we either assume or know before taking data D into account, $P(D|\theta, M)$ is the *likelihood* which is the probability of the data for a given model and parameter set, and $P(D|M)$ is the *evidence* for model M (the posterior marginalized over all parameters). Our detector follows a Poisson distribution which we defined to be the likelihood. We use Markov chain Monte Carlo (MCMC) to draw samples from the posterior and use an ensemble sampling method, see Foreman-Mackey *et al.* (2013) and Goodman *et al.* (2010) for details on the sampling algorithm. The evidence integral is often computationally expensive, the reason we approximated it with the *leave-one-out cross validation* (LOO-CV) likelihood (Bailer-Jones 2012). The base-10 logarithmic evidence ratios are compared and this way we identify the favored model. Although we cannot exclude any of the models, we favor the model with the highest evidence having log evidence ratio relative to the NL model larger than 0.5 dex (factor ≈ 3). If there is no model that is at least 3 times more likely than the NL model, we take the NL model as the favored one.

3. Results

We summarize the results in three ways: i) For every Voronoi bin and model we calculated posterior parameter distributions, and give their median and the highest-density 95% confidence intervals (later on 95% confidence interval) (Figure 1); ii) For every Voronoi bin we created evidence-weighted 1D-posterior of all models that feature the parameter of interest so that they contribute their marginalized posterior in proportion to their evidence, and calculated their median (Figure 2) and the 95% confidence interval (top row in Figure 3); iii) We consider the parameter's median across bins, sample its posterior (bottom row in Figure 3), and give this posterior distribution's median and confidence intervals as we did for the individual bins' posteriors. This provides us a measure of the global parameter values when considering the filament as a whole. We are particularly interested in the measured NL width, evidence for IL, its strength and width, evidence for a split in the NL, and the separation between the two NLs.

In Figure 1 we present two different bins, one favoring the NLIL model with log evidence ratio larger than 1 dex compared to the NL and NLNL model, and the other favoring

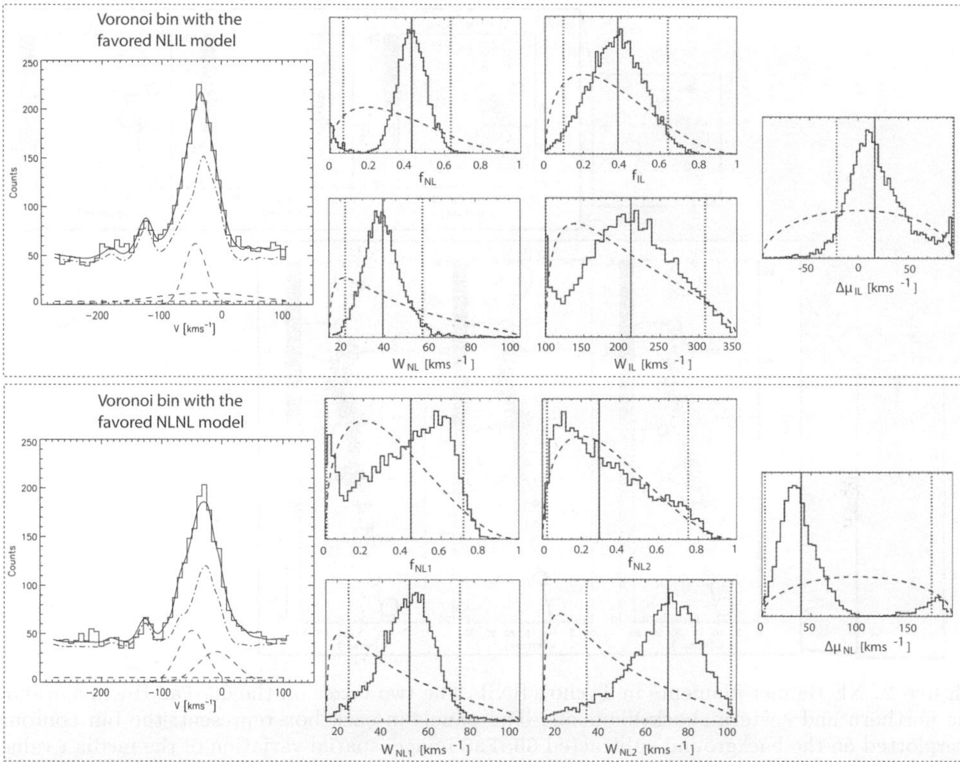

Figure 1. Voronoi bins in the NE filament of Tycho's SNR with the NLIL model (*top row*), and the NLNL model (*bottom row*) being the favored one. Left panels show the data as histograms, background models as dashed-dotted lines, dashed lines are the shock emission components, and the median models M are solid lines. The remaining five panels are marginalized posteriors of individual model parameters (solid lines): flux fractions in the lines ($f_{\rm NL}$, $f_{\rm IL}$, $f_{\rm NL1}$, $f_{\rm NL2}$), intrinsic line widths ($W_{\rm NL}$, $W_{\rm IL}$, $W_{\rm NL1}$, $W_{\rm NL2}$), IL offset from the NL centroid ($\Delta\mu_{\rm IL}$), and the separation between the two NLs ($\Delta\mu_{\rm NL}$). Overplotted dashed lines are prior distributions. Solid vertical lines are medians of the posterior distributions, and the vertical dotted lines are the 95% confidence intervals.

the NLNL model by >1 dex (factor 10) over the NL and >0.5 dex (factor 3) over the NLIL and NLNLIL models. Left panels show the data (histogram) and the median model (solid line), while the remaining panels show priors of chosen parameters (dashed lines), posteriors (solid line), the median (solid-vertical line) and the 95% confidence intervals (dotted-vertical line) of the posteriors. For the NL's full-width at half-maximum (FWHM) $W_{\rm NL}$ we used a prior range of $[15, 100]$ km s^{-1} that reflects the pre-shock temperature of ≈ 5000 K and the maximally predicted $W_{\rm NL}$ based on the Morlino *et al.* (2013) shock model that includes the emission from the CR precursor. We limited the $W_{\rm IL}$ prior to $[100, 350]$ km s^{-1} which is expected for the shock velocities around 2000 km s^{-1} (Morlino *et al.* 2012). We do not strongly prefer any parameter values within the model definition limits, which we ensured by using Dirichlet and Beta prior distributions with $\alpha(=\beta)$ $= 1.5$ shape parameter for the model parameters (flux fractions in the lines, log-line widths, NL centroid, NL-centroid separation and IL offset from the NL centroid). For the bin with the favored NLIL model we find $W_{\rm NL} \approx 40$ km s^{-1}, and an IL component with $W_{\rm IL} \approx 210$ km s^{-1} and $\approx 40\%$ of the total flux. The intrinsic widths of the two NLs in the bin with the favored NLNL model are also much larger than 20 km s^{-1}, nearly 52 km s^{-1} and 70 km s^{-1}, separated by 40 km s^{-1}.

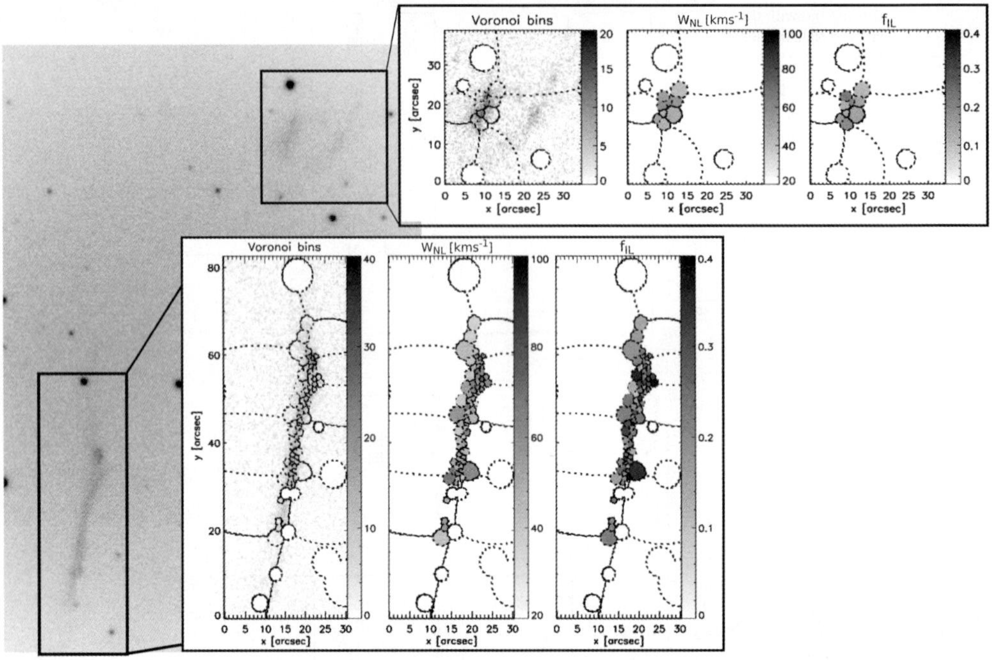

Figure 2. NE Balmer filaments in Tycho's SNR. The two boxes on the GHαFaS Hα image show the northern and eastern shock filaments. The panels in each box represent: the bin contours overplotted on the background subtracted GHαFaS image, spatial variation of the median values of evidence-weighted NL width (in km s^{-1}) and IL flux fraction posteriors.

In Figure 2 we present the Tycho's NE Balmer filaments that we observed and spatially resolved with GHαFaS, and show spatial variation of the median NL width ($W_{\rm NL}$ $\in$ [35, 72] km s^{-1}) and IL flux fraction ($f_{\rm IL} \in [0, 0.42]$), computed as described in ii). The distribution of all medians and their 95% confidence intervals are given in the top row in Figure 3 as solid and dotted histograms, respectively. In 24% of the bins we find a significant evidence for IL, while 18% of the bins have significant Bayesian evidence for a split in the NL. As expected, posteriors of the global (cross-bin median) parameters are significantly narrower than the constraints provided from an individual bin alone (bottom row in Figure 3). We found the median and the 95% confidence intervals of the cross-bin median posteriors on the following parameters: $W_{\rm NL} = (54.8 \pm 1.8)\,{\rm km\,s^{-1}}$, $f_{\rm IL}/f_{\rm NL} = (0.41 \pm 0.07)$, $W_{\rm IL} = (180.5 \pm 14.3)\,{\rm km\,s^{-1}}$, and $\Delta\mu_{\rm NL} = (38.5 \pm 5.1)\,{\rm km\,s^{-1}}$.

4. Conclusions and Summary

We presented observations of the Tycho's Balmer filaments in the NE rim, where for the first time, we spatially resolved the filaments, covered them in their entirety, and spectroscopically resolved the narrow Hα component. We also improved the analysis by applying Bayesian inference to obtain reliable parameter estimates and uncertainties, and to quantify the evidence for models with multiple line components.

Our results show that the $W_{\rm NL}$ is much larger than $20\,{\rm km\,s^{-1}}$ in the entire NE rim even when the NL is split and modeled accordingly. We find that 18% of the bins show significant evidence for double-NL models. These findings can currently only be explained by efficient CR acceleration in the NE rim confirming previous results in the Tycho's 'knot g' by Ghavamian *et al.* (2000) and Lee *et al.* (2007). The heating in the precursor broadens the NL widths to $\approx 55\,{\rm km\,s^{-1}}$, and the momentum transfer introduces two NLs

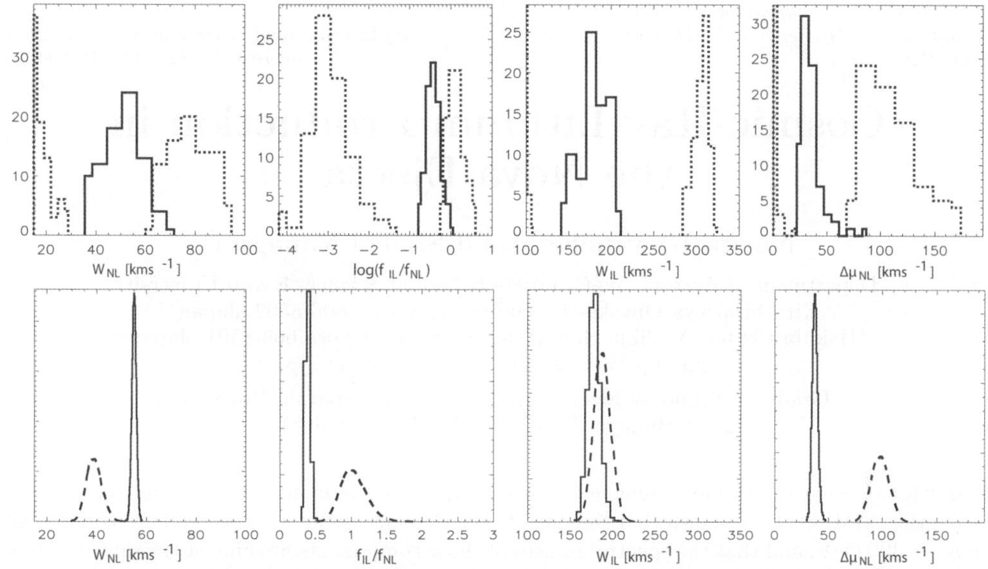

Figure 3. *Top row*: Distribution of median (solid) and highest-density 95%-confidence interval boundaries (dotted), quantifying the variation of narrow-line width ($W_{\rm NL}$, left panel), intermediate-to-narrow line flux fraction ($f_{\rm IL}/f_{\rm NL}$, centre-left), IL width ($W_{\rm IL}$, centre-right) and NL-centroid separation ($\Delta\mu_{\rm NL}$, right-most panel) across the filaments. *Bottom*: Posterior of the median across all bins (solid) created using all data to constrain the respective parameter values. Prior of the cross-bin median is overplotted with a dashed line.

in the observed projected shock emission separated by $\approx 38\,{\rm km\,s^{-1}}$ on average. As the amount of neutrals in the ambient medium around the remnant changes, we expect the NL width to vary across the filaments as we see in Figure 2. This effect occurs because the pre-shock gas is heated to higher temperature (larger $W_{\rm NL}$) if the medium has more neutrals as the ion-neutral damping of magnetic waves excited by CRs is more efficient.

Finally, significant evidence for an IL presence is measured in 24% of the bins, with $W_{\rm IL} \approx 180\,{\rm km\,s^{-1}}$ and $f_{\rm IL}/f_{\rm NL} \approx 0.41$. Detection of the IL indicates presence of a BN precursor. Therefore, our observations and analysis reveal the interplay of shock precursors in the NE rim of Tycho.

References

Bailer-Jones, C. A. L. 2012, *A&A*, 546, A89

Blasco-Herrera, J., Fathi, K., Beckman, J., *et al.*, 2010, 407, 2519

Cappellari, M. & Copin, Y. 2003, *MNRAS*, 342, 345

Chevalier, R. A., Kirshner, R. P., & Raymond, J. C. 1980, *ApJ*, 235, 186

Foreman-Mackey, D., Hogg, D. W., Lang, D., & Goodman, J. 2013, *PASP*, 125, 306

Ghavamian, P., Raymond, J. C., Hartigan, P., & Blair, W. P. 2000, *ApJ*, 535, 266

Ghavamian, P., Raymond, J. C., Smith, R. C., & Hartigan, P. 2001, *ApJ*, 547, 995

Goodman, J. & Weare, J., 2010, *Comm. App. Math. Comp. Sci.*, 5, 65

Heng, K. 2010, *PASA*, 27, 23

Hernandez, O., Fathi, K., Carignan, C., *et al.*, 2008, *PASP*, 120, 665

Lee, J-.J., Koo, B.-C., Raymond, J. C., *et al.*, 2007, *ApJ*, 659, L133

Lee, J. J., Raymond, J. C., Park, S., & Blair, W. P. 2010, *ApJ*, 715, L146

Morlino, G., Bandiera, R., Blasi, P., & Amato, E., 2012, *ApJ*, 760, 137

Morlino, G., Blasi, P., Bandiera, R., Amato, E., & Caprioli, D. 2013, *ApJ*, 100, 50

Supernova 1987A:30 years later - Cosmic Rays and Nuclei from Supernovae and their aftermaths
Proceedings IAU Symposium No. 331, 2017
A. Marcowith, M. Renaud, G. Dubner, A. Ray & A. Bykov, eds.

doi:10.1017/S1743921317004616

Cosmic-Ray Lithium Production in the Nova Ejecta

Norita Kawanaka[1,2] **and Shohei Yanagita**[3]

[1]Department of Astronomy, Graduate School of Science, Kyoto University, Kitashirakawa Oiwake-cho, Sakyo-ku Kyoto, 606-8502, Japan
[2]Hakubi Center, Yoshida Honmachi, Sakyo-ku, Kyoto 606-8501, Japan
email: `norita@kusastro.kyoto-u.ac.jp`

[3]Graduate School of Science and Engineering, Ibaraki University, 2-1-1 Bunkyo, Mito, Ibaraki 310-8512, Japan

Abstract. Recent direct measurements of cosmic-ray (CR) light nuclei (protons, helium, and lithium) by AMS-02 have shown that the flux of each element has an unexpected hard component above $\sim$ 300 GeV, and that the spectral indices of those components are almost the same ($\sim$ 2.5). This implies that there should be primary sources that produces CR lithium nuclei, which have been believed to be produced via spallation of heavier nuclei in the ISM (secondary origin). We propose the nearby Type Ia supernova following a nova eruption from a white dwarf as the origin of CR Li.

Keywords. supernovae: general – novae, cataclysmic variables – acceleration of particles – supernova remnants – cosmic rays

1. Introduction

Cosmic rays (CRs) below the knee ($\sim 10^{15.5}$ eV) are believed to be produced at the shock waves of Galactic supernova remnants (SNRs). Actually, γ ray emission detected from some SNRs interacting with molecular clouds seem to originate from hadronic interactions between high energy protons and neutral particles in molecular clouds, which implies that high energy CR particles are accelerated at these SNRs (Ackermann *et al.* 2013). On the other hand, recent experiments by PAMELA, CREAM, and AMS-02 have revealed the detailed CR spectra from nuclei to nuclei in the GeV-TeV range (Adriani *et al.* 2011; Ahn *et al.* 2010; Aguilar *et al.* 2015a; Aguilar *et al.* 2015b; Aguilar *et al.* 2016). They clearly show that a single power-law does not fit the data of CR protons and helium, and that above $\sim$ 300 GeV CR spectral indices become harder by $\simeq 0.13$. In addition, AMS-02 has recently reported that CR lithium nuclei also show such spectral hardening above $\sim$ 300 GeV, and that the hard component of CR lithium has the similar spectral index as that of CR protons and helium (Ting 2016). Since CR lithium nuclei are usually considered to be secondary particles (i.e. not produced at sources but produced via spallation in the interstellar medium), their spectrum should be softer than that of primary particles such as protons and helium. Therefore, the results of AMS-02 imply that there are CR sources that accelerate CR lithium primarily as well as protons and helium nuclei.

Recently, Tajitsu *et al.* (2015) reported on the detection of ^{7}Be in the post-outburst spectra of the classical nova V339 Del. Since ^{7}Be would decay into ^{7}Li with half-life of $\sim$ 53 days, this can be regarded as a direct observational evidence that lithium nuclei can be produced in nova ejecta. We propose that the CR lithium can be produced at a Type

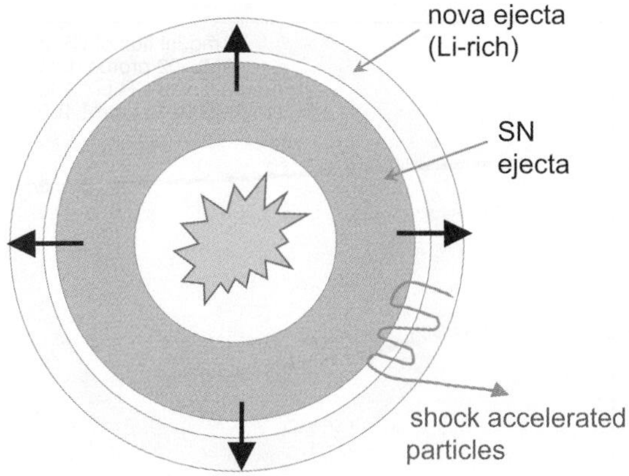

Figure 1. Schematic picture of a Type Ia supernova occurring after a nova eruption from a white dwarf. Since the nova ejecta is Li-rich, the subsequent supernova blast wave can accelerate a significant amount of lithium nuclei.

Ia supernova occurring after a nova eruption from a white dwarf (Fig. 1). In the context of the single degenerate scenario, the progenitor of a Type Ia supernova is a white dwarf accreting gas from its companion star. Therefore, it is possible that a Type Ia supernova occurs after a nova eruption, which can also be driven by the mass accretion onto a white dwarf. The blast wave of the supernova explosion would sweep the nova ejecta in the circumstellar space, making a shock, and producing high energy CRs via 1st order Fermi acceleration. Since the nova ejecta are Li-rich, we can expect that a significant amount of Li nuclei are accelerated as primary CRs.

2. Model

We assume that the extra components of CR particles are injected instantaneously into the interstellar medium at a single Type Ia supernova occurring at a distance r from the Earth. Then the distribution function of CR particles (i = proton, He, ...) at the distance r from a source and the time of t per unit energy bin can be described as

$$f_i(r, \epsilon, t) = \frac{Q_{i,0}(\epsilon)}{(4\pi Dt)^{3/2}} \exp\left(-\frac{r^2}{4Dt}\right), \tag{2.1}$$

whre ϵ is the energy of CR particles, $D(\epsilon)$ is the diffusion coefficient in the interstellar medium, and $Q_{i,0}$ is the energy spectrum of CRs injected from a source. Assuming the functional form of the diffusion coefficient as $D = D_0(\epsilon/1\ \mathrm{GeV})^\delta$ where D_0 and δ are constants, Eq. (2.1) can be approximated by a power-law with index $\alpha + (3/2)\delta$ in the high energy limit, where α is the spectral index at the source. Recent results of AMS-02 implies that the index of the diffusion coefficient is $\delta \simeq 1/3$ (Aguilar *et al.* 2016). Using these assumptions, we can fit the extra CR components with free parameters of r, t, α, and the total mass of CRs injected from a source. Note that, in order to fit the extra component by the function (2.1) without a spectral break, the peak energy of this function should be below $\lesssim 300$ GeV, above which the background flux becomes subdominant.

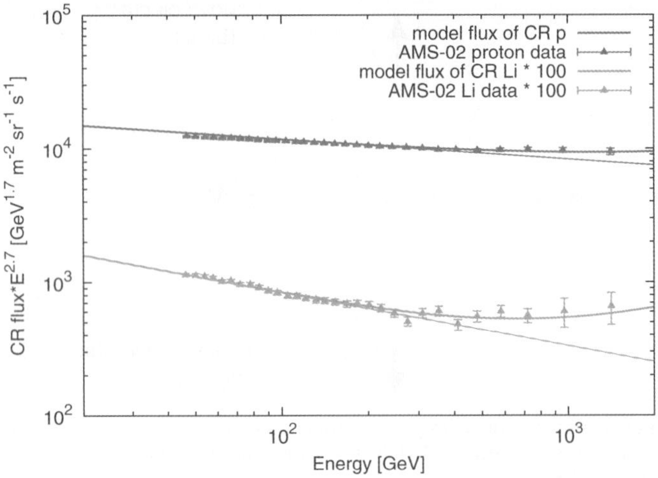

Figure 2. Comparison between CR protons and lithium nuclei spectra derived from our modeling and the data from AMS-02.

3. Results and Discussion

Fig 2 depicts the results of the spectral fitting of the extra flux components of CR protons and lithium nuclei provided by AMS-02 with our model. In this fitting, we assume the distance of the source (i.e. the Type Ia supernova occurring after a nova eruption), the total mass of CR protons, that of CR lithium, and the total energy of CRs as 400 pc, $10^{-6} M_{\rm odot}$, $2 \times 10^{-9} M_\odot$, and 2.0×10^{50} erg, respectively. Here we assume a diffusion coefficient at 1 GeV as 4×10^{28} cm^2 s^{-1}.

Let us discuss the relevance of the parameters adopted in the fitting presented above. First, the total CR energy, $\simeq 2 \times 10^{50}$ erg, is similar to the value that is often assumed as the CR energy available from a single supernova. Second, the total CR mass is about $\sim 1\%$ of the typical mass of a nova ejecta, $\sim 10^{-4} M_\odot$. Actually, if the nova ejecta is hot enough ($\sim 10^4$ K), the acceleration efficiency of the 1st-order Fermi process would be up to $\sim 1\%$ (Dogiel *et al.* 2017). Third, the mass ratio of CR Li to CR protons inferred from our fitting is $\sim 2 \times 10^{-3}$, which is ~ 20 times larger than that inferred from the observation of novae from which ^{7}Be lines are detected. However, the acceleration efficiency might be different from element to element. For example, Meyer (1985) shows how the overabundance of elements in Galactic cosmic rays with respect to local interstellar medium depends on the first ionization potential (FIP). According to his results, the abundance of sodium, whose FIP is ~ 5 eV, in Galactic cosmic rays is similar to that in the interstellar medium, while the abundance of hydrogen, whose FIP is ~ 13 eV, is a factor ~ 0.03 smaller than that in the interstellar medium. Since the FIP of lithium is about ~ 5 eV, the abundance ratio of lithium to hydrogen in cosmic rays would be enhanced by a factor of a few tens from that in the nova ejecta. In this sense, the abundance ratio in the extra component inferred from our fitting is relevant.

4. Summary and Future Work

From the fitting and discussion above, we can conclude that the unexpected CR flux component found in the recent AMS-02 data can be explained by assuming a nearby Type Ia supernova occurring after a nova as a CR source. In fact, the total energy, mass, and abundance ratio in those components derived from our fit are consistent with this model. In a nova ejecta, carbon nuclei are also efficiently synthesized and then can be accelerated by a subsequent supernova blast wave. According to the recent AMS-02 data, the ratio of CR carbon to CR helium is almost constant with respect to energy above ~ 300 GeV, which implies that CR carbon spectrum also has an extra hard component. In addition, since stable boron and beryllium are not synthesized in a nova, our model does not predict the enhancement of these nuclei in CRs, and actually one cannot see any spectral hardening of these elements in the recent CR measurement reported by AMS-02.

Our scenario will be supported or ruled out by more precise measurements of CR lithium, boron, and beryllium nuclei in near future. In addition, our hypothetical supernova remnant is quite close to the Earth, the CR flux from it would affect on the dipole anisotropy of CRs above ~ 300 GeV, which would be also measured precisely in future experiments.

References

Ackermann, M., Ajello, M., Allafort, A., *et al.* 2013, *Science*, 339, 807

Adriani, O., Barbarino, G. C., Bazilevskaya, G. A., *et al.* 2011, *Science*, 332, 69

Ahn, H. S., Allison, P. S., Bagliesi, M. G., *et al.* 2010, *ApJ*, 715, 1400

Aguilar, M., Aisa, D., Alpat, B., *et al.* 2015, *Physical Review Letters*, 114, 171103

Aguilar, M., Aisa, D., Alpat, B., *et al.* 2015, *Physical Review Letters*, 115, 211101

Aguilar, M., Ali Cavasonza, L., Ambrosi, G., *et al.* 2016, *Physical Review Letters*, 117, 231102

Ting, S. 2016, CERN Colloquium *The First Five Years of the Alpha Magnetic Spectrometer on the International Space Station*, https://indico.cern.ch/event/592392/

Tajitsu, A., Sadakane, K., Naito, H., *et al.* 2016, *ApJ*, 818, 191

Aguilar, M., Ali Cavasonza, L., Ambrosi, G., *et al.* 2016, *Phys. Rev. Lett.*, 117, 231102

Dogiel, V. A., Cheng, K. S., Chernyshov, D. O., *et al.* 2017, arXiv:1701.05481

Meyer, J.-P. 1985, *ApJS*, 57, 173

Supernova 1987A:30 years later - Cosmic Rays and Nuclei from Supernovae and their aftermaths
Proceedings IAU Symposium No. 331, 2017
A. Marcowith, M. Renaud, G. Dubner, A. Ray & A. Bykov, eds.

doi:10.1017/S1743921317004380

Bridging the gap between supernovae and their remnants through multi-dimensional hydrodynamic modeling

S. Orlando[1], M. Miceli[2,1] and O. Petruk[1,3]

[1]INAF-Osservatorio Astronomico di Palermo, Piazza del Parlamento 1, 90134 Palermo, Italy
[2]Dip. di Fisica e Chimica, Univ. di Palermo, Piazza del Parlamento 1, 90134 Palermo, Italy
[3]Inst. Appl. Probl. in Mechanics and Mathematics, Naukova Street, 3-b Lviv 79060, Ukraine

Abstract. Supernova remnants (SNRs) are diffuse extended sources characterized by a complex morphology and a non-uniform distribution of ejecta. Such a morphology reflects pristine structures and features of the progenitor supernova (SN) and the early interaction of the SN blast wave with the inhomogeneous circumstellar medium (CSM). Deciphering the observations of SNRs might open the possibility to investigate the physical properties of both the interacting ejecta and the shocked CSM. This requires accurate numerical models which describe the evolution from the SN explosion to the remnant development and which connect the emission properties of the remnants to the progenitor SNe. Here we show how multi-dimensional SN-SNR hydrodynamic models have been very effective in deciphering observations of SNR Cassiopeia A and SN 1987A, thus unveiling the structure of ejecta in the immediate aftermath of the SN explosion and constraining the 3D pre-supernova structure and geometry of the environment surrounding the progenitor SN.

Keywords. hydrodynamics - instabilities - shock waves - ISM: supernova remnants

1. Introduction

Supernova remnants (SNRs) are probably among the most beautiful objects which is possible to observe in the sky. Their beauty is mostly due to the complex morphology and highly non-uniform distribution of ejecta which characterize most of them. In fact the propagation of ejecta drives shocks back and forth in the interstellar medium (ISM) and through ejecta themselves, leading to features and complex structures virtually at all spatial scales. The physics governing SNRs is studied thanks to observations at all wavelength bands, from radio, to optical, to X- and γ-rays. Multi-wavelength observations allow us to study both the thermal and non-thermal emission arising from SNRs. The physical and chemical properties of the emitting plasma are inferred through the analysis of both low and high resolution spectra. The observations allow us to investigate the morphology and the projected distributions of properties in the remnants and to study the dynamics, energetics, and evolution of SNRs. Besides the possibility to investigate the physical processes at work in SNRs, there is a large consensus that the observations encode information also on the progenitor star and the SN explosion from which the remnant originates. Obvious questions are: What information might we hope to obtain by deciphering the observations? How can we decipher the observations?

In fact, the morphological properties of SNRs may reflect pristine structures and features of the progenitor SN explosion and the physical and chemical properties of the progenitor SN. Thus, investigating the intimate link that exists between the morphological properties of a SNR and the complex phases in the SN explosion may allow us to trace back the structure and chemical composition of SN ejecta, and the dynamics and energetics of the SN explosion. An example is the SNR Cassiopeia A (Cas A) whose complex morphology is mainly due to a highly inhomogeneous distribution of ejecta determined most likely in the immediate aftermath of the SN explosion. In the case of young remnants, the morphological properties of SNRs may also reflect the early interaction of the SN blast with the inhomogeneous circumstellar medium (CSM) formed during the latest stages of the stellar progenitors evolution. In this case, analyzing the morphology of a SNR may help us to probe the structure and geometry of the CSM immediately surrounding the SN, providing important clues on the final stages of stellar evolution. An example is the remnant of SN 1987A whose morphology reflects the interaction of the blast wave with the inhomogeneous CSM consisting mainly of an H II region and a dense equatorial ring.

All the above considerations point out that linking SNR morphology to their SN progenitors is a necessary step: 1) to probe the physics of SN engines by providing insight into the asymmetries that occur during the SN explosion, and 2) to investigate the final stages of stellar evolution by unveiling the structure of the medium immediately surrounding the progenitor star. This step is now essential to open new exploring windows on SN and SNR issues especially in view of the spatially resolved high-resolution spectroscopy capability of the forthcoming Athena satellite (Nandra *et al.* 2013). In the following, we review our recent achievements in this field.

2. The strategy to link supernovae to supernova remnants

How to link SNe to SNRs? In general existing hydrodynamic and magnetohydrodynamic models have described either the SN evolution or the expansion of the remnant. Former models describe the complex phases of the SN evolution and they do not follow its subsequent transition to the phase of SNR and the interaction between the SN blast wave and the CSM. Conversely, models of SNRs describe the interaction of the remnant with the environment; however they usually assume an initial parameterized ejecta profile, typically few years after the SN event, and leave out an accurate description of the ejecta structure and chemical stratification soon after the SN explosion. Unfortunately this sharp distinction between these two classes of models and the poor communication between the community studying the physics of SNe and that of SNRs have prevented so far to disentangle the effects of the initial conditions (namely the SN event) from those of the boundary conditions (namely the interaction of the SN blast wave with the environment).

In fact there are several criticalities in describing the whole phenomenon from the onset of the supernova to the full remnant development. One serious problem is to catch the very different scales in time and space of SNe and SNRs. This has hampered so far to study the SN-SNR connection in detail. The core collapse of a massive star occurs on time scales of the order of a second and on dimensions of the order of 10000 km (e.g. Janka *et al.* 2016). At the other end, a young SNR evolves on time scales of the order of hundreds of years (if not thousands of years in the case of more evolved remnants) and can be characterized by dimensions of the order of parsecs. In addition the phenomenon is inherently three-dimensional (3D), thus requiring for the modeling a huge amount of numerical resources. This is the reason why the rare attempts to study the SN-SNR connection have been made, in the past, by adopting a one-dimensional (1D) approach (e.g. Badenes *et al.* 2008, Yamaguchi *et al.* 2014, Patnaude *et al.* 2015). However, the 3D nature of SN explosions is well known, being these characterized by strong asymmetries (jets, knots, clumps; e.g. Li *et al.* 1993, Wang *et al.* 2002, Kifonidis *et al.* 2006, Wang & Wheeler 2008, Janka *et al.* 2016); hydrodynamic instabilities easily develop after the core-collapse, producing dense fingers of ejecta (e.g. Janka *et al.* 2016); and turbulence mixing of ejecta is most likely present (e.g. Janka *et al.* 2016). During the remnant expansion, the clumpy structure of the ejecta may induce complex interactions among the clumps and shrapnels of ejecta, enhancing the development of hydrodynamic instabilities, the mixing of initially chemically homogeneous layers, and some overturning of the ejecta due to hydrodynamic instabilities developing during clumps interaction (e.g. Orlando *et al.* 2012, Miceli *et al.* 2013). In addition the ambient medium (CSM or ISM) through which the SN blast wave expands is, in general, highly inhomogeneous and characterized, for instance, by nebulae, rings, clouds, or gradients of density. The ambient magnetic field is, in general, non-uniform; even in the case of a uniform magnetic field the obliquity angle between the magnetic field and the shock normal varies along the remnant outline (e.g. Petruk *et al.* 2009). Hydrodynamic instabilities develop at the contact discontinuity between the stellar debris and the shocked ambient medium and at the border of shocked CSM/ISM inhomogeneities (e.g. Orlando *et al.* 2005, Orlando *et al.* 2008).

In the attempt to study the SN-SNR connection, recently we have devised a strategy which is based on the coupling between a 1D model of core-collapse SN and a 3D model of SNR (Orlando *et al.* 2015, Orlando *et al.* 2016). The SN model describes the evolution of a core-collapse SN from the breakout of the shock wave at the stellar surface to the so-called nebular phase. The model is implemented through a relativistic, radiation-hydrodynamics Lagrangian code (Pumo & Zampieri 2011) which includes a full general relativistic treatment and the radiative transfer coupled to relativistic hydrodynamics at all regimes. The model includes also the gravitational effects of the central compact remnant on the evolution of the ejecta (fall-back of material onto the compact remnant; amount of ejected ^{56}Ni). The 1D SN model provides the initial radial distribution of ejecta for the 3D SNR model. The latter describes the transition from the phase of SN to that of SNR and the interaction of the remnant with the (inhomogeneous) ambient medium. The 3D SNR model is implemented with the FLASH code (Fryxell *et al.* 2000) and includes the effects of radiative losses from optically thin plasma, the deviations from equilibrium of ionization, the back reaction of accelerated cosmic rays, and the deviations from electron-proton temperature equilibration. The evolution of an appropriate post-explosion isotopic composition of the ejecta is accounted by performing multi-species simulations. The model describes the post-explosion structure of the ejecta by small-scale clumping of material and larger-scale anisotropies (see Fig. 1). The evolution of different plasma components (e.g. ejecta, inhomogeneities of CSM, etc.) is traced by considering passive tracers associated with each of them. From the model results, the thermal emission is synthesized in different bands. In this way, the model results can be compared with the observations to obtain a feedback on the SN-SNR models.

A major challenge in our approach is capturing the enormous range in time and space scales. The initial condition for the 3D simulations of the SNR is typically 1 day after the SN event. At that time the size of the ejecta distribution is of the order of tens of AU. Then the remnant evolution is followed for hundreds of years. At the final time, the size of the remnant can be of several pc. To catch this huge range of time and space scales, our approach exploits the adaptive mesh refinement capabilities of the FLASH code. About 20 nested levels of adaptive mesh refinement are typically used to reach an effective resolution of about 0.2 AU ($\approx 3 \times 10^{12}$ cm). This guarantees to have at least 100 computational cells per remnant radius during the whole evolution.

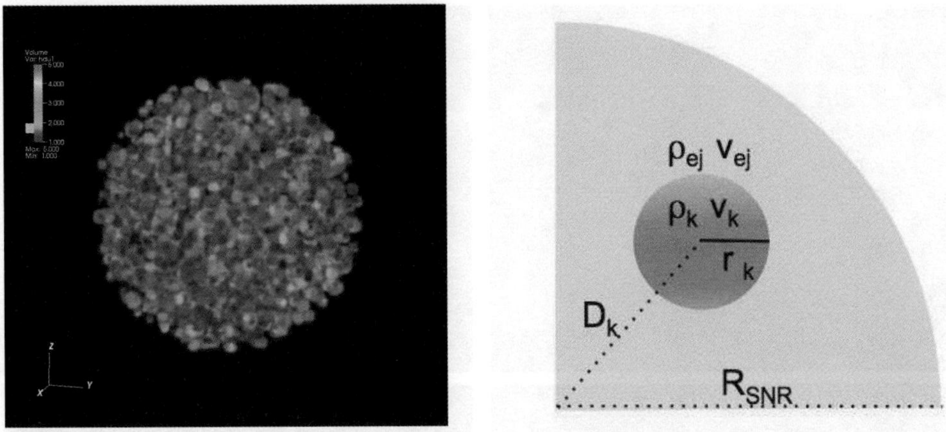

Figure 1. Initial 3D post-explosion structure of the ejecta. The panel on the left shows the small-scale clumping prescribed as in Orlando *et al.* (2012). The panel on the right shows a schematic view of the large-scale anisotropies described as overdense spherical knots characterized by radius, r_k, distance from the center of explosion, D_k, and density and velocity contrasts with respect to the surrounding ejecta, ρ_k/ρ_{ej} and v_k/v_{ej} respectively.

As first applications of our SN-SNR model, we have considered two study cases which are, in some way, complementary each other. The first is the SNR Cas A whose morphology is believed to reflect pristine structures and features developed in the immediate aftermath of the SN explosion. The second case is that of SN 1987A which is characterized by a very complex morphology which reflects the interaction of the SN blast wave with the strongly inhomogeneous CSM. These simulations required about 7.7 millions of CPU hours on the MareNostrum III cluster hosted at the Barcelona Supercomputing Center (Spain) and about 4 millions of CPU hours on the FERMI cluster hosted at CINECA (Italy).

3. Effects of supernova anisotropies: The case of SNR Cassiopeia A

The observations suggest that the morphology and expansion rate of Cas A are consistent with a remnant expanding through the wind of the progenitor red supergiant (Lee *et al.* 2014). Since the structure of the wind is expected to be spherically symmetric with the gas density proportional to r^{-2} (where r is the radial distance from the center of explosion), the complex morphology of the remnant and the highly non-uniform distribution of ejecta should originate from anisotropies developed soon after the SN explosion. For this reason, Cas A is considered a very attractive laboratory to bridge the gap between SNe and their remnants. In addition Cas A is one of the best studied SNR and its 3D structure has been characterized in good detail thanks to observations in the IR, optical, and X-ray bands (e.g. DeLaney *et al.* 2010, Hwang & Laming 2012, Milisavljevic & Fesen 2013). In particular, the observations show that the ejecta distribution is characterized by three extended Fe-rich regions and two Si-rich jets. In addition, rings of Si-, S-, and O-rich ejecta surround the Fe-rich regions (Milisavljevic & Fesen 2013).

Recently we have investigated the origin of the most pronounced features observed in the actual distribution of ejecta in Cas A following the approach outlined in Sect. 2 (Orlando *et al.* 2016). First the 1D SN model was used to provide the initial radial profile of ejecta for the 3D SNR model, about 1 day after the SN event. Then the 3D post-explosion structure of the ejecta was described by small-scale clumping of material and larger-scale anisotropies (Kifonidis *et al.* 2006, Wang & Wheeler 2008). The

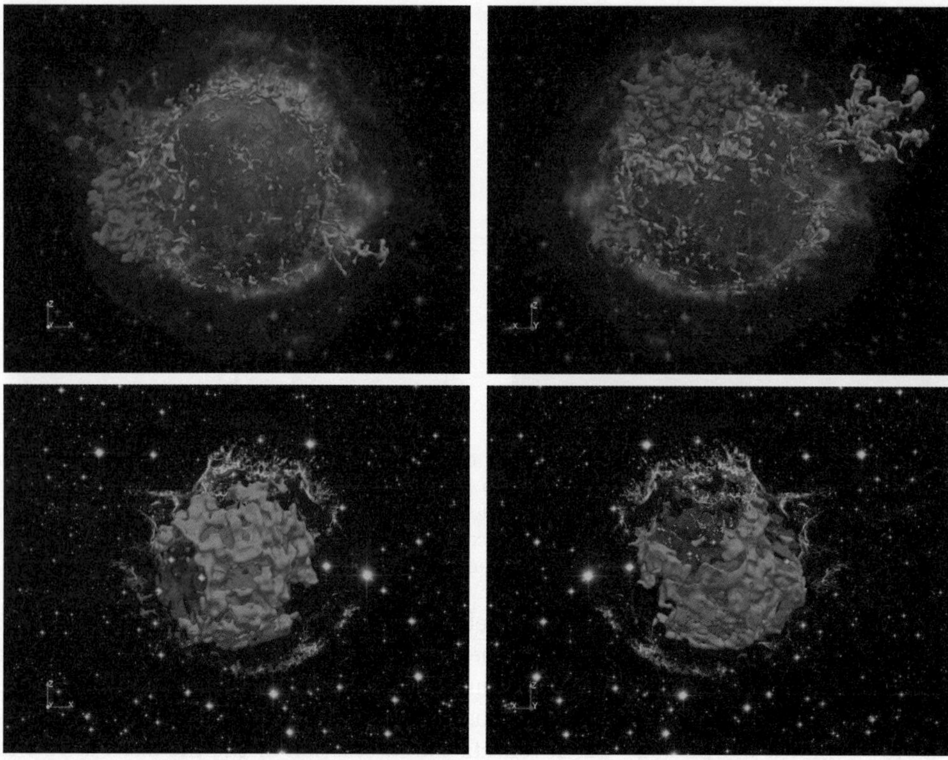

Figure 2. 3D spatial distribution of Cas A shocked (upper panels) and unshocked (lower panels) Fe (blue) and Si/S (green) rich ejecta derived from the best-fit model (adapted from Orlando *et al.* 2016). The figure shows the 3D distributions assuming the perspective in the plane of the sky (on the left) or from behind Cas A (on the right). The transparent red sphere marks the fiducial reverse shock. The transparent image in the upper panels is a Chandra observation showing the hot shocked plasma in the wavelength band [0.3, 10] keV (retrieved from www.nasa.gov); the transparent image in the lower panels is a composite Hubble Space Telescope (HST) image sensitive to emission in cold O and S lines (retrieved from www.spacetelescope.org).

small-scale clumping was prescribed as in Orlando *et al.* (2012, see left panel in Fig. 1). The large-scale anisotropies were described as overdense spherical knots characterized by the following parameters: radius of the knot, distance of the knot from the center of explosion, density and velocity contrasts of the knot with respect to the surrounding ejecta (see right panel in Fig. 1). By exploring the model parameters space, first we have derived the most likely mass of ejecta ($M_{\rm ej} \approx 4 M_{\odot}$) and explosion energy ($E_{\rm SN} \approx 2.3 \times 10^{51}$ erg) which are those able to reproduce simultaneously the radii and velocities of the forward and reverse shocks as observed at the current time (van Veelen *et al.* 2009) and the density of the shocked red supergiant wind inferred from observations (Lee *et al.* 2014). Then we have constrained the physical parameters (namely size, density, velocity, and position in the original onion-skin nucleosynthetic layering) characterizing the anisotropies developed soon after the SN explosion and responsible for the observed masses and distributions of Fe and Si/S in Cas A.

Figure 2 shows the distribution of both shocked and unshocked Fe-rich and Si/S-rich ejecta at the age of Cas A. We have found that post-explosion anisotropies (pistons) reproduce the observed distributions and masses of Fe and Si/S if their total mass was $\approx 0.25\,M_{\odot}$ (corresponding to $\approx 5\%$ of the total mass of ejecta) and their kinetic energy $\approx 1.5 \times 10^{50}$ erg (corresponding to $\approx 7\%$ of the total energy of explosion). The model has shown that the pistons produce a spatial inversion of ejecta layers at the epoch of Cas A, leading to the Si/S-rich ejecta physically interior to the Fe-rich ejecta, in nice agreement with the observations of Cas A (DeLaney *et al.* 2010, Milisavljevic & Fesen 2013). In addition the model explains the origin of the rings of Si/S-rich material encircling the Fe-rich regions (Milisavljevic & Fesen 2013): they form at the intersection between the reverse shock and the material accumulated around the pistons during their propagation. As for the unshocked ejecta, the model has shown that the distribution of Si/S is characterized by large cavities corresponding to the directions of propagation of the pistons/jets. This may explain why the cavities observed in near-IR observations are physically connected to the bright rings in the main-shell (Milisavljevic & Fesen 2015). The model, therefore, supports the idea that the bulk of asymmetries observed in Cas A are intrinsic to the explosion.

It is interesting to note that, very recently, 3D simulations of a neutrino-driven SN explosion were able to reproduce some basic properties of Cas A (Wongwathanarat *et al.* 2016). The initial conditions of this model are given immediately after the core bounce and the simulations follow the evolution until ≈ 1 day after the explosion, namely roughly the time when we have prescribed the post-explosion anisotropies in our 3D SNR simulations (Orlando *et al.* 2016). In one of their simulations, Wongwathanarat *et al.* (2016) have found three pronounced Fe-rich fingers that may correspond to the extended Fe-rich regions observed in Cas A. Interestingly their physical characteristics seem to be similar to those of initial pistons prescribed in our SN-SNR model, 1 day after the SN explosion.

4. Effects of inhomogeneous CSM: The case of SN 1987A

The effects of an inhomogeneous CSM on the morphology of a SNR can be investigated in SN 1987A. In fact, in this case, the surrounding medium (the nebula) is highly inhomogeneous as shown by optical images soon after the outburst. The nebula consists mainly of a dense central equatorial ring surrounded by an extended H II region. Two other rings lying in planes almost parallel above and below the equatorial one are also observed. This complex structure of the CSM is believed to originate from the interaction of a slow wind from the red supergiant phase with the faster wind from the blue supergiant phase (e.g. Morris & Podsiadlowski 2007). According to most recent observations (Frank *et al.* 2016), currently the blast wave from the explosion has left the inner equatorial ring.

In a previous work, we have investigated the effects of the CSM on the morphology of the remnant of SN 1987A and the imprint of the SN on the X-ray emission of its remnant through hydrodynamic simulations (Orlando *et al.* 2015). The strategy is again that described in Sect. 2. The model describes the complex structure of the CSM accounting, in particular, for the H II region and the dense equatorial ring. We have explored the space of model parameters, searching for those best fitting the morphology, light curves, and spectra observed in the X-ray band. We have found that a model with explosion energy in the range $1.2 - 1.4 \times 10^{51}$ erg and envelope mass in the range $15 - 17 M_{\odot}$ is able to reproduce the observed bolometric lightcurve of SN 1987A during the first 250 days after the SN event. The left panel of Fig. 3 shows the 3D volume rendering of the particle density of the shocked plasma at $t = 30$ yr for the best-fit model in Orlando *et al.* (2015). This model predicts a total mass of the ring $M_{\rm rg} = 0.062\, M_{\odot}$ of which $\sim 64\%$ is plasma with density $n \approx 1000$ cm^{-3} and $\sim 36\%$ is plasma with $n \approx 2.5 \times 10^4$ cm^{-3}. Interestingly, these values are in very nice agreement with those derived for the density structure of the ring from the analysis of optical observations (Mattila *et al.* 2010). The model also predicts that the blast wave has recently left the dense ring and is now propagating in

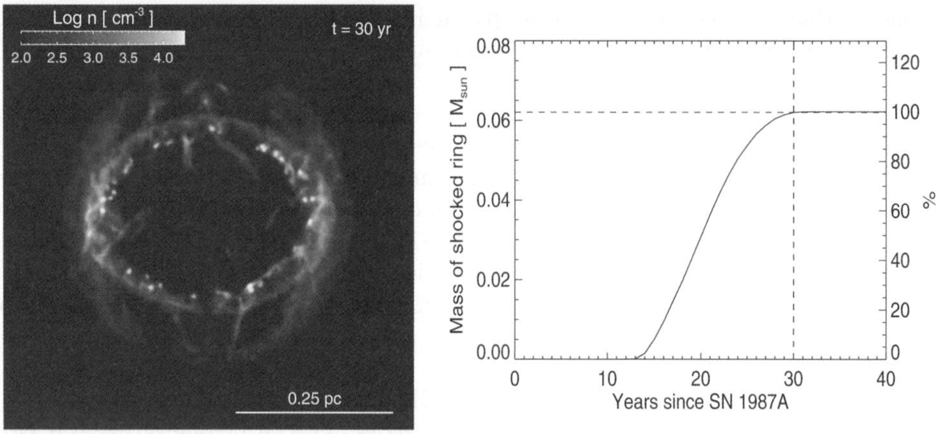

Figure 3. Left panel: three-dimensional volume rendering of the particle density of the shocked plasma during the interaction of the blast wave with the nebula, 30 years after the SN event (from the best-fit model of Orlando *et al.* 2015). Right panel: mass of the shocked ring vs. time; the vertical dashed line marks 30 years after the SN event; the horizontal dashed line indicates that the whole ring has been shocked.

a less dense environment (see right panel in Fig. 3). Again this is in excellent agreement with the recent findings from the analysis of X-ray observations (Frank *et al.* 2016) and, also, with the fading of the ring and implied destruction inferred from optical observations (Fransson *et al.* 2015).

From the model results, we have synthesized the X-ray emission, deriving light curves, images, and spectra to be compared with actual observations. Figure 4 shows the light curves derived from our best-fit model and the comparison with those observed. We have found that the same model fitting the bolometric lightcurve of the SN is also able to fit both the soft ([0.5, 2.0] keV) and hard ([3.0, 10] keV) X-ray light curves as derived from multi-epoch observations with ROSAT, ASCA, Chandra, and XMM-Newton X-ray telescopes. In particular, the predictions of the model fit the most recent Chandra observations (Frank *et al.* 2016), supporting the idea that the blast wave has now left the dense ring. This study has demonstrated, for the first time, the consistency between a physical model reproducing the observables of the SN (namely the cause) and that reproducing the observables of the resulting remnant (namely the effect).

The model includes also some passive tracers to follow the evolution of the different plasma components (namely ejecta, H II region, and dense ring) and their contribution to X-ray emission (see Figs. 4 and 5). We have identified three main phases in the evolution of the remnant of SN 1987A. The first starts when the ejecta reach the H II region and last for about 13 years. In this phase the emission is dominated by the shocked H II region with a smaller contribution from shocked ejecta. Then the blast wave starts to propagate through the dense ring ($\approx$ 13 yr after the SN event) and the X-ray emission becomes largely dominated by the shocked plasma from the ring (see Fig. 5). In this phase the contribution from shocked ejecta continue to increase faster than that of shocked plasma from the H II region, but still much lower than the contribution from shocked plasma from the ring (see Fig. 4). The model predicts that the blast wave has left the disk in the last few years and is now propagating through a less dense medium. As a consequence, the contribution from shocked ring will shortly decrease while the contribution from shocked ejecta will gradually increase becoming the dominant component in about $4-5$ years (see Fig. 4). During the first phase (H II region dominated), we have found that the

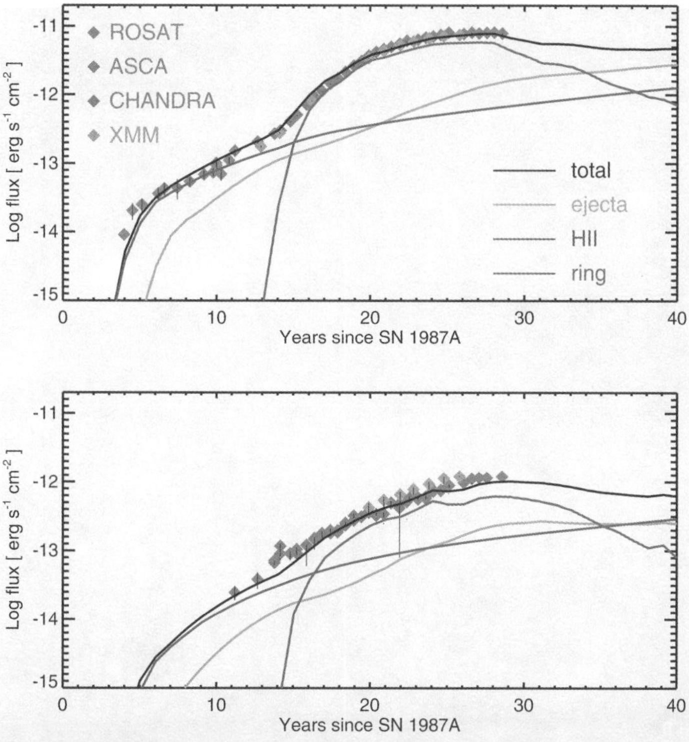

Figure 4. Observed and modeled X-ray light curves in the soft ([0.5, 2.0] keV; upper panel) and hard ([3.0, 10] keV; lower panel) bands (adapted from Orlando *et al.* 2015). Green, blue, and red lines mark the contribution to emission from the shocked ejecta, the shocked plasma from the H II region, and the shocked plasma from the ring, respectively.

shape of the X-ray lightcurve reflects the structure of the outer ejecta thus identifying the imprint of SN 1987A on the X-ray emission of its remnant. In the second phase (ring dominated), the comparison between model results and observations allowed us to constrain the structure and geometry of the CSM (see Fig. 5). In the third phase (ejecta dominated), expected in the next future, our model predicts that we will be able to investigate the structure and chemical stratification of the ejecta.

5. Conclusions

There is a growing interest in Astrophysics on studies aimed at linking the morphological properties of SNRs to their progenitor SNe. In fact SNRs morphology and properties are expected to reflect the physical and chemical properties of the progenitor SNe and of the environment in which blast waves travel. Multi-wavelength/multi-messenger observations of SNRs, therefore, should encode information about the physical and chemical properties of both the stellar debris and the surrounding CSM. In the first case, we could have the possibility to investigate the energetics and dynamics of the SN progenitor and the anisotropies developed during the complex phases of the SN explosion; in the latter we could derive important clues on the final stages of stellar evolution. Linking SNe to SNRs, therefore, has breakthrough potential to open new exploring windows on SN and SNR issues, especially in view of the spatially resolved high-resolution spectroscopy capability of the forthcoming Athena satellite (Nandra *et al.* 2013).

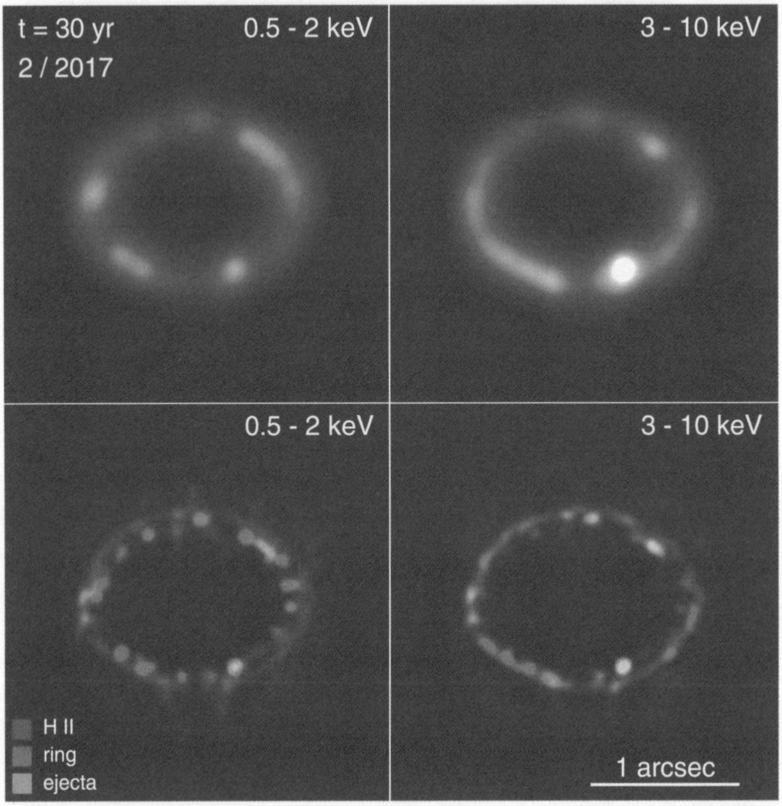

Figure 5. Synthetic maps of X-ray emission in the soft ([0.5, 2.0] keV; on the left) and hard ([3.0, 10] keV; on the right) bands integrated along the line of sight, 30 years after the SN event (from the best-fit model of Orlando *et al.* 2015); each image has been normalized to its maximum for visibility. In the upper panels the images have been convolved with a Gaussian of size 0.15 arcsec to approximate the spatial resolution of Chandra observations. In the lower panels, the figure shows three-color composite images of the X-ray emission smoothed with a Gaussian of size 0.025 arcsec; the colors in the composite show the contribution to emission from the different shocked plasma components, namely the ejecta (green), the ring (red), and the H II region (blue).

In this contribution we have reviewed our recent achievements in the study of the SN-SNR connection. In particular, we have devised a strategy based on 3D hydrodynamic models to link the morphological properties observed in SNRs to the structure of the ejecta soon after the progenitor SN explosions and to the structure of the environment in which the remnant expands. In the case of Cas A, we have shown that our approach was very effective in deriving the physical properties of the main anisotropies developed in the aftermath of the SN explosion and responsible for the observed non-uniform distribution of ejecta. In the case of SN 1987A, we have demonstrated that the physical model reproducing the main observables of the SN reproduces also the X-ray emission of the subsequent expanding remnant, thus bridging the gap between SNe and SNRs. The study was very effective also to identify the imprint of SN 1987A on the X-ray emission of its remnant and to constrain the structure and geometry of the surrounding nebula, formed in the latest stage of the star progenitor's evolution.

Our study has shown that detailed and complete numerical models accounting for the 3D nature of the phenomenon are necessary to successfully decipher the observations and, therefore, to connect the observed morphology of SNRs to the physics of the progenitor SN engine. However, deciphering observations might depend critically on the models. The latter should connect the stellar progenitor, the SN, and the SNR self-consistently. The models should be guided by observational facts and they should account for the dynamics, energetics, and spectral properties of SNe and SNRs. Finally it is highly desirable to strongly improve the synergy and communication between the SN and SNR communities. The IAU Symposium 331, bringing together theorists and observers from these communities, is an excellent example of how to accomplish this task.

Acknowledgements

The authors acknowledge partial support from the IAU to participate to the IAU Symposium 331. This paper was partially funded by the PRIN INAF 2014 grant "Filling the gap between supernova explosions and their remnants through magnetohydrodynamic modeling and high performance computing". The software used in this work was, in part, developed by the U.S. Department of Energy-supported Advanced Simulation and Computing/Alliance Center for Astrophysical Thermonuclear Flashes at the University of Chicago. We acknowledge that the results of this research have been achieved using the PRACE Research Infrastructure resource MareNostrum III based in Spain at the Barcelona Supercomputing Center (PRACE Award N.2012060993) and the FERMI cluster based in Italy at CINECA (ISCRA Award N. HP10BI36DG,2012).

References

Badenes, C., Hughes, J. P., Cassam-Chenaï, G., & Bravo, E. 2008, *ApJ*, 680, 1149
DeLaney, T. *et al.* 2010, *ApJ*, 725, 2038
Frank, K. A., Zhekov, S. A., Park, S., *et al.*, 2016, *ApJ*, 829, 40
Fransson, C., Larsson, J., Migotto, K., *et al.*, 2015, *ApJ*, 806, L19
Fryxell, B. *et al.* 2000, *ApJS*, 131, 273
Hwang, U. & Laming, J. M. 2012, *ApJ*, 746, 130
Janka, H.-T., Melson, T., & Summa, A. 2016, *Ann. Rev. Nucl. Part. Sci.*, 66, 341
Kifonidis, K., Plewa, T., Scheck, L., Janka, H.-T., & Müller, E. 2006, *A&A*, 453, 661
Lee, J.-J., Park, S., Hughes, J. P., & Slane, P. O. 2014, *ApJ*, 789, 7
Li, H., McCray, R., & Sunyaev, R. A. 1993, *ApJ*, 419, 824
Mattila, S., Lundqvist, P., Gröningsson, P., *et al.*, 2010, *ApJ*, 717, 1140
Miceli, M., Orlando, S., Reale, F., Bocchino, F., & Peres, G. 2013, *MNRAS*, 430, 2864
Milisavljevic, D. & Fesen, R. A. 2013, *ApJ*, 772, 134
Milisavljevic, D. & Fesen, R. A. 2015, *Science*, 347, 526
Morris, T. & Podsiadlowski, P. 2007, *Science*, 315, 1103
Nandra, K. *et al.* 2013, ArXiv e-prints: 1306.2307
Orlando, S., Peres, G., Reale, F., *et al.*, 2005, *A&A*, 444, 505
Orlando, S., Bocchino, F., Reale, F., Peres, G., & Pagano, P., 2008, *ApJ*, 678, 274
Orlando, S., Bocchino, F., Miceli, M., Petruk, O., & Pumo, M. L. 2012, *ApJ*, 749, 156
Orlando, S., Miceli, M., Pumo, M. L., & Bocchino, F. 2015, *ApJ*, 810, 168
Orlando, S., Miceli, M., Pumo, M. L., & Bocchino, F. 2016, *ApJ*, 822, 22
Patnaude, D. J., Lee, S.-H., Slane, P. O., *et al.*, 2015, *ApJ*, 803, 101
Petruk, O., Dubner, G., Castelletti, G., *et al.*, 2009, *MNRAS*, 393, 1034
Pumo, M. L. & Zampieri, L. 2011, Astrophys. J., 741, 41
van Veelen, B., Langer, N., Vink, J., *et al.*, 2009, *A&A*, 503, 495
Wang, L. *et al.* 2002, *ApJ*, 579, 671
Wang, L. & Wheeler, J. C. 2008, *ARA&A*, 46, 433
Wongwathanarat, A., Janka, H.-T., Mueller, E., *et al.*, 2016, *ArXiv e-prints*, 1610.05643
Yamaguchi, H. *et al.* 2014, *ApJ*, 785, L27

Supernova 1987A:30 years later - Cosmic Rays and Nuclei from Supernovae and their aftermaths
Proceedings IAU Symposium No. 331, 2017
A. Marcowith, M. Renaud, G. Dubner, A. Ray & A. Bykov, eds.

doi:10.1017/S1743921317004367

Linking supernovae and supernova remnants. Time-dependent injection in SN1987A and gamma-ray spectrum of IC443

O. Petruk[1,2], **S. Orlando**[1] **and M. Miceli**[1,3]

[1]INAF-Osservatorio Astronomico di Palermo, Piazza del Parlamento 1, 90134 Palermo, Italy
[2]Inst. Appl. Probl. in Mechanics and Mathematics, Naukova Street, 3-b Lviv 79060, Ukraine
[3]Dip. di Fisica e Chimica, Univ. di Palermo, Piazza del Parlamento 1, 90134 Palermo, Italy

Abstract. Acceleration times of particles responsible for the gamma-rays in supernova remnants (SNRs) are comparable with SNR age. If the number of particles starting acceleration was varying during early times after the supernova explosion then this variation should be reflected in the shape of the gamma-ray spectrum. In order to analyse this effect, we consider the time variation of the radio spectral index in SN1987A and solution of the non-stationary equation for particle acceleration. We reconstruct evolution of the particle injection in SN1987A, apply it to derive the particle momentum distribution in IC443 and model its gamma-ray spectrum. We show that: i) observed break in the proton spectrum around 50 GeV in IC443 is a consequence of the variation of the cosmic ray injection; ii) shape of the hadronic gamma-ray spectrum in SNRs critically depends on the temporal variation of the cosmic ray injection in the immediate post explosion phases.

Keywords. (stars:) supernovae: individual (SN1987A), (ISM:) supernova remnants: individual (IC443), acceleration of particles, gamma-rays

1. Introduction

The steady-state particle injection and the stationary acceleration are typically adopted in interpretation of the γ-ray spectra from supernova remnants (SNRs). Are these regimes justified to simulate the momentum distributions of the highest energy particles? Is it correct to consider the stationary solution for diffusive shock acceleration if injection is not constant or not impulsive? What consequences could be if the number of particles starting acceleration in the aftermath of the supernova event varies? May this early evolution be reflected after some time in the spectrum of the most energetic emission? In order to answer these questions, let us consider the general time-dependent equation.

2. Time-dependent particle acceleration at the SNR shock

The distribution function $f(t, x, p)$ for cosmic rays which accelerate diffusively on the shock of SNRs is a solution of the non-stationary equation (e.g., Skilling 1975, Jones 1990):

$$\frac{\partial f}{\partial t} + u\frac{\partial f}{\partial x} = \frac{\partial}{\partial x}\left[D\frac{\partial f}{\partial x}\right] + \frac{1}{3}\frac{du}{dx}p\frac{\partial f}{\partial p} + Q \tag{2.1}$$

where t is the time, x the spatial coordinate, p the momentum, D the diffusion coefficient, u the flow velocity in the shock reference frame, Q the injection term. We assume that

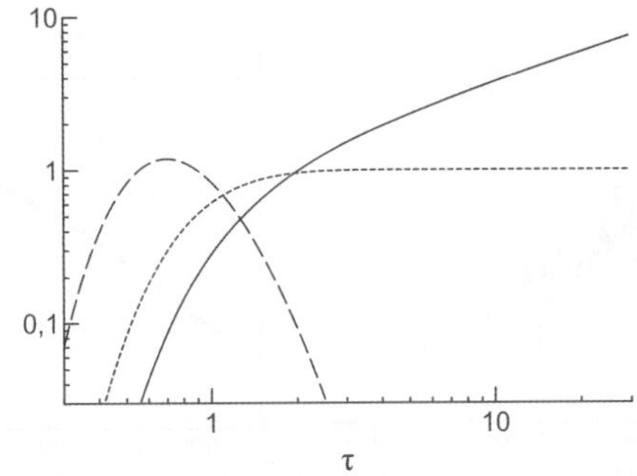

Figure 1. The distribution function φ_o (dashed line) and the integral in the solution (2.3) for $Q_t = 1$ (dotted line) and $Q_t = \tau^{0.7}$ (solid line) versus the dimensionless time τ.

particles are injected at the shock ($x = 0$) with the momentum p_i:

$$Q(t, x, p) = \frac{\eta n_1 u_1}{4\pi p_i^2} \delta(p - p_i) \delta(x) Q_t(t), \tag{2.2}$$

where the parameter η is the injection efficiency (the density fraction of accelerated particles); it is defined to be the value of the efficiency at the late steady-state regime. The term $Q_t(t)$ represents the time evolution of the injection; obviously, it is unity in the steady-state regime. The indices '1', '2' and 'o' correspond to upstream, downstream and to the shock position.

The test-particle solution of equation (2.1) for the steady-state injection $Q_t = 1$ was derived by Drury (1983), Forman & Drury (1983). The solution for the time-dependent injection was derived by Petruk & Kopytko (2016):

$$f_o(t, p) = \mathrm{f}_o(p) \int_0^\tau Q_t(\tau - \tau') \varphi_o(\tau') d\tau', \tag{2.3}$$

where

$$\mathrm{f}_o(p) = \frac{\eta n_1}{4\pi p_i^3} \frac{3\sigma}{\sigma - 1} \left(\frac{p}{p_i} \right)^{-s_f}, \tag{2.4}$$

is the solution of the stationary equation with the spectral index $s_f = 3\sigma/(\sigma - 1)$, $\sigma = u_1/u_2$ is the shock compression factor and the probability distribution is (Forman & Drury (1983), Petruk & Kopytko (2016))

$$\varphi_o(\tau) = \frac{e^{2A}}{2^{2A+1}\sqrt{\pi}} \frac{e^{-\xi(\tau)^2}}{\tau^{A/2+1}} \left(\mathrm{H}_{A+1}(\xi) - 2\tau^{1/2} \mathrm{H}_A(\xi) \right), \tag{2.5}$$

where $\mathrm{H}_m(x)$ is the Hermite polynomial, $\tau = t/t_1$, $t_1 = 4D_1/u_1^2$, $\xi(\tau) = \tau^{1/2} + A/(2\tau^{1/2})$, $A = s_f/\alpha$, α the index in the diffusion coefficient $D(p) \propto p^\alpha$.

It is clear from (2.3) and Fig. 1 that the widely-used stationary solution for the particle acceleration may actually be considered *only in the case when* $Q_t = 1$ and values $\tau \gg \tau_*$ where τ_* marks the maximum in the distribution $\varphi_o(\tau)$. Note that $\tau = t/t_1(p)$ depends on two variables, t and p. Therefore, the probability φ_o is a two-dimensional distribution.

In order to relate the unitless time τ to the physical time t and momentum p, the following expressions have to be used: $t = \tau t_{\rm age}$ and $p = \tau^{-1/\alpha} p_{\rm max}$ where $t_{\rm age}$ is the SNR age and $p_{\rm max}$ is the maximum particle momentum. Therefore, the larger τ corresponds to the larger t and the smaller p.

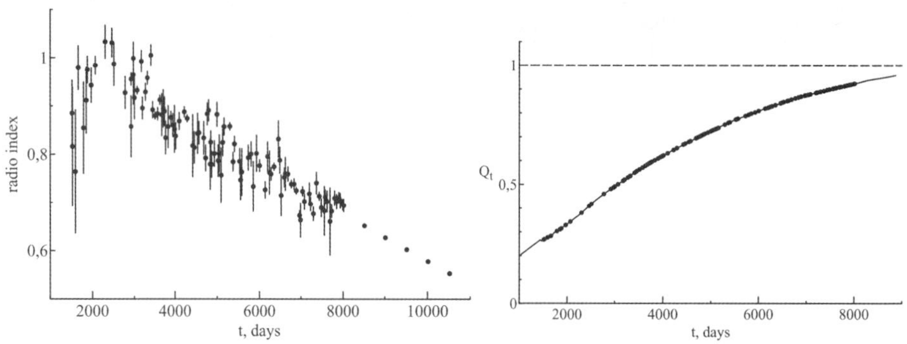

Figure 2. Observed evolution of the radio index in SN1987A (on the left) from Zanardo *et al.* (2013) and the temporal dependence of the particle injection derived from these data (on the right; dots corresponds to the period of the radio observations; dashed line represents the case of the steady-state injection). Adapted from Petruk *et al.* (2017).

It is important that the highest-energy particles has $\tau \ll 1$ where the integral in the solution is considerably less than unity. This means that the highest-energy particles (which actually are responsible for the γ-rays) cannot be described by the stationary solution.

Instead, the radio-emitting electrons has $\tau \gg \tau_*$ and may be described by the stationary solution *if the injection is characterized by* $Q_{\rm t} = 1$ (dotted line). However, if $Q_{\rm t} \neq 1$ then also the radio emission is affected by the temporal evolution of injection. It may be noted from Fig. 1 (solid line) that the injection term of the form $Q_{\rm t} \propto \tau^{\beta}$ affects, at large τ, the spectral index of the accelerated particle distribution. The solution in this case may be written in the form (Petruk & Kopytko (2016))

$$f_{\rm o}(t,p) \propto p^{-(s_{\rm f}+\alpha\beta)}. \tag{2.6}$$

The radio spectral index $\alpha_{\rm r}$ is known to be related to the index of the particle distribution. With that relation, we come to the simple formula

$$\alpha_{\rm r} = \frac{s_{\rm f} + \alpha\beta - 3}{2} \tag{2.7}$$

which directly connects the spectral index and β, with typical values of the adiabatic constant $\gamma = 5/3$ (therefore $\sigma = 4$ and $s_{\rm f} = 4$) and $\alpha = 1$ for the Bohm-like diffusion.

3. Radio index of SN1987A and evolution of particle injection

The formula (2.7) allows us to reconstruct the temporal evolution of the particle injection in SN1987A (see Petruk *et al.* (2017) for more details). We use the detailed observations of the radio index in SN1987A (Fig. 2 left) given by Zanardo *et al.* (2013). From these data we obtain a series of β, one for each observed $\alpha_{\rm r}$, and then represent the evolution of injection as a piecewise step function. This function is shown on Fig. 2 right and is represented by the following expression

$$Q_{\rm t}(t) = q_1 t^{\beta_{\rm i}} \prod_{m=2}^{i} t_{\rm m}^{\beta_{\rm m}-1-\beta_{\rm m}}, \qquad t_{\rm i} \leqslant t \leqslant t_{\rm i+1}. \tag{3.1}$$

It is derived from the condition that the function $Q_{\rm i} = q_{\rm i} t_{\rm i}^{\beta_{\rm i}}$ in a single step has to connect the two nearby points ($t_{\rm i}$ and $t_{\rm i+1}$) and the normalization $q_{\rm i+1}$ of each step is fixed by the function on the previous one: $q_{\rm i} t_{\rm i+1}^{\beta_{\rm i}} = q_{\rm i+1} t_{\rm i+1}^{\beta_{\rm i+1}}$. The value of $q_{\rm o}$ is given by requirement to provide $Q_{\rm t} = 1$ at large t.

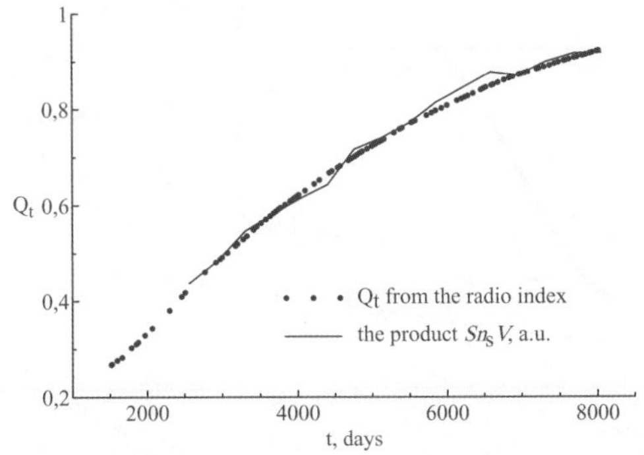

Figure 3. Evolution of Q_t during the period of the radio observations (dots) and of the product Sn_sV in 3-D model of SN1987A by Orlando *et al.* (2015) (line, in arbitrary units).

The observations are available from 1517 to 8014 days. After the last observed point, we extrapolate (Fig. 2 left, dots without errors) the behavior of the radio index using the fit suggested by Zanardo *et al.* (2013), up to the value $\alpha_r = 0.5$ and then keep it constant. Before the day 1517, we assume rather arbitrary Q_t to be constant till the day 100 and then $Q_t(t) \propto t^{0.8}$ after this day. Such a choice provides us with a good fit for the highest-energy part of the γ-ray spectrum of IC443 considered in the next section.

Right plot on Fig. 2 demonstrates clearly that the injection efficiency cannot be considered constant until some age of the remnant of supernova, when the radio index reaches a saturated level close to 0.5. This observational evidence needs to be understood from a theoretical point of view. Different factors is expected to influence the injection (e.g. shock strength, level of turbulence). In order to make initial insight, let us assume that the injected particles are a constant fraction η of the flux incoming through the shock. Then Q_t has to be proportional to the product of the shock surface S, post-shock density n_s and shock velocity V. In order to check this, we consider the three-dimensional model of this SNR by Orlando *et al.* (2015). Fig. 3 shows the product Sn_sV in comparison with the time dependence of Q_t. The model shows that, shortly after the day 2000, the forward shock enters the HII region with a constant density and freely expands there with roughly constant velocity. Therefore, the prominent change in the injection efficiency happens, to the large extent, because the surface of the shock from explosion increases considerably during the immediate aftermath of supernova (at later stages, the shock radius increases much slowly).

4. Non-stationary gamma-ray spectrum of IC443

We substitute the function $Q_t(t)$ derived above into the non-stationary solution (2.3) in order to obtain the particle spectrum in IC443. Having the proton spectrum, we calculate the hadronic γ-ray spectrum from this SNR (following the prescription of Kelner *et al.* 2006). The γ-ray spectrum from our model (Petruk *et al.* (2017)) is shown by the solid line on Fig. 4 (right plot). It fits the observations perfectly.

There are few conclusions from this figure. i) The spectrum of particles with the steady-state injection cannot explain the observed γ-ray spectrum. ii) The temporal evolution of the injection efficiency in the aftermath of the supernova *considerably affects* the γ-ray spectrum of SNR. iii) The break in the γ-ray spectrum of IC443 discovered by Ackermann *et al.* 2013 at the photon energies around ~ 2 GeV is a natural consequence of the smaller number of particles starting acceleration before the day 8000. iv) The early the particles were injected the more time they have for acceleration, the higher energies they may reach

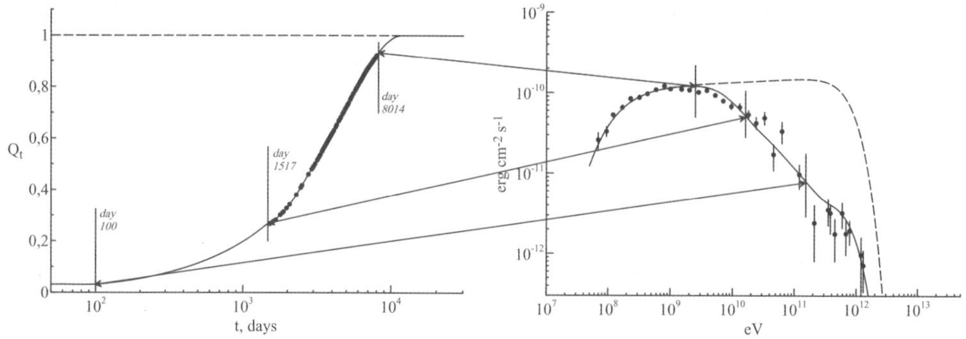

Figure 4. Evolution of Q_t (left) and the hadronic γ-ray spectrum of IC443 (right). The γ-ray observations of IC443 are reported by the Fermi (up to 10^{11} eV, Ackermann *et al.* 2013), MAGIC (Albert *et al.* 2007) and VERITAS (Acciari *et al.* 2009) collaborations. The solid line on the left plot shows $Q_\mathrm{t}(t)$ and on the right plot represents the spectrum from our model. The dashed line shows the case of the steady-state injection. Adapted from Petruk *et al.* (2017).

and the more energetic γ-rays they are able to emit. Fig. 4 uncovers the correspondence between the proton injection time and the maximum energies of γ-rays which may be generated by these protons. It is quite impressive that hadronic TeV γ-rays from SNR are emitted by protons injected just in the *first few months* after the supernova.

5. On the visibility of SN1987A in hadronic gamma-rays

We may use $Q_\mathrm{t}(t)$ derived from observations of SN1987A to estimate also the γ-ray visibility of this young SNR to the future experiment CTA.

If the acceleration time-scale in SN1987A is of the same order as in IC443, it is obvious that protons cannot be accelerated in SN1987A to the γ-ray emitting energies because of the difference in ages for these SNRs: time available for acceleration is small in SN1987A, $t_\mathrm{sn87a}/t_\mathrm{ic443} \simeq 0.01$. However, the right-hand part of the γ-ray spectrum (above $\sim 3\,\mathrm{GeV}$ on Fig. 4) shifts toward the right with increase of t/t_1. Therefore, if the acceleration in SN1987A is faster (i.e. the smaller diffusion coefficient and the larger shock velocity) to the extent which provides

$$\frac{t_\mathrm{sn1987a}}{t_\mathrm{ic443}}\frac{V^2_\mathrm{sn1987a}}{V^2_\mathrm{ic443}}\frac{D_\mathrm{ic443}}{D_\mathrm{sn1987a}} \geqslant 1. \tag{5.1}$$

and the density of target protons in vicinity of SN1987A is $n_\mathrm{pp,sn87a} \geqslant 0.005 n_\mathrm{pp,ic443}$ then the hadronic γ-rays from SN1987A could be detectable by CTA (Fig. 5).

6. Conclusions

We use the test-particle solution of the time-dependent equation for the diffusive particle acceleration on the strong shocks in order to develop the method to reconstruct the temporal evolution of the injection efficiency (fraction of particles which begin acceleration). For this purpose, the observational data on the radio index of SN1987A were adopted.

It is shown that the spectrum of the highest-energy particles, in general, cannot be described by the steady-state equation with the constant injection. It is obligatory therefore to consider the non-stationary particle spectrum for models of the hadronic γ-ray emission from SNRs. Variation of the injection during the early times after the supernova explosion is critical for interpretation of the TeV γ-ray spectra of SNRs.

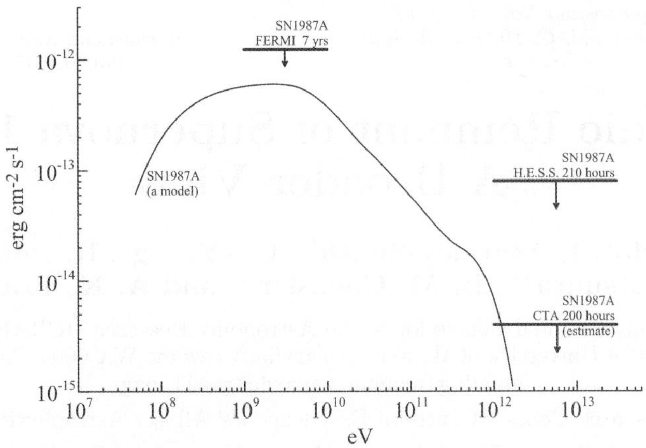

Figure 5. Eventual γ-ray spectrum of SN1987A. The upper limits on the γ-ray flux from SN1987A are from Ackermann *et al.* (2016) (FERMI), from H.E.S.S. Collaboration (2015) and the estimate for CTA is ours. Adapted from Petruk *et al.* (2017).

The radio observations of supernovae may be used to track the injection behavior. The main reason of the injection variation in the aftermath of the supernova is the rapid increase of the surface area of the forward shock.

In this approach, we have successfully fitted the GeV-TeV γ-ray spectrum of IC443, using SN1987A as a proxy of the parent supernova. The break in the proton spectrum around $50 - 200$ GeV is a natural consequence of the time-dependent particle injection.

Acknowledgements

We thank G. Zanardo for the radio data on the spectral index evolution in SN1987A. We acknowledge support from the IAU and organizers to participate in the IAU Symposium 331. This paper was partially funded by the PRIN INAF 2014 grant 'Filling the gap between supernova explosions and their remnants through magnetohydrodynamic modeling and high performance computing'.

References

Acciari V. *et al.* 2009, *ApJ*, 698, L133
Ackermann M. *et al.* 2013, *Science*, 339, 807
Ackermann M. *et al.* 2016 *A&A* 586, 71
Albert J. *et al.* 2007, *ApJ*, 664, L87
Drury L. 1983, *Rep. Prog. Phys.*, 46, 973
Forman M., Drury L. 1983, *Proc. 18th ICRC*, 2, 267
Jones F. 1990, *ApJ*, 361, 162
H. E. S. S. Collaboration 2015, *Science*, 347, 406
Kelner S., Aharonian F., Bugayov V. 2006, *Phys. Rev. D*, 74, 034018
Orlando S., Miceli M., Pumo M., Bocchino F. 2015, *ApJ*, 810, 168
Petruk O., Kopytko B. 2016, *MNRAS*, 462, 3104
Petruk O., Orlando S., Miceli M., Bocchino F. 2017, submitted
Skilling J. 1975, *MNRAS*, 172, 557
Zanardo G., Staveley-Smith L., Hg C.-Y. *et al.* 2013, *ApJ*, 767, 98

Supernova 1987A:30 years later - Cosmic Rays and Nuclei from Supernovae and their aftermaths
Proceedings IAU Symposium No. 331, 2017
A. Marcowith, M. Renaud, G. Dubner, A. Ray & A. Bykov, eds.

doi:10.1017/S1743921317005312

The Radio Remnant of Supernova 1987A – A Broader View

G. Zanardo[1], L. Staveley-Smith[1,2] C. -Y. Ng[3], R. Indebetouw[4], M. Matsuura[5,6], B. M. Gaensler[7,2] and A. K. Tzioumis[8]

[1]International Centre for Radio Astronomy Research (ICRAR), M468, The University of Western Australia, Crawley, WA 6009, Australia.
email: giovanna.zanardo@gmail.org

[2]Australian Research Council Centre of Excellence for All-sky Astrophysics (CAASTRO)

[3]Department of Physics, The University of Hong Kong, Pokfulam Road, Hong Kong

[4]Department of Astronomy, University of Virginia, P.O. Box 400325, Charlottesville, VA 22904-4325, USA

[5]School of Physics and Astronomy, Cardiff University, Queens Buildings, The Parade, Cardiff, CF24 3AA, UK

[6]Department of Physics and Astronomy, University College London, Gower Street, London, WC1E 6BT, UK

[7]CSIRO Astronomy and Space Science, Australia Telescope National Facility, PO Box 76, Epping, NSW 1710, Australia

[8]Dunlap Institute for Astronomy & Astrophysics, University of Toronto, Toronto, ON M5S 3H4, Canada

Abstract. Supernova remnants (SNRs) are powerful particle accelerators. As a supernova (SN) blast wave propagates through the circumstellar medium (CSM), electrons and protons scatter across the shock and gain energy by entrapment in the magnetic field. The accelerated particles generate further magnetic field fluctuations and local amplification, leading to cosmic ray production. The wealth of data from Supernova 1987A is providing a template of the SN-CSM interaction, and an important guide to the radio detection and identification of core-collapse SNe based on their spectral properties. Thirty years after the explosion, radio observations of SNR 1987A span from 70 MHz to 700 GHz. We review extensive observing campaigns with the Australia Telescope Compact Array (ATCA) and the Atacama Large Millimeter/submillimeter Array (ALMA), and follow-ups with other radio telescopes. Observations across the radio spectrum indicate rapid changes in the remnant morphology, while current ATCA and ALMA observations show that the SNR has entered a new evolutionary phase.

Keywords. circumstellar matter, ISM: supernova remnants, radio continuum: general, supernovae: individual (SN 1987A), acceleration of particles, radiation mechanisms: nonthermal

---

1. Background

Radio supernovae are most consistently identified with core-collapse supernovae (CC-SNe) and their transition into supernova remnants (SNRs). The radio emission results from the collision of the supernova (SN) shock and the progenitor's circumstellar medium (CSM). Many radio SNe can be associated with HII regions or regions of active star formation. Given the high ambient pressure of the surrounding HII region, radio SNe and their remnants are primarily identified by dense CSM environments. The SN shock wave modifies the CSM as it passes through it, both by compression and shock heating. The forward SN shock, i.e. the outer blast wave, and the reverse shock, which causes deceleration and heating of the inner ejecta, accelerate the shocked material, with some fractions of electrons and protons reaching cosmic ray energies. The acceleration mechanism usually associated with the shock propagation is diffusive shock acceleration (DSA), also known as the first-order Fermi mechanism (Drury 1983). According to DSA theory, energetic particles at shock fronts undergo spatial diffusion in a uniform magnetic field. If the magnetic field is not uniform, as in the case of young SNRs, the charged particles can get partially trapped in the magnetic field structure, while still gaining energy by scattering between the upstream and downstream regions of the forward shock (Kirk *et al.* 1996). This leads to a sub-diffusive particle transfer (sub-DSA), since the particle density at the shock front is lower than it is far downstream. In either case, as energetic electrons gyrate in the local magnetic field, they emit synchrotron emission. The synchrotron emission observed in SNRs can be well described by a power law $S_\nu \propto \nu^\alpha$, with S_ν the flux density measured at the observing frequency ν. Simplified synchrotron physics tells us that emission at the frequency ν is primarily by electrons with energy $E \propto 15\sqrt{\nu/B}$ GeV, with ν expressed in GHz and the magnetic field strength, B, expressed in μG (Reynolds 2011).

SN 1987A in the Large Magellanic Cloud, identified as Type IIP (Leibundgut & Suntzeff 2003) with progenitor of mass $\sim 20\, M_\odot$ (Smartt 2009), is the only nearby CCSN observed with a telescope since its early stages. Given its proximity, SN 1987A has allowed unique studies of the radio evolution of CCSNe and of the SN−CSM interaction. The complex CSM distribution in the remnant is believed to have originated from a red supergiant (RSG) which has evolved into a blue supergiant (BSG) about 20,000 years before the explosion (Crotts & Heathcote 2000). In its transition from supernova to SNR, SN 1987A has continued to brighten for the past three decades, thus making possible to monitor the rapid changes of the radio emission with a number of telescopes over an increasingly broad frequency range.

2. Evolution of the SN-CSM interaction

The interaction of the SN shock wave with the CSM in the equatorial plane (equatorial ring, ER) is the most notable feature of the radio remnant of SN 1987A. The ER defines the waist of the 'hourglass' environment created by the progenitor star prior to explosion. As the SN propagates outwards and impacts the denser CSM in the ER, the shocks expand above and below the ER plane (see Figure 1) and interact with high-latitude material (Potter *et al.* 2014), confined within the inner hourglass structure.

2.1. *Light curves*

The radio emission from SN 1987A was first detected two days after the arrival of the neutrinos with the Molonglo Observatory Synthesis Telescope (MOST) at 843 MHz, 1.4, 2.3, and 8.4 GHz (Turtle *et al.* 1987). The peak flux density was reached at ~ 130 mJy at 843 MHz four days after the explosion (Turtle *et al.* 1987), followed by a power-law decay of the emission, which became undetectable from September 1987 (Ball *et al.* 1995). It is understood that this radio outburst was due to the faster BSG wind, which produced a short-lived radio emission when hit by the SN shock (Chevalier & Fransson 1987). The radio emission re-appeared above noise ~ 1200 days after the explosion (Staveley-Smith *et al.* 1992), and has been monitored with the Australia Telescope Compact Array (ATCA) since then. Gaensler *et al.* (1997) calculated that, between the SN event and the second radio turn-on in mid-1990, the radio emitting regions expanded at an average velocity of $\sim 35,000$ km s^{-1}, which is possible in a very low-density inner CSM such as

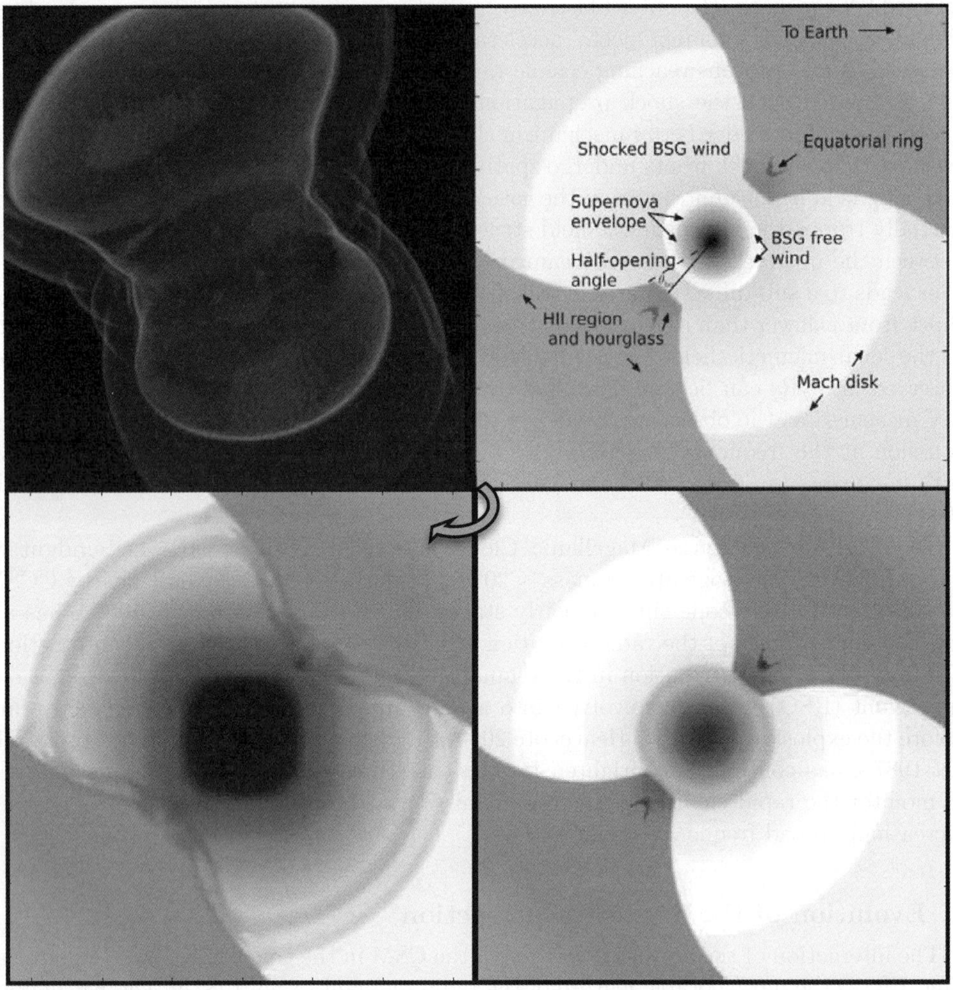

Figure 1. Stages of the shock expansion from HD simulations by Potter *et al.* (2014) (see their Figures 1, 6, and 7), as modelled around 1000 days (*top right*), at day ~2000 (*bottom right*), and at day ~8000 (*bottom left*) since the SN. By day ~8000, the forward shock has overtaken the equatorial ring (ER). The colour scheme varies from white to dark brown, indicating the increasing particle number density in $\log_{10}$ scale ($\sim 5 - 11 \times 10^7$ m^{-3}, see Potter *et al.* 2014). A 3D view of both the forward and reverse shocks expanding above and below the ER, is shown on the top left image, which is a rendering of the volumetric models developed by Potter *et al.* (2014) (see their Figure 9) at day ~8000. Source: Potter *et al.* (2014).

that surrounding Type Ib/Ic SNe. The expansion velocity sharply decreased to $v_s \lesssim 7000$ km s^{-1} from day ~1200 (Gaensler *et al.* 1997), when the radio emission started to increase (Ball *et al.* 2001). Since day ~ 5000 the radio emission has been rising at an exponential rate (Zanardo *et al.* 2010, Figure 2). A new phase of the SN−CSM interaction is observed at all frequencies after day ~ 8000 (Figure 2; Zanardo 2014), when the shock has likely become engulfed in the densest CSM in the ER, while still expanding above and below the ring.

2.2. *Morphology*

The morphology of the non-thermal radiation emitted by relativistic electrons accelerated in the remnant has been monitored with the Australia Telescope Compact Array (ATCA)

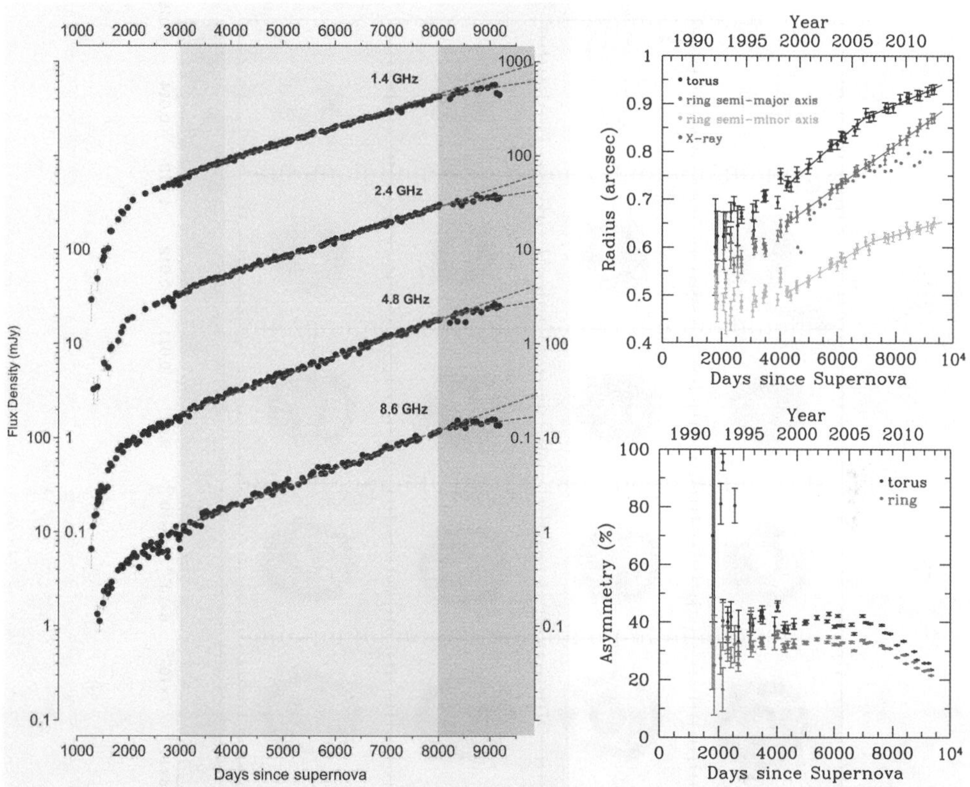

Figure 2. *Left* – Flux densities for SNR 1987A. Plots include data from observations at 1.4, 2.4, 4.8, and 8.6 GHz from August 1990 to June 2012 and are stacked with an offset of 0.1 mJy (Zanardo *et al.* 2010; Zanardo 2014). The red dashed lines are the exponential fits of data from November 2000 to February 2009, as derived by Zanardo *et al.* (2010). The blue dashed lines are 1500-day fits of data from February 2009. The colour bands are included to highlight different rates of the flux density increase. *Top Right* – Remnant radius resultant from images at 9 GHz, from August 1992 to March 2013, as derived by Ng *et al.* (2013) via fitting the SNR with a torus (*black*) and an elliptical ring (*red*: semi-major axis; *green*: semi-minor axis). The radius of the radio remnant is compared with that derived from X-ray images (*blue*, from Helder *et al.* 2013). The solid lines indicate the best-fit expansion. *Bottom Right* – Evolution of the remnant asymmetry at 9 GHz as derived from the torus (*black*) and ring (*red*) models (Ng *et al.* 2013). Source: Zanardo (2014), Ng *et al.* (2013).

since 1992 using images at 9 GHz (Staveley-Smith *et al.* 1992), with a spatial resolution of $0\farcs5$ (Gaensler *et al.* 1997; Ng *et al.* 2008). These images have provided insight into the trademarks of the radio emission (see Figure 3; Ng *et al.* 2013), such as: (*a*) the steady brightening; (*b*) the ring-like structure evolved from a two-lobe morphology; (*c*) the east-west asymmetry, with brightness peak on the eastern lobe. As discussed by Ng *et al.* (2013) (left plots in Figure 2), by day ∼8000 the faster eastbound shocks (Zanardo *et al.* 2013) have possibly overtaken the ER. This would have resulted in a reduction of the radio emission from the eastern rim and, thus, impacted the overall brightness asymmetry (see Figure 2). High-resolution images obtained at 44 GHz with the ATCA, and at 1.4 GHz via very-long baseline interferometry (VLBI) with the Australian Large Baseline Array (LBA; see Figure 4; Zanardo 2014), have shown small-scale structures more prominent on the eastern lobe, which are likely due to the shock wave interacting with discrete clumps near the inner surface of the SN shell (see Ng *et al.* 2011).

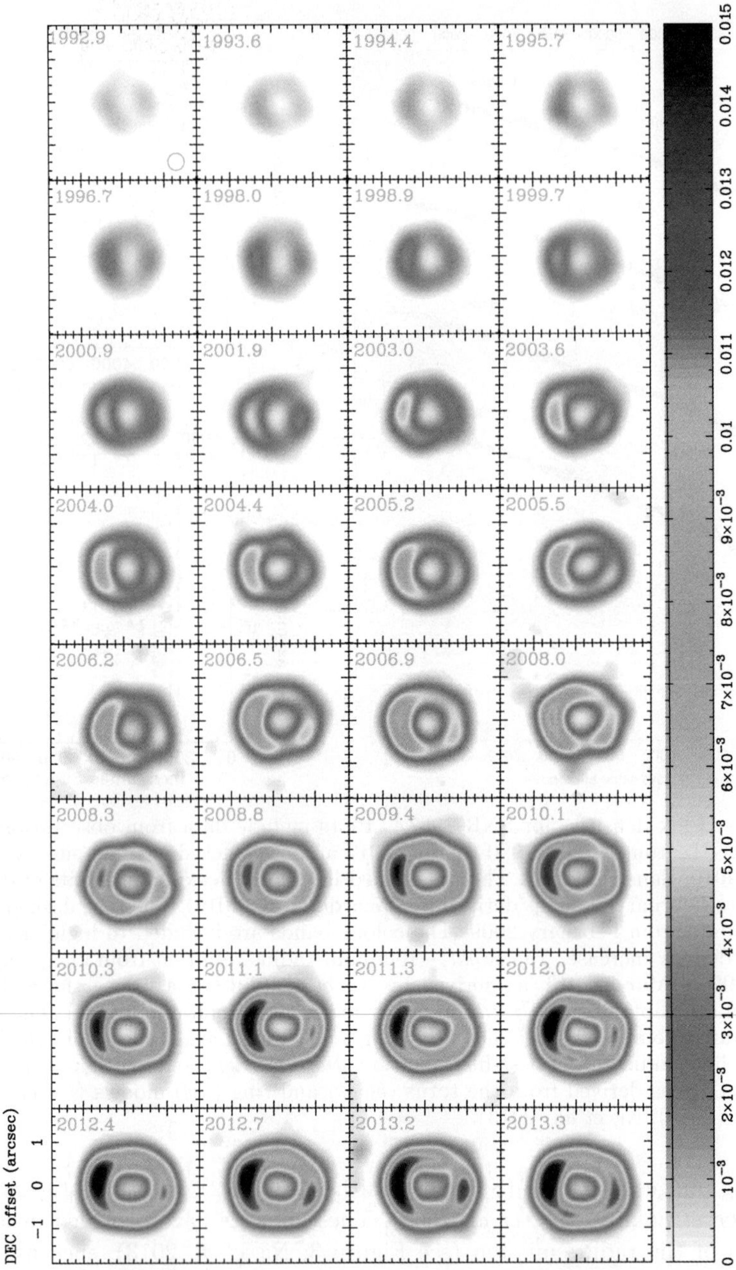

Figure 3. 9 GHz images of SNR 1987A from 1992 to 2013, adapted from Ng *et al.* (2013) (see Figure 1 therein). Images include data from Gaensler *et al.* (1997), Ng *et al.* (2008), and Ng *et al.* (2013). Colour-bar units are Jy beam^{-1}. Source: Ng *et al.* (2013), Zanardo (2014).

2.3. *Asymmetry*

The remnant asymmetry has been confirmed in all observations from radio to sub-mm (see Figure 5) and discussed for observations at contemporaneous epochs. The east-west asymmetry ratio is 1.4–1.5 at 18 and 44 GHz (Zanardo *et al.* 2013), and it shows a gradual

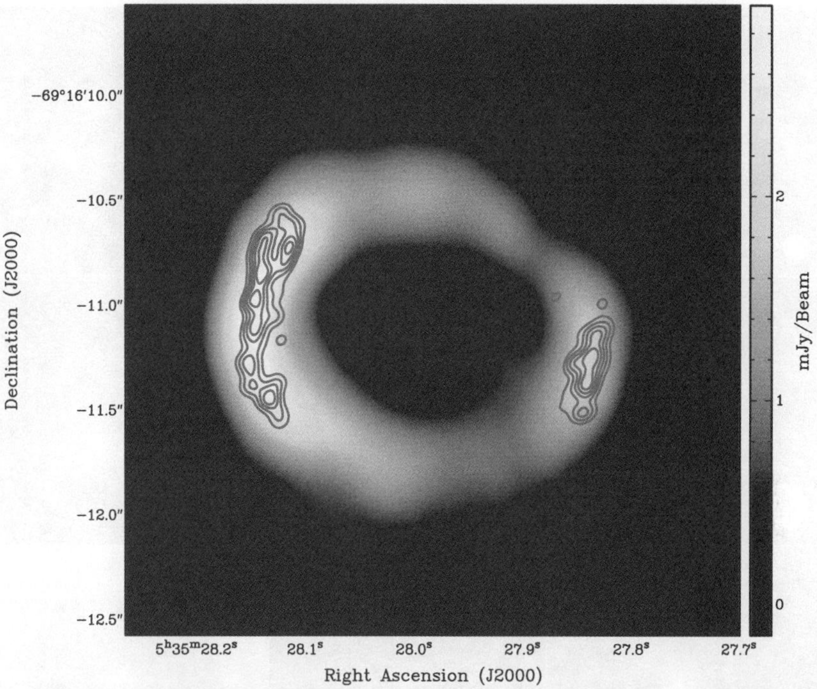

Figure 4. Contours of the 2010 VLBI image of SN 1987A at 1.4 GHz are overlaid with the 44 GHz image derived from 2011 observations (Zanardo *et al.* 2013). The 44 GHz image has a resolution of 0.″2, while the VLBI image has a restoring beam of 61 × 95 mas. The VLBI observations have been instrumental to align images prior to the derivation of spectral index maps (see Zanardo *et al.* 2013; Zanardo *et al.* 2014). Credit: Zanardo (2014).

decrease in images from 44 to 345 GHz (Zanardo *et al.* 2014), where there are hints of a reversed trend. As the synchrotron lifetime becomes shorter at higher frequencies, there would be regions in the eastern side of the remnant where electrons might be unable to cross the emission sites within their radiative life-time. In this scenario, the synchrotron emission would require the presence of relatively fresh injected and/or re-accelerated electrons to match the emission distribution at lower frequencies. Zanardo *et al.* (2014) estimated that, at frequencies higher than 213 GHz, the electrons might be unable to cross the emission regions on the eastern lobe.

Further, the asymmetry in the shock expansion velocities, with significantly faster eastbound shocks (Zanardo *et al.* 2013), has been related to an asymmetric explosion (Zanardo *et al.* 2013; Zanardo *et al.* 2014). The possibility of an asymmetric initial explosion was suggested by Chevalier & Soker (1989), while Podsiadlowski, Morris & Ivanova (2007) proposed a binary merger to reproduce the BSG progenitor and the rings formed from the BSG−RSG wind interaction. Although marked asymmetries have also been seen in the inner ejecta, (Kjær *et al.* 2010; Larsson *et al.* 2013), the recent detection of ^{44}Ti emission with the Nuclear Spectroscopic Telescope Array (*NuSTAR*) provides direct evidence of large-scale asymmetry in the explosion (Boggs *et al.* 2015).

2.4. *Spectral index*

A progressive hardening of the radio spectra obtained from ATCA multi-frequency flux monitoring has been observed since day ∼2500 (Zanardo *et al.* 2010), and up to day ∼8000. The results have pointed to sub-DSA particle transport (Kirk *et al.* 1996), which leads to increasingly efficient particle acceleration and cosmic ray production. On a local

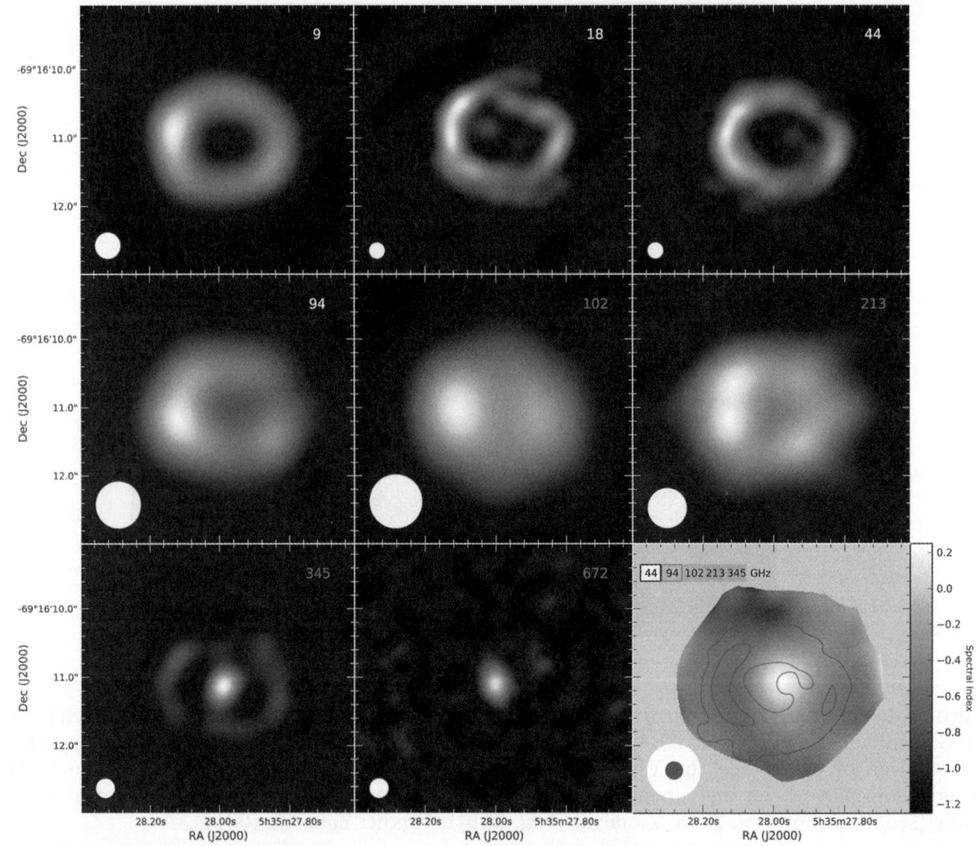

Figure 5. Mosaic of radio images of the remnant of SN 1987A at day 9280. Each image frequency is indicated on the top right corner, in GHz. Images below 100 GHz are derived from ATCA observations (Lakićević *et al.* 2012a; Ng *et al.* 2013; Zanardo *et al.* 2013), images above 100 GHz are from ALMA data (Zanardo *et al.* 2014). All images have been super-resolved with a circular beam (lower left corner). The bottom right image is the spectral index map obtained by combining images at five frequencies, i.e. 44, 94, 102, 213, and 345 GHz (see Zanardo *et al.* 2014). The spectral image is overlaid with contours (in blue) of the 15% and 60% emission levels at 44 GHz (Zanardo *et al.* 2013). Credit: Zanardo (2014).

scale, high-resolution spectral maps have consistently shown steeper synchrotron spectral indices and, thus, lower compression ratios, mainly localised on the eastern lobe of the remnant (see Figure 5; Zanardo *et al.* 2013; Zanardo *et al.* 2014). A softer particle spectrum on the eastern regions and the higher Mach numbers measured for the east-bound shocks (Zanardo *et al.* 2013), are likely to create the conditions for magnetic field amplification (Bell & Lucek 2001) with efficient cosmic ray production. Local magnetic field amplification is compatible with significant synchrotron losses at higher frequencies, which can explain the decrease of the east-west asymmetry ratio of the emission in the transition from radio to sub-millimetre wavelenghts (Zanardo *et al.* 2014).

The spectral energy distribution (SED) from 1.4 to 3000 GHz (see Figure 6) for 2011−2012 data yields a 'global' synchrotron spectral index $\alpha = -0.73$, which is confirmed in observations at low frequencies (72−230 MHz) carried out with the Murchison Widefield Array (MWA) in late 2013 (Callingham *et al.* 2016). The absence of a spectral turnover down to 72 MHz, places an upper limit of $n_e = 110$ cm^{-3} to the electron density of the medium in which the SN radio emission propagates (Callingham *et al.* 2016).

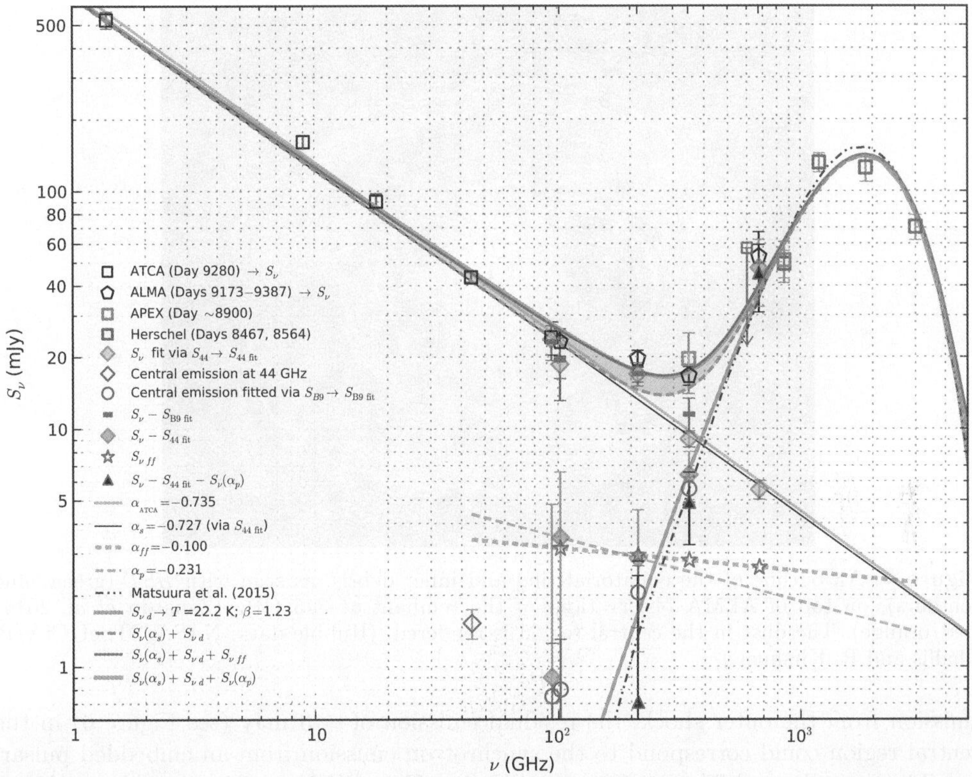

Figure 6. SED of SNR 1987A from radio to FIR, with data from: ATCA at 1.4 GHz (Zanardo 2014), 9 GHz (Ng *et al.* 2013), 18 and 44 GHz (Zanardo *et al.* 2013), and 94 GHz (Lakićević *et al.* 2012a); ALMA (Zanardo *et al.* 2014); the Atacama Pathfinder EXperiment (APEX) at 345 and 857 GHz (Lakićević *et al.* 2012b); the Herschel Space Observatory at 600−3000 GHz (Matsuura *et al.* 2011). The brown dash-dot-dotted curve is the amorphous carbon dust fit for ALMA data and *Herschel* observations carried out in 2012 (Matsuura *et al.* 2015). To match the average epoch of ALMA observations, ATCA data are scaled to day 9280, via exponential fitting parameters derived for ATCA flux densities from day 8000, as measured at 1.4, 8.6 and 9 GHz (Zanardo 2014) The hollow red diamond indicates the central emission measured at 44 GHz as reported by Zanardo *et al.* (2013) scaled to day 9280. The difference between possible spectral fits is highlighted in light orange. Credit: Zanardo *et al.* (2014).

3. Compact remnant

The search for a compact object in the inner regions of SNR 1987A has been undertaken on multiple fronts. It has been shown that flat spectral synchrotron indices can be associated either with accretion onto a black hole or with a pulsar wind nebula (PWN; Gaensler & Slane 2006). Spectral index maps of the remnant derived at different frequencies and epochs have consistently featured a flat central component (Lakićević *et al.* 2012a; Potter *et al.* 2009; Zanardo *et al.* 2013; Zanardo *et al.* 2014).

The discovery with ALMA (Indebetouw *et al.* 2014; Kamenetzky *et al.* 2013) and *Herschel* (Matsuura *et al.* 2011, 2015) of $\sim 0.8\ M_\odot$ of newly-forming dust grains in the SNR interior, complicates further the detection of a PWN expected to be expanding into the inner ejecta. The dust sits inside the ER and overlaps with the ejecta as seen with *HST* (Figure 7). Combined ALMA and ATCA observations have allowed one to separate the thermal and non-thermal components of the SNR emission (Indebetouw *et al.* 2014; Zanardo *et al.* 2014). After subtraction of the dust emission and the synchrotron

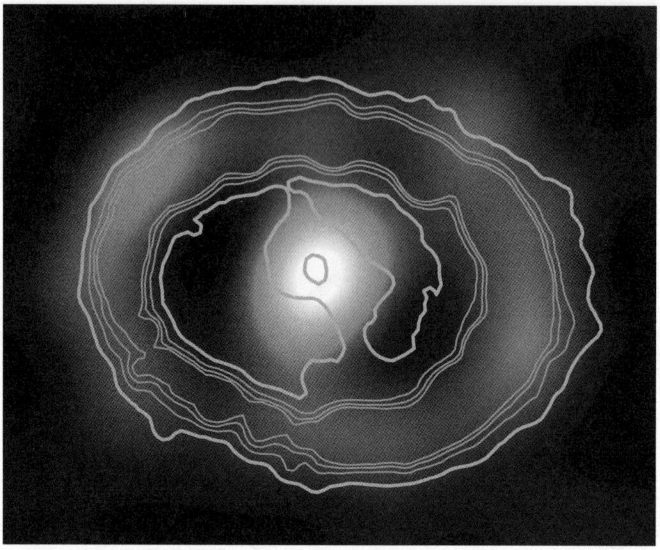

Figure 7. An outline of the equatorial ring and inner debris, as seen with *HST* (green/blue contours), on top of ALMA observations of the remnant at 345 GHz (Zanardo *et al.* 2014) (red/orange). The dust in the central region is rendered. (Hubble data: NASA/STScI/CfA/P. Challis and R. Kirshner.)

emission from the outer shock, the residual emission of $\sim$ 3 mJy (see Figure 6) in the central region could correspond to the synchrotron emission from an embedded pulsar, possibly located at $\sim$ 50 mas from the SN site. This PWN position would match the centre of the inner emission detected at 44 GHz (Figure 4), and would imply a natal kick of $\sim$ 400 km s^{-1}, for a possible pulsar period of $\sim$150 ms (Zanardo *et al.* 2014). An asymmetric explosion of the progenitor could have led to a high-velocity neutron star (Janka 2012) and, thus, to the formation of a PWN offset from the SN site.

After unsuccessful searches for the pulsar signal with the Parkes telescope (Manchester 1988, 2007), the preliminary analysis of data from an extended pulsar search campaign in 2013 has tentatively identified several candidates with period $30 \lesssim P \lesssim 300$ ms (Zanardo 2014). However, there is yet no convincing detection and the search continues.

References

Ball, L., Campbell-Wilson, D., Crawford, D. F., & Turtle, A. J. 1995, *ApJ.*, 453, 864

Ball, L., Crawford, D. F., Hunstead, R. W., Klamer, I., & McIntyre, V. J. 2001, *ApJ*, 549, 599

Bell, A. R. & Lucek, S. G. 2001, *MNRAS*, 321, 433

Bell, A. R. 2004, *MNRAS*, 353, 558

Boggs *et al.* 2015, *Science*, 348, 670

Callingham, J. R., Gaensler, B. M., Zanardo, G. *et al.* 2016, *MNRAS*, 462, 290

Chevalier, R. A. & Fransson, C. 1987, *Nature*, 328, 44

Chevalier, R. A. & Soker, N. 1989, *ApJ.*. 341, 867

Crotts, A. P. S. & Heathcote, R. S. 2000, *ApJ*, 528, 426

Drury, L. Oc. 1983, *Rep. Prog. Phys.*, 46, 973

Gaensler, B. M., Manchester, R. N., Staveley-Smith, L. *et al.* 1997, *ApJ*, 479, 845

Gaensler, B. M. 1998, *ApJ*, 493, 781

Gaensler, B. M. & Slane, P. O. 2006, *ARAA*, 44, 17

Helder, E. A., Broos, P. S., Dewey, D. *et al.* 2013, *ApJ*, 764, 11

Indebetouw, R., Matsuura, M., Dwek, E., *et al.* 2014, *ApJ.*, 782, L2

Janka, H.-T. 2012, *Annual Review of Nuclear and Particle Science*, 62, 407
Kamenetzky, J., McCray, R., Indebetouw, R., *et al.* 2013, *ApJ.*, 773, L34
Kjær K., Leibundgut, B., Fransson, C. *et al.* 2010, *A&A*, 517, A51
Kirk, J. G., Duffy P. & Gallant, Y. A. 1996, *A&A*, 314, 1010
Lakićević, M., Zanardo, G., van Loon, J. Th., *et al.* 2012a, *A&A*, 541, L2
Lakićević, M., van Loon, J. Th., Stanke, T. *et al.* 2012b, *A&A*, 541, L1
Larsson, J., Fransson, C., Kjær , K., *et al.* 2013, *ApJ*, 768, 89
Leibundgut, B. & Suntzeff, N. B. 2003, Supernovae and Gamma-Ray Bursters, edited by K. Weiler, Lecture Notes in Physics, vol. 598, p.77-90
Manchester, R. N. 1988, *PASA*, 7, 548-549.
Manchester, R. N. 2007, *AIP Conf. Proc.*, 937, 134-143
Matsuura, M., Dwek, E., Meixner, M., *et al.* 2011, *Science*, 333, 1258
Matsuura, M., Dwek, E., Barlow, M. J. *et al.* 2015, *ApJ.*, 800, 50
Matsuura, M., Indebetouw, R., Woosley, S. *et al.* 2017, *MNRAS*, in press, arXiv:1704.02324
Ng, C.-Y., Gaensler, B. M., Staveley-Smith, L. *et al.* 2008, *ApJ.*, 684, 481
Ng, C.-Y., Potter, T. M., Staveley-Smith, L. *et al.* 2011, *ApJ* (Letters), 728, L15
Ng, C.-Y., Zanardo, G., Potter, T. M., *et al.* 2013, *ApJ.*, 777,131
Podsiadlowski, Ph., Morris, T. S., & Ivanova, N. 2007, *AIP Conf. Proc.*, 937, 125-133
Potter, T. M., Staveley-Smith, L., Ng, C.-Y. *et al.* 2009, *ApJ*, 705, 261
Potter, T. M., Staveley-Smith, L., Reville, B., *et al.* 2014, *ApJ.*, 794, 174
Reynolds, S. P. 2011, *Ap&SS*, 336, 257
Smartt, S. J. 2009, *ARAA*, 47, 63
Staveley-Smith, L. *et al.* 1992, *Nature*, 355, 147
Turtle, A. J., Campbell-Wilson, D., Bunton, J. D. *et al.* 1987, *Nature*, 327, 38
Turtle, A. J., Campbell-Wilson, D., Manchester, R. N. *et al.* 1990, *IAU Circ.* 5086, 2
Zanardo, G., Staveley-Smith, L., Ball, Lewis *et al.* 2010 *ApJ*, 710, 1515
Zanardo, G., Staveley-Smith, L., Ng, C.-Y. *et al.* 2013, *ApJ*, 767, 98
Zanardo, G., Staveley-Smith, L., Indebetouw, R., *et al.* 2014, *ApJ.*, 796, 82
Zanardo, G. 2014, *Radio Evolution of the Remnant of Supernova 1987A*, PhD Thesis, University of Western Australia.

Supernova 1987A:30 years later - Cosmic Rays and Nuclei from Supernovae and their aftermaths
Proceedings IAU Symposium No. 331, 2017
A. Marcowith, M. Renaud, G. Dubner, A. Ray & A. Bykov, eds.

doi:10.1017/S1743921317004690

X-raying the evolution of SN 1987A

Vinay L. Kashyap[1,2], David van Dyk[3], Katy McKeough[4], Frank Primini[1], Diab Jerius[1], Akshay Gowrishankar[5], Aneta Siemiginowska[1] and Andreas Zezas[1,6]

[1]Harvard-Smithsonian Center for Astrophysics, 60 Garden St., Cambridge MA 02138, USA
[2]email: vkashyap@cfa.harvard.edu
[3]Dept. of Statistics, Imperial University, London, UK
[4]Dept. of Statistics, Harvard University, Cambridge MA, USA
[5]Acton-Boxborough Regional High, Acton MA, USA
[6]Dept. of Physics, University of Crete, Heraklion, Greece

Abstract. SN 1987A has been observed with the *Chandra* X-ray Observatory over the entire course of the mission. We have re-analyzed the archival data by constructing an empirical point spread function and reconstructing high-resolution images using a Bayesian multi-scale image reconstruction algorithm. We are able to resolve structure in the equatorial ring of SN 1987A with unprecedented detail, at scales of $\approx \frac{1}{4}$ arcsec. We describe how the point spread function is constructed, and the reconstruction method, and explore the evolution of the inner ring at different epochs and passbands.

Keywords. supernova remnants, supernovae: individual (SN 1987A), circumstellar matter, methods: data analysis, methods: statistical, techniques: image processing, X-rays: general

1. Introduction

SN 1987A is the closest supernova of the modern era (at ≈ 50 kpc in the Large Magellanic Cloud), and has served as a touchstone for models of Type II core-collapse supernovae and the subsequent evolution of the ejecta. In the thirty years since the event, it has been studied extensively in multiple wavebands. In the optical, it shows an evolving inner ring and two relatively stationary outer rings (see, e.g., Burrows *et al.* 1995). The outer rings have been interpreted as light echoes, and the inner ring as due to the blast wave running into dense material. The inner ring was first observed in X-rays with ROSAT (Hasinger *et al.* 1996). It was first resolved in X-rays with *Chandra* (Burrows *et al.* 2000) and its evolution has been continuously monitored since then (see, among many others, Park *et al.* 2006, Racusin *et al.* 2009, Helder *et al.* 2013, Frank *et al.* 2016). The X-ray intensity had increased continuously, faster since c.2004, but appears to have flattened out and started decreasing since c.2013, especially in the soft and medium bands (see Figure 1). There have been a number of morphological changes, with different parts of the ring brightening and fading. The expansion rate has shown a dichotomy, expanding faster in hard X-rays than in soft X-rays since c.2004, which has been interpreted as due to the blast wave hitting denser material, resulting in strong shocks that are brighter in soft X-rays and move more slowly (Frank *et al.* 2016).

We have developed new *Chandra* calibration products and devised image analysis methods that promise to radically improve the inferences that can be drawn from the data. In §2, we describe how an empirical PSF is made at sub-pixel resolution. In §3,

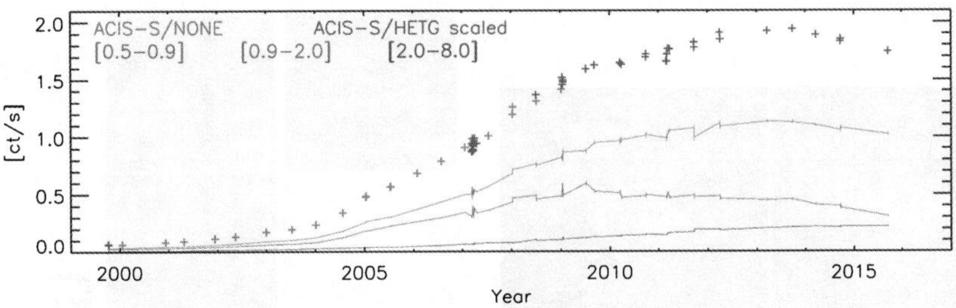

Figure 1. Count rates of SN 1987A in *Chandra*/ACIS-S. Observations made with bare ACIS-S (red '+') and those made with HETGS+ACIS-S (purple '+') scaled to match the bare ACIS count rates between 2007-2009 are shown, as well as the count rates in the soft (red line), medium (green line), and hard (blue line) passbands.

we describe a Bayesian multi-scale algorithm that we apply to the *Chandra* images. We then reanalyze *Chandra*/ACIS data of SN 1987A, and discuss preliminary results in §4.

2. *Chandra* Empirical PSF

SN 1987A has been observed predominantly with the *Chandra*/ACIS-S detector, a CCD with pixels $0''.49$ arcsec in size, comparable to the size of the point spread function (PSF). We have empirically constructed a PSF based on observations of on-axis, weak, isolated, point sources, and use it to analyze all ACIS-S data. The bare ACIS-S has accumulated 663 ksec of exposure, but most observations after c.2007 have been carried out with the high-energy transmission grating (HETG) in the optical path, accumulating 1.45 Msec of exposure. A few observations have utilized the low-energy transmission grating (LETG) and some have used the HRC-I, a microchannel plate detector; because these observations involve significantly different shapes of the PSF, we ignore them here.

Chandra data are acquired as photon events, with pixel location, arrival time, and the photon energy recorded. The CCD readout time varies between $2.4 - 3.2$ sec for observations of SN 1987A, and the energy resolution on the S3 chip ranges from $\approx 90 - 150$ eV between 1-5 keV. Pixel super-resolution in sky-coordinates is made possible primarily through spacecraft dither (the telescope slews during an observation in a Lissajous pattern with pitch period of 707 sec and yaw period of 1000 sec and an amplitude of 8 arcsec), which allows normal vectors to pixels to be mapped to the sky on average at a resolution of ≈ 0.07 pix.

Further improvements to photon location are made possible through more precise photon arrival locations. Li *et al.* (2004) have devised a method called EDSER (Energy Dependent Subpixel Event Resolution) to determine where on the pixel a photon landed based on the shape of the charge collected on the CCD (known as the charge bloom), recorded as the event grade. A fully symmetrical charge bloom centered on a pixel results in the assignment of grade=0, and no location correction is applied; for asymmetrical charge blooms, grade>0, and an energy dependent correction is applied to shift the photon location from pixel center towards one of the edges. EDSER is automatically applied to ACIS data in *Chandra* pipeline processing, and has been shown to substantively improve the sharpness of *Chandra* super-resolved images (see, e.g., `http://cxc.harvard.edu/ciao/why/acissubpix.html`). Unfortunately, the PSF of the super-resolved EDSER data has not yet been calibrated. As part of the *Chandra* calibration process, we have constructed an empirical PSF based on real data.

We constructed the empirical PSF by first identifying sources in the *Chandra* Source Catalog (CSC; `http://cxc.harvard.edu/csc`) that were flagged as not extended,

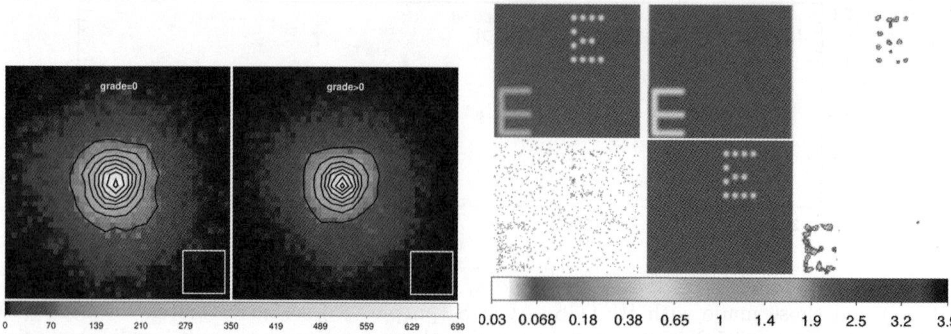

Figure 2. Left set: *The ACIS empirical PSF, constructed for grade=0 events (left) and for grade>0 events (right). The size of a detector pixel is shown as the square white box.*
Right set: *Feature detection using* `LIRA`, *showing simulation and analysis images. The true image (top left; of size 128×128) consists of two 'E' shaped sources of similar intensity, one extended and the other spotty, on a flat background and convolved with a Gaussian PSF ($\sigma = 6$ bins). A nominal simulated Poisson dataset is shown at bottom left. The images in the second column show baseline models that are input and are assumed to be known a priori: in the top panel, the diffuse "E" is taken to be known, and the unknown point sources are reconstructed; in the bottom panel, the point sources are taken to be known, and the unknown extended source is reconstructed. The images in the third column show the point sources (top) and the extended source (bottom) that are inferred from the multi-scale residuals (Connors et al. 2011).*

isolated (with no catalogued sources within $6''$ to avoid contamination of the wings), well-observed (>200 counts), on-axis (<1 arcmin), and weak (<0.1 ct sec^{-1}, to avoid pileup). The individual observations were downloaded from the archive in Jan 2017, reprocessed with CIAO v4.9 (see Fruscione *et al.* 2006) and CALDB v4.5.1.1 (Graessle *et al.* 2006), derolled so that all the datasets were aligned in the detector frame, and then stacked, resulting in a combined events list of >90 kcounts. We then split the accumulated events into separate grade=0 and grade>0 sets, and constructed images binned at $\frac{1}{8}$ pixel, corresponding to a sky resolution of $0''.06125$. The images were then locally adaptively smoothed to reduce statistical fluctuations in the wings. This was achieved by computing smoothed images at a number of smoothing scales, and choosing at each pixel the smallest smoothing scale which achieved a S/N of 2. A background was then estimated from an annulus of $4''.4 - 5''.9$ and subtracted out. Pixels that ended up with negative values were filled in by iteratively interpolating from surrounding pixels until no negative values were left. The final PSFs are shown in Figure 2 (*left*).

A number of future improvements to the PSF are still necessary: first, the current list of sources is not exhaustive, and with the release of version 2 of the CSC (with full RA coverage and newer data included), we expect to nearly double the number of sources included and improve S/N; second, the current PSF does not distinguish between the back- and front-illuminated ACIS chips; third, we assume that the HETGS does not introduce measurable distortions in the PSF shape, but this assumption has not been verified yet; and fourth, some of the sources selected still have nearby apparently astrophysical structures that are visible to the eye, and though their effect is minimized by derolling and stacking of multiple sources, they must be manually culled in order to obtain a clean PSF profile. Nevertheless, the empirical PSF is a higher fidelity representation of a *Chandra*/ACIS on-axis source than is achievable currently through raytraces.

3. `LIRA`

We apply the *Chandra*/ACIS PSF generated as above, weighted by the observed proportions of grade=0 and grade>0 events, to ACIS observations of SN 1987A to infer the true underlying intensity distribution. For this purpose, we use a Bayesian Markov chain Monte Carlo (MCMC) based multi-scale algorithm `LIRA` (Low-counts Image Reconstruction and Analysis; Esch *et al.* 2004, Connors & van Dyk 2007, Connors *et al.* 2011). `LIRA` is an open sourced R package, available at `https://github.com/astrostat/LIRA`.

A given image may be decomposed into a known "baseline" component and an unknown "multi-scale" component. `LIRA` constructs a Bayesian multi-scale model that is added to the baseline model to fit to the observed image. It can thus be thought of as modeling the residuals, and quantifies the structure in the residuals that exceeds what would be expected under the baseline model. Extended sources are identified in the multi-scale residuals.

We show an example of how `LIRA` works with two toy examples illustrated in Figure 2 (*right*). Two artificial sources, both shaped as 'E', one extended and one made up of 12 point sources, are analyzed. There are 132 counts in each of the 'E's, and there is a flat background of $\sim 10^3$ ct/image. In both cases, we recover the unknown component via multi-scale residuals, which easily pick out the relevant components at high significance. For each case, the 'sky-truth' (convolved with the PSF), the simulated Poisson data, the two different "baseline models", and the `LIRA` result, with the baseline modeled out are shown. Thus, we can extract the extended 'E' by adopting the set of 12 point source 'E' as the null model (top panels), or conversely extract the point sources by adopting the extended 'E' as the null model (bottom panels).

We note that `LIRA` is *not* a deconvolution algorithm, but rather draws from the posterior distribution of the intensity at each pixel. A Bayesian reconstruction algorithm has several advantages over regular deconvolution. It removes ambiguity about stopping rules, provides a measure of the uncertainty in the reconstruction via the posterior probability density distribution, and allows the daisy-chaining of analyses such that the result of analyzing one dataset may be easily incorporated into the next. Furthermore, because it is MCMC based, we can obtain error bars on all estimates of interest. The output of this MCMC algorithm is a series of images, one per iteration, describing the multiscale structure in the residuals. The uncertainty in the structures is defined by the distribution of the intensities in these images. For instance, the significance of an extended structure can be determined by comparing this distribution to that derived from images known to be free of structure, i.e., containing only the baseline model (Stein *et al.* 2015). We have recently used this to detect several jet features in X-rays from high-redshift Quasars (McKeough *et al.* 2016). Future improvements to the algorithm include accounting for uncertainties in the PSF (mainly based on the techniques pioneered by Lee *et al.* 2011 and Xu *et al.* 2014), and computing the intensities in different passbands without having to run `LIRA` independently on images made in each band.

4. SN 1987A Analysis Results

We have run `LIRA` on all the ACIS observations available in the *Chandra* public archive, but present only a subset of representative results here. In all cases, we adopt a flat baseline component that corresponds to the estimated background in the field, and run `LIRA` for 1000 iterations, and discard the first 400 to account for burn-in. We run `LIRA` on images in four separate bands: broad (B; $0.5-8$ keV), soft (S; $0.5-0.9$ keV), medium (M; $0.9-2$ keV), and hard (H; $2-8$ keV). In order to account for the possibility of systematic uncertainties in the PSF, we smooth the images output at each iteration with a Gaussian kernel of $\sigma = \frac{1}{4}$ image pixel. This is an *ad hoc* correction that disperses $\approx 9\%$ of the flux in each pixel into the surrounding pixels. We then compute the intensity maps of the X-ray emission from SN 1987A as the smoothed mean of the multi-scale residual

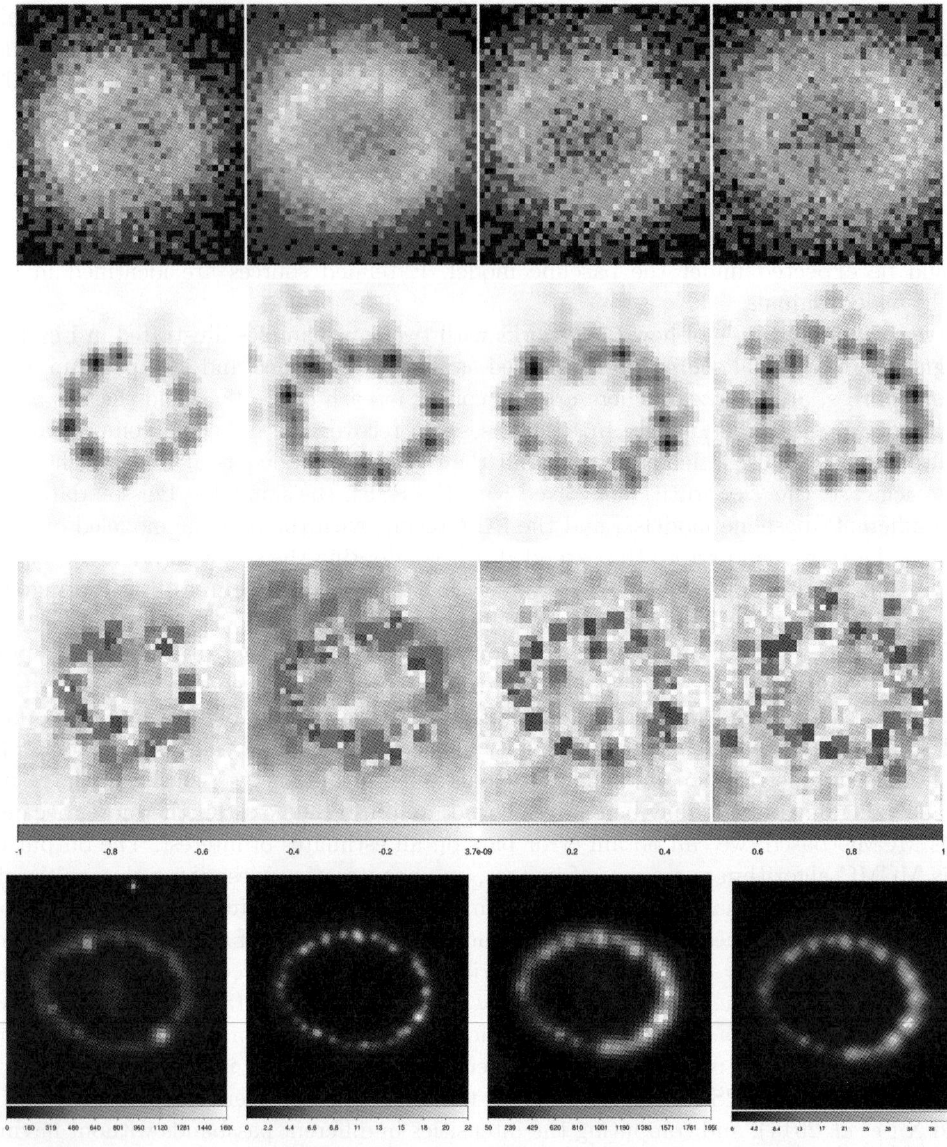

Figure 3. Comparing selected X-ray data and LIRA reconstructions with *Hubble*. The super-resolved *Chandra* broad band images are shown in the *top row* for data from 2000-Dec, 2007-Jul, 2011-Mar, and 2015-Sep. The corresponding LIRA reconstructions are shown in the *second row*; these images are further Gaussian smoothed to make the structure of the X-ray hot spots visually clearer. The X-ray color $HR_{HS} = \frac{H-S}{H+S}$ maps are showin in the *third row*, with red and blue shades indicating regions with soft and hard spectra respectively. Contemporaneous *Hubble* images (from 2000-Nov, 2006-Dec, 2011-Apr, and 2015-May) are shown for comparison in the *bottom* row (see also Fransson *et al.* 2015). All plots are $2.4'' \times 2.4''$ in size. Spatial registration across epochs is approximate.

images obtained in each iteration. We also compute the fractional hardness ratios in non-overlapping bands, e.g., $HR_{HS} = \frac{H-S}{H+S}$. We compute the uncertainties on the intensities and on the hardness ratios as the sample standard deviation.

Some results are shown in Figures 3 for a selected set of epochs. Two are for bare ACIS-S observations (ObsIDs 1967 and 7637 from 2000-Dec and 2007-Jul), and two are

HETGS+ACIS-S observations (ObsIDs 12539 and 16756 from 2011-Mar and 2015-Sep). The X-ray images show a clear increase in the size of the equatorial ring, as well as indications of emission outside the ring, indicating that the shock ejecta is ablating the material of the ring and punching through to the surrounding H II region. Comparison with near-contemporaneous HST images shows a remarkable coincidence of the optical knots with X-ray hot spots. There does not exist a one-to-one correspondence between the optical knots and the X-ray hot spots, but the matches and the morphology cannot be explained as chance coincidences. The ring of X-ray spots is more circular at earlier epochs, suggesting a large separation between the shock front and maximum thermalization, or a large variations in the density of the material in the equatorial ring. This is also borne out by the hardness ratio maps, which shows some X-ray spots dominated by soft emission and others by hard emission. A clear physical interpretation of these features is as yet lacking, and comparisons with hydrodynamic and MHD simulations (e.g., Orlando *et al.* 2015) are necessary.

In summary, we have developed a method for analyzing the high-resolution X-ray structure in SN 1987A, and have applied it to *Chandra* ACIS images. The combination of an improved PSF and a powerful reconstruction algorithm has unlocked the potential for significant new understanding in the evolution of the remnant. We plan further improvements to our analysis to use more accurate PSFs, to minimize Monte Carlo error in the reconstructions by running the chains 10x longer, to improve `LIRA` to better account for uncertainties in the PSF and to analyze the different passbands simultaneously, to evaluate the stability of the reconstructions by analyzing observations made close together, and to determine the statistical significances of X-ray features that have appeared outside the equatorial ring.

Acknowledgements

We acknowledge support for this work from NASA Contract NAS8-03060 to the *Chandra* X-ray Center. DvD and AZ acknowledge support from EU Horizon 2020 Project RISE-ASTROSTAT. DvD also acknowledges support from NSF Award number 1513492. We thank Xiao-Li Meng, Nathan Stein, and Salvatore Orlando for useful discussions.

References

Burrows, D., *et al.* 1995, *ApJ*, 452, 6808
Burrows, D., *et al.* 2000 *ApJ*, 543, L149
Connors, A. & van Dyk, D. A. 2007 *SCMA IV*, ASPC v371, p101
Connors, A., *et al.* 2011 *ADASS XX*, v442, p643
Esch, D., *et al.* 2004 *ApJ*, 610, 1213
Frank, K., *et al.* 2016 *ApJ*, 829, 40
Fransson, C., *et al.* 2015 *ApJL*, 806, L19
Fruscione, A., *et al.* 2006 *Proc. SPIE*, 6270, 62701V
Graessle, D., *et al.* 2006 *Proc. SPIE*, 6270, 62701X
Hasinger, G., *et al.* 1996 *A&A*, 312, 9
Helder, *et al.* 2013 *ApJ*, 764, 11
Lee, H., *et al.* 2011 *ApJ*, 731, 126
Li, *et al.* 2014 *ApJ*, 610, 1204
McKeough, K., *et al.* 2016 *ApJ*, 833, 123
Orlando, S., *et al.* 2015 *ApJ*, 810, 168
Park, T., *et al.* 2006 *ApJ*, 646, 1001
Racusin, D., *et al.* 2009 *ApJ*, 703, 1752
Stein, N., *et al.* 2015 *ApJ*, 813, 66
Xu, J., *et al.* 2014 *ApJ*, 794, 97

Supernova 1987A:30 years later - Cosmic Rays and Nuclei from Supernovae and their aftermaths
Proceedings IAU Symposium No. 331, 2017
A. Marcowith, M. Renaud, G. Dubner, A. Ray & A. Bykov, eds.

doi:10.1017/S1743921317004604

High-Resolution Observations of Dust in SN 1987A

Phil Cigan[1], Haley Gomez[1] and Mikako Matsuura[1,2]

[1]Cardiff University School of Physics and Astronomy
The Parade, Cardiff, CF24 3AA
email: ciganp@cardiff.ac.uk
[2]Department of Physics and Astronomy, University College
London, Gower Street, London, WC1E 6BT, UK

Abstract. The dust produced by supernovae is an important topic for understanding supernova physics and the chemical evolution of galaxies. Recent ALMA observations of SN 1987A have allowed us to peer into the inner ejecta to the cool dust, with spatial resolution from $0''.3$ at $\sim$300 GHz down to $0''.09$ at $\sim$680 GHz – an improvement over the previous 300 GHz Cycle 0 observations at $0''.69$. Comparison of the dust location and morphology with other multiwavelength emission presents an interesting picture of the role dust plays in the ejecta. The mm-FIR SED is compared to radiative models to study the dust composition 30 years after the initial explosion. Fits to the ring emission also probe the drift of the center of the system over time.

Keywords. dust, supernovae: individual (SN 1987A)

---

1. Introduction

The youngest known supernova (SN) remnant in the Milky Way Galaxy is G1.9+0.3, at roughly 100 years old (Reynolds *et al.* 2008). SN 1987A, located roughly 50 kpc away in the Large Magellanic Cloud, is the nearest supernova detected since then, and is the first supernova whose structure is able to be resolved with modern instrumentation less than 30 years after its initial explosion. Among many other areas in astrophysics, SN 1987A has made for an unprecedented target of study – new spatially resolved images of dust and molecules can reveal the chemistry resulting in SN dust formation.

Models in the literature predict that core-collapse supernovae (CCSN) can be an important source of dust in galaxies if they produce at least moderate (0.1–1.0 $M_\odot$) amounts of dust (Morgan & Edmunds 2003; Nozawa *et al.* 2003; Gomez *et al.* 2012; Gall *et al.* 2014; Dwek & Cherchneff 2011). Likewise, CCSN could be a major source of interstellar medium (ISM) dust in general (e.g., Matsuura *et al.* 2009; Dwek & Cherchneff 2011). Low-metallicity galaxies also provide insight into the requirements for dust sources. Their typically low dust-to-gas ratios can at least partially be due to low condensation rates in the ISM; subsequently, dust from stars could be responsible for a significant amount of dust on short timescales (Zhukovska 2014).

These nearby dwarfs are often taken as analogues to early galaxies in the universe, but direct observations of high redshift galaxies are also important for piecing together the puzzle of dust formation across cosmic timescales. The so-called dust budget crisis (e.g., Morgan & Edmunds 2003; Rowlands *et al.* 2014; Schaerer *et al.* 2015), similar to the case of low-Z dwarfs, highlights the need to be able to build up large amounts of dust relatively quickly. Sources beyond just AGB stars are needed to account for the dust observed, and supernovae such as SN 1987A are excellent candidates.

All of this leads to two big questions regarding the role of SN in dust production: How much dust do SNe really produce? And for matching any resolved observations to models, how is dust distributed in SN ejecta? The following work presents some developments in the effort to provide answers to these questions using new ALMA observations of SN 1987A.

2. History of Dust in SN 1987A

A brief history of the dust in SN 1987A is useful for context. From day 260–1316, the dust emission was dominated by mid-infrared (MIR) thermal emission (Bouchet *et al.* 1991; Wooden *et al.* 1993). During this time, $\sim 10^{-4}$ $M_{\odot}$ dust was inferred from the observed signal, presumed to be from the ejecta.

23 years later, between days 8467 and 8564, the dust had cooled down enough to be studied in the far-infrared (FIR). Thermal emission was detected with *Herschel* from 100–350 μm (Matsuura *et al.* 2011), where a quite massive quantity of 0.4–0.7 $M_{\odot}$ of dust was inferred. The dust emission was confirmed to originate in the ejecta by Indebetouw *et al.* (2014). This result raised questions, as the mass was quite a bit larger than those previously inferred from Spitzer MIR emission. However, *Herschel* observations of Cas A (Dunne *et al.* 2003, 2009; Barlow *et al.* 2010; De Looze *et al.* 2017) and the Crab Nebula (Gomez *et al.* 2012; Owen & Barlow 2015) also showed high masses of $M>0.1M_{\odot}$.

More recently during day 9090 in 2012, fresh *Herschel* observations presented by Matsuura *et al.* (2015) confirmed SN 1987A's large mass of dust of around 0.8 $M_{\odot}$. The best model of the emission was comprised of 0.3 $M_{\odot}$ of amorphous carbons and 0.5 $M_{\odot}$ of silicates. Crucially, these new observations used the 70 μm band to constrain the lower wavelength end of the thermal distribution and show that the dust temperature was cool, with $T < 30$ K.

3. Preliminary Results

The Atacama Large Millimeter Array (ALMA) brings to bear a game-changing capability to observe targets such as SN 1987A in the FIR to sub-millimeter. A full discussion of our observations, reduction, analysis, and implications is provided in our forthcoming paper (Cigan *et al.*, *in preparation*). The new ALMA observations in bands 7 (275–373 GHz) and 9 (602–720 GHz) give spatial maps at $0\farcs3$ and $0\farcs09$ resolution, respectively. This is an improvement over the previous Cycle-0 ALMA maps at $0\farcs69$ resolution presented by Indebetouw *et al.* (2014).

Qualitatively, the ring is visible in the ALMA submm windows in synchrotron radiation, while the ejecta follows a thermal greybody profile. The dust emission peaks near the notable gap in the optical HST map (Fransson *et al.* 2015). Figure 1 shows the SN 1987A 315 GHz image in comparison to several other wavelength bands.

Integrating the signal in each continuum window, it is possible to build spectral energy distributions (SEDs) for the separate ring and ejecta components. Unfortunately, the ALMA bands only cover the Rayleigh-Jeans tail of a cool thermal distribution, so modified blackbody fits would be relatively unconstrained. However, we can utilize the recent *Herschel* flux densities from Matsuura *et al.* (2015) as upper limits to the current SED (not as fully-weighted data points) and therefore constrain the fits. See Figure 2 for the integrated flux densities in the ring and ejecta. As an initial analysis of the ejecta dust, we fit a single modified blackbody function to the integrated ALMA ejecta flux densities. We use κ_{850}=0.07 m^2 kg^{-1} from James *et al.* (2002) for the absorption coefficient. The maximum likelihood estimation (MLE) fit yields a dust mass of 0.82 $\pm$0.40 $M_{\odot}$, consistent with the previous findings of a large dust mass in the system. The integrated fitted dust profile is also cool, at 22.8 $\pm$5.0 K.

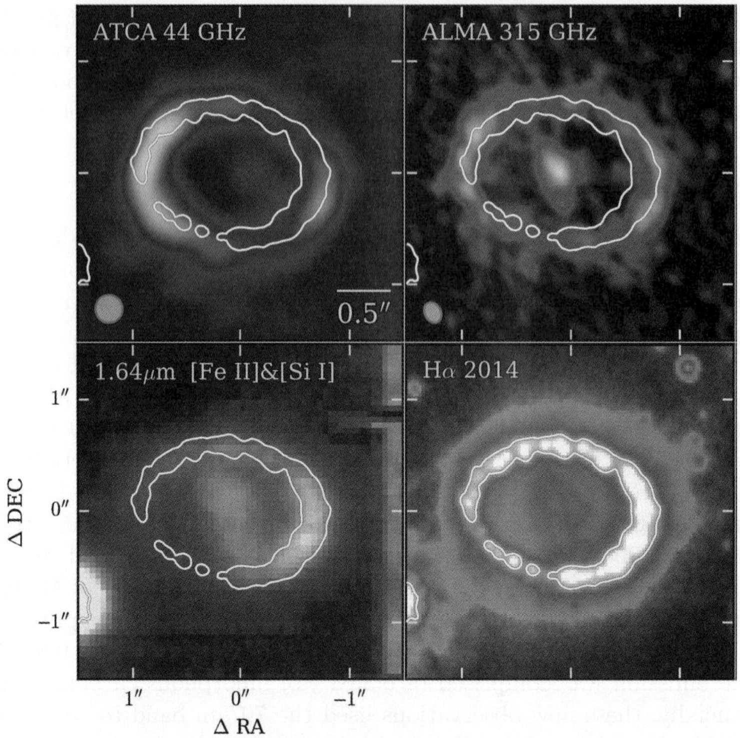

Figure 1. Multi-wavelength view of SN 1987A: ATCA 44 GHz (Zanardo *et al.* 2013), ALMA 315 GHz (this work), 1.64 μm [Fe II]& [Si I] (Fransson *et al.* 2016), and Hα (Fransson *et al.* 2015). The yellow contours trace the optical ring emission.

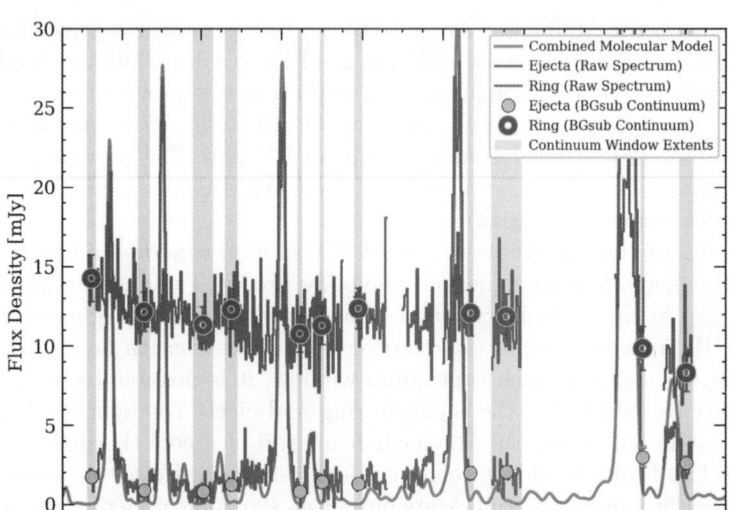

Figure 2. Raw (without background subtraction) band 6–7 spectra and integrated continuum flux densities in the ring and ejecta. Continuum windows selected to be relatively free of emission based on molecular models from Matsuura *et al.* 2017, *accepted*.

4. Conclusions

These new results confirm the previous findings of Matsuura *et al.* (2015) that SN 1987A has created a large amount of dust, ~0.8 $M_{\odot}$, within 30 years of its initial explosion. This lends evidence to the picture of CCSN being able to provide large amounts of dust in the ISM on short timescales. However, it should be noted that this is not the final word, as this will almost certainly not be the final yield of dust into the ISM from this object – that will depend on exactly how much ejecta dust is destroyed by the reverse shock, among other factors. Further monitoring and models will be required to verify the final dust yield. A spatially-resolved study of the thermal profiles in the ejecta is required for comparison with detailed models of the chemical evolution. We provide this spatially-resolved treatment in our second forthcoming paper, where we also compare the resolved dust emission to several molecular species from Abellan, Indebetouw, *et al.* (*submitted*).

Acknowledgments: PJC and HLG acknowledge support from the European Research Council (ERC) in the form of Consolidator Grant CosmicDust (ERC-2014-CoG-647939, PI H L Gomez). MM acknowledges support from the UK STFC Ernest Rutherford fellowship (ST/L003597/1).

References

Barlow, M. J., Krause, O., Swinyard, B. M., *et al.* 2010, *A&A*, 518, L138
Bouchet, P., Danziger, I. J., & Lucy, L. B. 1991, *AJ*, 102, 1135
De Looze, I., Barlow, M. J., Swinyard, B. M., *et al.* 2017, *MNRAS*, 465, 3309
Dunne, L., Eales, S., Ivison, R. J., *et al.* 2003, *Nature*, 424, 285
Dunne, L., Maddox, S. J., Ivison, R. J., *et al.* 2009, *MNRAS*, 394, 1307
Dwek, E. & Cherchneff, I. 2011, *ApJ*, 727, 63
Fransson, C., Larsson, J., Migotto, K., *et al.* 2015, *ApJL*, 806, L19
Fransson, C., Larsson, J., Spyromilio, J., *et al.* 2016, *ApJL*, 821, L5
Gall, C., Hjorth, J., Watson, D., *et al.* 2014, *Nature*, 511, 326
Gomez, H. L., Krause, O., Barlow, M. J., *et al.* 2012, *ApJ*, 760, 96
Indebetouw, R., Matsuura, M., Dwek, E., *et al.* 2014, *ApJL*, 782, L2
James, A., Dunne, L., Eales, S., & Edmunds, M. G. 2002, *MNRAS*, 335, 753
Matsuura, M., Barlow, M. J., Zijlstra, A. A., *et al.* 2009, *MNRAS*, 396, 918
Matsuura, M., Dwek, E., Meixner, M., *et al.* 2011, *Science*, 333, 1258
Matsuura, M., Dwek, E., Barlow, M. J., *et al.* 2015, *ApJ*, 800, 50
Morgan, H. L. & Edmunds, M. G., 2003, *MNRAS*, 343, 427
Nozawa, T., Kozasa, T., Umeda, H., Maeda, K., & Nomoto, K. 2003, *ApJ*, 598, 785
Owen, P. J. & Barlow, M. J. 2015, *ApJ*, 801, 141
Reynolds, S. P., Borkowski, K. J., Green, D. A., *et al.* 2008, *ApJL*, 680, L41
Rowlands, K., Gomez, H. L., Dunne, L., *et al.* 2014, *MNRAS*, 441, 1040
Schaerer, D., Boone, F., Zamojski, M., *et al.* 2015, *A&A*, 574, A19
Wooden, D. H., Rank, D. M., Bregman, J. D., *et al.* 1993, *ApJS*, 88, 477
Zanardo, G., Staveley-Smith, L., Ng, C.-Y., *et al.* 2013, *ApJ*, 767, 98
Zanardo, G., Staveley-Smith, L., Indebetouw, R., *et al.* 2014, *ApJ*, 796, 82
Zhukovska, S. 2014, *A&A*, 562, A76

Supernova 1987A:30 years later - Cosmic Rays and Nuclei from Supernovae and their aftermaths
Proceedings IAU Symposium No. 331, 2017
A. Marcowith, M. Renaud, G. Dubner, A. Ray & A. Bykov, eds.

doi:10.1017/S1743921317004719

ALMA observations of Molecules in Supernova 1987A

M. Matsuura[1,2], R. Indebetouw[3,4], S. Woosley[5], V. Bujarrabal[6], F. J. Abellán[7], R. McCray[8], J. Kamenetzky[9], C. Fransson[10], M. J. Barlow[2], H. L. Gomez[1], P. Cigan[1], I De Looze[2], J. Spyromilio[11], L. Staveley-Smith[12,13], G. Zanardo[12], P. Roche[14], J. Larsson[15], S. Viti[2], J. Th. van Loon[16], J. C. Wheeler[17], M. Baes[18], R. Chevalier[3], P. Lundqvist[10], J. M. Marcaide[7], E. Dwek[19], M. Meixner[20,21], C.-Y. Ng[22], G. Sonneborn[19] and J. Yates[2]

[1]School of Physics and Astronomy, Cardiff University, Queen's Buildings, The Parade, Cardiff, CF24 3AA, UK
[2]Department of Physics and Astronomy, University College London, Gower Street, London WC1E 6BT, UK
[3]Department of Astronomy, University of Virginia, P.O. Box 400325, Charlottesville, VA 22904-4325, USA
[4]National Radio Astronomy Observatory, 520 Edgemont Rd, Charlottesville, VA 22903, USA
[5]Department of Astronomy and Astrophysics, University of California, Santa Cruz, CA 95064, USA
[6]Observatorio Astronómico Nacional (OAN-IGN), Apartado 112, E-28803 Alcalá de Henares, Spain
[7]Universidad de Valencia, C/Dr. Moliner 50, E-46100 Burjassot, Spain
[8]Department of Astronomy, University of California, Berkeley, CA 94720-3411, USA
[9]Westminster College, Department of Physics, 1840 South 1300 East, Salt Lake City, UT 84105, USA
[10]Department of Astronomy, The Oskar Klein Centre, Stockholm University, Alba Nova University Centre, SE-106 91 Stockholm, Sweden
[11]European Southern Observatory (ESO), Karl-Schwarzschild-Strasse 2, 85748 Garching, Germany
[12]International Centre for Radio Astronomy Research (ICRAR), The University of Western Australia, Crawley, WA 6009, Australia
[13]ARC Centre of Excellence for All-Sky Astrophysics (CAASTRO), Australia
[14]Department of Physics, University of Oxford, Oxford OX1 3RH, UK
[15]KTH, Department of Physics, and the Oskar Klein Centre, AlbaNova, SE-106 91 Stockholm, Sweden
[16]School of Physical and Geographical Sciences, Lennard-Jones Laboratories, Keele University, Staffordshire ST5 5BG, UK
[17]Department of Astronomy, University of Texas, Austin, TX 78712-0259, USA
[18]Sterrenkundig Observatorium, Universiteit Gent, Krijgslaan 281 S9, B-9000 Gent, Belgium
[19]Observational Cosmology Lab, Code 665, NASA Goddard Space Flight Center, Greenbelt, MD 20771, USA
[20]Space Telescope Science Institute, 3700 San Martin Dr., Baltimore, MD 21218, USA
[21]Department of Physics and Astronomy, The Johns Hopkins University, 366 Bloomberg Center, 3400 N. Charles Street, Baltimore, MD 21218, USA
[22]Department of Physics, The University of Hong Kong, Pokfulam Road, Hong Kong, China

Abstract. Supernova (SN) 1987A has provided a unique opportunity to study how SN ejecta evolve in 30 years time scale. We report our ALMA spectral observations of SN 1987A, taken in 2014, 2015 and 2016, with detections of CO, ^{28}SiO, HCO^+ and SO, with weaker lines of ^{29}SiO.

We find a dip in the SiO line profiles, suggesting that the ejecta morphology is likely elongated. The difference of the CO and SiO line profiles is consistent with hydrodynamic simulations, which show that Rayleigh-Taylor instabilities causes mixing of gas, with heavier elements much more disturbed, making more elongated structure.

Using ^{28}SiO and its isotopologues, Si isotope ratios were estimated for the first time in SN 1987A. The estimated ratios appear to be consistent with theoretical predictions of inefficient formation of neutron rich atoms at lower metallicity, such as observed in the Large Magellanic Cloud (about half a solar metallicity).

The deduced large HCO^+ mass and small SiS mass, which are inconsistent to the predictions of chemical model, might be explained by some mixing of elements immediately after the explosion. The mixing might have made some hydrogen from the envelope to sink into carbon and oxygen-rich zone during early days after the explosion, enabling the formation of a substantial mass of HCO^+. Oxygen atoms may penetrate into silicon and sulphur zone, suppressing formation of SiS.

Our ALMA observations open up a new window to investigate chemistry, dynamics and explosive-nucleosynthesis in supernovae.

Keywords. supernovae: individual:Supernova 1987A — ISM: supernova remnants — ISM: molecules — ISM: abundances — radio lines: ISM

---

1. Introduction

Core-collapse supernovae (SNe) are important source of metal content in galaxies. Inside the stellar interior, metals are synthesised and newly synthesised elements are ejected from SNe, that enrich the interstellar medium of galaxies. So far, most abundance measurements of SN remnants have focused on detecting lines at X-ray, UV and optical wavelengths. Constraints obtained from these short wavelength spectra are limited to the atomic lines of the main isotope only. ALMA detection of molecules in SN 1987A provide a different angle in abundance measurements, by investigating isotope abundances in SNe, using molecular lines. Further, ALMA can investigate molecular chemistry occurring in early phase of evolution in SN 1987A. Here, we report our spectroscopic survey of SN 1987A from Matsuura *et al.* 2017.

2. Observations and data reduction

ALMA obtained spectra covering a continuous spectral range from 210 to 300 GHz, in August and September 2014. Additionally, CO J=3–2 and 357 GHz HCO^+ J=4–3 was obtained on 2015 July 25th A series of slightly overlapping ~2000 km s^{-1} wide segments were observed in sequence; whereas a typical SN 1987A ejecta line width is 2000 km s^{-1} (Kamenetzky *et al.* 2013), so that one emission line can fill one ALMA spectrum segment.

3. ALMA Spectrum

In Figure 1 we show the observed spectrum of SN 1987A's ejecta, showing a variety of broad molecular lines. The strongest molecular emissions are attributed to SiO and CO. The emission at 267 GHz is associated with HCO^+ J=3–2, with some possible contamination from SO_2. Other weak features are due to SO and ^{29}SiO.

It has been thought that supernovae are molecule-poor environments, because hydrogen, the key element for chemical reaction, is expected to be rather deficient in SN ejecta, and because He^+ destroys molecules formed within it (Lepp *et al.* 1990, Rawlings & Williams, Sarangi & Cherchneff 2013). For young (<10 years old) SNe, CO and SiO had been the only molecules firmly detected (Matsuura 2017), with a recent report of

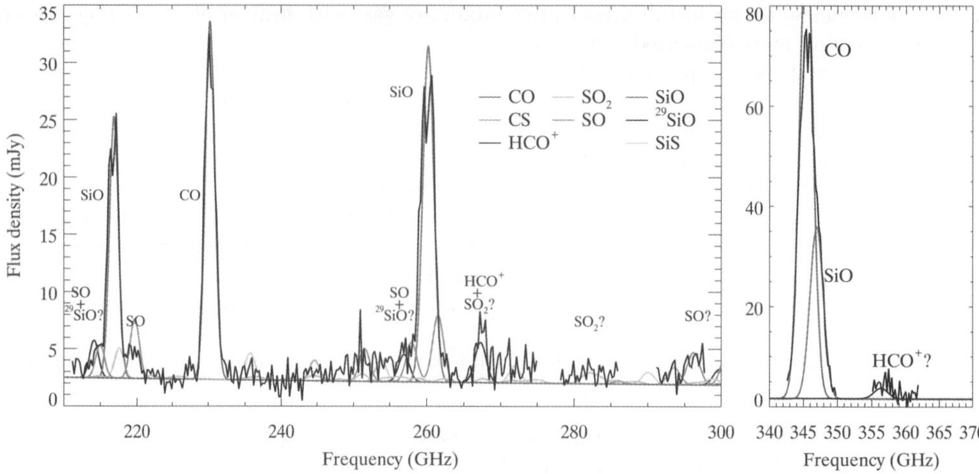

Figure 1. The ALMA 210–300 and 340-360 GHz spectra of SN 1987A's ejecta (black). Model molecular spectra with a 2000 km s^{-1} with FWHM Gaussian line profile and the excitation temperature of 40 K, are plotted in colour, guiding molecular identifications.

detecting near-infrared H_2 lines from SN 1987A (Fransson *et al.* 2016). Now from our spectral line survey with ALMA, we have discovered SO and HCO^+ in the ejecta of SN 1987A. These detections add to the already detected molecules at millimetre wavelengths, CO and SiO, and likely ^{29}SiO (Kamenetzky *et al.* 2013). The ejecta of young SN remnants may represent at a much more molecule-rich environment than had been thought before.

4. Discussion

4.0.1. *HCO^+ as a tracer of dense gas and possible requirement of mixing*

The detection of HCO^+ in SN 1987A was surprising. Chemical models by Sarangi & Cherchneff (2013) did not consider HCO^+ formation, while Rawlings & Williams (1990) did consider it but predicted very small HCO^+ mass.

Forming HCO^+ requires H_2, which is abundant only in the hydrogen envelope. Forming HCO^+ also requires H_2 in CO gas, as following reaction $H_3^+ + CO \rightarrow H_2 + HCO^+$, and $H_2^+ + H_2 \rightarrow H_3^+ + H$ (Williams & Viti 2014). There is a small fraction of C and O in the hydrogen envelope, because C and O were incorporated into the star at the time of star formation. However, the masses of intrinsic C and O are insufficient to account for the CO mass needed to generate substantial HCO^+ mass.

Forming a large mass of HCO^+ may have been enabled by macroscopic mixing in the ejecta early phases after the SN explosion, and that process might mix C and O in the stellar core and H in hydrogen envelope. A classic picture of stellar nucleosynthesis is that the progenitor undergoes a sequence of nuclear reactions in the stellar interiors, making onion-type shells composed of multiple zones with discrete elemental abundances. Figure 2 (a) shows the radial distributions of element mass without mixing. This figure is based on explosive nucleosynthesis models for SN 1987A (Sukhbold *et al.* 2016, Woosley 1988 and Woosley *et al.* 1997). Figure 2 (b) illustrate the effect of mixing; if there is some fraction of hydrogen mixed with carbon and oxygen. Instead of mixing elements completely, mixing might occur locally at the interfaces of the He and He-envelopes and the He and C+O zone, due to Rayleigh-Taylor instabilities (Müller *et al.* 1991,

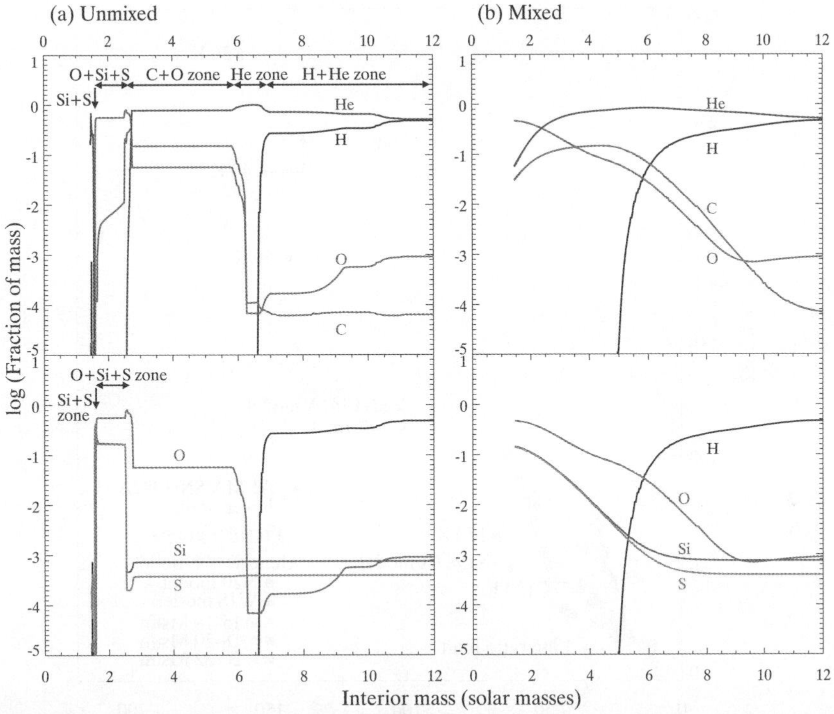

Figure 2. The modelled fractional abundance of atoms after SN explosion. The model is an $18\,M_\odot$ star with one third of the solar metallicity based on the model for Sk-69° 9202 (Sukhbold *et al.* 2016), but involving more extensive nuclear reaction network. Panel (a) shows SN 1987A model without mixing of nuclear burning zones. Panel (b) illustrates the effect of mixing, with the model with artificial mixing at $t = 10^7$ s. Having mixing triggered by the explosion can make elements from different nuclear burning zones mixed to some degree, though the scale of mixing is undetermined.

Wongwathanarat *et al.* 2015). The presence of a such large scale mixing at early times could have later resulted in a substantial mass of HCO^+.

4.0.2. *SiO and SiS chemistry*

Chemical models predict that the majority of Si is locked up in silicate dust grains at day 1500, and not so much SiO is present (Sarangi & Cherchneff 2013, Sarangi & Cherchneff 2015). Our measured SiO mass is consistent with this hypothesis qualitatively, but not quantitatively. Our estimated SiO mass is 4×10^{-5}–$2\times10^{-3}\,M_\odot$ which is a factor of 16–800 larger than the theoretically predicted SiO mass of $2.5\times10^{-6}\,M_\odot$(Sarangi & Cherchneff 2013). Nevertheless, Si atoms in SiO is only 3×10^{-5}–$1\times10^{-3}\,M_\odot$ which corresponds to less than 10 % of Si mass synthesised in the SN. Indeed, a large fraction of Si could be in silicate dust

Chemical models (Sarangi & Cherchneff 2013) predict that the majority of Si is in SiS rather than SiO, and the predicted SiS mass is $4.4\times10^{-2}\,M_\odot$. However, we find only small mass of SiS ($< 6\times10^{-5}\,M_\odot$).

We argue that a large SiO abundance and a small SiS abundance suggests the presence of some degree of mixing. Figure 2 (b) shows an example of a mixed case. At a time of 10^7 s (∼116 days) after the explosion, Si and S atoms are mixed with O atoms. The mixing enables one to form SiO more efficiently from co-existing Si and O atoms.

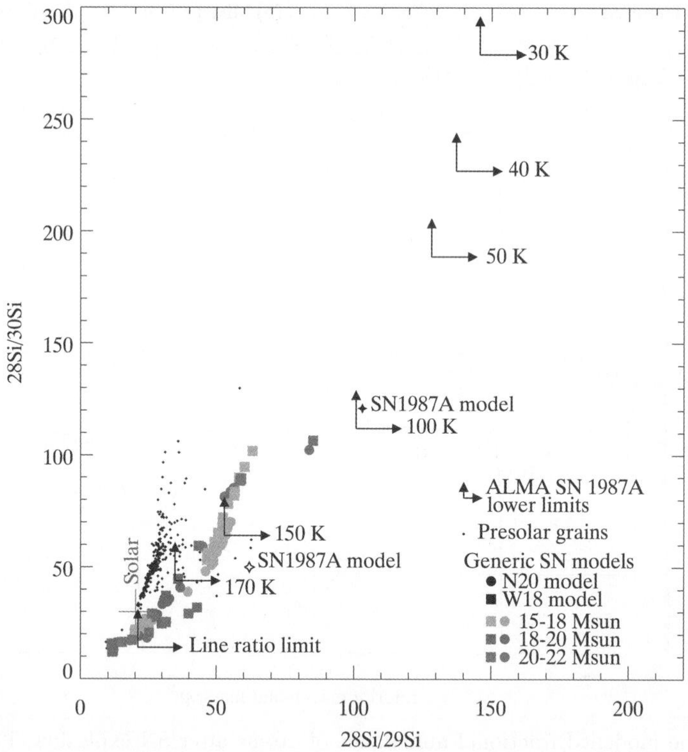

Figure 3. The Si isotope ratios. ALMA lower limits for SN 1987A are plotted with arrows, as a function of the assumed kinetic temperature of the molecule. The measured lower limits are compared with explosive nucleosynthesis models, including SN 1987A specific models (filled star: based on Woosley *et al.* 1988; open star: based on Nomoto *et al.* 2013). Generic solar abundance models are from Sukhbold *et al.* (2016), as a function of zero-age main sequence mass.

4.1. *Isotope ratios*

Our ALMA spectrum can cover the lines of SiO and CO isotopologues. That can be used to estimate lower limits for the isotope ratios of SN 1987A. Together with the Crab Nebula measurements of Ar isotope ratios (Barlow *et al.* 2013), SN 1987A is one of the first to estimate the isotope ratios for core-collapse SNe (Kamenetzky *et al.* 2013).

Fig. 3 shows our ALMA lower limits for ^{28}Si/^{29}Si and ^{28}Si/^{30}Si, as a function of the assumed RADEX kinetic temperature. These ALMA lower limits are compared with theoretically predicted ratios from explosive nucleosynthesis models based on Sukhbold *et al.* (2016).

The ratio of ^{30}Si/^{28}Si depends on the neutron excess, which depends on the initial metallicity of the star (Woosley 1988). Lower metallicity gives a lower neutron excess, resulting in a smaller production of neutron-rich isotopes. Because ^{30}Si requires more neutrons than ^{29}Si, and even more than ^{28}Si, ^{28}Si/^{30}Si becomes even larger than ^{28}Si/^{29}Si at lower metallicity, compared to the solar metallicity.

5. Acknowledgments

This paper makes use of the following ALMA data: ADS/JAO.ALMA#2013.1.00280.S, ADS/JAO.ALMA#2013.1.00063.S and ADS/JAO.ALMA#2012.1.00075.S. ALMA is a partnership of ESO (representing its member states), NSF (USA) and NINS (Japan), together with NRC (Canada), NSC and ASIAA (Taiwan), and KASI (Republic of Korea), in cooperation with the Republic of Chile. The Joint ALMA Observatory is operated by ESO, AUI/NRAO and NAOJ. MM and MJB acknowledge support from UK STFC grant (ST/J001511/1), and MM is further supported by an STFC Ernest Rutherford fellowship (ST/L003597/1). MJB acknowledges support from European Research Council (ERC) Advanced Grant SNDUST 694520. HLG acknowledges support from the European Research Council (ERC) in the form of Consolidator Grant CosmicDust (ERC-2014-CoG-647939). SW acknowledges support from supported by NASA grant (NNX14AH34G).

References

Barlow, M. J., Swinyard, B. M., Owen, P. J., Cernicharo, J., Gomez, H. L., Ivison, R. J., Krause, O., Lim, T. L., Matsuura, M., Miller, S., Olofsson, G., & Polehampton, E. T., 2013, *Science*, 342, 1343

Fransson, C., Larsson, J., Spyromilio, J., Leibundgut, B., McCray, R. A., & Jerkstrand, A., 2016, *ApJ*, 821, L5

Lepp, S., Dalgarno, A., & McCray, R. A., 1990, *ApJ*, 358, 262

Kamenetzky, J., McCray, R. A., Indebetouw, R., *et al.* 2013, *ApJ*, 773, L34

Matsuura, M., Indebetouw, R., Woosley, S. *et al.*, 2017, *MNRAS*, in press (arXiv/17)04.02324

Matsuura, M., "Handbook of Supernovae", Springer International Publishing, Ed. Athem W. Alsabti & Paul Murdin

Müller E., Fryxell, B., & Arnett, W. D., 1991, *A&A*, 251, 505

Nomoto, K., Kobayashi, C., & Tominaga, N., 2013, *ARA&A*, 51, 457

Rawlings, J. M. C. & Williams, D. A., 1990, *MNRAS*, 246, 208

Sarangi, A. & Cherchneff, I., 2013, *ApJ*, 776, 107

Sarangi, A. & Cherchneff, I., 2015, *A&A*, 575, A95

Sukhbold, T., Ertl, T., Woosley, S. E., Brown, J. M., & Janka, H. T., 2016, *ApJ*, 821, 38

Williams, D. A. & Viti Serena, S., 2014, *Observational Molecular Astronomy.* Cambridge University Press

Wongwathanarat, A., Muller, E., & Janka, H. T. T., 2015, *A&A*, 577, A48

Woosley, S. E., 1988, *ApJ*, 330, 218

Woosley, S. E., Pinto, P. A., & Weaver, T. A., 1988, *Australian Academy of Science*, 7, 355

Woosley, S. E., Heger, A., Weaver, T. A., & Langer, N., 1997, in Phillips M., Suntzeff N., eds, SN 1987A: Ten years after SN 1987A - Presupernova Evolution and the Progenitor Star. PASPC, pp (astro–ph/9705146)

Supernova 1987A:30 years later - Cosmic Rays and Nuclei from Supernovae and their aftermaths
Proceedings IAU Symposium No. 331, 2017
A. Marcowith, M. Renaud, G. Dubner, A. Ray & A. Bykov, eds.

doi:10.1017/S1743921317004562

Measuring the dust content and formation in SN 1987A using detailed radiative transfer modelling

Maarten Baes[1], Peter Camps[1], Phil J. Cigan[2], Christopher L. Fryer[3], Mikako Matsuura[2] and Sam Verstocken[1]

[1]Sterrenkundig Observatorium, Universiteit Gent
Krijgslaan 281 S9, 9000 Gent, Belgium
email: maarten.baes@ugent.be, peter.camps@ugent.be, sam.verstocken@ugent.be

[2]School of Physics and Astronomy, Cardiff University
The Parade, Cardiff 24CF 3YB, UK
email: ciganp@cardiff.ac.uk, matsuuram@cardiff.ac.uk

[3]Center for Theoretical Astrophysics, Los Alamos National Lab
Los Alamos, NM 87544, USA
email: fryer@lanl.gov

Abstract. Core-collapse supernovae are expected to be efficient producers of dust, and recent Herschel and ALMA observations have revealed up to 1 $M_\odot$ of cold dust in the inner ejecta of SN 1987A. The formation time scale, spatial distribution and clumpiness, and the importance of the different heating sources of the dust remain poorly understood. We have started a project to make detailed 3D dust radiative transfer models for SN 1987A, based on a combination of the latest observational constraints and input from 3D hydrodynamical models and dust formation models. Preliminary results seem to indicate the need for large, micron-sized dust grains, and a relatively large dust mass.

Keywords. radiative transfer, supernovae: individual (SN 1987A)

1. Introduction

Detailed observations and large-scale numerical modelling efforts over the past three decades have provided much insight into SN 1987A (and into core-collapse supernovae in general). Today, SN 1987A is making the transition to a supernova remnant, as the expanding supernova is colliding with circumstellar material that was released from the star prior to the explosion. It has long been expected that core-collapse supernovae are efficient producers of dust. They provide unique physical conditions for the production of dust grains: while the average metal mass fraction in galaxies is of the order of 1%, supernova ejecta can be 100% metal. Until recently, infrared observations have only detected very small amounts of warm dust in nearby supernova remnants (10^{-6} to 10^{-4} $M_\odot$: Gall *et al.* 2011). SN 1987A was the first core-collapse supernova with evidence for dust formation two years after the explosion. The reported dust masses in those early days were 10^{-6} to 10^{-5} $M_\odot$ (Bouchet *et al.* 1991; Wooden *et al.* 1993).

2. Cold dust in SN 1987A

SN 1987A was observed in 2010 with *Herschel* as part of the Magellanic Cloud survey, HERITAGE, and detected at wavelengths between 100 and 350 μm. The dust mass corresponding to these observations was 0.4–0.7 $M_\odot$, three orders of magnitude larger than previously reported (Matsuura *et al.* 2011). New pointed *Herschel* observations taken in 2012 confirmed these results, and suggest a total of 0.8 $M_\odot$ of dust (Matsuura *et al.* 2015). At the same time, SN 1987A was observed and spatially resolved with ALMA between 450 μm and 2.8 mm; these observations unambiguously show the formation of 0.2 to 1 $M_\odot$ of dust in the inner ejecta (Indebetouw *et al.* 2014).

One of the outstanding questions is how quickly and efficiently the dust has formed in SN 1987A (and by extension, in other core-collapse supernovae). Recent work by various teams has come to drastically different conclusions. Wesson *et al.* (2015) and Bevan & Barlow (2016) used radiation transfer calculations to simulate the evolution of the observed spectral energy distribution (SED) and emission line blueshifting of SN 1987A. According to their analyses, the dust mass increases as a power law in time, and 97.5% of the dust mass formed after 1300 days. In contrast, Dwek & Arendt (2015) applied an escape probability calculation on the same SED data, and inferred a sharply different, early and rapid dust mass evolution. According to their analysis, the dust mass had already reached 0.45 $M_\odot$ 615 days after the explosion, and it does not increase appreciably over the following two decades.

In a completely independent way, dust production in SN 1987A has been modelled from a chemical modelling perspective. In the past few years, theoretical computations of dust grain formation have become increasingly more realistic, and go far beyond the classical nucleation theory (e.g., Sarangi & Cherchneff 2015). Sluder *et al.* (2016) presented a detailed analysis of dust formation in SN 1987A based on the so-called molecular nucleation theory, which tracks the abundance of each molecular species with a non-equilibrium chemical reaction network, and which includes effects such as coagulation, grain charging, evaporation, accretion, and chemical weathering. They find a grain size distribution with a power law that is steeper than the one inferred by the studies of Wesson *et al.* (2015) and Dwek & Arendt (2015). Moreover, Sluder *et al.* (2016) predict a radically different dust composition compared to the one assumed by most previous studies.

3. 3D dust radiative transfer simulations of SN 1987A

3.1. *SKIRT simulations*

It is clear that there are some outstanding issues related to the dust in SN 1987A: the distribution and clumpiness, the evolution of the dust production rate, the importance of the different heating sources, etc. Detailed and self-consistent dust radiative transfer simulations in a realistic setting are necessary to address these issues. Our goal is to substantially improve on the existing dust radiative transfer models for SN 1987A in several ways, by using state-of-the-art self-consistent radiative transfer techniques, and by using a combination of observational constraints and input from 3D hydrodynamical models and dust formation models. Until recently, full-scale and self-consistent dust radiative transfer calculations in 3D were almost impossible, but thanks to new techniques and increased computing power, this is possible nowadays (for a review, see Steinacker *et al.* 2013).

A major tool we use for our investigation is the 3D Monte Carlo radiative transfer code SKIRT (Baes *et al.* 2011; Camps & Baes 2015). SKIRT has mainly been applied to investigate the dusty interstellar medium in galaxies (e.g., Baes *et al.* 2010; De Looze *et al.* 2014; Mosenkov *et al.* 2016), but it has also been applied to other systems including AGN accretion discs, molecular clouds, and binary systems (e.g., Stalevski *et al.* 2012; Hendrix

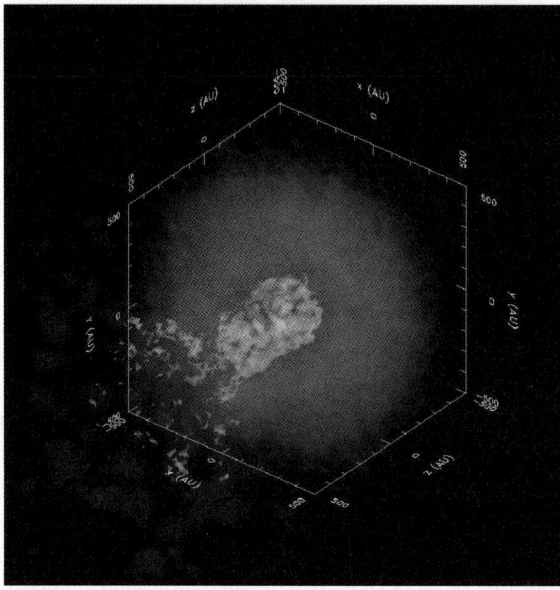

Figure 1. 3D rendering of an asymmetric core-collapse supernova explosion model from Hungerford *et al.* (2005), one year after the shock launch. The top-hat distribution of the imposed velocity asymmetry allows a significant portion of the cone material to expand without drag from fluid shear forces, resulting in the large splash of material at the outer ejecta. This and similar hydrodynamical models will be coupled with dust formation models and used as input for SKIRT radiative transfer simulations.

et al. 2016). One crucial aspect for the current project is that SKIRT solves the radiative transfer problem completely self-consistently, without any approximations. This implies that the effects of absorption, multiple anisotropic scattering and thermal emission (both equilibrium and transient) are taken into account. Recent work has shown there can be profound differences between such a complete 3D radiative transfer treatment and approximate solutions (Saftly *et al.* 2015).

3.2. *Parameterised models*

Concerning the input model for sources and dust, we use two different approaches. Our first approach is to use a (complex) parameterised model. We built a fitting routine on top of SKIRT that automatically determines the model parameters that best reproduce a given data set. The optimisation is based on genetic algorithms, which are ideally suited for fitting noisy Monte Carlo radiative transfer models (De Geyter *et al.* 2013, 2014). All model parameters can be varied, such as dust mass or geometry-related parameters, but also the grain size distribution, the dust clumpiness, etc. We have run initial test simulations for SN 1987A at the most recent epoch in which we vary 5 different parameters (luminosity, dust mass, silicate fraction, clumpiness and mean grain size). Preliminary results seem to indicate the need for large, micron-sized dust grains, and a relatively large dust mass ($> 0.5\ M_\odot$), in agreement with Wesson *et al.* (2015).

3.3. *Models based on hydrodynamical simulations*

A more advanced approach that we will explore in the near future consists of using state-of-the-art 3D hydrodynamical simulations of core-collapse supernovae as the starting point for our radiative transfer simulations. More specifically, we will use a suite of models with a large variation in the progenitor star, the explosion energy and asymmetry (Fryer & Warren 2002; Hungerford *et al.* 2005; Ellinger *et al.* 2012). These smoothed particle hydrodynamics calculations produce detailed tracers of the ejecta to calculate both detailed nucleosynthetic yields (including ^{44}Ti) and dust formation throughout the explosion. The Lagrangian nature of these calculations (Fryer *et al.* 2006) allows us to resolve the shock at $10^8 - 10^9$ cm at the beginning of the calculation and to follow the distribution of matter out to many parsecs at late times (Figure 1). These simulations will be coupled with the molecular nucleation calculations of Sluder *et al.* (2016) to obtain realistic chemical compositions and size distributions for the dust grains.

We will run SKIRT radiative simulations for each of the hydrodynamical models in our suite, calculate images and SEDs for different viewpoints at different time steps in the evolution, and compare the model results with specific observations of SN 1987A. Until very recently, there were essentially no direct observational constraints on the distribution of the dust in the ejecta of SN 1987A. The latest ALMA observations, however, resolve the dust and molecular line emission in the inner ejecta (Cigan *et al.* 2017; Abellan *et al.* 2017). The comparison between the synthetic model images and the ALMA dust observations will be a strong and novel test to discriminate between the different models.

Acknowledgements

We thank the organizers for a wonderful conference in Saint-Gilles-les-Bains.

References

Abellan, F., Indebetouw, R., Matsuura, M., *et al.* 2017, in prep.
Baes, M., Fritz, J., Gadotti, D. A., *et al.* 2010, *A&A*, 518, L39
Baes, M., Verstappen, J., De Looze, I., *et al.* 2011, *ApJS*, 196, 22
Bevan, A. & Barlow, M. J. 2016, *MNRAS*, 456, 1269
Bouchet, P., Phillips, M. M., Suntzeff, N. B., *et al.* 1991, *A&A*, 245, 490
Camps, P. & Baes, M. 2015, *Astronomy and Computing*, 9, 20
Cigan, P., Matsuura, M., Gomez, H. L., *et al.* 2017, in prep.
De Geyter, G., Baes, M., Fritz, J., & Camps, P. 2013, *A&A*, 550, A74
De Geyter, G., Baes, M., Camps, P., *et al.* 2014, *MNRAS*, 441, 869
De Looze, I., Fritz, J., Baes, M., *et al.* 2014, *A&A*, 571, A69
Dwek, E. & Arendt, R. G. 2015, *ApJ*, 810, 75
Ellinger, C. I., Young, P. A., Fryer, C. L., & Rockefeller, G. 2012, *ApJ*, 755, 160
Fryer, C. L. & Warren, M. S. 2002, *ApJL*, 574, L65
Fryer, C. L., Rockefeller, G., & Warren, M. S. 2006, *ApJ*, 643, 292
Gall, C., Hjorth, J., & Andersen, A. C. 2011, *A&AR*, 19, 43
Hendrix, T., Keppens, R., van Marle, A. J., *et al.* 2016, *MNRAS*, 460, 3975
Hungerford, A. L., Fryer, C. L., & Rockefeller, G. 2005, *ApJ*, 635, 487
Indebetouw, R., Matsuura, M., Dwek, E., *et al.* 2014, *ApJL*, 782, L2
Matsuura, M., Dwek, E., Meixner, M., *et al.* 2011, *Science*, 333, 1258
Matsuura, M., Dwek, E., Barlow, M. J., *et al.* 2015, *ApJ*, 800, 50
Mosenkov, A. V., Allaert, F., Baes, M., *et al.* 2016, *A&A*, 592, A71
Saftly, W., Baes, M., De Geyter, G., *et al.* 2015, *A&A*, 576, A31
Sarangi, A. & Cherchneff, I. 2015, *A&A*, 575, A95
Sluder, A., Milosavljevic, M., & Montgomery, M. H. 2016, arXiv:1612.09013
Stalevski, M., Fritz, J., Baes, M., Nakos, T., & Popović, L. Č. 2012, *MNRAS*, 420, 2756
Steinacker, J., Baes, M., & Gordon, K. D. 2013, *ARA&A*, 51, 63
Wesson, R., Barlow, M. J., Matsuura, M., & Ercolano, B. 2015, *MNRAS*, 446, 2089
Wooden, D. H., Rank, D. M., Bregman, J. D., *et al.* 1993, *ApJS*, 88, 477

Supernova 1987A:30 years later - Cosmic Rays and Nuclei from Supernovae and their aftermaths
Proceedings IAU Symposium No. 331, 2017
A. Marcowith, M. Renaud, G. Dubner, A. Ray & A. Bykov, eds.

doi:10.1017/S1743921317004355

The role of the diffusive protons in the gamma-ray emission of SNR RX J1713.7−3946

Xiao Zhang[1] and Yang Chen[1,2]

[1]Department of Astronomy, Nanjing University, 163 Xianlin Avenue, Najing 210023, China
email: xiaozhang@nju.edu.cn

[2]Key Laboratory of Modern Astronomy and Astrophysics, Nanjing University, Ministry of Education, Najing 210023, China
email: ygchen@nju.edu.cn

Abstract. RX J1713.7−3946 is a prototype in the γ-ray-bright supernova remnants (SNRs) and is in continuing debates on its hadronic versus leptonic origin of the γ-ray emission. We explore the role played by the diffusive relativistic protons that escape from the SNR shock wave in the γ-ray emission, apart from the emission of high energy particles from the inside of the SNR. In the scenario that the SNR shock propagates in a clumpy molecular cavity, we consider that the γ-ray emission from the inside of the SNR may either arise from the IC scattering or from the interaction between the trapped energetic protons and the shocked clumps. The dominant origin between them depends on the electron-to-proton number ratio. The surrounding molecular cavity wall is considered to also produce γ-ray emission due to the "illumination" by the diffusive protons that escaped from the shock wave during the expansion history. The broad-band spectrum can be well explained by this two-zone model, in which the γ-ray emission from the inside governs the TeV band, while the outer emission component substantially contributes to the GeV γ-rays. The two-zone model can also explain the TeV γ-ray radial brightness profile that significantly stretches beyond the nonthermal X-ray emitting region.

Keywords. radiation mechanisms: nonthermal, ISM: individual (G347.3−0.5), (ISM:) supernova remnants, gamma rays: theory

---

1. Introduction

SNR RX J1713.7−3946 (G347.3−0.5) was discovered by *ROSAT* X-ray observation (Pfeffermann & Aschenbach 1996) and suggested to be the remnant of the historical supernova AD393 (Wang *et al.* 1997, hereafter an age of the remnant $t_{\rm age} \sim 1620\,$yr will be used). Its shell-like X-ray emission is dominated by non-thermal component, lack of thermal line features (e.g., Koyama *et al.* 1997; Slane *et al.* 1999; Cassam-Chenaï *et al.* 2004). It is found to be confined in a molecular cavity and the elevated gas temperature and broad molecular line wings in a few molecular cloudlets are ascribed to the high-energy events of the SNR (Fukui *et al.* 2003; Moriguchi *et al.* 2005). In radio band, it shows faint emission and has an average angular $60'$ in diameter, corresponding to an average radius $R_s \sim 9\,$pc at distance $d \approx 1$ kpc (e.g. Fukui *et al.* 2003; Moriguchi *et al.* 2005). The SNR is suggested to be in the free expansion evolutionary phase (e.g. Moriguchi *et al.* 2005; Sano *et al.* 2010), and that the forward shock has not yet reached the wall enclosing the wind-blown cavity in view of the high velocity ($V_s \sim 4000$ km s^{-1}) currently measured (Federici *et al.* 2015).

Since the detection of TeV γ-ray emission from it (Enomoto *et al.* 2002; Aharonian *et al.* 2004, 2007), SNR RX J1713.7−3946 has absorbed enormous amount of attention and aroused constant debates on its hadronic versus leptonic origin of the γ-ray emission (e.g. Aharonian *et al.* 2006; Berezhko & Völk 2008; Liu *et al.* 2008; Morlino *et al.* 2009; Zirakashvili & Aharonian 2010). The debates seemed to have been concluded when the 2-yr *Fermi*-LAT observation revealed a hard GeV spectrum with a power-law photon index $\Gamma = 1.5 \pm 0.1$ (Abdo *et al.* 2011), which appears to support the leptonic scenario (e.g., Abdo *et al.* 2011; Yuan *et al.* 2011; Li *et al.* 2011; Ellison *et al.* 2012; Lee *et al.* 2012; Finke & Dermer 2012). Subsequently, however, a strong correlation of azimuthal distribution was found between the TeV γ-ray flux and the column density of total interstellar protons, which is in favor of hadronic origin of the γ-rays (Fukui *et al.* 2012). Some hadronic interaction models have also been elaborated to interpret the hard GeV emission (Inoue *et al.* 2012; Gabici & Aharonian 2014; Federici *et al.* 2015). Both the scenarios in terms of relativistic leptons and protons accelerated via standard DSA can explain the hard spectrum with their advantages, but meanwhile also leave difficulties, respectively.

Here we explore the role played by the diffusive relativistic protons that escape from the SNR shock and hit the surrounding dense gas in the γ-ray emission of RX J1713.7−3946, considering that the blast wave is propagating in a molecular cavity, in which there may be an accumulation of shocked clumps. In the following, we show that the GeV-TeV γ-ray spectrum of RX J1713.7−3946 with updated 5-yr *Fermi*-LAT data can be perfectly interpreted with a two-zone scenario, in which the π^0-decay γ-rays resulted from the diffusive protons will substantially contribute to the GeV γ-rays while the leptonic process or the p-p interaction in the shocked clumps will be responsible for the TeV γ-rays.

2. A two-zone Model

Considering the large shock velocity and the low ambient density, we assume that the shock of SNR RX J1713.6−3946 doesn't extensively hit the surrounding MC and is still evolving in the molecular cavity, but may sweep up some dense clumps (see Figure 1). The first (outer) emission zone is the cavity wall (the surrounding dense matter at the cavity boundary, which may be clumpy even without a sharp interface), which is bombarded by the protons that escaped from the shock wave during the history of expansion. The second (inner) emission zone is inside the SNR, and the emission could be either IC scattering off the accelerated electrons or the hadronic emission from the shocked clumps. Due to the two possible channels of generating γ-rays inside SNR, two cases will be considered within inner zone: (I) it is dominated by the IC process of the accelerated electron; and (II) it is dominated by the collision of the accelerated protons with the shocked clumps. For comparison, we also only use the inner zone to fit the broadband data (referred to as Model Ib and Model IIb respectively). All models are summarized in Figure 1.

Distribution of particles in the inner zone. The shock accelerated particles are assumed to obey a power-law distribution with an high energy cut-off:

$$dN_i/dE_i = A_i E_i^{\alpha_i} \times \exp(-E_i/E_{\mathrm{c},i}) \tag{2.1}$$

where $i = \mathrm{e}, \mathrm{p}$, E_i is the particle kinetic energy, α_i is the power-law spectral index and is taken as $\alpha_\mathrm{e} = \alpha_\mathrm{p} = \alpha$, and $E_{\mathrm{c},i}$ is the cut-off energy and will be set to the CR "knee" energy for the protons, leaving $E_{\mathrm{c,e}}$ as free parameter. The normalization parameter A_i is determined from the total kinetic energy of particles above 1 GeV. In our calculation, two physical parameters are adopted instead of the normalization parameters: the energy conversion efficiency η, namely the faction of explosion energy

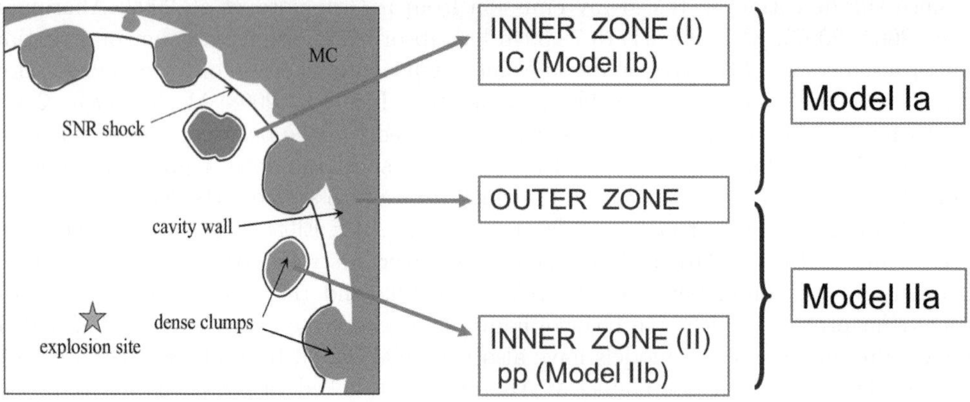

Figure 1. Schematic of the two-zone model.

converted to CR energy, and the number ratio between the accelerated electrons and protons at 1 GeV, $K_{\rm ep} = A_{\rm e}/A_{\rm p}$.

Distribution of particles in the outer zone. For the continuous injection of the accelerated protons that escape from the propagating shock wave surface, Li & Chen (2010) established an "accumulative diffusion" model, in which the protons at given position is a collection of the diffusive protons escaping from the entire shock surface at different radii throughout the history of the SNR expansion. Instead of the treatment in Li & Chen (2010), the distribution function of the escaping protons can be simplified by directly solving the diffusion equation of the escaping protons as:

$$f(E_{\rm p}, R, t) = \int_0^t \frac{Q_{\rm p}(E_{\rm p})}{4\pi^{3/2} R_s(t_i) R_{\rm d} R} \left\{ \exp\left[-\left(\frac{R - R_s(t_i)}{R_{\rm d}} \right)^2 \right] - \exp\left[-\left(\frac{R + R_s(t_i)}{R_{\rm d}} \right)^2 \right] \right\} {\rm d}t_i \tag{2.2}$$

where $f(E_{\rm p}, R, t)$ is the distribution of the escaping protons with energy $E_{\rm p}$ at a given distance R from the SNR center at a given time t from the explosion time, $Q_{\rm p}(E_{\rm p})$ is the injection rate from the spherical surface $4\pi R_s^2$, $R_s(t)$ is the shock radius, and $R_{\rm d} = 2\sqrt{D(t - t_i)}$ is diffusion radius. Here, we assume that the diffusion coefficient has the form of $D(E_{\rm p}) = 10^{28}\chi(E_{\rm p}/10\,{\rm GeV})^{\delta}$ cm^2 s^{-1}, where χ is the correction factor of slow diffusion around SNR, and $\delta \approx 0.3 - 0.7$ (Berezinskii 1990) is the index of diffusion coefficient.

Mass of the gas in the inner zone II. Following Inoue *et al.* (2012), the total mass of the gas in the clumps directly bombarded by the accelerated protons, $m_{c,tot}$, can be obtained via three steps: (1) Estimate the average mass density ρ_c for each clump with physical parameters given by Sano *et al.* (2013, see Table 3 and Table 4 therein). (2) Estimate the mass in the outer layer of each clump that is penetrated by relativistic protons using $m_c = 4\pi\rho_c a_c^2 l_{\rm pd}$ (see Inoue *et al.* 2012), where a_c is the clump radius and $l_{\rm pd} \simeq \sqrt{\kappa_d t_{\rm pd}}$ is the penetration depth of the accelerated protons into the clumps during diffusion time $t_{\rm pd}$ with the diffusion coefficienct $\kappa_d = 4\eta_{\rm B} c E_{\rm p}/3\pi e B_{\rm MC}$ ($\eta_{\rm B}$ is the degree of magnetic field fluctuations and $B_{\rm MC}$ is the magnetic field in the dense region of MCs). The diffusion time is estimated as $t_{\rm pd} \sim 0.2\,(l_m/1\,{\rm pc})(V_s/4000\,{\rm km\,s^{-1}})^{-1}$ kyr, where $V_s = 4000\,{\rm km\,s^{-1}}$ (e.g., Cassam-Chenaï *et al.*2004; Acero *et al.* 2009) is the shock velocity and $l_m \sim 1\,{\rm pc}$ is the mean length that the shock moves forward from the clumps, estimated from the spatial distribution of the clumps (Sano *et al.* 2013). (3) Scale the mass of the gas collided by protons according to the area fraction covered by X-rays for

Table 1. The SED fitted parameters with 1σ statistical error.

Model	K_{ep}	η (%)	α	B_{SNR} (μG)	$E_{c,e}$ (TeV)	$E_{c,p}$ (TeV)	$\chi/0.01$	δ	M_t ($10^3\ M_\odot$)
Ia	0.01[a]	$2.5^{+0.6}_{-0.9}$	$2.09^{+0.03}_{-0.04}$	$15.9^{+4.2}_{-1.3}$	$34.2^{+1.7}_{-4.0}$	—	$1.2^{+1.0}_{-0.4}$	$0.67^{+0.03}_{-0.14}$	$8.8^{+6.6}_{-3.1}$
Ib	0.01[a]	$4.1^{+0.5}_{-0.5}$	$2.11^{+0.02}_{-0.02}$	$12.7^{+0.3}_{-0.3}$	$38.6^{+0.7}_{-0.8}$	—	—	—	—
IIa	0.001[a]	$6.0^{+2.9}_{-3.3}$	$2.09^{+0.08}_{-0.06}$	$37.9^{+6.2}_{-5.5}$	$22.2^{+2.9}_{-5.0}$	$81.3^{+13.7}_{-11.9}$	$3.1^{+0.7}_{-0.4}$	$0.55^{+0.15}_{-0.25}$	$2.6^{+1.8}_{-1.0}$
IIb	0.001[a]	$10.6^{+1.8}_{-1.5}$	$2.10^{+0.02}_{-0.03}$	$27.6^{+1.5}_{-2.0}$	$26.2^{+1.0}_{-0.8}$	$60.4^{+08.8}_{-10.1}$	—	—	—

Notes:
[a] Fixed in the model fit.

Table 2. The best-fit χ^2 value for each set of data.

Model	radio	X-ray	GeV	TeV	χ^2_ν (d.o.f.)
Ia	3.8	364.6	10.2 (13.5[a])	37.1	1.78(233)
Ib	0.09	374.1	36.3 (5.5[a])	120.0	2.25(236)
IIa	3.7	364.6	9.1 (7.0[a])	36.4	1.78(232)
IIb	0.8	368.6	23.9 (3.0[a])	41.7	1.85(235)

Notes:
[a] Corresponding to the old *Fermi* data (Abdo *et al.* 2011).

each clump and sum the mass of the clumps. Hence, we obtain

$$m_{c,tot} \approx 13 \left(\frac{\eta_B}{1}\right)\left(\frac{E_p}{10\,\mathrm{GeV}}\right)^{1/2}\left(\frac{B_{MC}}{100\,\mu\mathrm{G}}\right)^{-1/2}\left(\frac{t_{pd}}{0.2\,\mathrm{kyr}}\right)^{1/2} M_\odot. \tag{2.3}$$

3. Modelling and Results

Spectral behavior. We use the two-zone model to fit the broadband spectra of SNR RX J1713.7−3946 and employ the MCMC approach to constrain the model parameters. The best-fit parameters with 1σ statistical uncertainties and the χ^2 values are listed in the Table 1 and Table 2, respectively. The model γ-ray spectral energy distribution (SED, see Figure 2) shows that this model perfectly explains the γ-ray data and that the hadronic γ-rays from the escaping protons play an important role on the explanation of the spectral data, especially in the GeV band. In the energy range from few GeV to several 100 GeV, the γ-rays are mainly contributed by the hadronic emission produced within the outer emission zone. Given the presence of the shocked clumps, in the remaining γ-ray band the dominant mechanism of γ-ray emission in the downstream depends on the electron-to-proton number ratio, K_{ep}: it was found that the hadronic emission from the inside clumps dominates for $K_{ep} < \sim 4\times10^{-3}$ while the leptonic component dominates for $K_{ep} > \sim 4\times10^{-3}$. In addition, we also calculate the χ^2 value in GeV band using our best-fit SED and the previous *Fermi* data (Abdo *et al.* 2011), which are listed in Table 2 with parentheses.

Spatial behavior. According to the flux ratio between the outer and inner components derived from the above model calculation, we fit the TeV profile of the sector-shaped "region 3" given in de Naurois (2015) and the entire SNR given in Aharonian *et al.* (2006), assuming uniform emissivities in the two emission zones. Then, the TeV γ-ray profile is determined by the positions (R_s, R_w) and thicknesses (ΔR_s, ΔR_w) of the two mission zones. It should be noted that R_s represents the outer radius of the inner emission zone while R_w is the inner radius of the outer emission zone. In Figure 3, we

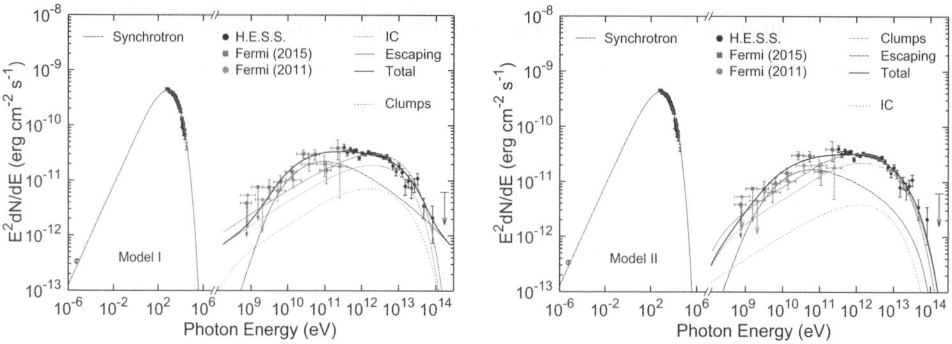

Figure 2. The best fitted broad-band SED of SNR RX J1713.7−3946 according to the radio (Acero *et al.* 2009), X-ray (Tanaka *et al.* 2008), GeV (*Fermi*: Abdo *et al.* 2011; Federic *et al.* 2015) and TeV (H.E.S.S.: Aharonian *et al.* 2011) data for Model Ia (left) and Model IIa (right). The grey solid lines represent the fitting SED for Model Ib (left) and Model IIb (right).

Table 3. The fitted parameters used in Figure 3.

Model	Ratio[a]	R_s [pc]	R_w [R_s]	ΔR_s [R_s]	ΔR_w [R_s]
Ia	4.1:5.9 (3.9:6.1[b])	7.0 (9.0[b])	1.13 (1.11[b])	0.12 (0.23[b])	0.12 (0.22[b])
IIa	2.8:7.2 (2.7:7.3[b])	7.0 (9.0[b])	1.20 (1.11[b])	0.10 (0.18[b])	0.09 (0.20[b])

Notes:
[a] Flux ratio (above 0.25 TeV for "region 3" and between 0.3 and 40 TeV for the entire remnant) between the outer and inner zone derived from the model calculations.
[b] Results in parentheses are corresponding to the entire remnant.

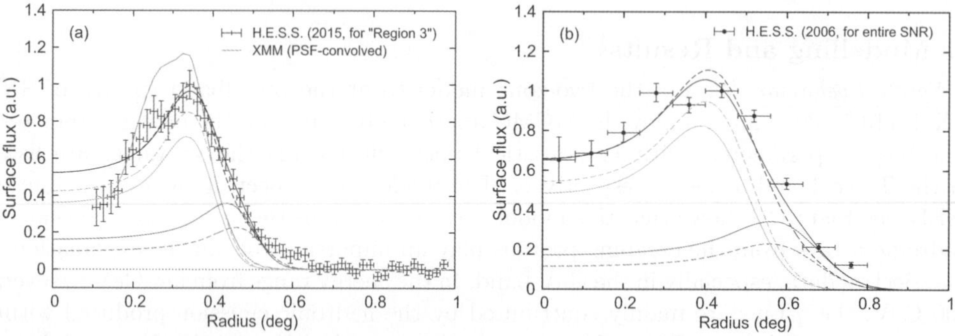

Figure 3. (a) TeV γ-ray radial surface brightness profile of sector "region 3". The H.E.S.S. flux data and the *XMM*-Newton X-ray brightness profile convolved with the H.E.S.S. PSF of $3'$ (grey line) are adopted from de Naurois (2015). The radial profile of "region 3" is fitted with a two-zone emission (in black), which is the sum of the inner (in green) and outer (in blue) components. The solid and dashed lines represent the profiles in Model Ia and Model IIa, respectively. All of the fitting profile lines have been convolved with the H.E.S.S. PSF of $3'$. (b) TeV γ-ray radial surface brightness profile for the entire remnant. The H.E.S.S. flux data are adopted from Aharonian *et al.* (2006). The meanings of all lines convolved with the H.E.S.S. PSF of $5'$ are the same as those in (a).

display the fitting profile lines (black) which are the sum of the inner (green) and outer (blue) components can explain the observational data. For example, the profile lines in left pannel can reproduce both the TeV brightness peak, which is coincident with the non-thermal X-ray peak, and the broad profile wing that extends outside the X-ray emitting region. For the entire remnant, the fitting profiles match the observational data at well. All parameters used to plot Figure 3 can be found in Table 3.

4. Summary

In the scenario that the SNR shock is propagating in a molecular cavity, we have explored the role played by the diffusive relativistic protons that escaped from the SNR shock wave in the γ-ray emission, apart from the emission of high energy particles from the inside of the SNR. The broad-band fluxes can be well explained by the two-zone model, in which the γ-ray emission from the inside governs the TeV band, while the outer emission component substantially contributes to the GeV γ-rays. Meantime, we show that the two-zone model can also reproduce the TeV γ-ray radial brightness profile that significantly extends outside the nonthermal X-ray emitting region.

References

Abdo, A. A., Ackermann, M., Ajello, M., *et al.* 2011, *ApJ*, 734, 28
Acero, F., Ballet, J., Decourchelle, A., *et al.* 2009, *A&A*, 505, 157
Aharonian, F., Akhperjanian, A. G., Bazer-Bachi, A. R., *et al.* 2006, *A&A*, 449, 223
Aharonian, F., Akhperjanian, A. G., Bazer-Bachi, A. R., *et al.* 2007, *A&A*, 464, 235
Aharonian, F., Akhperjanian, A. G., Bazer-Bachi, A. R., *et al.* 2011, *A&A*, 531, C1
Aharonian, F. A. & Atoyan, A. M. 1996, *A&A*, 309, 917
Aharonian, F. A., Akhperjanian, A. G., Aye, K.-M., *et al.* 2004, *Nature*, 432, 75
Berezhko, E. G. & Völk, H. J. 2008, *A&A*, 492, 695
Berezinskii, V. S., Bulanov, S. V., Dogiel, V. A., & Ptuskin, V. S. 1990, *Astrophysics of cosmic rays*
Cassam-Chenaï, G., Decourchelle, A., Ballet, J., *et al.* 2004, *A&A*, 427, 199
de Naurois, M. 2015, ArXiv e-prints: 1510.00635
Ellison, D. C., Slane, P., Patnaude, D. J., & Bykov, A. M. 2012, *ApJ*, 744, 39
Enomoto, R., Tanimori, T., Naito, T., *et al.* 2002, *Nature*, 416, 823
Fan, Z. H., Liu, S. M., Yuan, Q., & Fletcher, L. 2010, *A&A*, 517, L4
Federici, S., Pohl, M., Telezhinsky, I., Wilhelm, A., & Dwarkadas, V. V. 2015, *A&A*, 577, A12
Finke, J. D. & Dermer, C. D. 2012, *ApJ*, 751, 65
Fukui, Y., Moriguchi, Y., Tamura, K., *et al.* 2003, *PASJ*, 55, L61
Fukui, Y., Sano, H., Sato, J., *et al.* 2012, *ApJ*, 746, 82
Gabici, S. & Aharonian, F. A. 2014, *MNRAS*, 445, L70
Inoue, T., Yamazaki, R., Inutsuka, S.-i., & Fukui, Y. 2012, *ApJ*, 744, 71
Koyama, K., Kinugasa, K., Matsuzaki, K., *et al.* 1997, *PASJ*, 49, L7
Lee, S.-H., Ellison, D. C., & Nagataki, S. 2012, *ApJ*, 750, 156
Li, H. & Chen, Y. 2010, *MNRAS*, 409, L35
Li, H., Liu, S., & Chen, Y. 2011, *ApJL*, 742, L10
Liu, S., Fan, Z.-H., Fryer, C. L., Wang, J.-M., & Li, H. 2008, *ApJL*, 683, L163
Moriguchi, Y., Tamura, K., Tawara, Y., *et al.* 2005, *ApJ*, 631, 947
Morlino, G., Amato, E., & Blasi, P. 2009, *MNRAS*, 392, 240
Pfeffermann, E. & Aschenbach, B. 1996, in Roentgenstrahlung from the Universe, ed. H. U. Zimmermann, J. Trümper, & H. Yorke, 267–268
Sano, H., Sato, J., Horachi, H., *et al.* 2010, *ApJ*, 724, 59
Sano, H., Tanaka, T., Torii, K., *et al.* 2013, *ApJ*, 778, 59
Slane, P., Gaensler, B. M., Dame, T. M., *et al.* 1999, *ApJ*, 525, 357
Tanaka, T., Uchiyama, Y., Aharonian, F. A., *et al.* 2008, *ApJ*, 685, 988
Wang, Z. R., Qu, Q.-Y., & Chen, Y. 1997, *A&A*, 318, L59
Yuan, Q., Liu, S., Fan, Z., Bi, X., & Fryer, C. L. 2011, *ApJ*, 735, 120
Zirakashvili, V. N. & Aharonian, F. A. 2010, *ApJ*, 708, 965

Supernova 1987A:30 years later - Cosmic Rays and Nuclei from Supernovae and their aftermaths
Proceedings IAU Symposium No. 331, 2017
A. Marcowith, M. Renaud, G. Dubner, A. Ray & A. Bykov, eds.

doi:10.1017/S174392131700432X

The GeV Gamma-Ray Emission Detected by *Fermi*-LAT Adjacent to SNR Kesteven 41

Bing Liu[1], Yang Chen[1,2], Xiao Zhang[1], Gao-Yuan Zhang[1], Yi Xing[3] and Thomas G. Pannuti[4]

[1]Department of Astronomy, Nanjing University, 163 Xianlin Avenue, Nanjing 210023, China

[2]Key Laboratory of Modern Astronomy and Astrophysics, Nanjing University, Ministry of Education, Nanjing 210093, China

[3]Key Laboratory for Research in Galaxies and Cosmology, Shanghai Astronomical Observatory, Chinese Academy of Sciences, 80 Nandan Road, Shanghai 200030, China

[4]Space Science Center, Department of Earth and Space Sciences, Morehead State University, 235 Martindale Drive, Morehead, KY 40351, USA

Abstract. Gamma-ray observations for Supernova remnant (SNR)-molecular cloud (MC) association systems play an important role in the research on the acceleration and propagation of cosmic-ray protons. Through the analysis of 5.6 years of *Fermi*-Large Area Telescope observation data, here we report on the detection of a gamma-ray emission source near the SNR Kesteven 41 with a significance of 24σ in 0.2–300 GeV. The best-fit location of the gamma-ray source is consistent with the MC with which the SNR interacts. Several hypotheses including both leptonic and hadronic scenarios are considered to investigate the origin of these gamma-rays. The gamma-ray emission can be naturally explained by the decay of neutral pions produced via the collision between high energy protons accelerated by the shock of Kesteven 41 and the adjacent MC. The electron energy budget would be too high for the SNR if the gamma-rays were produced via inverse Compton (IC) scattering off the Cosmic Microwave Background (CMB) photons.

Keywords. acceleration of particles, gamma rays: observations, supernova remnants ,ISM: individual (Kes 41)

1. Introduction

Cosmic rays (CRs) are highly energetic particles comprised primarily of protons and atomic nuclei with a small fraction ($\lesssim 1\%$) of leptons. Since CRs were discovered by Victor Hess in 1912, the origin of CRs remains one of the most debated issues in high energy astrophysics. As results of stellar explosions, supernova remnants (SNRs) are widely believed to be the main accelerators of Galactic CRs (e.g., Ginzburg & Syrovatskii 1969). In recent years, a lot of breakthroughs have been made in the study of the SNRs and the origin of Galactic CRs through gamma-ray observations. For instance, gamma-ray emissions produced through π^0-decays have been detected from several SNRs close to MCs (e.g., Ackermann *et al.* 2013). There are generally two scenarios on the hadronic gamma-rays associated with these SNR-MC systems. In one scenario, the π^0-decay emission is suggested to arise from shock-crushed dense clouds where the accelerated protons frozen in the clouds efficiently collide with target cloud gas (e.g., Blandford & Cowie 1982;

Uchiyama *et al.* 2010; Tang & Chevalier 2014). In the other scenario, the gamma-rays are ascribed to the hadronic interaction between the relativistic protons escaping from the SNR shock and the adjacent MCs (e.g.,Aharonian & Atoyan 1996; Gabici *et al.* 2009; Li & Chen 2010; Ohira *et al.* 2011). In both scenarios, SNRs interacting with MCs play an important role in the research on acceleration of protons. The hadronic gamma-ray emission from the SNR-MC systems is usually bright in the GeV domain. Several GeV-bright SNRs interacting with MCs have recently been discovered with the Large Area Telescope (LAT) on board the *Fermi Gamma-ray Space Telescope*, such as W51C (Abdo *et al.* 2009), W28 (Abdo *et al.* 2010a), W44 (Abdo *et al.* 2010b), IC 443 (Abdo *et al.* 2010c), W41 (Castro *et al.* 2013), RCW 103 (Xing *et al.* 2014). Additional GeV observations continue to enlarge the sample of hadronic interaction between SNRs and MCs, and here we present a GeV study of another SNR, namely, Kesteven 41 (G337.8−0.1).

Kesteven 41 (here after Kes 41) is a southern-sky SNR interacting with MC. It is centrally brightened in X-rays within a distorted radio shell, therefore, is classified as a thermal composite (or mixed-morphology) SNR. Kes 41 has also been found to be interacting with an adjacent MC, as indicated by the 1720 MHz hydroxyl radical (OH) maser emission detected in the northern radio shell (Koralesky *et al.* 1998, Caswell 2004). Recently, we found that Kes 41 is associated with a giant MC at a systemic local standard of rest (LSR) velocity of $-50\,\mathrm{km\,s^{-1}}$ and is confined in a cavity delineated by a northern molecular shell, a western concave MC, and a southeastern HI cloud (Zhang *et al.* 2015). It is logical and meaningful to search for GeV emission due to the interaction of high-energy particles.

2. Fermi-LAT Data Analysis and Results

Using 5.6 years **Pass 7** data from *Fermi*-LAT (Atwood *et al.* 2009), we select the LAT events in the energy range of 0.2–300 GeV, and perform binned likelihood analyses inside a $14° \times 14°$ region of interest (ROI) centered at the position of Kes 41.

Source detection . A gamma-ray source (hereafter source A) is detected with a significance of 24σ in the 0.2-300GeV energy range. Its best-fit position is found to be R.A. (J2000) = $16^h\ 38^m\ 36^s.00$, decl. (J2000) = $-46°\ 55'\ 06''.96$.

Spectral analysis . Its spectra ranging from 0.2 to 300 GeV can be fitted with a simple power-law model with a photon index of $\Gamma = 2.38 \pm 0.03$. The flux is $(9.2 \pm 1.0) \times 10^{-11}\,\mathrm{erg\,cm^{-2}\,s^{-1}}$, and the corresponding luminosity is $\sim 1.6 \times 10^{36} d_{12}^2$ erg s^{-1}, where $d_{12} = d/12$kpc is the distance to the MC associated with SNR Kes 41. See also section 3.1 for an estimate of the flux and luminosity with an exponential cutoff.

Timing analysis. We also apply long-term variability test on source A, using the method introduced in section 3.6 of Nolan *et al.* (2012). Its Variability Index (TS_{var}), equal to 65.2, doesn't reach the threshold (98.0 for 69 time bins) to claim a long-term variability. And no periodic signal is detected through our *Fermi*-LAT aperture photometry analyses.

3. The Origin of the Detected Gamma-Ray Emission

3.1. *Emission from a pulsar?*

We fit the spectrum of source A with an exponential cutoff power law model,

$$dN_{\mathrm{ph}}/dE_{\mathrm{ph}} = KE_{\mathrm{ph}}^{-\Gamma}\exp(-E_{\mathrm{ph}}/E_{\mathrm{ph,cut}}),$$

and obtain the cutoff energy, $E_{\mathrm{ph,cut}} = 4.0 \pm 0.9$ GeV and the spectral index, $\Gamma = 1.9 \pm 0.1$, similar to those of the detected gamma-ray pulsars (Abdo *et al.* 2013). The corresponding

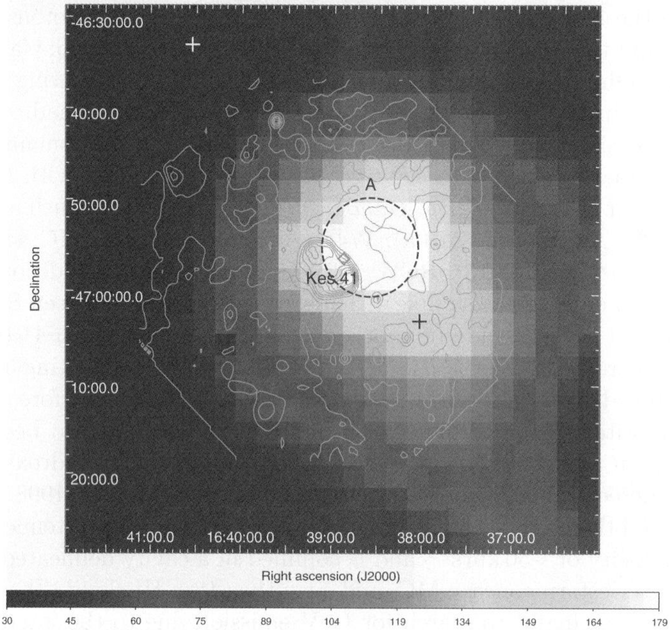

Figure 1. TS map (2–300 GeV) of $1^\circ \times 1^\circ$ region centered on Kes 41. All sources have been subtracted, except 2FGL J1638.0−4703c (see Liu *et al.* (2015) for more details). The white cross presents the position of a 2FGL source, the black cross shows the position of 2FGL J1638.0−4703c, and the dashed black circle indicates the 3σ error range of the best-fit position for the residual emission found in the Kes 41 region. The image is overlaid with the MOST 843 MHz radio contours (from Whiteoak & Green (1996)). The magenta diamond indicates the OH (1720 MHz) maser spot (Koralesky *et al.* 1998).

flux and luminosity in the energy range 0.2–300 GeV is $(7.5 \pm 0.9) \times 10^{-11}\,\mathrm{erg\,cm^{-2}\,s^{-1}}$, and $(1.3 \pm 0.2) \times 10^{36} d_{12}^2$ erg s^{-1} respectively. But no known pulsar is within the 3σ error circle of the best-fit position of source A. Nonetheless, we can not rule out the possibility that the emission comes from a pulsar.

3.2. *Emission from Particles Accelerated by Kes 41?*

We search the SIMBAD Astronomical Database (Wenger *et al.* 2000) within the 3σ error circle of the best-fit source location, looking for the possible counterpart of the gamma-ray emission. In addition to Kes 41, only several stars, dark clouds, a young stellar object candidate and an infrared source are known in this region. As shown in Fig.2, the location of source A is consistent with the shock-MC interaction region. Therefore, relativistic particles accelerated by the SNR shock waves are very likely to be responsible for the gamma-ray emission.

Case A. First, we consider the leptonic scenario (case A): gamma-rays are produced by the IC emission of relativistic electrons accelerated by the SNR. We assumed a power-law electron spectrum with an exponential cutoff,

$$dN_\mathrm{e}/dE_\mathrm{e} \propto E_\mathrm{e}^{-\alpha_\mathrm{e}} \exp(-E_\mathrm{e}/E_\mathrm{e,cut}).$$

In this case, only the cosmic microwave background is considered as the seed photons during the fitting process.

Case B. Next, we consider the hadronic scenario in which these gamma-rays are produced by the π^0-decay after the inelastic collision of the shock accelerated protons with

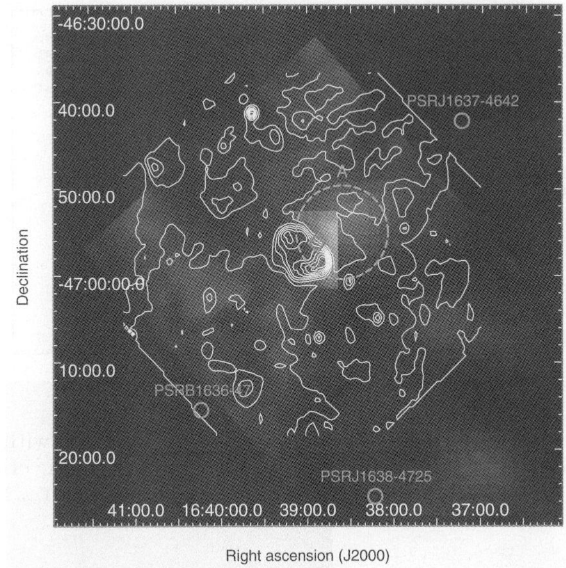

Figure 2. Tricolor image of Kes 41. Red: *Fermi*-LAT 2–300GeV residual counts map with the emissions from the Galactic and extragalactic diffuse background as well as the 2FGL sources (except 2FGL J1638.0-4703c) deducted. Green: ^{12}CO (J=1–0) integrated emission ($V_{\rm LSR}$ = −70 to $-40\,{\rm km\,s^{-1}}$) with the field of view $11' \times 10'$. Blue: HI line emission from SGPS integrated map ($V_{\rm LSR}$ = −55 to $-50\,{\rm km\,s^{-1}}$). The green circles label the positions of known pulsars. The dashed circle, the diamond and the contours are the same as in Fig.1.

dense molecular gas (case B). We assumed the target gas with average number density $n_{\rm t}$, and the protons with a broken power-law distribution,

$$dN_{\rm p}/dE_{\rm p} \propto E_{\rm p}^{-\alpha_{\rm p}}(1+(E_{\rm p}/E_{\rm b})^2)^{-\Delta\alpha_{\rm p}/2}.$$

The total energy in protons with energy above 1 GeV is

$$W_{\rm p}(>1\,{\rm GeV}) \sim 0.7\times 10^{50} E_{51}(n_{\rm t}/100\,{\rm cm}^{-3})^{-1}\ \textbf{erg},$$

where $E_{51} = E_{\rm SN}/10^{51}$ erg is the dimensionless SN explosion energy.

Case C. As shown in Fig.1, the centroid of the gamma-ray emission is outside the northwestern boundary of Kes 41. Thus, the interaction of high-energy protons escaped from Kes 41 with the nearby MC (case C), is a very reasonable scenario for the origin of the gamma-rays. In this case, the energy distribution of protons in the finite volume of the adjacent cloud is obtained following the work of Li & Chen (2012).

Case D. According to Zhang *et al.* (2015), Kes 41 is enclosed in a cavity that may send a reflected shock backward when the blast wave collides with it. It is necessary to consider the case that the reflected shock could accelerate particles effectively after the forward shock becomes radiative . We simplify this situation as a continuous proton injection from the SNR center and follow the prescription of Aharonian & Atoyan (1996). In addition, the MC is regard as a point and the energy conversion fraction $\eta = 0.1$ is adopted.

Comparison among different models. The parameters obtained for the above four cases via fitting the spectra of Kes 41 are listed in Table 1 (for the definition of the parameters and calculation details, see Liu *et al.* (2015), Li & Chen (2012) and the references therein). For the leptonic scenario (case A), the total energy required in electrons with energy above 1 GeV, $W_{\rm e}(>1{\rm GeV}) \sim 1.3\times10^{51}$ erg, is unreasonably high, i.e. of the order of the canonical SN explosion energy. As for the hadronic scenarios, the variation ranges of these parameters are reasonable given Kes 41's own conditions and its ambient environment.

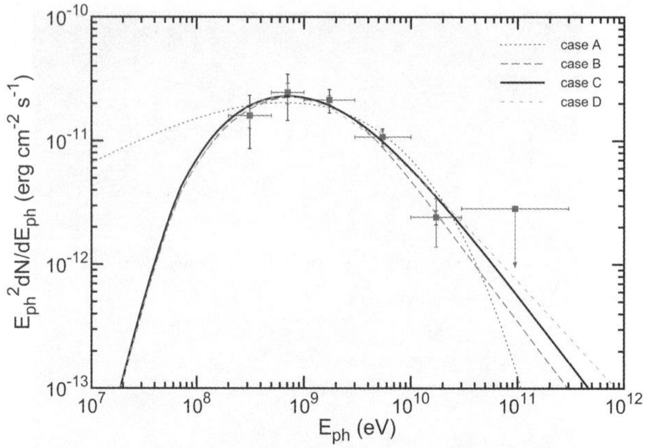

Figure 3. *Fermi* γ-ray spectral energy distribution of source A, fitted with various models (see text). The statistical errors are indicated by red bars and the systematic errors are indicated by black bars. For case C, t_{age} = 10 kyr and R_c = 20 pc and for case D, t_{age} = 10 kyr (see Table 1).

Table 1. Model parameters for the emissions of SNR accelerated particles

		α_e		$E_{e,cut}$ (GeV)	$W_e(>1\mathrm{GeV})$ (10^{51} erg)	
case A		2.0		400	1.3	
		α_p	$\Delta\alpha_p$	E_b (GeV)	$n_t E_{51}^{-1} W_p(>1\mathrm{GeV})$ (10^{51} erg cm^{-3})	
case B		2.0	1.2	18	7	
	t_{age} (kyr)	α_p	δ	χ	ΔR_c (pc)	M_{cl} ($10^4\,M_\odot$)
case C	4	2.4	0.7	0.07	5	4.5
	10	2.4	0.7	0.03	10	11
	100	2.4	0.7	0.004	13	18
	t_{age} (kyr)	α_p	δ	χ	R_c (pc)	M_{cl} ($10^4\,M_\odot$)
case D	10	2.4	0.7	0.25	15	18
	10	2.4	0.7	0.45	20	40
	100	2.4	0.7	0.05	20	40

Taking the case B as an example, if the mean target density $n_t \sim 100\,\mathrm{cm}^{-3}$, **then the total energy in protons with energy above 1 GeV, $W_p(>1\,\mathrm{GeV})$ is $\sim 1 \times 10^{50} E_{51}$ erg, i.e. the fraction, η, of the SN explosion energy converted into protons is of the order of 0.1.** As shown in Fig. 3, the fitting effect of case A is less satisfactory, and the escape cases (case C/D) have harder model spectra above 10 GeV than the interaction-at-the-shock case (case B). Further TeV observations will likely be helpful to distinguish the two scenarios. Although both the interaction at the shock and the illumination by escaping protons, can generally explain the gamma-ray properties of source A, the latter cases (C/D) are preferred as the GeV emission is offset from the SNR itself and centered on the SNR-MC interaction region.

4. Summary

Using 5.6 years of *Fermi*-LAT observation data, we have detected a gamma-ray source to the northwest of SNR Kes 41 with a significance of 24σ in 0.2–300 GeV. Neither significant long-term variability nor periodicity is detected from the timing analysis of source A in the same energy range. The best-fit location of the gamma-ray emission is within the region of the MC with which the SNR interacts. Although the spectrum is similar to those of pulsars, there is no known radio pulsar within the 3 σ location uncertainty that could be responsible for the high GeV luminosity. The energy budget would be too high if the gamma-ray emission was produced by high energy electron inverse Compton scattering off the CMB photons. The decay of neutral pions produced by the collision between relativistic protons accelerated by the shock of Kes 41 and the adjacent MC, can naturally explain the origin of the gamma-ray emission. Interaction of high-energy protons escaped from the SNR shock front with the adjacent MC seems more consistent with the *Fermi*-LAT observations presented in this article.

References

Abdo, A. A., Ackermann, M., Ajello, M., *et al.* 2009, *ApJ* (Letters), 706, L1
Abdo, A. A., Ackermann, M., Ajello, M., *et al.* 2010a, *ApJ*, 718, 348
Abdo, A. A., Ackermann, M., Ajello, M., *et al.* 2010b, *Science*, 327, 1103
Abdo, A. A., Ackermann, M., Ajello, M., *et al.* 2010c, *ApJ*, 712, 459
Abdo, A. A., Ackermann, M., Ajello, M., *et al.* 2013, *ApJS*, 208, 17
Ackermann, M., Ajello, M., Allafort, A., *et al.* 2013, *Science*, 339, 807
Aharonian, F. A. & Atoyan, A. M. 1996, *A&A*, 309, 917
Atwood, W. B., *et al.* 2009, *ApJ*, 697, 1071
Blandford, R. D. & Cowie, L. L. 1982, *ApJ*, 260, 625
Castro, D., Slane, P., Carlton, A., & Figueroa-Feliciano, E. 2013, *ApJ*, 774, 36
Caswell, J. L. 2004, *MNRAS*, 349, 99
Gabici, S., Aharonian, F. A., & Casanova, S. 2009, *MNRAS*, 396, 1629
Ginzburg, V. L. & Syrovatskii, S. I. 1969, The origin of cosmic rays (New York: Gordon and Breach)
Koralesky, B., Frail, D. A., Goss, W. M., Claussen, M. J., & Green, A. J. 1998, *AJ*, 116, 1323
Li, H. & Chen, Y. 2010, *MNRAS*, 409, L35
Li, H. & Chen, Y. 2012, *MNRAS*, 421, 935
Nolan, P. L., Abdo, A. A., Ackermann, M., *et al.* 2012, *ApJS*, 199, 31
Ohira, Y., Murase, K., & Yamazaki, R. 2011, *MNRAS*, 410, 1577
Tang, X. & Chevalier, R. A. 2014, *ApJ* (Letters), 784, L35
Uchiyama, Y., Blandford, R. D., Funk, S., Tajima, H., & Tanaka, T. 2010, *ApJ* (Letters), 723, L122
Wenger, M., Ochsenbein, F., Egret, D., *et al.* 2000, *A&AS*, 143, 9
Whiteoak, J. B. Z. & Green, A. J. 1996, *A&AS*, 118, 329
Xing, Y., Wang, Z., Zhang, X., & Chen, Y. 2014, *ApJ*, 781, 64
Zhang, G.-Y., Chen, Y., Su, Y., Zhou, X., Pannuti, T. G., & Zhou, P. 2015, *ApJ*, 799, 103

Supernova 1987A:30 years later - Cosmic Rays and Nuclei from Supernovae and their aftermaths
Proceedings IAU Symposium No. 331, 2017
A. Marcowith, M. Renaud, G. Dubner, A. Ray & A. Bykov, eds.

doi:10.1017/S1743921317004677

Investigating the region of 3C 397 in High Energy Gamma rays

Pooja Bhattacharjee[1], Pratik Majumdar[2], Tulun Ergin[3], Lab Saha[4] and Partha S. Joarder[5]

[1]Centre for Astroparticle Physics and Space Science, Bose Institute, Block EN, Sector V, Salt Lake, Kolkata 700091, India,
Department of Physics, Bose Institute, 93/1 A.P.C. Road, Kolkata 700009, India
email: pooja.bhattacharjee@jcbose.ac.in

[2]Saha Institute of Nuclear Physics, HBNI, 1/AF Bidhannagar, Kolkata 700064
email: pratik.majumdar@saha.ac.in

[3]TUBITAK Space Technologies Research Institute, ODTU Campus, 06800 Ankara, Turkey
email: ergin.tulun@gmail.com

[4]Universidad de Complutense, E-28040 Madrid, Spain
email: labsaha@ucm.es

[5]Centre for Astroparticle Physics and Space Science, Bose Institute, Block EN, Sector V, Salt Lake, Kolkata 700091, India,
Department of Physics, Bose Institute, 93/1 A.P.C. Road, Kolkata 700009, India
email: partha@jcbose.ac.in

Abstract. We investigate the supernova remnant (SNR) 3C 397 and its neighboring pulsar PSR J1906+0722 in high energy gamma rays by using nearly six years of archival data of *Large Area Telescope* on board *Fermi Gamma Ray Space Telescope* (Fermi-LAT). The off-pulse analysis of gamma-ray flux from the location of PSR J1906+0722 reveals an excess emission which is found to be very close to the radio location of 3C 397. Here, we present the preliminary results of this gamma-ray analysis of 3C 397 and PSR J1906+0722.

Keywords. Supernova Remnant, 3C 397, Pulsar, PSR J1906+0722, High Energy Gamma Rays

1. Introduction

3C 397 (G41.1-0.3) is a radio and X-ray bright young Galactic Supernova Remnant (SNR) located 1° away from the location of PSR J1906+0722 (Safi-Harb *et al.* 2000). The estimated age of 3C 397 is $\sim$ 5300 year (Safi-Harb *et al.* 2005). Whether 3C 397 is a "mixed-morphology" SNR (Safi-Harb *et al.* 2005) or not, is a matter of recent investigations. The *Suzaku* X-ray spectrum of 3C 397 showed a strong K-shell emission from stable Fe-peak elements. The high Ni/Fe and Mn/Fe mass ratios in the hot plasma component that dominates the K-shell emission lines may result from exploding white dwarfs with a near-Chandrasekhar mass (Yamaguchi *et al.* 2015). The radio location (Green 2014) of 3C 397 is very close to PSR J1906+0722, which is detected as part of a blind search of unidentified Fermi-LAT sources by the computing system, Einstein@Home1. This pulsar is associated with the Fermi 3rd catalog source (3FGL) (Acero *et al.* 2015), 3FGL J1906.6+0720. For past several years, 3FGL J1906.6+0720 had been known as one of the brightest undefined Fermi-LAT sources but its association was not detected in radio or in X-ray observations conducted before and after the gamma-ray observations. PSR

1 www.einsteinathome.org

J1906+0722 is a young, energetic, isolated pulsar with a spin frequency of 8.9 Hz, a characteristic age of 49 kyr and spin down power 1.0×10^{36} erg s^{-1} (Clark *et al.* 2015). No pulsar wind nebula (PWN) associated with the PSR J1906+0722, is, however, detected so far (Xing *et al.* 2014).

2. Gamma-ray Analysis

The large effective area and wide field of view of *Large Area Telescope* on board *Fermi Gamma Ray Space Telescope* (Fermi-LAT) (Atwood *et al.* 2009) allow for the detection and analysis of the gamma-ray emission from pulsars and PWNe. In our analysis, we have selected nearly six years (i.e from 04/08/2008 to 01/10/2014) of Pass 8 source class LAT events in about 100 MeV to 300 GeV energy range within a 20° region of interest (ROI) around PSR J1906+0722. We have used `gtselect` of the *Fermi Science Tools* (FST) analysis package[2] to select photons of energies greater than 100 MeV with a 90° zenith angle cut to remove contaminations from the Earth's limb. In order to perform the pulsed analysis of PSR J1906+0722, we have used an updated ephemeris for this pulsar (Clark *et al.* 2015). To extract the information about PWN or any other possible associations of the PSR J1906+0722, we have to eliminate the possible contamination due to the strong emissions from the pulsar itself. We have applied the pulsar gating technique[3] for such a purpose. Figure 1 shows the off-pulse phase interval of PSR J1906+0722 between the phases 0.36 and 0.68. We have then performed the FST binned likelihood analysis[4] on this off-pulse phase interval data. For the gamma-ray background model, we have used the diffuse Galactic (*gll_iem_v6.fits*) and the isotropic (*iso_P8R2_SOURCE_V6_v06.txt*) emission components in our analysis. Moreover, we have also incorporated all the 3FGL (Acero *et al.* 2015) point sources within a $10° \times 10°$ quadratic region centered at the pulsar location. The spectral and normalisation parameters of the sources within 4° of the ROI center and the normalizations of the two diffuse background components were left free. After performing this initial analysis, we have added 3C 397 as a new point source with a power-law (PL) type spectrum at the radio location of G41.1-0.3 (Green 2014) to the background model considered above. We have repeated the analysis with this new background model. The results are summarized in the next section.

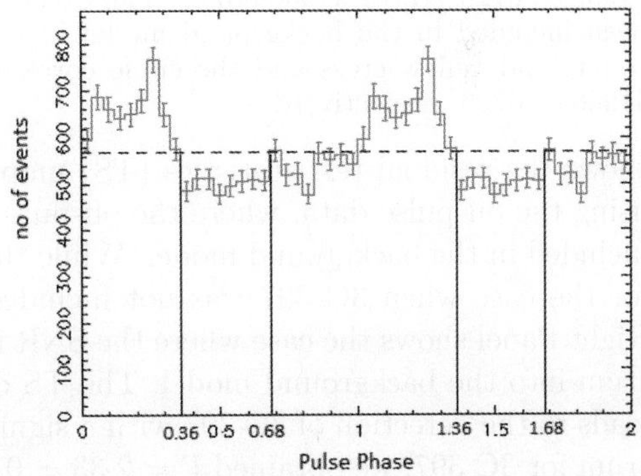

Figure 1. The gamma-ray phasogram of PSR J1906+0722.

3. Preliminary Results

3.1 3FGL J1906.6+0720 & 3C 397

From the binned-likelihood analysis of the off-pulse data, we found that by assuming a Log-parabola type spectrum, the off-pulse emission of 3FGL J1906.6+0720 had a significance of $\sim 31\sigma$. The spectral indices were found to be $\alpha = 2.34 \pm 0.05$ and $\beta = 0.27 \pm 0.04$. The total photon flux and energy flux were obtained to be $(4.66 \pm 0.33) \times 10^{-8}$ photons cm^{-2} sec^{-1} and $(3.84 \pm 0.16) \times 10^{-5}$ MeV cm^{-2} sec^{-1}, respectively. Figure 2

2 `fermi.gsfc.nasa.gov/ssc/data/analysis/software/`
3 `https://fermi.gsfc.nasa.gov/ssc/data/analysis/scitools/pulsar_gating_tutorial.html`
4 `https://fermi.gsfc.nasa.gov/ssc/data/analysis/scitools/binned_likelihood_tutorial.html`

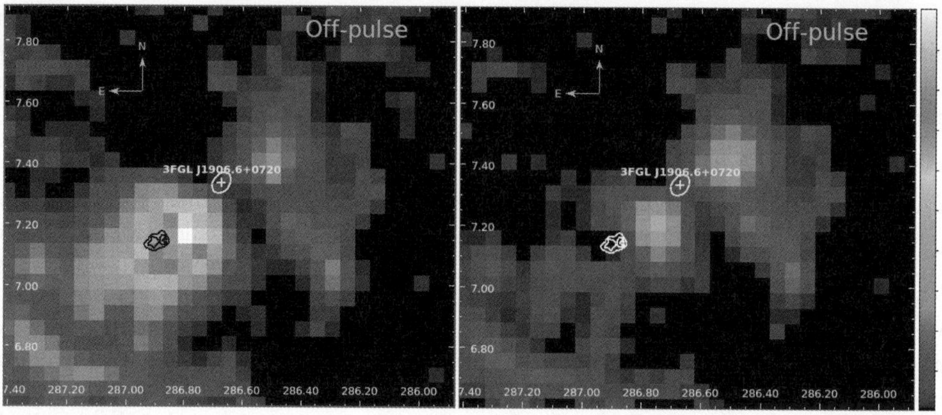

Figure 2. $10^\circ \times 10^\circ$ residual TS map showing the analysis region around the location of 3FGL J1906.6+0720 / PSR J1906+0722. The excess emission near 3C 397 seen in the Left panel has been included in the background model in the Right panel. Contours represent the *ROSAT* counts and yellow cross and the circle correspond to the best-fit position and error of 3FGL J1906.6+0720, respectively.

shows the residual test-statistics (TS) map of the $10^\circ \times 10^\circ$ analysis region created by using the off-pulse data, where the off-pulse emission model of 3FGL J1906.6+0720 was included in the background model. While the Left Panel of Figure 2 displays the TS map for the case when 3C 397 was not included in the gamma-ray background model, the Right Panel shows the case where the SNR is added as a point source with PL-type spectrum into the background model. The TS difference between the Right and Left Panels leads to the detection of 3C 397 with a significance of $\sim 8\sigma$. By assuming a PL-type spectrum for 3C 397, we obtained $\Gamma = 2.33 \pm 0.09$ and the total photon flux and energy flux of 3C 397 were found to be $(1.59 \pm 0.30) \times 10^{-8}$ photons cm^{-2} sec^{-1} and $(1.17 \pm 0.16) \times 10^{-5}$ MeV cm^{-2} sec^{-1}, respectively. In Figure 2, the contours of the *ROSAT* data are representing the X-ray counts of 8.78, 17.57, 26.36, and 35.15 in black and white colors on the Left and Right Panels, respectively. After modeling out the off-pulse emission of the 3FGL J1906.6+0720 and the excess emission coming from 3C 397, Figure 2 shows that there is still some amount of excess in the region between the pulsar and the SNR. There are various scenarios to explain such excess emission in terms of their possible hadronic or leptonic origins. For example, Clark *et al.* 2015 suspected that this excess gamma-ray emission could be due to the interaction of 3C 397 with a Molecular Cloud as observed by (Jiang *et al.* 2010).

3.2 PSR J1906+0722 & its Possible PWN Association Figure 3 shows the residual plot of the region around 3FGL J1906.6+0720. In the Upper Panel of this figure, we have displayed the full phase data from PSR J1906+0722. In the Lower panel of Figure 3, we have displayed the off-pulse phase data from PSR J1906+0722, where we applied a phase cut on our dataset. This cut reduces possible contaminations due to strong emissions from the pulsar PSR J1906+0722.

It is quite evident from both of the panels of Figure 3 that there is small bump at energies > 10 GeV. It is important to note that such an excess emission, that appears in the off-pulse phase data from PSR J1906+0722 (displayed in the Lower Panel of Figure 3), is not likely to be related to 3C 397 itself as we have already included 3C 397 in our background model. As PSR J1906+0722 is a young pulsar, it is likely to be associated with a PWN. This excess gamma-ray emission above 10 GeV in Figure 3 may, perhaps, be the result of such a possible association of PWN with PSR J1906+0722.

4. Conclusion

We investigated the region of PSR J1906+0722 and 3C 397 in high energy gamma rays by using seventy-four months of Fermi-LAT data. Our preliminary analysis shows that 3FGL J1906.6+0720 has an off-pulse gamma-ray emission of $\sim 31\sigma$ significance. After eliminating the component due to the pulsar and taking into account the off-pulse emission of PSR J1906+0722 in the background model, we obtained an excess emission of $\sim 8\sigma$ at the location of 3C 397. After including 3C 397 into the background model, there is still some excess emission left in the region between SNR and the pulsar. Since 3C 397 is supposed to be a Type Ia supernova (Yamaguchi *et al.* 2015), it is rather unlikely that it is associated with the pulsar PSR J1906+0722. As seen in Figure 2 Right Panel and Figure 3 Lower Panel, we still find some excess gamma-ray emission above 10 GeV in close neighbourhood of the SNR and the pulsar. As PSR J1906+0722 is a young pulsar, there is a possibility that the aforesaid excess emission might arise from an associated PWN. In view of these results, we need to disentangle the SNR from the pulsar and its possibly associated PWN by carefully studying the spectra and the systematic effects. In future, we plan to investigate these points in greater detail.

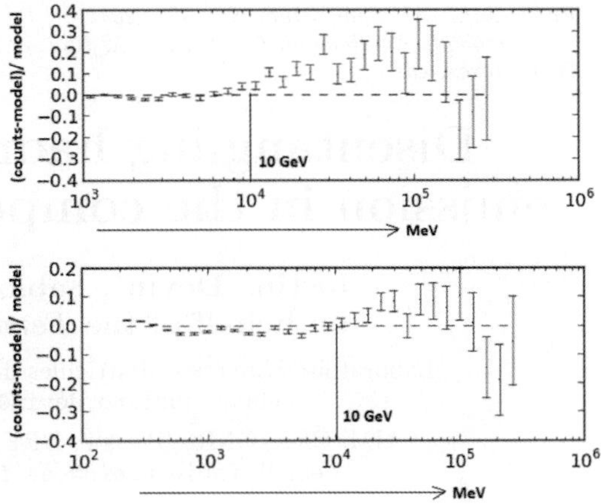

Figure 3. Residual plots of 10° ROI around the location of PSR J1906+0722 shown for two cases: The full-phase data (Upper Panel) and the off-pulse phase data (Lower Panel). The line at 10 GeV is the energy where the excess of gamma-ray emission starts.

5. Acknowledgement

The authors are grateful to Dr. C. J. Clark, Albert-Einstein-Institut, MPIG, Germany for making the Ephemeris of PSR J1906+0722 accessible to us. PB is particularly grateful to the International Astronomical Union (IAU) for the allocation of a grant thus enabling her to attend this IAU S331 symposium. We are grateful to all the organizers of the IAUS331 symposium for their warm hospitality. TE thanks the Science Academy, Turkey for the support under the Science Academy's Young Scientists Award Program (BAGEP-2015), Turkey. We would like to thank the referee for carefully reading the manuscript and for giving very constructive comments.

References

Acero, F., *et al.* 2015, *ApJS*, 218, 23
Atwood, W. B., *et al.* 2009, *ApJ*, 697, 1071
Clark, C. J., *et al.* 2015 *ApJ* (Letters), 809, L2
Green, D. A. 2014, *BASI*, 42, 47
Jiang, B., Chen, Y., Wang, J., Su, Y., Safi-Harb, S., & DeLaney, T. 2010, *ApJ*, 712, 1147
Safi-Harb, S., Petre, R., Arnaud, K. A., Keohane, J. W., Borkowski, K. J., Dyer, K. K., Reynolds, S. P., & Hughess, J. P. 2000, *ApJ*, 545, 922
Safi-Harb, S., Dubner, G., Petre, R., Holt, S. S., & Durouchoux, P. 2005, *ApJ*, 618, 321
Xing, Y. & Wang, Z. 2014, *PASJ*, 66, 72
Yamaguchi, H., Badenes, C., Foster, A. R., Bravo, E., William, B. J., Maeda, R., & Koyama, K. 2015, *ApJ* (Letters), 801, L31

Supernova 1987A:30 years later - Cosmic Rays and Nuclei from Supernovae and their aftermaths
Proceedings IAU Symposium No. 331, 2017
A. Marcowith, M. Renaud, G. Dubner, A. Ray & A. Bykov, eds.

doi:10.1017/S1743921317004823

Disentangling hadronic from leptonic emission in the composite SNR G326.3-1.8

Justine Devin[1], Fabio Acero[2], Jean Ballet[2]
on behalf of the *Fermi*-LAT collaboration

[1]Laboratoire Univers et Particules de Montpellier, CNRS/IN2P3, France
email: justine.devin@etu.umontpellier.fr

[2]AIM, CEA/CNRS/Paris 7, DRF/IRFU/DAp, CEA Saclay, France
email: fabio.acero@cea.fr or jean.ballet@cea.fr

Abstract. G326.3-1.8 (also known as MSH 15-56) has been detected in radio as a middle-aged composite supernova remnant (SNR) consisting of a SNR shell and a pulsar wind nebula (PWN) which has been crushed by the reverse shock. With the recent *Fermi*-LAT data release Pass 8 providing increased acceptance and angular resolution, we investigate the morphology of this SNR to disentangle the PWN from the SNR contributions and understand the nature of the γ-ray emission. We thus perform a morphological and spectral analysis from 300 MeV to 300 GeV which highlights the contributions from these two components. The simplest interpretation is hadronic emission from the SNR and harder leptonic emission from the PWN.

Keywords. supernova remnants, gamma-rays: observations, acceleration of particles, cosmic rays

1. Introduction

Supernova remnants (SNRs) and pulsar wind nebulae (PWNe) have long been considered potential sources of Galactic cosmic rays and have therefore been investigated over a wide range of energies. With a SNR shell and a PWN seen at radio wavelengths, the Galactic SNR G326.3-1.8 is a prototype of the so-called composite SNRs (Mills *et al.* 1961). Its distance is estimated between 3.1 kpc (Goss *et al.* 1972) and 4.1 kpc (Rosado *et al.* 1996) as established by the HI absorption profile and Hα velocity measurements respectively. Temim *et al.* (2013) estimated the SNR to be 16,500 years old with a shock velocity of 500 km s^{-1}. Figure 1, obtained from radio observations (Whiteoak & Green 1996), shows a symmetric SNR shell with a 0.3° radius and a PWN which has been displaced from the geometric center of the SNR and likely crushed by the reverse shock. The associated pulsar has not been detected but *Chandra* maps have revealed a point source embedded in the X-ray PWN located in the South West of the radio nebula (Temim *et al.* 2013). At higher energies, the origin of the γ-ray emission remains uncertain (Temim *et al.* 2013) and SNR G326.3-1.8 has recently been found to be extended with the *Fermi* Large Area Telescope (LAT) data (Acero *et al.* 2016). In this proceeding, we revisit the morphology of this SNR, with the latest data release Pass 8 of the *Fermi*-LAT (Atwood *et al.* 2013), and we also report a spectral analysis of our best model using two spatial components for the γ-ray emission.

2. Data analysis

The LAT on board the *Fermi* satellite is a pair conversion instrument sensitive to γ-rays in the energy range from 30 MeV to more than 300 GeV. The new data release Pass

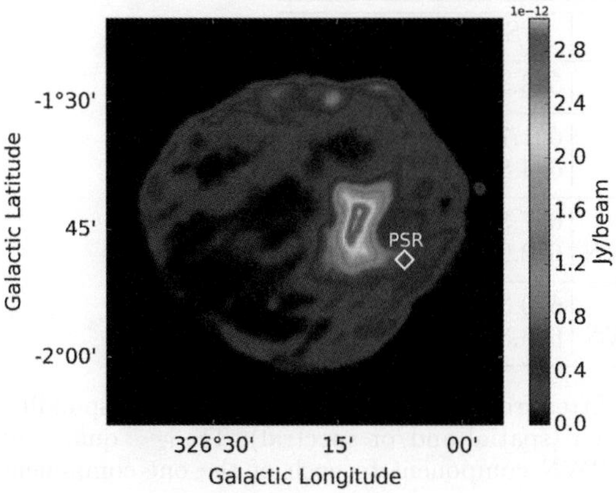

Figure 1. 843 MHz MOST radio image of SNR G326.3-1.8 (Whiteoak & Green 1996). The position of the pulsar candidate is represented by a white diamond.

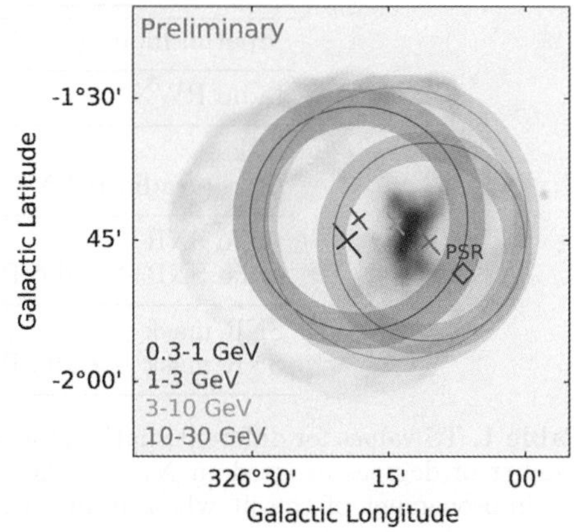

Figure 2. Best-fit Gaussians in 4 energy bands. Crosses: position uncertainties, circles: r_{68} of the Gaussian (if the source is extended), shaded areas: errors on size.

8 allows for event selection based on the quality of the reconstruction of their direction, resulting in four different Point Spread Function (PSF) classes. To avoid as much as possible the contamination between the PWN and the SNR, we keep the data having the best angular resolution (0.4° at 1 GeV compared to 0.8° with no selection) at the cost of a factor four reduction of the effective area. We perform a binned analysis from 300 MeV to 300 GeV using 6.5 years of data collected from August 4, 2008 to January 31, 2015, within a 10° × 10° region around the position of SNR G326.3-1.8. We use a Test Statistic (TS) which is defined to test the likelihood of one hypothesis $\mathcal{L}_1$ (including a source) against the null hypothesis $\mathcal{L}_0$ (absence of source), such as: $\text{TS} = 2 \times (\log \mathcal{L}_1 - \log \mathcal{L}_0)$. This can be directly interpreted in terms of significance of hypothesis 1 with respect to the null hypothesis 0 since the TS follows a χ^2-law for n additional parameters. When adding a second component, the model with one component becomes the null hypothesis to test the significance of the second component. We first model the region starting with the 3FGL sources catalog (Acero *et al.* 2015) based on the previous Pass 7 data, and we add eleven sources at the positions where the significance exceeds 5σ.

2.1. *Morphological analysis*

To investigate the source morphology, we divide the data into four logarithmically spaced energy bins that we subsequently fit individually using a 2D-symmetric Gaussian. Since the PSF changes over half a decade, we need to specify the source spectrum. We parametrize it with a power law with free spectral index. The normalizations of the source, the Galactic and isotropic diffuse emissions are let free while the other sources are fixed to their best-fit values. Figure 2 depicts the results of the fitting procedure in the four energy bands. At low energies (300 MeV - 1 GeV), the PSF (0.4° at 1 GeV) is larger than the SNR radius and we cannot measure the extension. The centroid of the Gaussian lies outside the nebula. Between 1 and 3 GeV, the significance of the extension is more than 5σ and the position of the Gaussian appears to be fairly consistent with the center of the radio SNR. At higher energies (from 3 to 30 GeV), the γ-ray morphology is still significantly extended (more than 5σ) and shows a clear trend toward the position of the radio PWN.

This energy-dependent source morphology clearly requires a more detailed investigation, beyond a one-component modeling. Since the PSF below 1 GeV is not good enough

Spatial models	TS	N_{dof}	TS_{PWN}
radio PWN	593.38	2	—
disk	681.78	5	—
disk + radio PWN	694.81	7	13.03
radio SNR	667.32	2	—
radio SNR + radio PWN	682.97	4	15.65
SNR mask	670.32	2	—
SNR mask + radio PWN	696.36	4	26.04

Table 1. TS values for different spatial models fitted from 1 GeV to 300 GeV. The corresponding number of degrees of freedom N_{dof} is also given (spatial and/or spectral). TS_{PWN} quantifies the improvement of the fit when adding the PWN component to each of the one-component models.

to resolve the SNR, the following morphological analysis uses data between 1 and 300 GeV and we describe the spectral models associated to each component by a power law. We first model the γ-ray emission assuming it comes only from the PWN, with its radio template (see Figure 1) and we perform a likelihood fit where the spectral parameters of the PWN as well as those of the Galactic and isotropic emissions are free during the fit. Table 1 reports the corresponding TS value. To model the contribution of an additional component which seems to be responsible for the low energy part (as depicted in Figure 2), we test several templates using first a simple disk component and then physically motivated templates (derived from the radio map of the SNR). We thus use the radio map replacing the contribution of the nebula by the average value of the radio emission around it (labeled "radio SNR" in Table 1). From that, we also create another template, following the radio shock and filled homogeneously (called "SNR mask" in Table 1). The results from our maximum likelihood fit are given in Table 1. The TS values obtained using the one-component models (disk, radio SNR and SNR mask) alone are clearly higher than the one obtained using only the radio PWN, indicating that the fit prefers a model more extended than the radio nebula. In Table 1, we compare the TS values obtained with each of the one-component models to the two-component models testing the significance of adding the radio template of the nebula (TS_{PWN} value for two degrees of freedom). For all our models, the significance of adding the PWN is between 3 and 4σ and the maximum TS value is obtained for the model including the radio PWN with either the disk or the SNR mask. In terms of significance, our best model involves the SNR mask and the radio PWN since it requires fewer free parameters during the fit than the disk whose spatial components have been optimized.

2.2. *Spectral analysis*

To understand the underlying emission processes, we perform a spectral analysis from 300 MeV to 300 GeV using our best two-component model: the SNR mask with the radio PWN. Modeling both by a power law, the fit leads to a spectral separation between the two components: a softer spectrum for the remnant ($\Gamma = 2.17 \pm 0.06$ being the photon index) and a harder spectrum for the nebula ($\Gamma = 1.79 \pm 0.12$). The associated Spectral Energy Distributions (SEDs), shown in Figure 3, clearly emphasize that two distinct morphologies give rise to two distinct spectral signatures. The TS values of the components in each energy bin highlight their different contributions: at low energy ($E < 10$ GeV), the emission is dominated by the SNR while the contribution of the PWN protrudes above 10 GeV. If both the SNR and the PWN contribute to the γ-ray

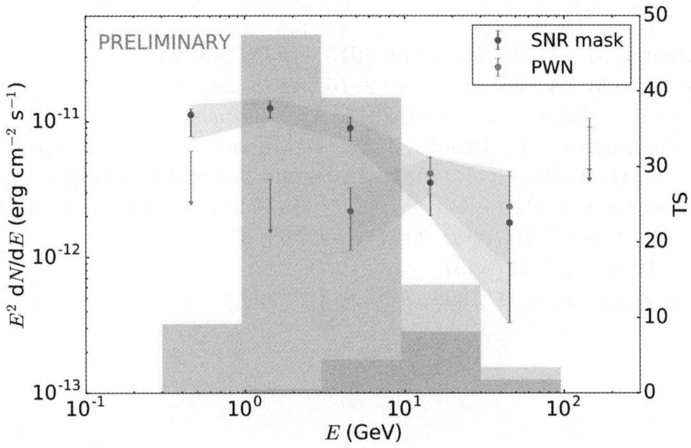

Figure 3. SED (data points) and TS values (histograms) in individual energy bands obtained with the model including the SNR mask and the radio PWN. The shaded areas represent the quadratic sum of statistical and systematic errors. The systematic errors are calculated using 8 alternative Galactic diffuse emissions as explained in the first *Fermi*-LAT supernova remnant catalog (Acero *et al.* 2016).

emission, their spectral shapes are consistent with two different underlying processes, namely hadronic and leptonic emission, respectively. Even if no molecular cloud has been reported close to this SNR, the detection of Hα emission in several regions (Van den Bergh 1979) is an evidence of interaction of the shock with dense material, therefore providing a large density of targets to produce a significant amount of hadronic emission.

3. Conclusions

We perform an analysis from 300 MeV to 300 GeV of the composite SNR G326.3-1.8 with the *Fermi*-LAT Pass 8 data. We find that the γ-ray morphology evolves with energy and shrinks to the radio nebula at high energies (E $>$ 10 GeV). We thus investigate a more detailed morphology using the radio map of the PWN and other models for an additional contribution such as an uniform disk, the radio map of the remnant and its radio template filled homogeneously, called here the SNR mask. Using the maximum likelihood fitting procedure, our best model involves the SNR mask and the radio PWN for which we obtain two different spectral signatures: a softer spectrum for the remnant ($\Gamma = 2.17 \pm 0.06$) and a harder spectrum for the nebula ($\Gamma = 1.79 \pm 0.12$). Their spectral shapes are consistent with hadronic and leptonic origins for the SNR and the PWN respectively and we are currently investigating the physical properties of this source.

Acknowledgements

The *Fermi*-LAT Collaboration acknowledges support for LAT development, operation and data analysis from NASA and DOE (United States), CEA/Irfu and IN2P3/CNRS (France), ASI and INFN (Italy), MEXT, KEK, and JAXA (Japan), and the K.A. Wallenberg Foundation, the Swedish Research Council and the National Space Board (Sweden). Science analysis support in the operations phase from INAF (Italy) and CNES (France) is also gratefully acknowledged. This work performed in part under DOE Contract DE-AC02-76SF00515.

Many thanks to the organizing committees and for the financial support obtained through an IAU grant.

References

Acero, F., Ackermann, M., Ajello, M., *et al.* 2015, *ApJS*, 218, 23
Acero, F., Ackermann, M., Ajello, M., *et al.* 2016, *ApJS*, 224, 8
Atwood, W., Albert, A., Baldini, L., *et al.* 2013, ArXiv e-prints
Goss, W. M., Radhakrishnan, V., Brooks, J. W., & Murray, J. D. 1972, *ApJS*, 24, 123
Mills, B. Y., Slee, O. B., & Hill, E. R. 1961, *Australian Journal of Physics*, 14, 497
Rosado, M., Ambrocio-Cruz, P., Le Coarer, E., & Marcelin, M. 1996, *A & A*, 315, 243
Temim, T., Slane, P., Castro, D., *et al.* 2013, *ApJ*, 768, 61
Van den Bergh, S. 1979, *ApJ*, 227, 497
Whiteoak, J. B. Z. & Green, A. J. 1996, *A&AS*, 118, 329

Supernova 1987A:30 years later - Cosmic Rays and Nuclei from Supernovae and their aftermaths
Proceedings IAU Symposium No. 331, 2017
A. Marcowith, M. Renaud, G. Dubner, A. Ray & A. Bykov, eds.

doi:10.1017/S1743921317004628

Upper Limits on Gamma-ray Emission from Supernovae Serendipitously Observed with H.E.S.S.

Rachel Simoni[1], Nigel Maxted[2], Mathieu Renaud[3] and Jacco Vink[1]
On behalf of the H.E.S.S. Collaboration

[1]GRAPPA/Anton Pannekoek Institute, Universiteit van Amsterdam,
Postbus 94249, 1090 GE, Amsterdam, the Netherlands
email: r.c.simoni@uva.nl

[2]School of Physics, University of New South Wales, 2052, Sydney, Australia
email: n.maxted@unsw.edu.au

[3]Laboratoire Univers et Particules, Université Montpellier, 34095, Montpellier, France

Abstract. It is hypothesized that some young supernovae might have the correct properties to accelerate cosmic rays, which in turn might generate gamma-ray emission by-products. We search for gamma-ray excesses towards supernovae in nearby galaxies which were serendipitously within the field of view of the HESS telescopes within a year of the supernova event. HESS cherenkov air-shower data collected between December 2003 and March 2015 were considered and compared to recent catalogs. Nine candidate supernovae were identified and analysed. No significant emission from these supernovae has been found, and upper limits for their very high energy emission are reported.

Keywords. gamma rays: observations, supernovae: general, cosmic rays

1. Introduction

It is well established that supernova remnants (SNRs) can be emitters of high-energy and very-high-energy (E>100 GeV) gamma-rays and that cosmic-ray (CR) protons and nuclei accelerated locally are responsible for this gamma-ray emission in some of these objects (e.g. Ackermann *et al.* 2013, Aharonian *et al.* 2008), while other objects remain probable CR accelerators (e.g. Abdalla *et al.* 2016, Lau *et al.* 2017). Nevertheless, the SNRs under strong scrutiny from a very-high-energy gamma-ray perspective are often 10^2-10^4 years old and have never been conclusively shown to be PeV (10^{15} eV) CR acceleration sources, despite some shell-type SNRs still being potential candidates (e.g. Fukui *et al.* 2012, Gabici *et al.* 2014).

It has been theorized that young supernovae (SNe) might have the correct properties to accelerate CRs to PeV energies within a few months to a few years after the progenitor explosion (e.g. Cardillo *et al.* 2015, Katz *et al.* 2011, Murase *et al.* 2011, Marcowith *et al.* 2014).

This can occur only in a very dense circumstellar medium (CSM) as this results in sufficient number of particles entering the shock, as well as creation and growth of plasma instabilities. This last process is driven by the particle acceleration itself, leading to a fast magnetic field amplification. The amplified, turbulent magnetic fields allow CR particles to be accelerated to PeV energies or beyond. Accelerated hadrons will interact with the dense surrounding matter and produce very high energy gamma rays and neutrinos via pion production/decay. Dense CSM is expected around core collapse SNe with high

Table 1. The list of SN positions tested for H.E.S.S. gamma-ray excess emission. The list was compiled using a system of cuts described in Section 2.2. The name, host galaxy, coordinates, estimated distance, SN type and discovery date is shown for each SN.

SN Name	Host galaxy	RA [J2000]	DEC [J2000]	Dist. [Mpc]	Type	Disc. date
SN 2004cx	NGC 7755	23h47m52.86s	$-30°31'32.6''$	26 ± 5	II	2004-06-26
SN 2005dn	NGC 6861	20h11m11.73s	$-48°16'35.5''$	38.4 ± 2.7	?	2005-08-27
SN 2008bk	NGC 7793	23h57m50.42s	$-32°33'21.5''$	4.0 ± 0.4	IIP	2008-03-25
SN 2008bp	NGC 3095	10h00m01.57s	$-31°33'21.8''$	29 ± 6	IIP	2008-04-02
SN 2008ho	NGC 922	02h25m04.00s	$-24°48'02.4''$	41.5 ± 2.9	IIP	2008-11-26
SN 2009hf	NGC 175	00h37m21.79s	$-19°56'42.2''$	53.9 ± 3.8	IIP	2009-07-09
SN 2009js	NGC 918	02h25m48.28s	$+18°29'25.8''$	16 ± 3	IIP?	2009-10-11
SN 2011ja	NGC 4945	13h05m11.12s	$-49°31'27.0''$	5.28 ± 0.38	IIP	2011-12-18
SN 2012cc	NGC 4419	12h26m56.81s	$+15°02'45.5''$	$\sim 20 \pm$	II	2012-04-29

mass-loss rates and slow wind speeds; this applies particular to SN type IIp, IIb and IIn (Chevalier *et al.* 2016).

This possibility of CR acceleration in young SNe motivates a search for signatures of particle acceleration beyond TeV energies in H.E.S.S. Cherenkov air-shower data.

2. Analysis

2.1. *H.E.S.S. observations*

H.E.S.S., the High Energy Sterescopic System, is an array of five imaging atmospheric Cherenkov telescopes (IACTs) located in the Khomas Highland of Namibia at an altitude of 1800 m above sea level. Four 12 m-diameter telescopes were operating from December 2003 and a 5th telescope of 28 m-diameter became operational in September 2012. In our analysis, the data of 3-4 of the 12 m telescopes has been utilised. This generally results in field of view and beam full-width half-maximum of 5° and ~0.1°, respectively, and an energy threshold and energy resolution of ~ 100 GeV and ~ 15%, respectively (Aharonian *et al.* 2006). The SNe in our sample all occur within ~3° of the centre of the field of view.

2.2. *Candidate Selection*

The online IAU Central Bureau of Astronomical Telegrams (CBAT) supernova catalogue † was downloaded to compile an initial, extensive list of SN candidates. The NASA/IPAC Extragalactic Database (NED) ‡ was then queried for the redshift of each SN host galaxy to compile a short-list of SNe with redshift $z < 0.01$ to ensure that only SNe near enough to plausibly be observable at gamma-ray energies were considered. If a host galaxy was not stated for a given SN in the CBAT SN catalogue, the SN was discarded from the short-list. The H.E.S.S. database was then queried for air shower measurements in the direction of each short-listed SN, within a time-range beginning 7 days before the SN discovery date to a year after the SN discovery date. The SNe with coincidental H.E.S.S. data are presented in Table 1. All H.E.S.S. data taken from December 2003 until the 31st of December 2014 was searched, and all SNe presented on CBAT on the 30th of March, 2015 were considered. We removed type Ia and Ic SNe from our sample, because these types are unlikely to occur in an interstellar density large enough to accelerate CRs up to TeV energies.

† www.cbat.eps.harvard.edu/lists/Supernovae.html
‡ ned.ipac.caltech.edu/forms/z.html

Table 2. Observed statistics for each SNe (see text) and preliminary upper limits on the integrated flux above the energy threshold E_{Th}, and 1 TeV. These upper limits are derived with 95% confidence level assuming a power law spectrum of index 2

SNe	N_{on}	N_{off}	α	N_{excess}	Sig	Exp hrs	Delay day	E_{Th} (TeV)	$F(> E_{Th})$	$F(> 1\,\text{TeV})$
									(10^{-13} cm^{-2}s^{-1})	
SN 2004cx	169	10387	65	8.7	0.7	39.9	-6	0.18	10	1.9
SN 2005dn	571	11452	19	-38.6	1.5	53.1	-3	0.21	2.2	0.41
SN 2008bk	50	3652	54	-18.0	-2.3	9.6	98	0.21	6.0	4.8
SN 2008bp	32	1860	60.	1..1	0.2	4.7	272	0.21	29	5.5
SN 2008ho	9	369	33	-2.3	-0.7	1.4	36	0.33	16	7.7
SN 2009hf	43	1404	35	3.3	0.5	4.0	0	0.21	20	5.3
SN 2009js	14	711	67	3.4	1	4.8	94	0.63	15	11
SN 2011ja	56	834	15	-1.2	-0.2	3.8	91	0.23	13	3.9
SN 2012cc	7	660	74	-1.9	-0.7	3.0	53	0.72	15	10

Notes: Exp gives the lifetime of the observations, Delay the number of days beetween first observation and SN discovery and Sig, the significance.

2.3. *Data Analysis*

Standard quality cuts were applied to remove bad-quality data from each data set, and the Model analysis outlined in (de Naurois *et al.* 2009) was employed for each candidate SN. For a given SN, multiple OFF regions were selected via the reflected-region method (Berge *et al.* 2007) to estimate a background of gamma-ray-like events which to subtract from the gamma-ray-like events detected in the ON-region (0.6 deg radius), selected on the SN position as for a point source. Results were confirmed using the H.E.S.S. Analysis software version hap_ICRC2015, with the ImPACT analysis framework (Parsons *et al.* 2014) and standard cuts.

For each 28-minute observation run which passed the criteria outlined in Section 2.2, the excess was computed (using $N_{excess}=N_{on}$ - αN_{off}) as well as the cumulative statistical significance using Equation 17 of Li & Ma 1983. Given the serendipitous nature of the observations presented in this report, the live time varies between ~1 and ~50 hours (see Table 2).

2.4. *Results*

Table 1 lists the observed supernovae along with corresponding host galaxies, coordinates, distance, type and discovery date. Distances range between 4 and 54 Mpc and discovery dates range between 2004 and 2012.

In Table 2 we report the relevant statistics of the gamma-ray observations and the TeV gamma-ray upper limits set by H.E.S.S.. No significant TeV emission is found towards any of the SNe within one year of the initial explosion.

Plots of >1 TeV gamma-ray emission vs time binned on a nightly basis are presented in Fig. 1. All SNe are consistent with having zero TeV emission within the sensitivity limits of our observations.

3. Conclusion

We search for TeV gamma-ray emission from supernovae serendipidously observed with the H.E.S.S. gamma-ray telescope, amending previous efforts (Lennarz *et al.* 2013). We find no significant detection above >1 TeV, a result that can be linked to other recent non detections (Ackermann *et al.* 2015, Ahnen *et al.* 2017). Another related result is the non

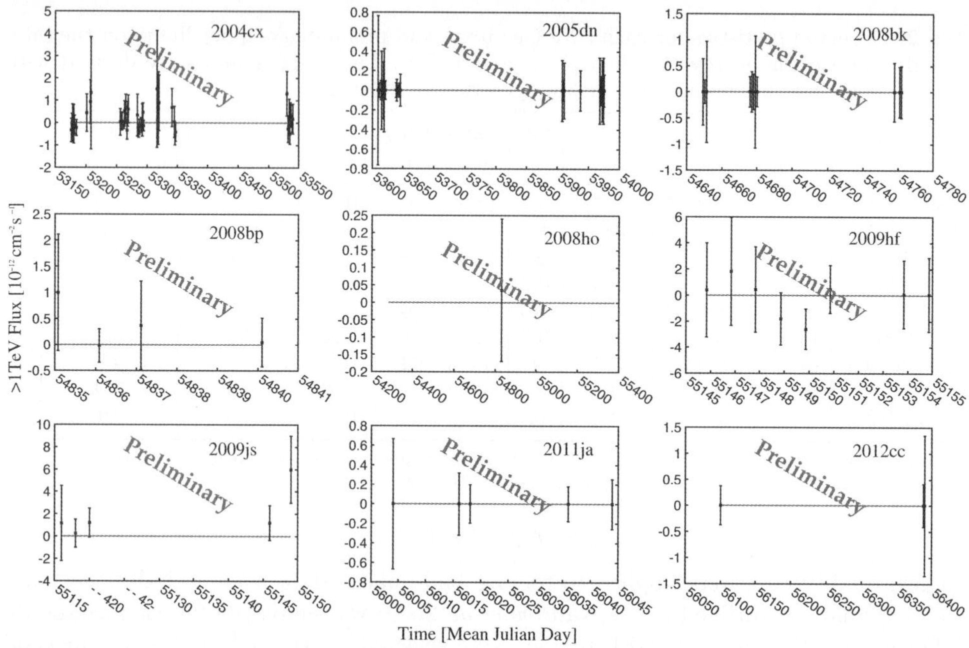

Figure 1. Light curves of the 9 observed SNe.

detection of nearby SN1987A (Aharonian *et al.* 2015), despite predictions (Völk *et al.* 2011).

These limits on gamma-ray emission can translate into limits on the physical environment of these SN events - something to be investigated in future papers.

References

Abdalla, H., *et al.* (H. E. S. S. Collaboration) 2016, arXiv1609.00600
Ackermann, M., *et al.* (Fermi Collaboration) 2013, *Science*, 339, 807
Ackermann M, *et al.* (Fermi Collaboration) 2015, *ApJ*, 807, 169
Aharonian, F., *et al.* (H. E. S. S. Collaboration) 2006, *A&A*, 457, 899
Aharonian, F, *et al.* (H. E. S. S. Collaboration) 2008, *A&A*, 481, 401
Aharonian, F., *et al.* (H. E. S. S. Collaboration) 2015, *Science*, 347, 406
Ahnen, M., *et al.* (Magic Collaboration) 2017, arXiv1702.007677
Berge, D., Funk, S., & Hinton, J. 2007, *A&A*, 466, 1219
Cardillo,M , Amato E., & Blasi P. 2015, *Astroparticle Physics* 69 1-10.
Chevalier, Roger A., & Claes Fransson 2016 arXiv preprint arXiv:1612.07459
Fukui, Y., Sano, H., Sato, J. *et al.* 2012, *ApJ*, 746, 82
Gabici, S. & Aharonian, F., 2014, *MNRAS*, 445, L70
Katz *et al.* 2011, arXiv:1106.1898
Lau J, Rowell G, Burton G, Fukui Y *et al.* 2017, *MNRAS*, 464, 3757
Lennarz, D., *et al.* (H. E. S. S. Collaboration) 2013, arXiv1307.7727
Li, T.-P. & Ma, Y.-Q. 1983, *ApJ*, 272, 317
Marcowith, A., Renaud, M., Dwarkadas, V., & Tatischeff, V. 2014, *NuPhS*, 256, 94M
Murase, Thompson *et al.*, 2011, *PhRvD*, 84, 3003
de Naurois, M. & Rolland, L. 2009, *Aph*, 32, 231
a Parsons, R. D. & Hinton, J. A. 2014, *Astroparticle Physics*, 56, 26
Völk, H. J, Berezhko, E. G., & Ksenofontov, L. T. 2011, *ApJ*, 732, 58

Supernova 1987A:30 years later - Cosmic Rays and Nuclei from Supernovae and their aftermaths
Proceedings IAU Symposium No. 331, 2017
A. Marcowith, M. Renaud, G. Dubner, A. Ray & A. Bykov, eds.
doi:10.1017/S174392131700446X

Multimessenger Predictions from 3D General-Relativistic Core-Collapse Supernovae Models

Kei Kotake[1], Takami Kuroda[2] and Kazuhiro Hayama[3]

[1]Department of Applied Physics, Fukuoka University, Jonan, Nanakuma, Fukuoka 814-0180, Japan,

[2]Institute of Kernphysik, Technische Universtät Darmstadt, D-64289 Darmstadt, Germany,

[3]Institute for Cosmic Ray Research, University of Tokyo, 5-1-5 Kashiwa-no-Ha, Kashiwa City, Chiba 277-8582, Japan

Abstract. In this contribution, we present results from fully general-relativistic three-dimensional (3D) simulations of a non-rotating $15M_{\odot}$ star using different nuclear equations of state (EOSs). We show that the SASI (standing-accretion-shock-instability) activity occurs much more vigorously in models with softer EOS. By performing detailed analysis of the gravitational-wave (GW) emission, we find a new GW signature that is produced predominantly by the SASI-induced downflows to the proto-neutron star. We discuss the detectability of the GW signal by performing a coherent network analysis where multiple detectors including LIGO Hanford, LIGO Livingston, VIRGO, and KAGRA are considered. We point out that the GW signal, whose typical frequency is in the best sensitivity range of the laser-interferometers, could potentially provide the *live broadcast* that pictures how the supernova shock is dancing in the core. The detection horizon of the signal is estimated as 2∼3 kpc for the current generation detectors, which can extend up to ∼100 kpc for the third generation detectors like Cosmic Explorer. We furthermore perform a correlation analysis between the SASI-modulated GW and neutrino signals. Our results show that the time correlation of the two signals becomes highest when we take into account the travel timescale of adverting material from the (average) neutrino-sphere to the proto-neutron star surface.

Keywords. supernovae:general, hydrodynamics, instabilities, neutrinos, gravitational waves

1. Introduction

Accumulating observational evidence from precise, multi-wavelength electromagnetic-wave (EM) observations of ejecta morphologies, spatial distributions of nucleosynthetic yields have all pointed towards core-collapse supernovae (CCSNe) being generally aspherical (e.g., Maeda *et al.*(2008), Grefenstette *et al.* (2015), Tanaka *et al.* (2017) and references therein). Unquestionably important as these discoveries are, these EM signals are secondary for probing the multidimensionality inside the supernova engine because they are only able to provide images of optically thin regions far away from the central core.

Much more direct information is delivered to us by neutrinos and gravitational waves (GWs). This year is 30 years after SN1987A, which has opened up neutrino astronomy (Hirata *et al.* (1987)). In 2015, the centurial year after Einstein's theory of general relativity (GR), the twin detectors of the Laser Interferometer Gravitational-wave Observatory (LIGO) coined the *first direct detection* of GWs. It turns out that this (potentially) Nobel Prize-winning event is produced by the merger of two black holes (Abbott *et al.* (2016)).

The second generation detectors like KAGRA (Aso *et al.* (2013)), advanced VIRGO (Hild *et al.* 2009) will be on-line in the coming years. Next to compact binary mergers, CCSNe are among one of the most promising astrophysical sources of GWs (e.g., Kotake (2013) for reviews).

From a theoretical point of view, understanding the origin of the explosion multi-dimensionalities is also indispensable for clarifying the yet uncertain CCSN mechanisms. After a half-century of continuing efforts, theory and neutrino radiation-hydrodynamics simulations are now converging to a point that multi-dimensional hydrodynamics instabilities play a pivotal role in the neutrino mechanism (Bethe (1990)), the most favoured scenario to trigger explosions. In fact, a number of self-consistent simulations in two or three spatial dimensions (2D, 3D) now report shock revival of the stalled bounce shock into explosion by the neutrino mechanism. These successful models have strengthened our confidence in the multi-D neutrino-driven paradigm (see Müller (2016), Janka *et al.* (2016), Foglizzo *et al.* (2015), Burrows (2013), Kotake *et al.* (2012) for recent reviews).

In this contribution, we present results of 3D-GR simulations of a non-rotating $15M_\odot$ star using two representative EOSs based on our recent work (Kuroda *et al.* (2016)). We start to briefly summarize our numerical method for GR-radiation hydrodynamics and initial models in Section 2. In Section 3, we report results, which includes new analysis on the neutrino emission from the 3D-GR models and the correlation analysis between the SASI-modulated neutrino and GW signals.

2. Numerical Method and Initial Models

In our full GR radiation-hydrodynamics simulations, the evolution equations of metric, hydrodynamics are solved based on the BSSN formalism (Shibata and Nakamura (1995), Baumgarte and Shapiro (1999)) and the evolution of neutrino radiation field based on the M1 scheme (see collective references in Kuroda *et al.* (2012), Kuroda *et al.* (2014)). To follow the 3D hydrodynamics up to $\lesssim 400$ ms postbounce, we shall omit the energy dependence of three-flavor neutrino radiation fields in this work (see, however, the multi-energy transport scheme in Kuroda *et al.* (2016)).

We employ a frequently used 15 $M_\odot$ star of Woosley and Weaver (1995). The 3D computational domain (15000 km width) consists of nested cubic boxes with the minimum grid size near the origin being $\Delta x = 458$m (the effective angular resolution is $\sim 1°$). We use two kinds of EOSs based on different nuclear interaction treatments, which is TM1 of Hempel and Schaffner-Bielich (2010) and SFHx of Steiner *et al.* (2013). We label the two models simply by the EOS used (SFHx or TM1). From the mass-radius relation of a cold neutron star (NS), the maximum gravitational mass and the radius is $M_{\rm max} =$ 2.1 and 2.2 $M_\odot$, $R = 12$ and 14.5 km for SFHx and TM1, respectively (Fischer *et al.* (2014)). Note that SFHx is softer than TM1 because of its smaller NS radius and mass relative to TM1.

Extraction of GWs is done by the conventional quadrupole formula (Misner *et al.* (1973)), in which the transverse and the traceless gravitational field h_{ij} is expressed as,

$$h_{ij}(\theta, \phi) = \frac{A_+(\theta, \phi)e_+ + A_\times(\theta, \phi)e_\times}{D}. \tag{2.1}$$

In Eq.(2.1), $A_{+/\times}(\theta, \phi)$ represent amplitude of orthogonally polarized wave components with emission angle (θ, ϕ) dependence, $e_{+/\times}$ denote unit polarization tensors and D is the source distance where we set $D = 10$ kpc in this contribution, unless otherwise stated.

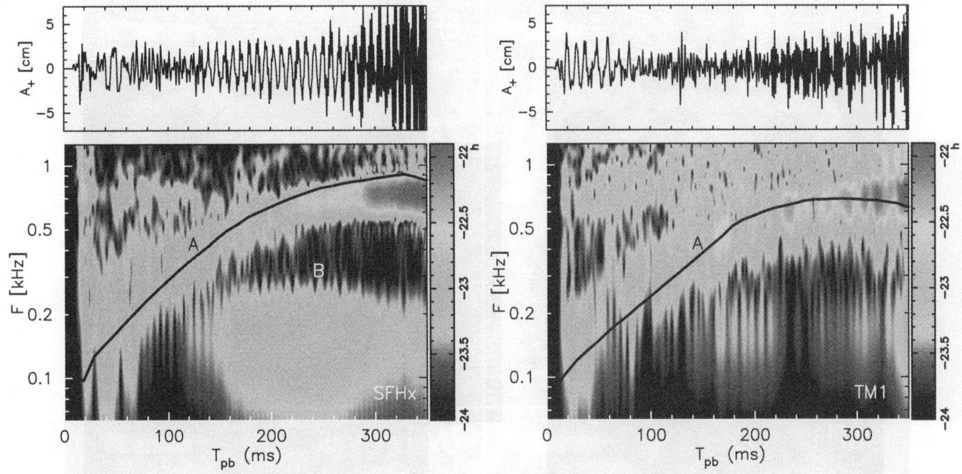

Figure 1. In each set of panels, we plot, top; gravitational wave amplitude of plus mode A_+ [cm] (black), bottom; characteristic wave strain in frequency-time domain $\tilde{h}$ in a logarithmic scale which is over plotted by the expected peak frequency $F_{\rm peak}$ (black). Left and right panels are for SFHx and TM1, respectively.

3. Results from 3D GR Simulations of a Non-rotating 15 $M_\odot$ Star

3.1. *3D hydrodynamics and GW signatures*

Figure 1 shows time evolution of the GW amplitude (only plus mode is shown $A_+(\theta,\phi)$, black line) in the top panel and the characteristic wave strain in time-frequency domain $\tilde{h}(\theta,\phi,F)$ (see Eq.(44) in Kuroda *et al.* (2014)) in the bottom panel. Here F denotes the GW frequency. We extract GWs along the north pole $(\theta,\phi)=(0,0)$ because the GW amplitude is less sensitive to the viewing angle.

The GW amplitude (A_+, upper two panels) shows a consistent behavior as reported in Müller *et al.* (2013), Andresen *et al.* (2016), and Yakunin *et al.* (2017). It shows initial low frequency and slightly larger amplitude till $T_{\rm pb}\sim 60$ ms ($T_{\rm pb}$ represents postbounce time), which comes from the so-called prompt convection. This is followed by a quiescent phase with higher frequency till $T_{\rm pb}\sim 150$ ms. Afterward the amplitude and frequency become larger with time as non-sphericity of fluid motion becomes chaotically bigger with the growth of neutrino-driven convection and the SASI.

From the GW spectrograms (bottom panels in Figure 1), we see a narrow band spectrum (labeled as "A" in both models) which shows an increasing trend in its peak frequency. As previously identified (Müller *et al.* (2013), Murphy *et al.* (2009)), this peak shift can be explained by properties of PNS, such as its compactness and surface temperature. By following Eq.(17) in Müller *et al.* (2013), we superimpose $F_{\rm peak}$ in the bottom panels of Figure 1 (black line). One can see that in both models, $F_{\rm peak}$ indeed tracks spectral peak quite well, although there is some exception in late phase of SFHx ($T_{\rm pb}\gtrsim 200$ ms) when the other strong component appears at $100\lesssim F\lesssim 200$ Hz (labeled as "B"). The component "A" is thus actually originated from the g-mode oscillation of the PNS surface.

Before going into detail to explain the origin of the (low-frequency) component "B", we touch on several key differences in the hydrodynamics evolution between SHFx and TM1. Figure 2 shows that SFHx experiences violent sloshing (top left) and spiral motions of the SASI (top right), before neutrino-driven convection dominates over the SASI (bottom

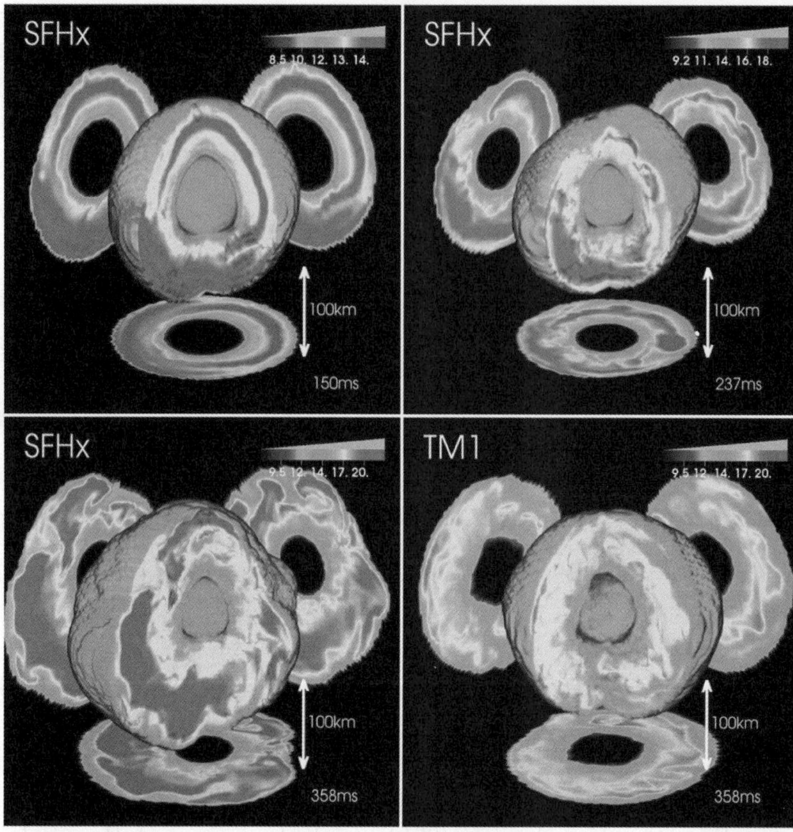

Figure 2. Snapshots of the entropy distribution ($k_{\rm B}$ baryon^{-1}) for models SFHx and TM1 (top left; $T_{\rm pb}$ = 150 ms of SFHx, top right; $T_{\rm pb}$ = 237 ms of SFHx, bottom left; $T_{\rm pb}$ = 358 ms of SFHx, bottom right; $T_{\rm pb}$ = 358 ms of TM1). The contours on the cross sections in the $x = 0$ (back right), $y = 0$ (back left), and $z = 0$ (bottom) planes are, respectively projected on the sidewalls of the graphs. The 90° wedge on the near side is excised to see the internal structure. Note that to see the entropy structure clearly in each dynamical phase, we change the maximum entropy in the colour bar as $s_{\rm max} = 16$, 20 and 22 $k_{\rm B}$ baryon^{-1} for $T_{\rm pb} = 150$, 237 and 358 ms, respectively.

left), whereas the SASI activities are less developed in TM1. For SFHx, the clear SASI motions are observed after the prompt convection phase ceases at $T_{\rm pb} \sim 50$ ms.

So how are these hydrodynamical evolutions related to the GW emission "B" in Fig. 1 ? For clarity, we present a back-of-the-envelope estimation of the GW amplitude as

$$D|h| \sim 2\epsilon M R^2/T_{\rm dyn}^2 \sim 2\epsilon M^2/R \sim 2\epsilon R^2 \dot{M}^2/M, \tag{3.1}$$

where M, R and $T_{\rm dyn}$ represent the mass, size and dynamical time scale of the system, respectively, in the geometrized unit. Here we have used the following reasonable assumptions

$$T_{\rm dyn} \sim M/\dot{M} \tag{3.2}$$

or

$$T_{\rm dyn} \sim R/V \sim \sqrt{R^3/M}, \tag{3.3}$$

with $V \sim \sqrt{M/R}$ being the velocity derived by the energy conservation. From the last

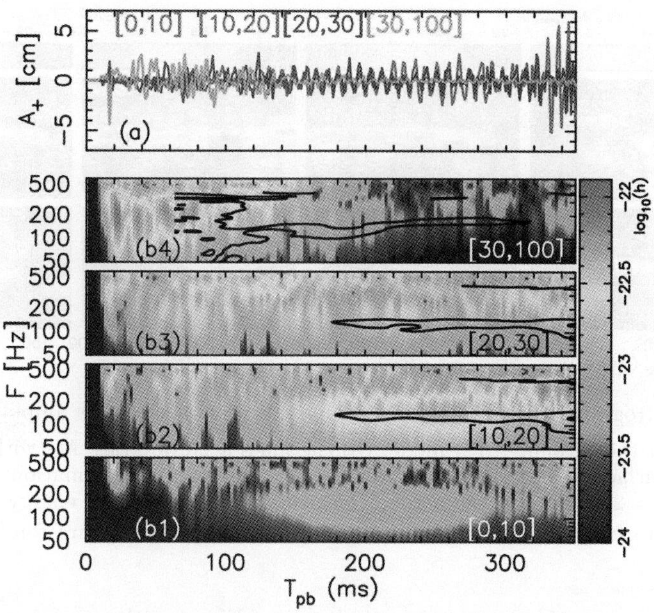

Figure 3. Rough measurement of contribution from each spherical shell to (a) the GW amplitude and (b1-4) their spectrogram $\tilde{h}$ in a logarithmic scale. We show the contributions from four spherical shells with interval of [0,10], [10,20], [20,30] and [30,100] km. Black contours over plotted on spectrograms for $\tilde{h}$ represent half maximum of spectrograms for mass accretion rate measured at $R = 17$ km (panel b2), 23 km (panel b3), and 48 km (panel b4).

relation in Eq.(3.1), we expect that significant time variation in the mass accumulation onto the PNS can potentially lead to the GW emission.

In Figure 3, we superimpose spectrogram of the mass accretion rate $\dot{M}(R)$ (the black contour at half maximum) measured at $R = 17$, 23, and 48 km on top of the GW spectrogram. While $\dot{M}(R = 48\text{km})$ starts quasi-periodic oscillation at $F \sim 100 - 200$Hz around $T_{\rm pb} \sim 120$ ms, we find a time delay of ~ 60 ms for their appearance at deeper region ($R = 17$ and 23 km). Since the density averaged mean radial velocity between the lepton driven ($10 \lesssim R \lesssim 20$ km) and the entropy driven ($R \gtrsim 40$ km) convection layers is $\sim 5 \times 10^7$ cm s^{-1}, the time delay is consistent with the advection time scale over the stable layer ($20 \lesssim R \lesssim 40$ km). Furthermore, coincidence of time modulation in $\dot{M}(R)$ and the GW component "B" is obvious from panel (b2).

Finally, we are trying to connect the SASI activities with the GW component B. Figure 4 shows spectrograms of normalized mode amplitude of the sloshing-SASI mode $|\tilde{A}_{10}|$, the mass accretion rate $|\tilde{\dot{M}}|$ measured at $R = 17$ km, normalised quadrupole deformation of the isodensity surface $\tilde{\epsilon}_l$ for $l = 2$. As a guide to measure the GW energy spectrum, $\tilde{\epsilon}_l$ denotes a Fourier component of normalised mode amplitude ϵ_l defined by

$$\epsilon_l \equiv \sqrt{\sum_{m=-l,l} \left(R^{14}_{l,m}\right)^2} \Big/ R^{14}_{0,0}\,, \tag{3.4}$$

where $R^{14}_{l,m}$ is evaluated by the spherical polar expansion of the isodensity surface R^{14} extracted at $\rho = 10^{14}$ g cm^{-1} as the same way as for the shock surface. Although several other modes are excited at the surface, only the leading contribution ($l = 2$ mode) to the GW emission is shown in the panel. As a reference, the isodensity surface R^{14} locates ~ 13.5 km during $150 \lesssim T_{\rm pb} \lesssim 300$ ms in SFHx. From the last relation in Eq.(3.1), we

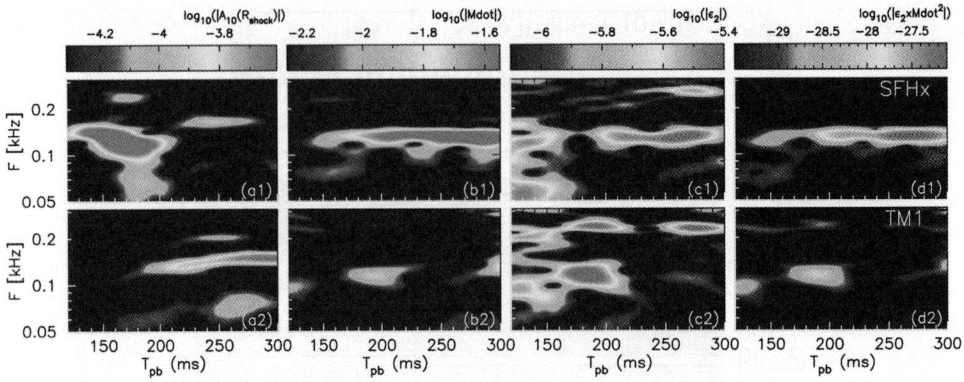

Figure 4. Spectrograms of; (a) Fourier decomposed normalized mode amplitude $|\tilde{A}_{10}|$ of the shock surface for the sloshing-SASI mode, (b) the mass accretion rate $\tilde{\dot{M}}$ (with a dimension of $M_\odot$), through surface of a sphere with radius of $R = 20$ km, (c) deformation of the isodensity surface $\tilde{\epsilon}_l$ for $l = 2$ mode and (d) a rough measurement of the GW energy spectrum which is proportional to $\sim \epsilon R^2 \dot{M}^2 M^{-1}$ (see text). Top and bottom rows are for SFHx and TM1, respectively.

plot $\log_{10}|h| \sim \log_{10}\epsilon\dot{M}^2 + \text{const.}$ in panels (d) of Fig. 4 with assuming $M = 0.5M_\odot$, a mass contained in $10 \lesssim R \lesssim 20$ km, and $R^{14} = 13.5$ km stay nearly constant.

During $140 \lesssim T_{\rm pb} \lesssim 180$ ms in SFHx, we see a strong sloshing motion which has its peak frequency at $100 \lesssim F \lesssim 200$ Hz (a1). With some time delay (~ 50 ms) from the appearance of it, the mass accretion rate $\dot{M}$ starts showing a quasi-periodic oscillation at the same frequency range $100 \lesssim F \lesssim 200$ Hz (b1) and it excites oscillation on the isodensity surface (c1). A combination of large $\dot{M}$ and ϵ_2 expect GW emissions appearing in panel (d1) and it can well explain the feature in Figure 1. During $200 \lesssim T_{\rm pb} \lesssim 300$ ms, ϵ_2 stays $\sim 3 \times 10^{-4}$ in SFHx. A rough measurement of the GW amplitude due to this deformation, $A \sim 2\epsilon_2 M^2 R^{-1}$, deduces $A \sim 2$ cm which is consistent with the actual amplitude (Figure 3).

3.2. *Detectability of the SASI-induced GW signals*

As we have discussed above, the GW spectrogram is characterized by the two components, which comes from the high-frequency g-mode PNS (surface) oscillation (e.g., A in Figure 1, hereafter we shortly call it shortly as "g-mode" GW) and from the low-frequency ($100 \sim 200$ Hz) SASI-induced downflows (e.g., B in Figure 1, we call it as "SASI-mode" GW). In order to discuss detectability of the two distinct features in the spectrogram, we perform a coherent work analysis where multiple detectors including LIGO Hanford, LIGO Livingston, VIRGO, and KAGRA are considered (e.g., Hayama *et al.* (2015) for more details).

The left panel of Figure 5 shows the spectrogram of the reconstructed waveform of SFHx at a distance of 3 kpc. Setting an optimal threshold of detection as 10 (see the color scale), one could see that the g-mode GW in the fully non-linear phase (at $T_{\rm pb} \sim 350$ ms) is most likely to be detectable, whereas the SASI-mode GW (in the low-frequency range) is second largest (compare the (pre-processed) original spectrogram in Figure 1 (left panel)). Our results suggest that LIGO-class detectors could marginally detect and hopefully differentiate the two signatures (g-mode vs. SASI-mode GWs) only for a very close-by event (a few kpc). But with the advent of the third generation detectors (right panel of Figure 5), the detection horizon is expected to extend up to 100 kpc (out to LMC). For a *Galactic* source, the prompt convection (red region $T_{\rm pb} \lesssim 50$ ms),

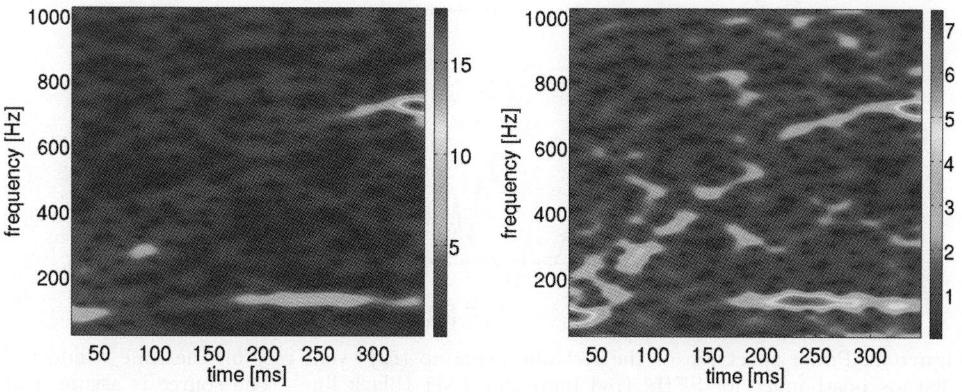

Figure 5. The left panel is the spectrogram of the reconstructed waveform of SFHx at a distance of 3 kpc. The color scale represents the signal-to-noise ratio (SNR) in the spectrogram (e.g., $\mathrm{SNR_{TF}}$, see the definition in Hayama *et al.* (2015)). The right panel is similar to the left panel but the distance to the source is taken as 100 kpc and the design sensitivity of Cosmic Explore (Abbott *et al.* (2017)) is also included in the coherent network analysis.

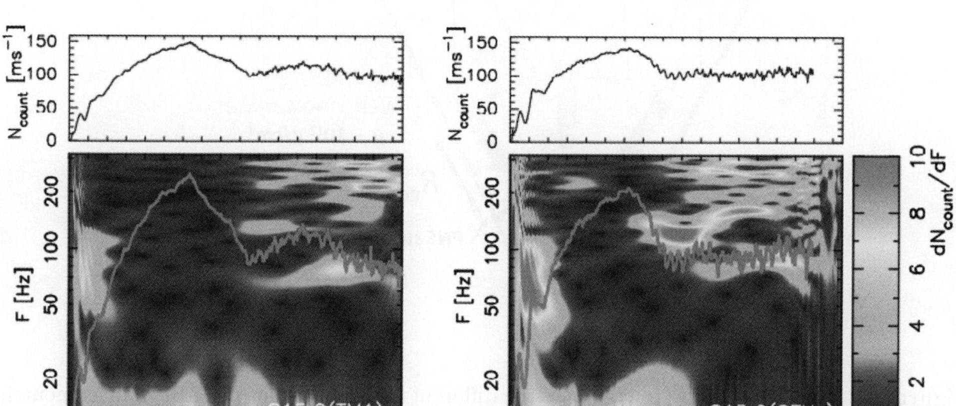

Figure 6. The top panel shows detection rate of $\bar{\nu}_e$ for TM1 (left panel) and SFHx (right panel) in Hyper-Kamiokande at 10 kpc. The bottom panel shows the spectrogram where the detection rate is superimposed by the red line.

the running-up of the g-mode GW, and the SASI-induced island (red region at $200 \lesssim T_{\mathrm{pb}} \lesssim 300$ ms) might be clearly visible by the third generation detectors as if it were the noise-free theoretical spectrogram (e.g., Figure 1 (left panel)).†

3.3. *SASI-induced Modulation in Neutrino Signals*

Figure 6 shows detection rate of anti-electron neutrino events ($\bar{\nu}_e$) at (the future) Hyper-Kamiokande (Hyper-K)(Abe *et al.* (2011)). Here the fiducial volume of Hyper-K is taken as 440 kton. Note that from the fiducial volume of Super-K (32 kton), one can get the number in Super-K from Figure 6.

As previously identified (Tamborra *et al.* (2013)), the SASI-induced signal modulation is both seen in the top panels of Figure 6. Reflecting the stronger SASI activity in SFHx

† The Cosmic Explore is proposed to start operation in the 2030's (https://dcc.ligo.org/public/0120/T1500290/003/T1500290.pdf).

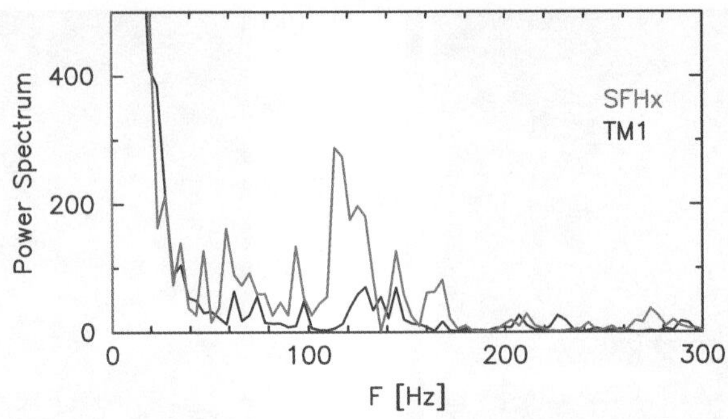

Figure 7. Power spectrum of the IceCube neutrino ($\bar{\nu}_e$) event rate on the time window of 0 - 300 ms postbounce for SFHx (red line) and TM1 (black line). The source is assumed at a distance of 10 kpc. The shot noise caused by the background of IceCube of 1.5×10^3 ms^{-1} is assumed (e.g., Tamborra *et al.* (2014)).

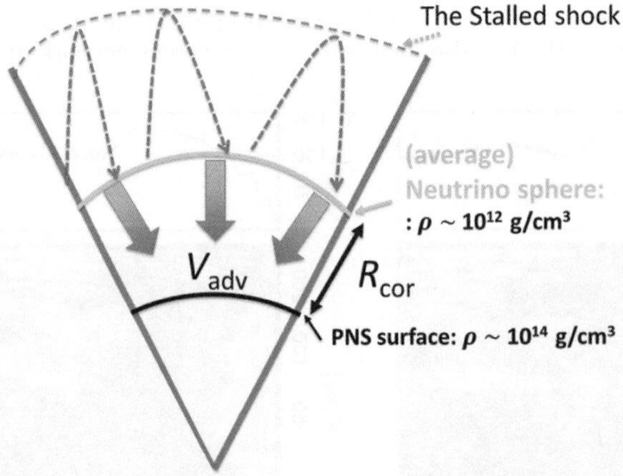

Figure 8. Schematic drawing to illustrate the different radial positions of SASI-induced neutrino and GW emission in the postbounce core. Below the stalled shock (dashed blue line, labeled as "The stalled shock"), non-spherical flows (red line with arrow) hit first the (average) neutrino sphere then penetrates into the PNS surface. $R_{\rm cor}$ represents the distance between the neutrino sphere ($\bar{\nu}_e$ in this case) and the PNS. $V_{\rm adv}$ is the typical velocity of the downflows there.

(right panel) compared to TM1 (left panel), more clearer excess (bottom panels) is seen for SFHx in the frequency range of 100 - 200 Hz in accordance with the SASI-modulation (red thick line). Note that the detection rate in Figure 6 is angle-averaged one. If one could take into account the viewing angle correction, the modulation amplitude could become more pronounced as shown in Tamborra *et al.* (2014).

IceCube by its superior statistics compared to Hyper-K could help differentiate smaller modulation amplitudes in the two models. In fact, Figure 7 supports this, where one can see the more pronounced peak for SFHx (red line) compared to TM1 (black line).

3.4. *Correlation between SASI-induced GW and Neutrino Signals*

From Figures 1 and Figure 7, one can see that both of the modulation frequency of the GW and neutrino signals is relatively close (i.e., in the range of 100 $\sim$ 200 Hz). Is there any correlation between them ? Figure 8 illustrates how the two signals could have a correlation. During the simulation timescale, the radial distance ($R_{\rm cor}$) between the neutrino sphere and the PNS can be crudely estimated as 50 km, and $V_{\rm adv}$ as 10^8 cm/s,

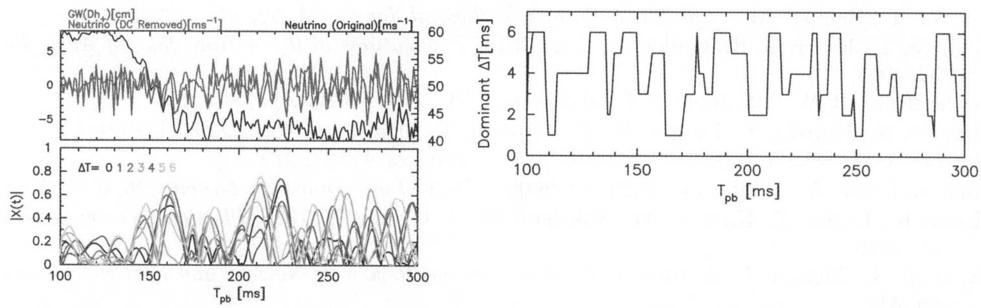

Figure 9. The left panel (*top*) shows the gravitational waveform ($h(t)$, blue line) and the neutrino count rate ($n_\nu(t)$, red line at Hyper-K), where the count rate with the DC component is also shown (black line) as a reference. The left-bottom panel shows the correlation coefficient $0 \leqslant X(t) \leqslant 1$ between the two signals (top panel, blue and red curves) when we shift the time of the two waveforms by ΔT (ms) ($X(t) \equiv \langle h(t+\Delta T) n_\nu(t) \rangle$). Note $\Delta T = 0$ means that we take the correlation using the data in the top panel. The right panel shows the evolution of ΔT where the correlation becomes maximum at a given time.

then the timescale of downflows travelling from the neutrino-sphere to the PNS reads $\Delta T \sim R_{\rm cor}/V_{\rm adv} \sim 5$ ms.

Figure 9 shows that the correlation between the SASI-modulated neutrino and GW signals (left panels) takes maximum for the time delay of $\Delta t = 4 \sim 6$ ms (right panel), which fits well with the schematic understanding in Figure 8. More careful analysis is apparently needed if we could or could not distinguish the signal modulation from noises. But if it could be possible, we could extract the information regarding the distance ($R_{\rm cor}$ in Figure 8) from the neutrino-sphere and the PNS, which is otherwise very hard to measure by other means.

4. Perspectives; where are we going ?

First we need to investigate the dependence of numerical resolution on the presented results. For enhancing the predictive power of the multi-messenger predictions, we have to upgrade from the gray to multi-energy neutrino transport including more detailed neutrino opacity (Kotake *et al.* in preparation). Here we have only presented a snapshot of our continuing endeavour for making more precise multi-messenger predictions based on more sophisticated 3D-GR models.

We are thankful to stimulating discussions/conversations with R. Diehl, H.T. Janka, G. Raffelt, C. Fryer, A. Heger, R. Hirshi, S. Katsuda, A. Wongwathanarat, E. O'Connor, J. Guilet, and R. Kazeroni during the conference. This study was supported by JSPS (Nos. 24244036, 26707013 and 26870823) and by MEXT (Nos. 15H00789, 15H01039, 15KK0173) and JICFuS as a priority issue to be tackled by using Post 'K' Computer.

References

Abe, K., Abe, T., Aihara, H., *et al.* 2011, arXiv:1109.3262

Abbott, B. P., Abbott, R., *et al.* 2016, *Physical Review Letters*, 116, 061102

Abbott, B. P., *et al.* 2017, *Classical and Quantum Gravity*, 34, 044001

Aso, Y., Michimura, Y., Somiya, K., *et al.* 2013, *Phys. Rev. D*, 88, 043007

Andresen, H., Mueller, B., Mueller, E., & Janka, H.-T. 2016, arXiv:1607.05199

Baumgarte, T. W. & Shapiro, S. L. 1999, *Phys. Rev. D*, 59, 024007

Bethe, H. A. 1990, *Reviews of Modern Physics*, 62, 801

Burrows, A. 2013, *Reviews of Modern Physics*, 85, 245

Fischer, T., Hempel, M., *et al.* 2014, *European Physical Journal A*, 50, 46
Foglizzo, T., Kazeroni, R., Guilet, J., *et al.* 2015, *Publications of the Astron. Soc. of Australia*, 32, e009
Grefenstette, B. W., Reynolds, S. P., Harrison, F. A., *et al.* 2015, *ApJ*, 802, 15
Hayama, K., Kuroda, T., Kotake, K., & Takiwaki, T. 2015, *Phys. Rev. D*, 92, 122001
Hempel, M. & Schaffner-Bielich, J. 2010, *Nuclear Physics A*, 837, 210
Hild, S., Freise, A., Mantovani, M., *et al.* 2009, *Classical and Quantum Gravity*, 26, 025005
Hirata, K., Kajita, T., Koshiba, M., Nakahata, M., & Oyama, Y. 1987, *Physical Review Letters*, 58, 1490
Janka, H.-T., Melson, T., & Summa, A. 2016, *Annual Review of Nuclear and Particle Science*, 66, 341
Kotake, K., Sumiyoshi, K., Yamada, S., *et al.* 2012, *Progress of Theoretical and Experimental Physics*, 2012, 01A301
Kotake, K. 2013, *Comptes Rendus Physique*, 14, 318
Kuroda, T., Kotake, K., & Takiwaki, T. 2012, *ApJ*, 755, 11
Kuroda, T., Takiwaki, T., & Kotake, K. 2014, *Phys. Rev. D*, 89, 044011
Kuroda, T., Takiwaki, T., & Kotake, K. 2016, *ApJS*, 222, 20
Kuroda, T., Kotake, K., & Takiwaki, T. 2016, *ApJL*, 829, L14
Maeda, K., Kawabata, K., Mazzali, P. A., *et al.* 2008, *Science*, 319, 1220
Misner, C. W., Thorne, K. S., & Wheeler, J. A. 1973, Gravitation
Müller, B. 2016, *Publications of the Astron. Soc. of Australia*, 33, e048
Müller, B., Janka, H.-T., & Marek, A. 2013, *ApJ*, 766, 43
Murphy, J. W., Ott, C. D., & Burrows, A. 2009, *ApJ*, 707, 1173
Shibata, M. & Nakamura, T. 1995, *Phys. Rev. D*, 52, 5428
Steiner, A. W., Hempel, M., & Fischer, T. 2013, *ApJ*, 774, 17
Takiwaki, T., Kotake, K., & Suwa, Y. 2016, *MNRAS*, 461, L112
Tamborra, I., Hanke, F., Müller, B., Janka, H.-T., & Raffelt, G. 2013, *Physical Review Letters*, 111, 121104
Tamborra, I., Raffelt, G., Hanke, F., Janka, H.-T., & Müller, B. 2014, *Phys. Rev. D*, 90, 045032
Tanaka, M., Maeda, K., Mazzali, P. A., Kawabata, K. S., & Nomoto, K. 2017, *ApJ*, 837, 105
Woosley, S. E. & Weaver, T. A. 1995, *ApJS*, 101, 181
Yakunin, K. N., Mezzacappa, A., Marronetti, P., *et al.* 2017, arXiv:1701.07325

Discussion

JANKA: In your plot showing the SNR (signal-to-noise ratio) as a function of distance (not shown in this contribution), the GW detection horizon extends out to 10 kpc. But the SASI-modulated signature (Figure 5 (left panel)) can be seen only visible to 2∼3 kpc. What is the different between the two statements ?

KOTAKE: This is an important question. Let me first clarify this. The first plot you are mentioning is the so-called *optimal* SNR where the matched filtering method is employed using the GW signals over the entire simulation time. On the other hand, the SNR on the spectrogram (Figure 5 (left)) is estimated only over the time when the SASI activity is active, and the SNR is estimated after the signal reconstruction. So the SNR becomes much smaller compared to the optimal SNR.

JANKA: I'm asking this because we have also a paper by Andresen *et al.* (2016), where we pointed out that the detection horizon of the SASI-modulated GW signal is a few kpc by the currently running detectors.

KOTAKE: All right. So, our results are consistent in this regard.

Supernova 1987A:30 years later - Cosmic Rays and Nuclei from Supernovae and their aftermaths
Proceedings IAU Symposium No. 331, 2017
A. Marcowith, M. Renaud, G. Dubner, A. Ray & A. Bykov, eds.

doi:10.1017/S1743921317004495

ANTARES and KM3NeT programs for the supernova neutrino detection

Vladimir Kulikovskiy
on behalf of the ANTARES and KM3NeT collaborations
Aix Marseille Univ, CNRS/IN2P3, CPPM, Marseille, France
email: kulikovskiy@cppm.in2p3.fr

Abstract. The currently working ANTARES neutrino telescope has capabilities to detect neutrinos produced in astrophysical transient sources. Neutrino alerts are regularly generated to trigger multi-wavelength observatories. Potential sources include gamma-ray bursts, core-collapse supernovae, and flaring active galactic nuclei. In particular, the neutrino detection together with the multi-wavelength observations may reveal hidden jets in the supernova explosions.

Supernovae remnants are currently the most promising acceleration sites of the cosmic rays in our Galaxy. The neutrino emission is expected during the cosmic ray interaction with the surrounding matter. The neutrino telescopes in the Northern hemisphere have excellent visibility to the most of the galactic supernovae remnants. Recent results on the search for point-sources with the ANTARES detector and the prospects for the future KM3NeT detector are presented.

Although ANTARES and KM3NeT detectors are mainly designed for high energy neutrino detection, the MeV neutrino signal from the supernova can be identified as a simultaneous increase of the counting rate of the optical modules in the detector. The noise from the optical background due to ^{40}K decay in the sea water and the bioluminescence can be significantly reduced by using nanosecond coincidences between the nearby placed photomultipliers. This technique has been tested with the ANTARES storeys, each one consisting of three 10-inch photomultipliers, and it is further optimized for the KM3NeT telescope where the directional optical modules containing 31 3-inch photomultipliers provide very promising expectations.

Keywords. neutrinos, supernovae: general, supernova remnants, instrumentation: detectors

---

1. Introduction

ANTARES and KM3NeT are neutrino telescopes presenting mainly three-dimensional photomultiplier (PMT) arrays in the deep sea water well shielded from atmospheric muons (more technical details will be given in Sec. 2). Such telescopes can be used to study different evolution stages of the SuperNova (SN) explosions:

• Prompt SN emission (20 MeV neutrinos) should produce an increase of PMT rates in the detector. No event reconstruction but high sensitivity in time domain is expected for such observations. Simulations and search optimization for KM3NeT are presented in Sec. 3.

• Transient high energy neutrino emission (~GeV–TeV e.g. neutrinos from GRB-SN, hidden jets) can be promptly detected by the neutrino telescopes (fast event reconstruction, neutrino direction and energy estimation) and the alerts can be forwarded to the followers. The existing and future programs are described in Sec. 4.

• Supernova Remnants can be efficient high energy hadron accelerators (~GeV–TeV neutrinos). Searches of unknown steady point sources on the sky are performed with the long-term ANTARES data and planned with KM3NeT. They are enriched with searches through candidate catalogs (from gamma-ray observations mainly). Some ANTARES results and KM3NeT perspectives are summarized in Sec. 5.

2. Technology and performance

The ANTARES detector is the first undersea neutrino telescope. It is installed at 2475 m depth; its effective mass is about 10 Mt and it is optimized for $E_\nu \gtrsim 10\,\mathrm{TeV}$ (Ageron M. *et al.*(2011)). KM3NeT is a distributed research infrastructure that will host a high energy neutrino telescope (ARCA), offshore of Capo Passero, and an atmospheric neutrino detector (ORCA), offshore of Toulon in France (close to the ANTARES site), for the determination of the neutrino mass hierarchy (Adrián-Martínez S. *et al.* (2016)). ARCA has two detector blocks instrumenting about 1 Gt and it is optimized for $E_\nu \gtrsim$ 10 TeV. Each detector block contains 115 detection units. Detection unit present a string with 18 optical modules. The ORCA detector has one detector block that instruments about 5.7 Mt optimized for 5–500 GeV neutrinos. The total number of strings and optical modules in ORCA is the same as in one ARCA block, but the detector elements are placed closer to each other to improve the detector sensitivity at lower energies.

The KM3NeT detectors use new directional optical modules (DOMs) that host 31 3-inch PMTs. Each DOM has a total PMT photocathode effective area similar to the ANTARES storey with three 10-inch PMTs. Multi-PMT technology allows better light direction estimation and better optical background suppression by using time coincidence between PMT signals. PMTs and their digitization electronics as well as positioning/calibration instrumentation for each DOM are stored inside a 17-inch glass sphere (the same one that is used to store one ANTARES PMT). This concept allows furling of the detection unit for the deployment, which means more units deployment during each marine operation. Both detector sites have the underwater infrastructure ready for the detection units connection. Data taking at the ARCA site is ongoing with two detection units already.

Although being designed for the high energy neutrinos, both KM3NeT detectors are sensitive to the prompt MeV neutrinos from supernova explosions by observing the global light rate increase in the detector. More details are provided in the next section.

3. Prompt emission detection

Supernova fluxes simulation. The simulation of the supernova explosions are well advanced nowadays and reproduce the detected spectra and time distributions of the only so far existing SN1987A detection (for review see, for example, Mirizzi A. *et al.* (2016)). Variety of the models and parameters is present and used by different experiments for sensitivity estimations. Often, there is no overlap between the models used by different experiments, which complicates the direct comparison. In order to overcome this complexity a pragmatic approach is proposed in this work. In fact, it is very similar to the approach used by the JUNO collaboration (An F. *et al.* (2016)).

The SN neutrino spectrum is parametrized with the following function:

$$f = E_\nu^\alpha e^{-(\alpha+1)E_\nu/\tilde{E_\nu}}, \tag{3.1}$$

which is defined with average energy $\tilde{E_\nu}$ and spectral shape/pinching parameter α. As it was shown by Tamborra I. *et al.* (2012), the neutrino spectra are always fitted with parameters in the range of $2 \lesssim \alpha \lesssim 5$ or $0 \lesssim \eta \lesssim 4$.

Results of different supernova simulations predict about $15\% - 25\%$ of the $\bar{\nu}_e$ emission in the first 100 ms. The total ν flux considered by JUNO and generally accepted as a benchmark is 3×10^{53} erg divided equally by flavours and $\nu/\bar{\nu}$. The more precise fraction of the $\bar{\nu}_e$ flux in total flux corresponds to $1/5 \cos^2 \theta_{12} \approx 0.14$, which includes oscillations in supernova mantle (Strumia A. and Vissani F. (2003)) and it is similar to 1/6. The shape parameter α is around 3 during this interval and as also chosen by JUNO. The average energy for $\bar{\nu}_e$ is about 15 MeV. In JUNO three values are considered: 12 MeV, 14 MeV and 16 MeV.

The cross-sections for inverse beta decay are taken from Strumia A. and Vissani F. (2003). The positron energy is calculated as:

$$E_{e^+}^{CM} = \frac{m_p^2 - m_n^2 + m_e + 2E_\nu m_p}{2\sqrt{m_p^2 + 2m_p E_\nu}} \approx E_\nu - \frac{m_n^2 - m_p^2 - m_e^2}{2m_p} \approx E_\nu - 1.239\,\text{MeV}, \qquad (3.2)$$

where $E_\nu \ll m_p$ is considered. As the Lorentz factor of the central mass frame is small for $E_\nu \ll m_p$ as well, $E_{e^+}^{CM} \approx E_{e^+}$ (in the lab frame).

The positron distribution as a differential number of particles in 1 kton detector volume from as SN at $D = 10$ kpc was calculated according to the following formula:

$$dN_{e^+}/dE = \frac{1}{F} E_\nu^\alpha e^{-(\alpha+1)E_\nu/\tilde{E}_\nu} \{f_{100\text{ ms}} \frac{E_{\bar{\nu}_e}^{\text{tot}}}{\tilde{E}_\nu}\} \frac{1}{4\pi D^2} \sigma_{IBD} \{M_{\text{det}}/\mu_{\text{H}_2\text{0}} N_{\text{A}} \times 2\} \qquad (3.3)$$

Where $1/F$ is the normalization factor for the neutrino energy distribution, $f_{100\text{ ms}} =$ 25% used, $E_{\bar{\nu}_e}^{\text{tot}} = 3 \times 10^{53}/6$ erg. The first brackets present the calculation of the number of neutrinos at the source.

The previous analysis done for ANTARES (Kulikovskiy V. (2011)) used model 57 from Burrows A. (1988). According to this model, there are 65.7 positrons per kiloton of the detector in the first 100 ms and their spectrum follows the distribution described by the following parameterization:

$$dN/dE \sim \frac{E[\text{MeV}]^2}{1 + e^{\frac{E[\text{MeV}]-16}{6}}}. \qquad (3.4)$$

The distribution has shape similar to the previously described model with $\tilde{E}_\nu = 16$ MeV. Also the total numbers of positrons in 1 kton are similar: 45.7, 53.6 and 61.1 for 12 MeV, 14 MeV and 16 MeV correspondingly.

The positrons generator with the previously described parameterization was added to the GEANT 4 simulation of the optical modules (Hugon C. M. F. (2016)).

Background suppression with coincidences. The main optical background in the deep sea water arises from the ^{40}K decay that goes mostly with 1.3 MeV electron production. This background is well reproduced in GEANT 4 simulations and used for time and efficiency calibration of the detectors. Atmospheric muons represent also background for this measurement. Their contribution was taken from the data directly (Adrián-Martínez S. *et al.* (2014) and Adrián-Martínez S. *et al.* (2016b)). For simplicity, the muon rate for the ARCA site was taken as 1/4 of the muon flux at the ORCA site.

To suppress the background the coincidence method was developed for ANTARES (Kulikovskiy V. (2011)). Instead of all PMT signals, only coincident signals between two or three nearby PMTs (from the same storey) in a short time window (~25 ns) are used. The sensitivity estimations using these double and triple coincidence selections yield similar results (the sensitivity is about 5σ for SN at 5 kpc). Since there are 31 nearby PMTs located in each DOM, higher coincidence level selections can be applied. The rate of hit selections with different coincidence levels in a single DOM is shown in Fig. 1 for ^{40}K and muon backgrounds. For SN the constant emission during 100 ms is considered on the same plot for comparison.

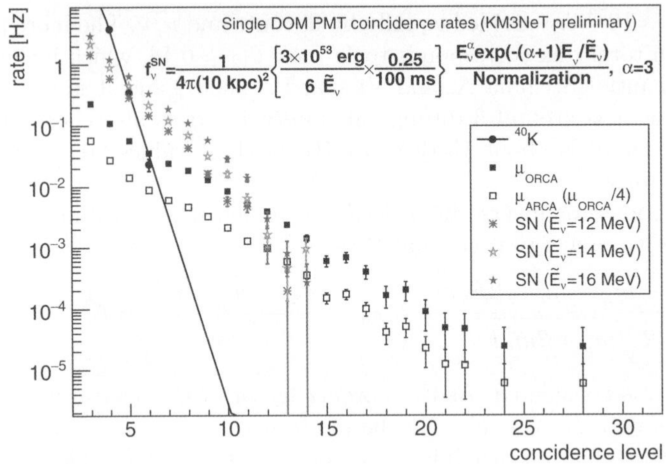

Figure 1. Coincidence event rates for the first 100 ms of the SN explosion (assuming constant emission during this period) for different mean neutrino energies. Main backgrounds (^{40}K decays and atmospheric muons) are also shown.

Table 1. Summary of the results on the SN prompt emission detection with KM3NeT.

$\tilde{E}_{\bar{\nu}_e}$ **MeV**	N_{events} **per block**	$D_{5\sigma/3\sigma}$ **(kpc) ARCA**	$D_{5\sigma/3\sigma}$ **(kpc) ARCA**
12	60	23/30	16/20
14	100	29/37	19/25
16	150	37/47	24/31

As it can be seen, the ^{40}K decays produce in general events with fewer coincident hits in a DOM comparing to SN events. Muons produce even harder coincidence spectrum. This correlates with the lepton energies (1.3 MeV e^- for ^{40}K, $\gtrsim$ 10 MeV e^+ for SN and MeV–GeV energies for μ) and suggests that coincidence level can be a rough event energy estimator.

Sensitivity estimation. For known signal and noise rates during a limited period sensitivity can be estimated as following:

$$S = \frac{N_{\mathrm{DOMs}} \times T \times f_{\mathrm{SN}}}{\sqrt{N_{\mathrm{DOMs}} \times T \times f_{\mathrm{bg}}}} = \frac{f_{\mathrm{SN}}}{f_{\mathrm{bg}}}\sqrt{T \times N_{\mathrm{DOMs}}} \tag{3.5}$$

The sensitivity decreases with source-detector distance as $1/D^2$ following the positron production decrease with a distance (Eq. 3.3). For $\tilde{E}_\nu = 16$ MeV model one can estimate the 5 sigma sensitivity till 36 kpc for ARCA (all Milky Way) and 24 kpc for ORCA (95% of the Galactic progenitors). The complete set of results is summarized in Table 1.

These results are rather preliminary and conservative. In the next future time window optimization is planned; muon veto using coincidences between DOMs and topological features should increase sensitivity by increasing the signal/noise ratio.

4. Transient high energy sources

For transient and periodic sources the correlations in space and time between X-ray, optic and high energy gamma-ray telescopes and gravitational waves detectors restricts the background and boosts the sensitivities. A multi-wavelength follow-up program, denoted as TAToO (Telescopes-ANTARES Target of Opportunity), has operated within the

ANTARES Collaboration since 2009. It is based on optical and, since mid 2013, X-ray follow-ups of selected high-energy neutrino events very shortly after their detection. The network is composed of the small robotic optical telescopes TAROT and ROTSE, and the Swift X-ray telescope (Adrián-Martínez S. *et al.* (2016b)). To be sensitive to all types of time variability in the astrophysical sources, the observational strategy is composed of a real-time observation for rapidly fading sources, such as Gamma Ray Burst afterglows, complemented by several observations during the following month specially adapted to detect the rising light curve of core collapse SNe. Unprecedented in this domain, the ANTARES telescope is able to generate alerts within a few seconds after the neutrino detection, and provides a precision on the reconstructed direction to better than 0.5° at high energies ($E > 1\,\mathrm{TeV}$).

The exchange of alerts from the gamma-ray detectors is established through GCN†. The ANTARES detector receives the GCN alert, which contains the position of the burst and its main features. The typical delay between the detection of a Gamma Ray Burst by the satellite and the time of the alert message distributed is around 10 s. Buffered unfiltered data of about 2 minutes before the alert can be saved and analyzed afterwards using more efficient event reconstructions with a known direction (Albert A. *et al.* (2016)). That improves ANTARES sensitivity up to two times for 100 GeV – 1 TeV neutrinos.

The future KM3NeT programs will follow the ANTARES developments extending them thanks to particular attention to the topic by the European Union‡.

5. Steady sources detection

Super novae remnants (SNRs) are one of the most promising candidates for the cosmic rays acceleration sites in our Galaxy. Debates about hadronic or leptonic acceleration mechanisms are still ongoing for many SNRs. Neutrino detection or strict upper limits can be an ultimate argument in such debates.

The structures of various SNRs defined with gamma ray telescopes cannot be resolved with current neutrino telescopes; they should appear as point-like or slightly extended spots in some cases. For the detection of neutrinos from the point-like sources, the most suited are the track-like events thanks to their sub-degree angular resolution. Atmospheric muon background should be suppressed. The event selection optimization was performed for the general point sources with E_ν^{-2} spectrum. For 3 years of ARCA 5σ discovery potential fluxes are $E^2\Phi \approx 2 \cdot 10^{-9}\,\mathrm{GeV\,cm^{-2}s^{-1}}$ for sin(declination) < 0.6 (Fig. 2). This is more than one order of magnitude below the same discovery potential fluxes for IceCube with similar exposure at sin(declination) < -0.2. This difference grows up for the softer neutrino spectrums since the IceCube selection prefers higher event energies to suppress the atmospheric muon background in the Southern hemisphere. At high declinations track analysis loses its sensitivity due to atmospheric muon veto. Cascade events can recover sensitivity in this area. Event selection optimization including vertex containment inside the detector volume reaches 5σ discovery potential at the level of $E^2\Phi \leqslant 5 \cdot 10^{-9}\,\mathrm{GeV\,cm^{-2}s^{-1}}$ which is rather independent on the declination (Adrián-Martínez S. *et al.* (2016a)).

A separate analysis was done for the young shell-type SNR RXJ1713.7-3946. The neutrino spectrum used for the sensitivity estimation is derived from the measured

† gcn.gsfc.nasa.gov
‡ ASTERICS-CLEOPATRA www.asterics2020.eu/cleopatra

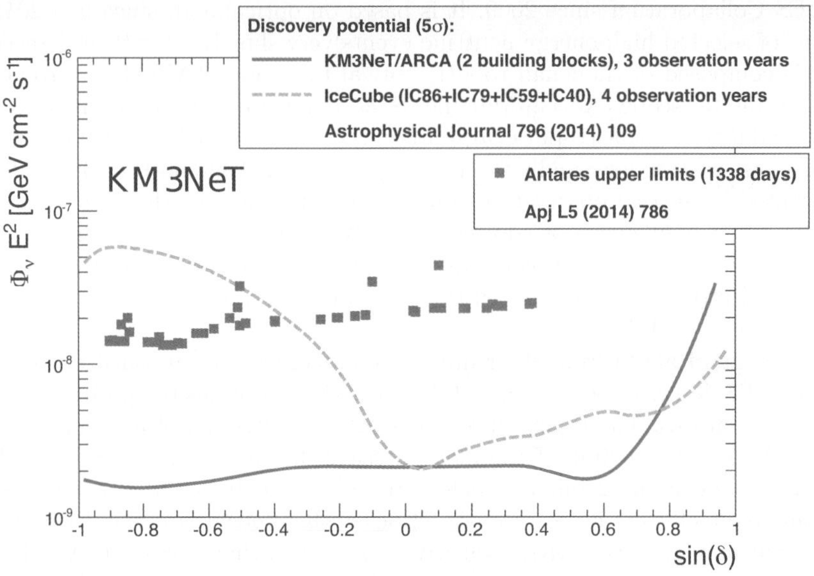

Figure 2. 5σ discovery potential as a function of the source declination (red line) for one neutrino flavour, for point-like sources with a spectrum $\propto E^{-2}$ (ARCA, IceCube) and upper limits on particular sources for the ANTARES detector.

gamma-ray spectrum, which extends up to 100 TeV. The analysis adapted for the track-like events reaches about 3σ sensitivity in 4 years (Adrián-Martínez S. *et al.* (2016a)).

Recently, a new track reconstruction algorithm was developed. It provides more well reconstructed events and better neutrino angular resolution: $< 0.1^\circ$ for $E > 100$ TeV and almost reaches the intrinsic angle limit for $E < 10$ TeV. Also, different multivariate analysis tools are tested to discriminate the signal from two background components (atmospheric neutrinos and muons) which have different energy and angular spectra. In particular, Random Forests algorithms of Boosted Decision Trees perform well for this task. The first very preliminary results show that observation time may be reduced by one year to reach 3σ discovery for RXJ1713 (Coniglione R. (2017)).

References

Adrián-Martínez, S. *et al.* 2016, *J. of Phys. G* 43 084001
Ageron, M. *et al.* 2011, *Nucl. Instrum. Meth. A* 656 11
Mirizzi, A. *et al.* 2016, *Riv. Nuovo Cimento* 39 1
An, F. *et al.* 2016, *Phys. G 43* 030401
Tamborra, I. *et al.* 2012, *Phys. Rev. D 86* 125031
Strumia, A. & Vissani F. 2003, *Phys. Lett. B* 564 42
Burrows, A. 1988, *APJ* 334 891
Kulikovskiy, V. 2011, *Proceedings of ICRC 2011* dx.doi.org/10.7529/ICRC2011/V07/1083
Hugon, C. M. F. 2016, *PoS ICRC2015* 1106
Adrián-Martínez, S. *et al.* 2014, *Eur. Phys. J. C* 74 3056
Adrián-Martínez, S. *et al.* 2016, *Eur. Phys. J. C* 76 54
Adrián-Martínez, S. *et al.* 2016, *JCAP* 02 062
Albert, A. *et al.* 2016, *Accepted for publication to MNRAS*, arXiv:1612.08589
Coniglione, R. 2017, *Proceedings of XXV European Cosmic Ray Symposium* arXiv:1701.05849

Supernova 1987A:30 years later - Cosmic Rays and Nuclei from Supernovae and their aftermaths
Proceedings IAU Symposium No. 331, 2017
A. Marcowith, M. Renaud, G. Dubner, A. Ray & A. Bykov, eds.

doi:10.1017/S1743921317004574

Synergy SKA - CTA: Supernova remnants as cosmic accelerators

Adriano Ingallinera, Corrado Trigilio, Grazia Umana, Paolo Leto, Carla Buemi, Francesco Schillirò, Filomena Bufano, Simone Riggi and Francesco Cavallaro

INAF - Osservatorio Astrofisico di Catania,
Via Santa Sofia 78, 95123 Catania, Italy
email: ingallinera@oact.inaf.it

Abstract. Supernova remnants (SNRs) are one of the most important sites where particles are accelerated with high efficiency and in a wide range of energies, becoming an important component of cosmic rays. A good test for this hypothesis will be possible using the data collected by next-generation radio and gamma-ray observatories, like the Square Kilometre Array (SKA) and the Cherenkov Telescope Array (CTA). Radio emission is fundamental to explore the SNR environment and to shed light on the physical processes involved in particle acceleration, providing direct links to high-energy physics. Two cases of SNRs recently studied in radio are presented, showing the importance of high-resolution radio images. An overview of SKA and its precursors is given with our ongoing preparation work. In particular, we present the EMU survey and the pathfinder project SCORPIO. Finally a direct view of the tight connection between SKA and CTA future studies of SNRs is provided.

Keywords. acceleration of particles, masers, instrumentation: interferometers, supernova remnants

---

1. Introduction

Radio is probably the best suited band at which searching for Galactic SNRs. About 95% of all of the known Galactic SNRs are detected in radio (Dubner & Giacani 2015) and many of them are detected only in this band. The radio emission originates from synchrotron processes involving electrons accelerated up to relativistic energies. Radio is therefore a unique tool to locate the potential particle acceleration sites and probe the local plasma conditions. Radio spectral index maps, for example, allow us to reveal the interaction of the remnant with molecular clouds (Giacani *et al.* 2011) or the presence of a pulsar wind nebula (PWN; Gaensler & Slane 2006). In these cases a significant spatial variation of the spectral index is expected. Furthermore, radio polarimetric study can give us hints on the intensity and direction of the magnetic field disclosing the observation of turbulence in plasma (DeLaney *et al.* 2002). Finally the observation of the OH maser transition at 1720 MHz is an extremely powerful tracer of shocked medium, since this line is inverted only collisionally (Frail, Goss & Slysh 1994). These phenomena (interaction with molecular clouds, presence of a PWN, turbulence and shocks) are directly associated with sites where particle acceleration can take place. These sites are therefore usually characterized also by γ-ray emission, from GeV to PeV. In this sense radio observations are a valuable probe of those environments where high-energy emission originates.

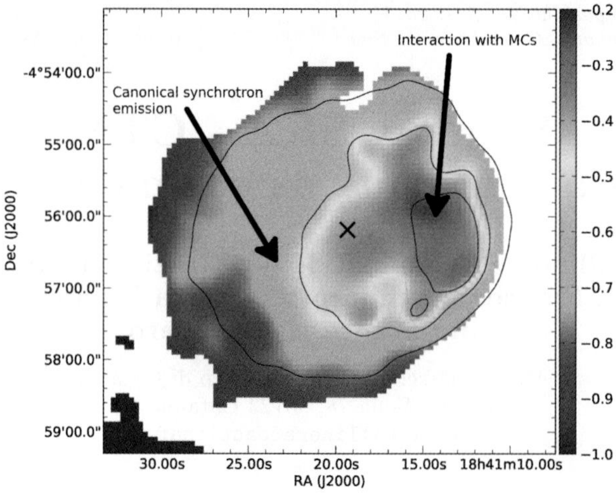

Figure 1. Spectral index map of SNR G027.4+00.0 (adapted from Ingallinera *et al.* 2014).

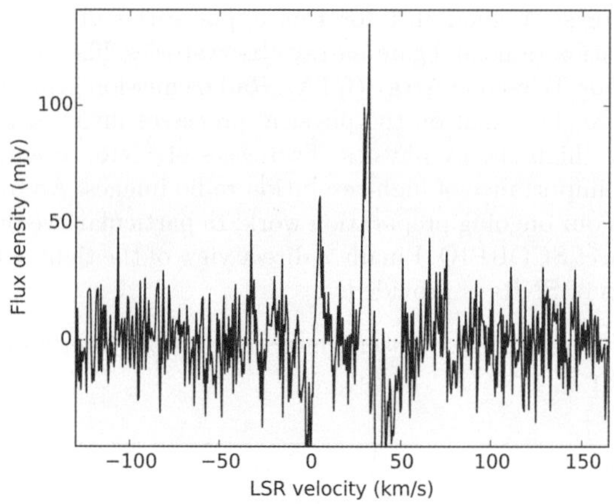

Figure 2. OH maser line at 1720 MHz detected toward SNR G011.2-00.3.

2. Case studies: SNR G027.4+00.0 and SNR G011.2-00.3

In this Section we provide two examples of SNRs recently studied by our group (Ingallinera *et al.* 2014), using the Very Large Array (VLA) and the Green Bank Telescope (GBT). The data from the interferometer and from the single-dish telescope were combined in the *uv*-plane ("feathering" process) to achieve both a very high resolution and the full sensitivity to large spatial scales. The images of the two SNRs produced in this way do not suffer from those flux-loss issues typical of interferometers, and accurate flux density measurements can be obtained. The first remnant is SNR G027.4+00.0, observed at two different frequencies, 1.4 and 5 GHz, in order to derive spectral information. In Figure 1 we show the spectral index map of SNR G027.4+00.0, which clearly shows a flatter spectral index toward the west region, interpreted as the sign of an interaction with molecular clouds.

The second remnant is SNR G011.2-00.3, observed in continuum at 5 GHz and in spectral line at 1.7 GHz. For this remnant we were able to detect the OH maser line at 1720 MHz (Figure 2), signature of an ongoing shock.

The two SNRs presented in this Section have a radius of about 2 arcmin. SNRs with such an angular extension are too small to be investigated only with single-dish radio

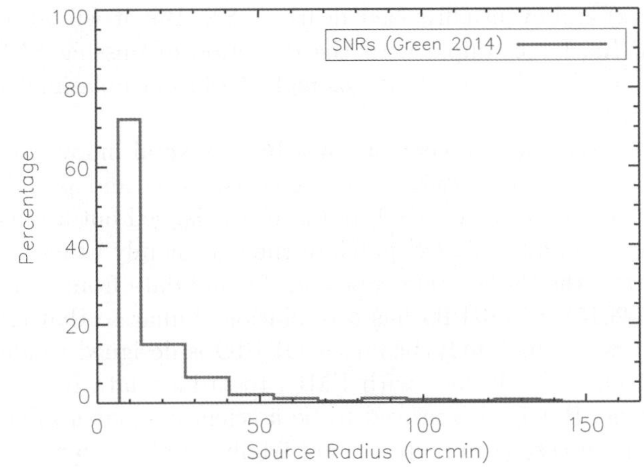

Figure 3. Distribution of Galactic SNRs in Green (2014) according to their radius.

telescopes and interferometric data are absolutely necessary. On the other hand, remnants with small angular extension are very common. In Figure 3 we show a histogram plotting the percentage of the Galactic SNRs included in the catalogue by Green (2014) against their radius. It is possible to notice that the majority of known Galactic SNRs have radii of order of 5 arcmin or less. Furthermore, it is likely that many missing SNRs could be even smaller. So, in order to increase the sample of Galactic SNRs that can be studied in detail, achieving a good angular resolution in radio data will be mandatory.

3. Science with SKA and its precursors

In the next future radio astronomers will be provided with have access to the most advanced astronomical facility ever designed, SKA. SKA will consist of two different arrays, LOW and MID, located in Australia and South Africa, respectively, and will cover a frequency range between 50 MHz and 14 GHz. Its construction is divided in two phases. SKA-1 phase (10% of the array) will start in 2018 and its early-science is expected around 2022. SKA-2, that is the full array as originally designed, will start to be built just after SKA-1 and its completion is scheduled in 2030. SKA-1 angular resolution will span from about 20 arcsec down to tens of milliarcseconds, and a two-order of magnitude improvement is expected at the fully completion of the array (SKA-2). Since SKA is a revolutionary telescope, before its construction several minor observatories have been built as precursors and are now operational, among which MWA, MeerKAT and ASKAP. The task of SKA precursors is not only to address scientific and technological challenges of the SKA but also to test innovative elements (for example the Phased Array Feeds) and huge data sets processing.

ASKAP (Australian SKA Pathfinder) is testing the Phased Array Feeds technology, by which large-area very-sensitive surveys will be feasible in reasonable amounts of time. The Evolutionary Map of the Universe (EMU) is one of the large programs already approved for ASKAP (Norris *et al.* 2011). It is a wide-field survey that will observe the entire southern sky up to $\delta = +30^{\circ}$, with a resolution of about 10 arcsec. About 75% of the Galactic plane will be imaged. EMU will cover approximately the same sky extension as NVSS, the current reference all-sky survey, but it will be 100 times more sensitive (down to 0.01 mJy/beam at 1σ level). The impact of EMU on Galactic science and in particular on Galactic SNRs will be extraordinary. It will allow a detailed study of the entire evolution of massive stars, from their formation to their last phases (e.g.

LBV and WR stars) and eventually their death as SN. But it will also provide a direct contribution to SNRs, for example with the detection of missing SNRs, in particular the young SNRs, or with the possibility of high-resolution polarimetric study of their filamentary structures.

ASKAP, and the other SKA precursors, will bring several innovations, so a long verification phase has been undertaken where extensive tests and pathfinders have been employed. In this context, between 2011 and 2016 we carried out a blind survey, named SCORPIO, conducted in a 2×2 deg^2 patch of the sky, mainly using the Australia Telescope Compact Array, the Parkes Telescope and the first data from ASKAP commissioning (Umana *et al.* 2015). SCORPIO has a resolution similar to that expected in EMU, and a sensitivity down to 0.03 mJy/beam. SCORPIO is designed to identify in advance what kind of problems we will face with EMU, regarding data reduction and analysis in the Galactic plane. But it also proved to be a scientific golden mine. A catalogue of point and extended sources are in preparation (Trigilio *et al. in prep.*; Ingallinera *et al. in prep.*), a method for extended source extraction has already been published (Riggi *et al.* 2016) and work on the spectral index analysis is submitted (Cavallaro *et al. submitted*). The SCORPIO field harbours a SNR (G344.7-0.1) and two SNR candidates. Studies are ongoing to characterize and classify these three sources.

4. Connection between SKA and CTA

Radio observations of SNRs with next-generation instruments will have a tight link to those high-energy phenomena that will be probed by future γ-ray instruments like CTA. CTA is a high-energy observatory that will be fully operational in 2024. It will cover an energy range from 20 GeV to 300 TeV, a window suitable to directly observe many energetic phenomena taking place in SNRs. Radio data with SKA, thanks to its sensitivity, can help addressing and refining CTA observations, providing the possibility to study the counterparts of γ-ray sources in a completely different region of the electromagnetic spectrum.

In this scenario, SNRs are the most interesting Galactic sources that can be studied simultaneously with the SKA and the CTA. In fact, though the hypothesis that SNRs are by far the main contributor of Galactic cosmic rays has gained an enormous consensus among the scientific community (e.g. Bell 2013), a conclusive evidence still lacks (Cristofari *et al.* 2013). In particular it must be proven that SNR shocks are able to accelerate particle with an energy spectrum compatible to what we observe for cosmic rays. The most direct way to link the SNRs to the cosmic rays is γ-ray observations. In fact, the charged particle constituting the cosmic rays are deflected by magnetic fields and their origin cannot be inferred by their detection. But γ-rays associated with sites where charged particle are accelerated travel in straight lines and their origin can be reconstructed. However, SNRs are predicted to emit γ-rays also via inverse Compton scattering of cosmic microwave background photons by relativistic electrons (e.g. Aharonian *et al.* 2008). These two γ-ray emission mechanisms, referred to as 'hadronic' and 'leptonic' respectively, are concurrent and, so far, it has been difficult to asses which one dominates over the other. SNRs can be the main (or only) source of Galactic cosmic rays only if their γ-ray emission is dominated by the hadronic process. Otherwise SNRs cannot take into account the measured cosmic ray fluxes and other sources must be searched. It is then fundamental to have a more accurate picture of the acceleration sites and more reliable and extended spectral energy distributions (SEDs) of as many SNRs as possible, and SKA and CTA are the ideal facilities to achieve these goals. The two observatories will provide complementary information on the local conditions and phenomena characterizing each remnants. But they will also be able to observe different aspects of the same feature, concurring to give us the most detailed picture ever on the physics of these celestial objects, especially thanks to their unprecedented resolution and sensitivity.

The best candidate regions for particle acceleration are the filaments, where magnetic field amplification takes place at scales < 0.1 pc. SKA-MID will be able to observe in great detail the filamentary structures of the SNRs, achieving a very high resolution and, simultaneously, a very high sensitivity. But it will be also sensitive to large scale structures (up to several arcminutes), providing an extremely accurate picture of these regions of SNRs. Its high resolution will significantly reduce depolarization phenomena occurring when different polarization angles are averaged together because of poor resolution. This prevents the reconstruction of the magnetic field, both in direction and in intensity, heavily affecting studies on the local dynamics of plasma. The other main depolarization agent is the Faraday rotation. Different rotations are produced by different plasma depths, so it is very difficult to estimate its contribution to total depolarization. However, provided enough resolution and a wide observing band, in principle the Faraday rotation can be determined by studying the position angles of polarization vectors at different frequencies (e.g. DeLaney *et al.* 2002). Once these issues will be overcome, radio observations will disclose us the finest details of the plasma interacting with magnetic fields, allowing us to make accurate predictions on particle acceleration location and efficiency.

At lower frequencies, SKA-LOW will observe radio emission from low-energy electrons, those with energies of order of GeV, resulting a mirror of the γ-ray emission at this energy band. Deviations from the canonical spectral behaviour of synchrotron emission are expected at this very low frequency when the hadronic process dominates over the leptonic one as the origin of γ-ray emission. In fact in this case radio emission is mainly due to secondary electrons generated in *pp* collisions (Wang *et al.* 2015). An analogue difference in the spectral energy distribution is expected to be observed with the CTA, resulting in an independent test of the same phenomenon.

Therefore the results of the two facilities will have a great impact in answering the open questions on the cosmic ray origins. In order two achieve these goals observations are already planned and included in key science projects of both instruments. For CTA the main possibility for SNR observations will be supplied by the Galactic Plane Survey. Designed to achieve a sensitivity of at least 4.2 mCrab, the survey will observe the entire Galactic plane, using both northern and southern arrays, in about 1600 hours. One of its scientific goal is to provide a census of Galactic SNRs and pulsar wind nebulae. On the other hand, one possibility for the study of SNRs with SKA, already in phase 1, are the so-called 'generic surveys', a small number of SKA1 surveys that provide maximum commensality to a wide range of scientific objectives, at an early date. The concept is still under debate among the SKA Scientific Working Groups and among those proposed there is the survey of the Galactic Plane to be conducted with SKA1-MID at band 5 (5 to 14 GHz), with a planned sensitivity of 0.04 mJy/beam and a resolution of 1 arcsec. If conducted, this survey will enable the study of all visible SNRs, likely including many missing ones. Analogue surveys are proposed for SKA1-LOW. In this case, the better *uv* coverage at short spatial frequency will allow a very accurate measurement of the SNR flux density, hence a precise reconstruction of its SED.

5. Summary and conclusions

The advent of the SKA and the CTA will allow a great result in a major improvement in the study of Galactic SNRs as sites where cosmic rays are produced. Both radio and

γ-ray emissions are a probe of the local plasma conditions and provide us insights on the fundamental phenomena resulting in particle acceleration. The work on two Galactic SNRs presented here suggests that high-resolution radio observations, possible with next-generation radio instruments, will be a formidable tool to enlarge the sample of Galactic SNRs that can be studied in great detail. Such an opportunity will be naturally offered by SKA and its precursors. The science preparation phase with these new facilities is currently ongoing and many results will come shortly. In the next future, radio and γ-ray observatories will provide us a clear portrait of SNRs, better refining their role as cosmic accelerators. In particular with radio and γ-ray observations we will able to determine if SNRs are indeed the main contributors for Galactic cosmic rays or if other sources must be considered.

References

Aharonian, F., Buckley, J., Kifune, T., & Sinnis, G. 2008, *RPPh*, 71, 096901

Bell, A. R. 2013, *APh*, 43, 56

Cristofari, P., Gabici, S., Casanova, S., Terrier, R., & Parizot, E. 2013, *MNRAS*, 434, 2748

DeLaney T., Koralesky, B., Rudnick, L., & Dickel, J. R. 2002, *ApJ*, 580, 914

Dubner, G. & Giacani, E. 2015, *A& AR*, 23, 3

Frail, D. A., Goss, W. M., & Slysh, V. I. 1994, *ApJ*, 424, L111

Gaensler, B. M. & Slane, P. O. 2006, *ARAA*, 44, 17

Giacani, E., Smith, M. J. S., Dubner, G., & Loiseau, N. 2011, *A& A*, 531, A138

Green, D. A. 2014, *BASI*, 42, 47

Ingallinera, A., Trigilio, C., Umana, G., Leto, P., Agliozzo, C., & Buemi, C. 2014, *MNRAS*, 445, 4507

Norris, R. P., *et al.* 2011, *PASA*, 28, 215

Riggi, S., *et al.* 2016, *MNRAS*, 460, 1486

Umana, G., *et al.* 2015, *MNRAS*, 454, 902

Wang, L., Cui, X., Zhu, H., & Tian, W. 2015, *PoS: aska.conf*, 64

Supernova 1987A:30 years later - Cosmic Rays and Nuclei from Supernovae and their aftermaths
Proceedings IAU Symposium No. 331, 2017
A. Marcowith, M. Renaud, G. Dubner, A. Ray & A. Bykov, eds.

doi:10.1017/S1743921317004410

The e-ASTROGAM space mission: a major step forward for supernova physics

Vincent Tatischeff[1], Roland Diehl[2] and Alessandro De Angelis[3], on behalf of the e-ASTROGAM Collaboration †

[1]CSNSM, IN2P3-CNRS/Univ. Paris-Sud, Université Paris-Saclay, F-91405 Orsay Campus, France
email: vincent.tatischeff@csnsm.in2p3.fr

[2]Max Planck Institut für extraterrestrische Physik, D-85748 Garching, Germany
email: rod@mpe.mpg.de

[3]INFN Padova, via Marzolo 8, 35141, Padova, Italy; LIP/IST, Av. Elias Garcia 14, 1000 Lisboa, Portugal
email: alessandro.deangelis@pd.infn.it

Abstract. e-ASTROGAM is a gamma-ray observatory operating in a broad energy range, 0.15 MeV – 3 GeV, recently proposed as the M5 Medium-size mission of the European Space Agency. It has the potential to revolutionize the astronomy of medium-energy gamma-rays by increasing the number of known sources in this domain by more than an order of magnitude and providing gamma-ray polarization information for many of these sources. In these proceedings, we discuss the expected capacity of the mission to study the physics of supernovae, both thermonuclear and core-collapse, as well as the origin of cosmic rays in SN shocks.

Keywords. Gamma rays: observations, telescopes, radiation mechanisms: nonthermal, supernovae: general, supernova remnants, cosmic rays

1. Introduction

e-ASTROGAM is a proposal for an open gamma-ray observatory dedicated to the exploration of the Universe with unprecedented sensitivity in the energy range 0.15 – 100 MeV, extending up to about 3 GeV. Many of the most-spectacular objects in the Universe have their peak emissivity in this energy band (e.g. gamma-ray bursts, blazars, pulsars, etc.), which is thus essential to study the physical properties of these objects. This energy range is also known to feature a characteristic spectral turn-over associated to hadronic emission from pion decay, which makes it paramount for the study of cosmic ray acceleration, by making it possible to distinguish hadronic and leptonic processes of nonthermal radiation. Moreover, this energy domain covers the crucial range of nuclear gamma-ray lines produced in various astrophysical sites by radioactive decay, nuclear collision, positron annihilation, or neutron capture.

The core mission science of e-ASTROGAM addresses three major topics of modern astrophysics:

• Processes at the heart of the extreme Universe: jet and outflow astrophysics (active galactic nuclei, gamma-ray bursts, compact binaries) and the link to new astronomies (gravitational waves, neutrinos, ultra-high energy cosmic rays);

• Origin and impact of high-energy particles on the evolution of the Milky Way, from cosmic rays to antimatter;

† See http://eastrogam.iaps.inaf.it/

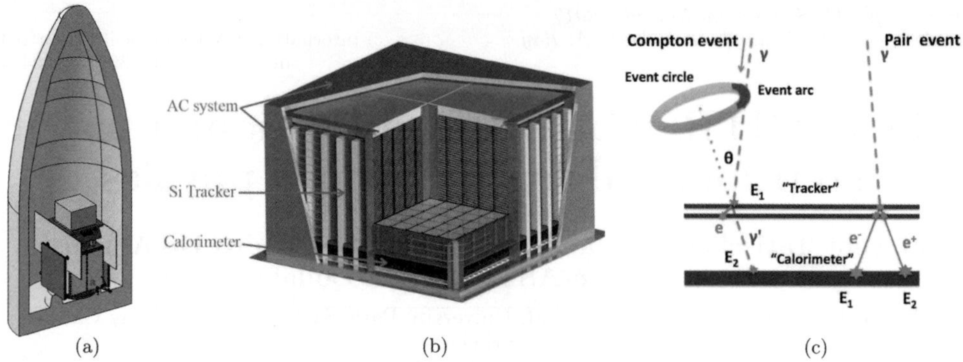

Figure 1. (a) The e-ASTROGAM satellite under Ariane 6.2 fairing in upper position. (b) Overview of the e-ASTROGAM telescope with its three main detectors. (c) Representative topologies for a Compton event and a pair event in the telescope. Photon tracks are shown in pale blue, dashed, and electron and/or positron tracks in red, solid.

- Supernovae, nucleosynthesis and the chemical evolution of our Galaxy.

In addition, e-ASTROGAM is ready for many serendipitous discoveries (the "unknown unknowns") through its combination of wide field of view and much improved sensitivity compared to previous missions in this domain, measuring in three years the spectral energy distributions of thousands of Galactic and extragalactic sources. e-ASTROGAM will also provide a groundbreaking capability for measuring gamma-ray polarization, thus giving access to a new observable that can provide valuable information on the geometry and emission processes of various high-energy sources.

The e-ASTROGAM payload consists of a single telescope for the simultaneous detection of Compton and pair-producing gamma-ray events (see Figure 1). It is made up of three detection system: a silicon Tracker in which the cosmic gamma rays undergo a Compton scattering or a pair conversion, a scintillation Calorimeter to absorb and measure the energy of the secondary particles, and an anticoincidence system to veto the prompt-reaction background induced by charged particles. The telescope has a size of $120 \times 120 \times 78$ cm^3 and a mass of 1.2 tons (including maturity margins plus an additional margin of 20% at system level).

The e-ASTROGAM mission is presented at length in De Angelis *et al.* (2017). Here, we discuss the breakthrough capabilities of the proposed observatory to study the physics of both thermonuclear and core-collapse supernovae (SNe), as well as the process of cosmic-ray acceleration in SN shocks.

2. Thermonuclear supernovae

SNe Ia are the outcome of a thermonuclear disruption of a white dwarf, either through centrally-ignited energy release from carbon fusion reactions, or from external triggering such as, e.g., collision/merging with a white dwarf companion star. But exactly how the ignition conditions are obtained, and on which white dwarfs, and more so how the thermonuclear runaway proceeds through the white dwarf and turns it into a variety of isotopes that are ejected, are all questions that are subject to considerable debate (e.g. Hillebrandt & Niemeyer (2000), Hillebrandt *et al.* (2013) and references therein). Such uncertainties are troublesome for cosmology since the use of SN Ia as standard candles depends on an empirical relationship between the shape and the maximum of the light curve (Phillips 1993), which is closely related to the mass of synthesized ^{56}Ni. Factors like

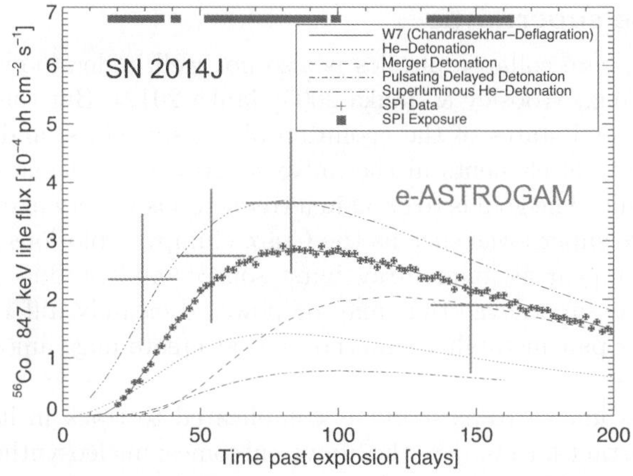

Figure 2. Light curve of the 847 keV line from ^{56}Co decay in SN 2014J. INTEGRAL data (adapted from Fig. 4 in Diehl *et al.* (2015); red data points) are compared to various models of Type Ia SN (The & Burrows 2014). A simulation of the e-ASTROGAM response to a time evolution of the 847 keV line such as in the W7 model (Nomoto *et al.* 1984) shows that the sensitivity improvement by e-ASTROGAM (blue points) will lead to a much better understanding of the SN progenitor system and explosion mechanism.

the progenitor evolution, ignition density, flame propagation, mixing during the burning, completeness of burning in outer, expanding regions, all lead to different amounts of synthesized ^{56}Ni, which, however, can be measured directly through gamma-ray lines (see contribution by Diehl, this volume).

The first detection of a thermonuclear SN in gamma rays was realized with SN 2014J and INTEGRAL's instruments (Churazov *et al.* 2014, 2015; Diehl et 2015). The detected lines of ^{56}Co (lifetime of ~111 days) at 847 and 1238 keV provided good constraints on the mass of radioactive material and the expansion velocity of the ejecta. Moreover, possible signatures of the radioactive ^{56}Ni (mean lifetime ~8.8 days) lines early-on have been reported (Diehl *et al.* 2014; Isern *et al.* 2016), suggesting either a surface explosion triggered from the companion or very special morphology of the runaway, in contrast to any conventional model.

e-ASTROGAM will achieve a major gain in sensitivity compared to INTEGRAL for the main gamma-ray lines arising from ^{56}Ni and ^{56}Co decays. Thus with an expected sensitivity of 3.5×10^{-6} ph cm^{-2} s^{-1} for the 847 keV line in 1 Ms of integration time, which represents an improvement in sensitivity by a factor of about 70 compared to INTEGRAL/SPI, e-ASTROGAM will be able to detect events like SN 2014J up to a distance of 35 Mpc. In the corresponding volume, that includes, in particular, the Virgo cluster of galaxies, one can expect about 10 type Ia SN explosions in 3 years (nominal mission lifetime). As illustrated in Figure 2, e-ASTROGAM will provide much better data than we have now with INTEGRAL for SN 2014J from similarly nearby events. These data will allow us to probe the explosion mechanism in detail, and compare with astrophysical models for each event to better understand the progenitor system(s) and the thermonuclear explosion process. Moreover, the detected sample of thermonuclear SNe would provide a clean and fundamental test of the Phillips (1993) relation for a dominant population of Branch Normal type Ia SNe. The contribution of positrons to energy deposits from radioactivity will also be addressed by e-ASTROGAM, due to its excellent sensitivity for tracking the light curve of the 511 keV annihilation line.

3. Core-collapse supernovae

Similar to SN Ia, core collapse physics is also not well understood in terms of an astrophysical model (e.g., Woosley & Janka 2005; Janka 2012). But these events are more common, being the end states of the evolution of massive stars, and are key to understanding the diversity of elements in the universe. Also here, deviations from spherical symmetry are the rule. The goal is to explain a tremendous variety of core collapse events, e.g. electron capture supernovae such as the Crab, clumpy explosions such as Cassiopeia A, collapsars that appear as gamma-ray burst sources and produce stellar mass black holes, superluminous supernovae that may be powered entirely differently by magnetar rotational energy, or pair instability supernovae that create huge amounts of radioactive ^{56}Ni.

Stellar rotation is known to exist but is complicated to track in its effects on stellar evolution, yet important for many of the above outcomes: nucleosynthesis, pre-supernova structure, core collapse. Measuring nucleosynthesis products such as ^{44}Ti, ^{56}Co, and ^{56}Ni is one of the more direct ways to extract information on the inner processes triggering the explosion near the newly forming compact stellar remnant (e.g., Grefenstette *et al.* 2014, 2017) – other observables are indirect, and mostly reflect interactions within the envelope, or with circumstellar, pre-explosively ejected, or ambient gas.

With a gain in sensitivity for the ^{44}Ti line at 1157 keV by a factor of 27 compared to INTEGRAL/SPI, e-ASTROGAM should detect the radioactive emission from ^{44}Ti (half-life of 58.9 ± 0.3 yr) from most of the young (age $\lesssim$ 500 yr) SNRs in the Milky Way (e-ASTROGAM expected sensitivity for the 1157 keV line: 3.6×10^{-6} ph cm^{-2} s^{-1} in 1 Ms of integration time). The observatory thus will measure precisely the amount of ^{44}Ti in the remnant of SN 1987A, which is currently disputed in the literature (Grebenev *et al.* 2012; Boggs *et al.* 2015; Tsygankov *et al.* 2016). These observations will give new insights on the physical conditions of nucleosynthesis in the innermost layers of a supernova explosion and the dynamics of core collapse near the mass cut.

e-ASTROGAM will also detect the signatures of ^{56}Ni and ^{56}Co decay from several core-collapse SNe in nearby galaxies. Comparing γ-ray characteristics of different classes of core-collapse SNe, possibly including the pair instability SNe with their order of magnitude higher ^{56}Ni production (e.g., Gal-Yam *et al.* 2009), will probe potentially large variations in their progenitors and offer a direct view of their central engines. Rare core collapse events are predicted to have gamma-ray line brightnesses orders of magnitude above typical supernovae: pair instability and magnetar-powered jet explosions will reveal much larger amounts of ^{56}Ni. e-ASTROGAM will reach out to a larger part of the nearby universe to constrain the rate of such events, if not detect them.

4. Cosmic-ray acceleration in supernova shocks

Galactic cosmic rays are believed to be accelerated in expanding shock waves initiated by SN explosions, at least up to the "knee" energy of the cosmic-ray spectrum at $\sim 10^{15}$eV (e.g., Axford 1994). But several fundamental questions remain partly unanswered, in particular on the injection of particles in the acceleration process, the efficiency of the process as a function of time after explosion, and the maximum energy achieved by accelerated particles in SN shocks.

Here also, gamma-ray observations are crucial. Thus, spatially and spectrally resolved observations of a handful of SNRs with *Fermi*-LAT have revealed important spectral modifications with the age of the system: young SNRs tend to be harder and fainter at GeV energies than older (Acero *et al.* 2016). The performance of e-ASTROGAM will

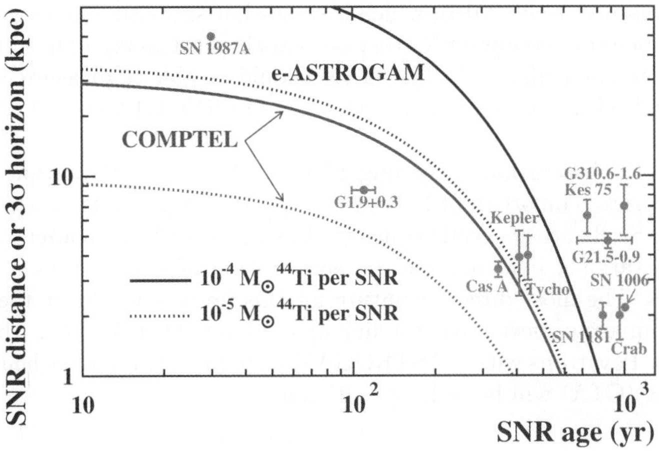

Figure 3. Horizon of detectability of the ^{44}Ti line at 1157 keV as a function of SNR age for CGRO/COMPTEL (blue lines) and e-ASTROGAM (red lines). The plotted sensitivities are for an effective source exposure of 1 year (COMPTEL: 9.0×10^{-6} ph cm^{-2} s^{-1}; e-ASTROGAM: 6.4×10^{-7} ph cm^{-2} s^{-1}), assuming two different yields of ^{44}Ti production per SNR: 10^{-5} $M_\odot$ (typical of SNe Ia; dotted lines) and 10^{-4} $M_\odot$ (Cas A-like events; solid lines). Data for the age and distance of the known, young SNRs are from Tsygankov *et al.* (2016).

open the way for spectral imaging of a score of SNRs, spanning ages from 10^3 to 10^5 years. The 0.15 MeV – 3 GeV gamma-ray band covered by e-ASTROGAM is essential to separate the emission from relativistic electrons and nuclei above 100 MeV, so the new data can constrain how electrons and protons are differentially injected into the shock, how large and sometimes highly intermittent magnetic fields build up near the shock, and how the acceleration efficiency and the total cosmic-ray content of a remnant evolves as the shockwave slows down. Moreover, the bremsstrahlung emitting electrons seen with e-ASTROGAM have energies close to the radio synchrotron emitting ones, and lower than those seen in synchrotron X-rays, thus permitting tomographic reconstruction of the magnetic field and electron distributions inside the remnant.

Older remnants often interact with molecular clouds that provide target gas for cosmic rays escaping the remnant (see, e.g., Uchiyama *et al.* 2012). Resolving the diffuse pion emission produced in those clouds against the bright Galactic background is essential to probe the cosmic-ray spectra that are actually injected into the interstellar medium. Imaging the remnant and shocked clouds both require an angular resolution better than 0.2° around one GeV and a sensitivity below 10^{-11} erg cm^{-2} s^{-1} above 50 MeV that e-ASTROGAM will achieve.

The instrument may also detect the gamma-ray line emission between 0.3 and 10 MeV from nuclear excitation caused by low-energy (i.e. sub-GeV) cosmic-ray nuclei, opening a new and unique way to remotely measure the flux and elemental composition of these elusive particles. This long awaited detection (e.g. Ramaty *et al.* 1979) will be essential to better understand the role of low-energy cosmic rays in the Galactic ecosystem.

5. Recent supernova history and Galactic Pevatrons

Using the gamma-ray line emission from ^{44}Ti decay, e-ASTROGAM is expected to uncover about 10 new, young SNRs in the Galaxy presently hidden in highly obscured clouds (see Figure 3), as well as the youngest SNRs in the LMC. Among the youngest Galactic SNRs, only Cassiopeia A has been firmly detected in ^{44}Ti surveys carried out up to now (Tsygankov *et al.* 2016), which is somewhat surprising in view of an otherwise inferred rate of one core-collapse SN every 50 years in our Galaxy (The *et al.*2006; Dufour & Kaspi 2013). In particular, e-ASTROGAM should be able to measure for the first time the amounts of ^{44}Ti in the remnants of the Type Ia SNe G1.9+0.3, Tycho, and Kepler (Figure 3).

e-ASTROGAM observations of young SNRs will be very important for cosmic-ray acceleration physics. The origin of PeV (1 PeV = 10^{15} eV) cosmic rays is still uncertain, although young SNRs are generally considered as the most likely sources. However, there is still no observational evidence for the acceleration of cosmic rays in SN shocks up to PeV energies. The most direct signature for this process would be the detection of a nonthermal gamma-ray spectrum extending up to about 100 TeV. Thus, the identification of potential new Pevatrons with e-ASTROGAM would be invaluable when the Cherenkov Telescope Array (CTA) will be fully operational.

References

Acero, F., Ackermann, M., Ajello, M., *et al.* 2016, *ApJS*, 224, 8
Axford, W. I. 1994, *ApJS*, 90, 937
Boggs, S. E., Harrison, F. A., Miyasaka, H., *et al.* 2015, *Science*, 348, 670
Churazov, E., Sunyaev, R., Isern, J., *et al.* 2014, *Nature*, 512, 406
Churazov, E., Sunyaev, R., Isern, J., *et al.* 2015, *ApJ*, 812, 62
De Angelis, A., Tatischeff, V., Tavani, M., *et al.* 2016, *Exp. Ast.*, in press, arXiv:1611.02232
Diehl, R., Siegert, T., Hillebrandt, W., *et al.* 2014, *Science*, 345, 1162
Diehl, R., Siegert T., Hillebrandt W., *et al.*, 2015, *A&A*, 574, A72
Dufour, F. & Kaspi, V. M. 2013, *ApJ*, 775, 52
Gal-Yam, A., Mazzali, P., Ofek, E. O., *et al.* 2009, *Nature*, 462, 624
Grebenev, S. A., Lutovinov, A. A., Tsygankov, S. S., & Winkler, C. 2012, *Nature*, 490, 373
Grefenstette, B. W., Fryer, C. L., Harrison, F. A., *et al.* 2017, *ApJ*, 834, 19
Grefenstette, B. W., Harrison, F. A., Boggs, S. E., *et al.* 2014, *Nature*, 506, 339
Hillebrandt, W., Kromer, M., Röpke, F. K., & Ruiter, A. J. 2013, *Frontiers of Physics*, 8, 116
Hillebrandt, W. & Niemeyer, J. C. 2000, *ARAA*, 38, 191
Isern, J., Jean, P., Bravo, E., *et al.* 2016, *A&A*, 588, A67
Janka, H.-T. 2012, *Annual Review of Nuclear and Particle Science*, 62, 407
Nomoto, K., Thielemann F.-K., Yokoi K., 1984, *ApJ*, 286, 644
Phillips, M. M. 1993, *ApJ*, 413, L105
Ramaty, R., Kozlovsky, B., & Lingenfelter, R. E. 1979, *ApJS*, 40, 487
The, L.-S. & Burrows, A. 2014, *ApJ*, 786, 141
The, L.-S., Clayton, D. D., Diehl, R., *et al.* 2006, *A&A*, 450, 1037
Tsygankov, S. S., Krivonos, R. A., Lutovinov, A. A., *et al.* 2016, *MNRAS*, 458, 3411
Uchiyama, Y., Funk, S., Katagiri, H., *et al.* 2012, *ApJL*, 749, L35
Woosley, S. & Janka, T. 2005, *Nature Physics*, 1, 147

Author index